Surface and Interfacial Aspects of Biomedical Polymers

Volume 1

Surface Chemistry and Physics

Surface and Interfacial Aspects of Biomedical Polymers

Volume 1
Surface Chemistry and Physics

Edited by
Joseph D. Andrade
University of Utah
Salt Lake City, Utah

PLENUM PRESS • NEW YORK AND LONDON

Library of Congress Cataloging in Publication Data

Main entry under title:

Surface and interfacial aspects of biomedical polymers.

Bibliography: v. 1. p.
Includes index.
Contents: v. 1. Surface chemistry and physics—1. Biomolecules. 2. Polymers and polymerization. 3. Surface chemistry. I. Andrade, Joseph D., 1941–
QD419.S94 1985 574.19′24 84-26601
(v. 1)

ISBN 978-1-4684-8612-4 ISBN 978-1-4684-8610-0 (eBook)
DOI 10.1007/978-1-4684-8610-0

Softcover reprint of the hardcover 1st edition 1985
A Division of Plenum Publishing Corporation
233 Spring Street, New York, N.Y. 10013

Contributors

J. D. Andrade ● College of Engineering, University of Utah, Salt Lake City, Utah 84112

F. M. Fowkes ● Department of Chemistry, Lehigh University, Bethlehem, Pennsylvania 18015

D. E. Gregonis ● Department of Materials Science and Engineering, College of Engineering, University of Utah, Salt Lake City, Utah 84112

B. Hupfer ● Institut fur Organische Chemie, Johannes Gutenberg-Universitat, Johann-Joachim-Bercher-Weg 18-20, D-6500 Mainz, FRG

M. S. Jhon ● Korean Advanced Institute of Science, P.O. Box 150, Chongyangni, Seoul, Korea

K. Knutson ● Department of Pharmaceutics, College of Pharmacy, University of Utah, Salt Lake City, Utah 84112

J. Lyklema ● Laboratory for Physical and Colloid Chemistry, Agricultural University, De Dreijen 6/6703 BC Wageningen, The Netherlands

D. J. Lyman ● Department of Materials Science and Engineeering, University of Utah, Salt Lake City, Utah. 84112

Y. G. Oh ● Korean Advanced Institute of Science, P.O. Box 150, Chongyangni, Seoul, Korea

B. D. Ratner ● Department of Chemical Engineering, University of Washington, Seattle, Washington 98195

M. Reichert ● Department of Bioengineering, College of Engineering, University of Utah, Salt Lake City, Utah. 84112

H. Ringsdorf ● Institut fur Organische Chemie, Johannes Gutenberg Universitat, Johann-Joachim-Bercher-Weg 18-20, D-6500 Mainz, FRG

L. M. Smith ● Department of Bioengineering, College of Engineering, University of Utah, Salt Lake City, Utah 84112

Preface

This book is intended to provide a fundamental basis for the study of the interaction of polymers with living systems, biochemicals, and with aqueous solutions.

The surface chemistry and physics of polymeric materials is a subject not normally covered to any significant extent in classical surface chemistry textbooks. Many of the assumptions of classical surface chemistry are invalid when applied to polymer surfaces. Surface properties of polymers are important in the development of medical devices and diagnostic products. Surface properties are also of vital importance in fields such as adhesion, paints and coatings, polymer-filler interactions, heterogeneous catalysis, composites, and polymers for energy generation.

The book begins with a chapter considering the current sources of information on polymer surface chemistry and physics. It moves on to consider the question of the dynamics of polymer surfaces and the implications of polymer surface dynamics on all subsequent characterization and interfacial studies. Two chapters are directed toward the question of model polymers for preparing model surfaces and interfaces. Complete treatments of X-ray photoelectron spectroscopy and attenuated total reflection infrared spectroscopy are given. There is a detailed treatment of the contact angle with particular emphasis on contact angle hysteresis in aqueous systems, followed by chapters on interfacial electrochemistry and interface acid–base charge-transfer properties. The very difficult problem of block and graft copolymer surfaces is also discussed. The problem of theoretical calculations of surface and interfacial tensions is presented. Raman spectroscopy is considered as an analytical technique for polymer surface characterization. The final chapter treats a number of techniques which have not been treated in previous chapters as well as some conclusions and expectations.

It is hoped that the book will serve as a useful text in the general area of polymer surface analysis and characterization and make its way into advanced undergraduate and graduate courses in surface chemistry and physics as well as serve as a reference source for researchers and engineers in the general area of polymer surfaces and interfaces. The treatment is by no means complete. Understanding of the properties of polymer surfaces is minimal at the present time, and considerable advances in theory and experimental techniques are required before we can expect to have a thorough understanding of the surface properties of organic systems. Let us hope that this book will help those who decide to enter this stimulating field.

When I began the study of polymer surfaces as biomaterials some fifteen years ago, I lacked any formal training in the general area of surface chemistry. Thus, I am particularly indebted to a number of classic textbooks which are highly suitable for self-study, many of which are listed in Chapter One. The texts by Bikerman; Davies and Rideal; Aveyard and Haydon; Kipling; Kaelble; DeBoer; Adam; and Adamson have been particularly helpful. I owe an enormous debt to my students and coworkers over the years as we attempted to learn and understand the basic principles of polymer surfaces. Although it was often the blind leading the blind, with many hesitations and frustrations, the exercise was always challenging, stimulating and generally productive. I am also in debt to a number of stimulating and even controversial colleagues in the area of surface chemistry of polymers who have provided hypotheses, conclusions, and discussions: R. E. Baier, A. W. Neumann, F. M. Fowkes, B. D. Ratner, A. S. Hoffman, F. Holly, L. Vroman, O. Wichterle, D. Lim, P. M. Sawyer, D. J. Lyman, S. W. Kim, and others.

I wish to thank and to dedicate this volume to W. J. Kolff, pioneer in artificial organs and supporter, promoter, and encourager of often inexperienced people with new and interesting ideas.

J. D. Andrade, Salt Lake City, December 1983

"*All we can create and cry is interface.*
It forms and folds to mate and die and
by its grace
Life's clutch can hold a cloudy sky
'till water laps the coast.
That is perhaps why what we touch
is uppermost"

Leo Vroman

Contents

4. Polymeric Oriented Monolayers and Multilayers as Model Surfaces

B. Hupfer and H. Ringsdorf

5. X-ray Photoelectron Spectroscopy (XPS)

J. D. Andrade

7. The Contact Angle and Interface Energetics

J. D. Andrade, L. M. Smith, and D. E. Gregonis

8. Interfacial Electrochemistry of Surfaces with Biomedical Relevance

J. Lyklema

9. Interface Acid-Base/Charge-Transfer Properties

F. M. Fowkes

10. Graft Copolymer and Block Copolymer Surfaces
B. D. Ratner

11. Interfacial Tensions at Amorphous Polymer–Water Interfaces: Theory
M. S. Jhon and Y. Oh

12. Surface Raman Spectroscopy
W. M. Reichert and J. D. Andrade

Introduction to Surface Chemistry and Physics of Polymers

Joseph D. Andrade

1. SOURCES

The surface properties of synthetic polymers are important in many industrially significant processes and applications. Because of the critical importance of polymer surface properties in many major industries and in technology in general, good textbooks and monographs are readily available, including several focusing on biomedical problems (Table 1).

In recent years, there have been a number of symposia specifically on polymer surfaces and related areas (Table 2) and symposia specifically relating to biomedical phenomena and polymer surfaces (Table 2). There are many periodicals and serials focusing on surface chemistry and physics which are vitally important to scientists and engineers working on polymer surface problems (Table 3). Some of the surface science journals which emphasize non-organic surfaces contain occasional papers on polymer surfaces. These include:

Applications of Surface Science (North-Holland)
Applied Physics, A. Solids and Surfaces (Springer-Verlag)
J. Electroanalytical Chemistry and Interfacial Electrochemistry (Elsevier)
J. Vacuum Science and Technology (American Institute of Physics)
Surface Science (North-Holland)
Surface Technology (Elsevier Sequoia)
Thin Solid Films (Elsevier Sequoia)

A number of professional scientific and engineering societies are heavily involved with polymer surface chemistry, physics, and engineering (Table

Joseph D. Andrade • Departments of Bioengineering and Materials Science, College of Engineering, University of Utah, Salt Lake City, Utah 84112.

TABLE 1
Key Textbooks and Monographs on Polymer Surfaces, Surface Chemistry, and Biomedical Aspects

Author	Year of publication	Title	Emphasis	Reference
Adamson	1983	*Physical Chemistry of Surfaces* (*4th ed.*)	General textbook, basic reference	1
Bikerman	1970	*Physical Surfaces*	General text, critical, challenges assumptions	2
Cherry	1981	*Polymer Surfaces*	Fracture, mechanical aspects	3
Fowkes, ed.	1964	*Contact Angle, Wettability, and Adhesion*	Contact angle	22
Hair, ed.	1971	*Chemistry of Biosurfaces*	Solution properties	5
Harkins	1952	*Physical Chemistry of Surface Films*	Liquid interfaces and monolayers	6
Hench and Ethridge	1983	*Biomaterials: An Interfacial Approach*	Biomaterials text	7
Jones	1975	*Biological Interfaces*	Membranes and monolayers	8
Kaelble	1971	*Physical Chemistry of Adhesion*	Adhesion, polymer surfaces	9
Laslo and Quintana, eds.	1973	*Surface Chemistry and Dental Integuments*	Dental aspects	10
Manly, ed.	1970	*Adhesion in Biological Systems*	Cell adhesion	11
Prince and Sears, eds.	1973	*Biological Horizons in Surface Science*	Biophysics, interfaces	12
Szycher and Robinson, eds.	1980	*Synthetic Biomedical Polymers*	Biomaterials	13
Vold and Vold	1983	*Colloid and Interface Chemistry*	Colloid and gels	34
Wu	1982	*Polymer Interface and Adhesion*	Surface and interfacial tension, adhesion	14

TABLE 2
Recent Key Symposium Proceedings Related to Polymer Surfaces and to Their Biomedical Aspects

Editor	Year of publication	Title	Emphasis	Reference
Andrade	1976	*Hydrogels for Medical and Related Applications*	Hydrogels, hydrophilic interfaces	15
Baier	1975	*Applied Chemistry at Protein Interfaces*	Proteins at interfaces	16
Casper and Powell	1982	*Industrial Applications of Surface Analysis*	Mainly surface analysis	33
Clark and Feast	1978	*Polymer Surfaces*	All areas of polymer surfaces	17
Cooper and Peppas	1982	*Biomaterials: Interfacial Phenomena and Applications*	Blood and protein interactions	18
Copley and Seaman	1984	*Surface Phenomena in Hemorheology*	Rheology	19
Dwight *et al.*	1981	*Photon, Electron, and Ion Probes of Polymer Structure and Properties*	Instrumental surface analysis	20
Goldberg and Nakajima	1980	*Biomedical Polymers*	Compatibility, drug delivery	21
Lee	1975	*Adhesion Science and Technology*	Adhesion	23
Lee	1977	*Characterization of Metal and Polymer Surfaces, Vol. 2, Polymer Surfaces*	Surface characterization	24
Lee	1980	*Adhesion and Adsorption of Polymers*	Polymer adsorption and adhesion	25
Mittal	1979	*Surface Contamination*	Contamination and cleaning	27
Mittal	1983	*Physicochemical Aspects of Polymer Surfaces*	All areas of polymer surfaces	28
Sedlacek *et al.*	1979	*Medical Polymers: Chemical Problems*	Blood interactions, protein adsorption, surface modification	31
Vroman and Leonard	1977	*Behavior of Blood and Its Components at Interfaces*	Blood interactions, protein adsorption	26
Winter *et al.*	1979	*Evaluation of Biomaterials*	Biomaterials—all areas	29
Winter *et al.*	1982	*Biomaterials—1980*	Biomaterials—all areas	30

TABLE 3
Serials and Periodicals Important to Polymer Surface Science and Engineering

Title	Publisher	Editor(s)	Frequency	Focus
Advances in Colloid and Interface Science	Elsevier	A. C. Zettlemoyer	Bimonthly	General review journal
Colloid and Polymer Science	Steinkopff, Darmstadt	H. G. Kilian	Monthly	Polymer science, colloids and surfaces
Colloids and Surfaces	Elsevier	P. Somasundaran	8 issues/yr	Colloids and general surface science
J. Adhesion	Gordon and Breach	L. H. Sharpe	Quarterly	Adhesion
J. Applied Polymer Sci.	Wiley	H. Mark	Monthly	All aspects of applied polymer science
J. Colloid Interface Sci.	Academic	A. C. Zettlemoyer	Monthly	All aspects of surface and colloid science
J. Electron Spectroscopy	Elsevier	C. R. Brundle	Monthly	Photoelectron spectroscopy
J. Electrostatics	Elsevier	J. C. Giddings	Bimonthly	Electrostatics, electrical properties
Polymer Preprints	American Chemical Society	B. M. Culbertson	Biannually	American Chemical Society meeting proceedings
Polymeric Materials: Science and Engineering	American Chemical Society	R. S. Bauer	Biannually	American Chemical Society meeting proceedings
Progress in Surface Science	Pergamon	S. G. Davison	Monthly (Volumes 1–9)	General review journal
Recent Progress in Surface and Membrane Science	Academic	D. A. Cadenhead, J. F. Danielli	Annually (Volumes 1–15)	General review journal
Surface and Colloid Science	Wiley	E. Matijevic	Annually (Volumes 1–12)	General review journal
Surface and Interface Analysis	Wiley–Heyden	D. Briggs	Bimonthly	Instrumental surface analysis—emphasis on photoelectron spectroscopy

4). There are many medical, bioengineering, and biomaterials journals which often publish articles related to polymer surface properties and, in particular, their biomedical consequences (Table 5).

Much of the literature on biomedical aspects of polymer surface properties is scattered throughout the entire journal literature, including the medical, chemical, physical, and engineering literatures. The serious investigator in this field must accept the fact that he/she must monitor a very wide range of journals on a regular basis. This is generally optimally done by some sort of computer-based literature retrieval service based on a custom search profile, such as provided by the Institute for Scientific Information, the American Chemical Society, and others.

A variety of tutorial short courses are available for the beginning worker and specialty meetings are often held for the experienced worker. In addition to the tutorial and research symposia, such as provided by the meetings of the key societies in this field (Table 4), other specialty meetings include Gordon Research Conferences (Department of Chemistry, University of Rhode Island, Providence, RI) which are offered annually or every other year; conferences offered by the New York Academy of Sciences (2 East 63rd St., N.Y., NY 10021); and meetings organized by the British Plastics and Rubber Institute (11 Hobart Place, London SW1W 0HL).

Although there is considerable information on polymer surface characterization, properties, and applications, there is no modern textbook specifically focusing on those aspects of polymer surfaces of importance to biomedicine.

2. OBJECTIVES

The purpose of this volume, *Surface Chemistry and Physics*, is to provide a basic source of information for working scientists and engineers dealing with the surface and interfacial aspects of biomedical polymers.

Classical surface chemistry of polymers is based on a series of assumptions, many of which are not valid for certain polymer classes and/or for the particular environments in which biomedical polymers are applied. Many of the surface characterization methods used in classical surface chemistry depend on some of these assumptions in order to interpret the experimental results obtained. These methods are critically evaluated in this volume, and the assumptions are generally challenged and tested. It is important to point out that a detailed understanding of polymer surface properties, particularly in physiologically and biologically relevant environments, is far from complete. We shall see that many of the existing methods are not fully suited to the task. It is hoped and expected that this volume will serve as a concise base and source of information on the properties

TABLE 4
Societies and Professional Organizations Related to Polymer Surfaces and Their Biomedical Applications

Society/address	Division	Focus	Meetings	Other
Adhesion Society c/o G. P. Hamed, University of Akron, Whitby Hall, Akron, Ohio 44325		Subjects related to adhesion and adhesives	Annually	Publishes newsletter of the Adhesion Society
American Chemical Society (*ACS*) 1155 16th St., NW, Washington, DC 20036	Division of Colloid and Surface Chemistry; Division of Organic Coatings and Plastics (publishes biannual preprints); Division of Polymer Chemistry (publishes biannual preprints)	All areas of chemistry: surfaces, polymers, and coatings	Biannually and regional meetings	Publishes meeting proceedings via *Advances* and *Symposium* Series, also many chemical journals, including *Macromolecules* and *Analytical Chemistry*
American Physical Society (*APS*) 335 East 45th St., New York, NY 10017	Division of Biological Physics; Division of High Polymer Physics; Division of Solid State Physics; others	All areas of physics	Biannually and specialty meetings	Publishes abstracts for meetings in *Physical Rev. Abst.*, also several physics journals
American Society for Artificial Internal Organs (*ASAIO*) P.O Box C, Boca Raton, FL 33432	—	All aspects of artificial organs	Annually	Publishes *Trans ASAIO*, proceedings of annual meet.; publishes *ASAIO Journal* quarterly
American Society of Testing and Materials (*ASTM*) 1916 Race Street, Philadelphia, PA 19103	Special committees on specific subjects	Development of voluntary standards and protocols	Special meetings on an occasional basis	Publishes *ASTM Standards*, annually updated

Biological Engineering Soc. c/o K. Copeland, Royal College of Surgeons, Lincoln's Inn Fields, London WCZA 3PN, United Kingdom	Biomaterials Group	6–8 subgroups in specialty areas, including biomaterials	Occasional special topics conferences	Abstracts, books of all meetings, publishes *J. of Biomedical Engineering* and *Biomaterials* quarterly
Biomedical Engineering Society (*BMES*) P.O. Box 2399, Culver City, CA 90230	—	General biomedical engineering—not very strong on materials or surfaces	Annually, a part of FASEB (Fed. of Amer. Soc. for Expt. Biology)	Publishes *Annals of Biomedical Engineering*, 6 issues/year
European Society for Artificial Organs (*ESAO*) # 10, Route des Jeuves, 1227 Geneva—La Prarille, Switzerland	—	All areas of artificial organs	Annual congress	Abstract booklet
European Society for Biomaterials c/o Dr. F. Burny, Department Orthopedic Surgery, Hospital Universitaire Brugmann, Place Van Gehuchten 4, B 1020 Brussels, Belgium	—	General biomaterials	Annual meeting	Sponsor of *Adv. in Biomaterials*, proceedings of annual meeting
International Society for Artificial Organs (*ISAO*) 8937 Euclid Ave, Cleveland, OH 44106	—	All areas of artificial organs	International congress annually—abstract booklet	Publishes *Int. J. Artificial Organs*, quarterly

TABLE 4 (*continued*)

Society/address	Division	Focus	Meetings	Other
International Assoc. of Colloid and Interface Scientists (*IACIS*) c/o Lab. for Physical and Colloid Chem., Agricultural University, 6703 BC Wageningen, The Netherlands	—	General colloid and surface chemistry	Every 3 years—Potsdam, USA, 1985; Japan, 1988	Publishes IACIS newsletter
Microbeam Analysis Society (*MAS*) P.O. Box 502, Fairport, NY 14450	—	Microarea analysis, electron microscopy, Raman microprobe, microprobe techniques	Annually	Publishes *Microbeam Analysis,* D. E. Newbury, ed. proceedings of annual meeting
New York Academy of Sciences (*NYAS*) 2 East 63rd St., New York, NY 10021	—	All areas of science	Specialty meetings on specific topics	Publishes *Annals NYAS,* proceedings of specialty meetings
Society for Biomaterials c/o Southwest Research Institute, 6220 Culebra Road, San Antonio, TX 78284	—	Biomedical materials and implants, mainly synthetics rather than natural biomaterials	Publishes annually a preprint "Trans. of the—Annual Meet."	Sponsor of *J. Biomedical Materials Res.*, 6 issues/year

and methods of characterization of polymer surfaces at this point in time, and that it will serve to stimulate and encourage workers in this general area.

3. OVERVIEW

The volume begins with a chapter on polymer surface dynamics, which provides a conceptual background for several of the chapters to follow. Many of the polymers used in biomedicine are dynamic; that is, side chains and large sections of the main chain may be in motion at normal use temperatures. These motions are present, in modified form, at the surface and are expected to influence the surface properties. This effect has been largely ignored in classical polymer surface science.

Chapters 3 and 4 discuss the development and preparation of model polymers. These are very important because many of the phenomena to be discussed are not fully developed nor understood and model systems are necessary to probe those phenomena and their biological analogues. Most of the practical literature on biomedical aspects of polymer surfaces has utilized commercial or relatively common polymers. In very few cases have model systems been used.

Once one has a polymer sample, for example, a polymer thin film deposited on some appropriate substrate, the first and most basic characterization to be performed is to look at it. An introduction to the optical and electron microscopic examination of polymer surfaces is given in Chapter 13.

Following a detailed optical examination, the next question one may ask is what is the surface chemical composition of the polymer surface? There are two key techniques presently widely used for this purpose, X-ray photoelectron spectroscopy, the subject of Chapter 5, and surface infrared spectroscopy, the subject of Chapter 6.

Biomedical polymers are generally used in an aqueous environment and there are relatively few techniques available for the direct, *in situ* characterization of the polymer–water interface. One of these methods, based on water contact angle dynamics, is presented and discussed in Chapter 7.

The surface charge and interface potential characteristics of polymer surfaces and polymer–water interfaces is of considerable importance and significance and is presented in Chapter 8. In addition to ionic and electrostatic interactions, it is becoming increasingly evident that the partial acid–base or electron donor–acceptor properties of polymer surfaces are of vital importance in practically all applications, including biomedical. This area is presented in Chapter 9, together with methods for characterizing the charge transfer properties of surfaces.

TABLE 5
Biomaterials/Bioengineering Journals Which Publish Papers Related to Polymer Surfaces

Title	Editor(s)	Publisher	Frequency	Focus
Advances in Biomaterials	—	J. Wiley	Sporadic	Proceedings of symposia
Annals of Biomedical Engineering	P. Albrecht, 1708 Meadowbrook Ave., Ann Arbor, MI 48103	Pergamon Press	Bimonthly	General bioengineering
Artificial Organs	Managing Editor, Y. Nose, Cleveland Clinic Foundation 10300 Carnegie Ave., Cleveland, OH 44106	Inter. Society for Artificial Organs, 10300 Carnegie Ave., Cleveland, OH 44106	Quarterly	Artificial organs
ASAIO Journal	P. M. Galletti, Div. of Medicine, Brown University, Providence, RI	J. B. Lippincott, Co. and American Society for Artificial Internal Organs	Quarterly	Artificial organs
Biomaterials	See list following table[a]	Butterworth Scientific, Ltd. GU2 5BH, UK	Quarterly	Biomaterials
Biomaterials, Medical Devices, Artificial Organs	T. F. Yen, University of Southern California, PCE-201, Los Angeles, CA 90089	Marcel Dekker	Quarterly	Biomaterials

Devices and Technology Branch (DTB), Contractors Meeting, NIH	DTB, NHLBI, NIH, Bethesda, MD	U.S. Government Printing Office	Annually	Biomaterials, instrumentation and devices in cardiovascular medicine
International J. of Artificial Organs	Coordinating Editor, D. Brancaccio, Policlinico Renal Unit, Milan, Italy	Wichtig Editore s.r.l., Milan, Italy	Bimonthly	Artificial organs
J. Biomedical Materials Res.	N. Cranin, Brookdale Hospital Medical Centre, Linden Blvd. at Brookdale Place, Brooklyn, NY 11212	J. Wiley and Sons Monthly, 605 3rd Ave., New York, NY 10158	Bimonthly	Biomaterials
Polymers in Medicine	H. Kus (Warsaw)	PWN, Polish Scientific Publ.	Quarterly	Biomedical polymers
Trans. American Society Artificial Internal Organs	G. E. Schreiner, Dept. of Medicine, Georgetown Univ. Hospital, Washington, DC 20007	ASAIO (see Table 4)	Annually	Artificial organs

[a] *Editors list for Biomaterials*: Dr. G. W. Hastings, Biomedical Engineering Unit, Medical Institute, Hartshill, Stoke-on-Trent, Staffs, ST4 7NY, UK; Professor R. S. Langer, Department of Nutrition and Food Science, Massachusetts Institute of Technology, Cambridge, MA 02139, USA; Professor N. A. Peppas, School of Chemical Engineering, Purdue University, West Lafayette, Indiana 47907, USA.

Polymers which have the possibility of phase separation, such as blends, block copolymers, and graft systems, are particularly difficult to characterize because of their multiphase nature. This subject is addressed in Chapter 10.

Theoretical methods for probing and understanding polymer surfaces and interfaces have been developing rapidly, particularly in the area of estimation of surface and interfacial tensions, presented in Chapter 11.

Newer methods for polymer surface characterization, and indeed for surface science in general, are constantly being developed. An area which is developing very rapidly, and for which considerable application is already evident, is surface Raman spectroscopy of polymers, discussed briefly in Chapter 12.

Chapter 13 briefly summarizes the volume and reviews other methods for polymer surface and interface characterization. It also discusses future expectations in this area.

4. LIMITATIONS

There are many important areas of polymer surface chemistry and physics which are not covered in this volume. Many of the electronic aspects of polymer surfaces have not been treated. These are available in a number of other sources, particularly the book edited by Clark and Feast.[17] The surface and interfacial aspects of polymer fracture and adhesion have not been discussed. These are covered in the texts by Kaelble,[9] Cherry,[3] and Wu.[14] The surface properties of polymers considerably above room temperatures, such as in the melt, are discussed in detail by Wu[14] and are not included here. Surface and interfacial properties related to nucleation and crystallization of polymers, as well as orientation and related aspects seen in extrusion and related processes, are not included. The question of polymers at solution/air interfaces, such as polymer monolayers,[32] although discussed briefly in Chapter 3, are not generally discussed in this book. This subject is, of course, very important for understanding polymer surface properties. The entire question of polymer surface modification is not generally discussed here although there is some discussion in the chapter on graft copolymers, Chapter 10. Such processes as Corona discharge, radiofrequency glow discharge, and surface chemical reactions have not been treated. These are discussed in some detail in the book edited by Clark and Feast.[17]

A particularly important omission is that of highly porous surfaces, including fibers, fabrics, microporous surfaces, sponges, etc. This is a very important area, particularly in the case of vascular grafts, but it is also highly complex and even less completely understood than the relatively smooth polymer surfaces emphasized in this book. There is considerable information in the textile and fiber literature on these topics as well as in basic texts.[1]

Another geometric omission is that of small particles, including colloidal particles. Although much of what is said in this book applies to fibers, porous materials, and microparticles, each of these geometries has peculiarities which are different from those of the smooth, flat surface. There are a number of recent treatments in the colloid science literature which focus on these topics (see Reference (1)).

Aspects of polymer surfaces related to adsorption phenomena are treated in Volume 2.

5. REFERENCES

1. A. W. Adamson, *Physical Chemistry of Surfaces*, 4th ed., Wiley, New York (1983).
2. J. J. Bikerman, *Physical Surfaces*, Academic Press, New York (1970).
3. B. W. Cherry, *Polymer Surfaces*, Cambridge University Press, Cambridge (1981).
4. J. T. Davies and E. K. Rideal, *Interfacial Phenomena*, 2nd ed., Academic Press, New York (1963).
5. M. L. Hair, ed., *Chemistry of Biosurfaces*, Marcel Dekker, New York (1971).
6. W. D. Harkins, *Physical Chemistry of Surface Films*, Reinhold Publ., New York (1952).
7. L. L. Hench and E. C. Ethridge, *Biomaterials: An Interfacial Approach*, Academic Press, New York (1983).
8. M. N. Jones, *Biological Interfaces*, Elsevier, New York (1975).
9. D. H. Kaelble, *Physical Chemistry of Adhesion*, Wiley, New York (1971).
10. A. Lasslo and R. P. Quintana, eds., *Surface Chemistry and Dental Integuments*, C. C. Thomas, Publ., Springfields, Illinois (1973).
11. R. S. Manly, ed., *Adhesion in Biological Systems*, Academic Press, New York (1970).
12. L. M. Prince and D. F. Sears, eds., *Biological Horizons in Surface Science*, Academic Press, New York (1973).
13. M. Szycher and W. J. Robinson, eds., *Synthetic Biomedical Polymers*, Technomic Publ. Co., Lancaster, Pennsylvania (1980).
14. S. Wu, *Polymer Interface and Adhesion*, Marcel Dekker, New York (1982).
15. J. D. Andrade, ed., *Hydrogels for Medical and Related Applications*, *Am. Chem. Soc. Symp. Ser.* **31** (1976).
16. R. E. Baier, ed., *Applied Chemistry at Protein Interfaces*, *Adv. Chem. Ser.* **145** (1975).
17. D. T. Clark and W. J. Feast, eds., *Polymer Surfaces*, Wiley, New York (1978).
18. S. L. Cooper and N. A. Peppas, eds., *Biomaterials: Interfacial Phenomena and Applications Adv. Chem. Ser.* **199** (1982).
19. A. L. Copley and C. V. F. Seaman, eds., *Surface Phenomena in Hemorheology*, *Ann. N.Y. Acad. Sci.*, in press (1984).
20. D. H. Dwight, T. J. Fabish, and H. R. Thomas, *Photon, Electron, and Ion Probes of Polymer Structure and Properties*, *Am. Chem. Soc. Symp. Ser.* **162** (1981).
21. E. P. Goldberg and A. Nakajima, eds., *Biomedical Polymers*, Academic Press, New York (1980).
22. F. W. Fowkes, ed., *Contact Angle, Wettability, and Adhesion*, *Adv. Chem. Ser.* **43** (1964).
23. L. H. Lee, ed., *Adhesion Science and Technology*, Plenum Press, New York (1975).
24. L. H. Lee, ed., *Characterization of Metal and Polymer Surfaces*, Vol. 2, *Polymer Surfaces*, Academic Press, New York (1977).

25. L. H. Lee, ed., *Adhesion and Adsorption of Polymers*, Plenum Press, New York (1980).
26. L. Vroman and E. Leonard, eds., *Behaviour of Blood and Its Components at Interfaces, Ann. N.Y. Acad. Sci.*, **283** (1977).
27. K. L. Mittal, ed., *Surface Contamination*, Plenum Press, New York (1979).
28. K. L. Mittal, ed., *Physicochemical Aspects of Polymer Surfaces*, Plenum Press, New York (1983).
29. G. D. Winter, J. L. Leray, and K. de Groot, eds., *Evaluation of Biomaterials*, Wiley, New York (1979).
30. G. D. Winter, D. F. Gibbons, and H. Plenk, Jr., *Biomaterials—1980*, Wiley, New York (1982).
31. B. Sedlacek, C. G. Overberger, and H. F. Mark, eds., *Medical Polymers: Chemical Problems J. Polymer Sci., Polymer Symp.* **66** (1979).
32. M. Breton, *J. Macromol. Sci., Rev. Macromol. Chem.* **C21**, 61 (1981).
33. L. A. Casper and C. J. Powell, eds., *Industrial Applications of Surface Analysis, Am. Chem. Soc. Symp. Ser.* **199** (1982).
34. R. E. Vold and M. J. Vold, *Colloid and Interface Chemistry*, Addison-Wesley Publ. Co., Reading, Massachussetts (1983).

Note added in proof: A new journal should be added to Table 3: *Langmuir*, published by the American Chemical Society, and edited by A. W. Adamson, Department of Chemistry, University of Southern California, Los Angeles, California.

2

Polymer Surface Dynamics

J. D. Andrade, D. E. Gregonis, and L. M. Smith

1. INTRODUCTION

Classical surface chemistry assumes that solid surfaces are rigid, immobile, and at equilibrium. These assumptions allow one to probe adsorption and wetting or contact angle processes purely from the point of view of the liquid phase, because one assumes that the solid phase does not in any way respond, reorient, or otherwise change in the different liquid environments. Although such assumptions may be partially correct for truly rigid solids, they are generally inappropriate for polymers (see also Chapter 7).

Polymer structures and properties are, in general, time and temperature dependent. Because of the relatively large size and high molecular weights of synthetic polymer molecules, it is unlikely that most polymeric solids ever achieve a true equilibrium. Solid polymers are, therefore, inherently nonequilibrium structures and as such exhibit a range of relaxation times and properties under normal conditions and in response to changing environments.[1] This situation is well known in the area of bulk polymer properties but, with few exceptions,[44] has been largely neglected or ignored in polymer surface chemistry and physics. There is now considerable evidence that the surface properties of polymers are also time, temperature, and environment dependent.

Transition and relaxation phenomena in solid polymers are treated in practically all polymer science and polymer materials textbooks.[2] In this chapter we will be concerned only with the case of amorphous, noncrystalline polymer systems. A preliminary report of some of these topics has appeared.[3]

J. D. Andrade, D. E. Gregonis, and L. M. Smith • Department of Bioengineering, University of Utah, Salt Lake City, Utah 84112.

TABLE 1
Suggested Relationship between Polymer Surface Motions and Blood or Biocompatibility

Investigator(s)	Year	Idea	Comment	Reference
Nyilas and Ward	1970–1977	Fluctuating arrays of methyl groups—randomized dispersion force field leads to weak, reversible protein adsorption	Rationale for compatibility of poly(dimethylsiloxane) (PDMSO) and grafts or blends containing PDMSO	4
Holly and Refojo	1975	Side chain reorientation influences wetting properties of soft contact lenses	Considerable evidence now exists for this concept—see remainder of this chapter	5
Merrill	1977	Surface motions result in a surface entropy which should produce a negative entropy of mixing, leading to decreased protein adsorption	There is now evidence for decreased adsorption with increased surface entropy or surface mobility for gel-like interfaces.[45] This paper stimulated much activity on polymer surface motions	6
Barenberg *et al.*	1979	Side chain surface motions influence protein adsorption and thrombogenesis; polymer block morphology influences adsorption and thrombogenesis	No characterization of model surfaces *in* water; some correlation between motions, morphology, and thrombogenesis was observed. Important contribution is their suggestion of the role of surface bound ions in protein adsorption	7–9

Sakurai *et al.*	1980	Surface dynamic molecular fluctuations (micro-Brownian motion) influence bio- and blood interactions	No direct evidence, although this group has studied the effect of surface morphology in block systems on protein and platelet interactions	10
Brier-Russell *et al.*	1981	Surface motions result in a surface entropy which should produce a negative entropy of mixing, leading to decreased protein adsorption	Main chain T_g motions in alkyl methacrylates correlate with platelet retention and platelet release	11
Coleman	1980	T_g and side chain length of alkyl methacrylates correlates with platelet adhesion and clotting time measurements	Platelet adhesion increases with T_g; whole blood clotting time shows no simple correlation with T_g	12
Nagaoka *et al.*	1983	PEO side chain mobility is important; as mobility increases, blood interactions and protein adsorption decrease	Platelet adhesion and plasma protein adsorption decrease as surface mobility increases	45

A number of hypotheses and suggestions relating polymer surface motions to blood and biocompatibility have appeared and are reviewed in Table 1.

2. POLYMER TRANSITIONS AND RELAXATIONS

Polymer relaxation has been defined by North[13] as "A time dependent return to equilibrium of the system which has recently experienced a change in the constraints acting upon it." If one in any way changes or perturbs the polymer system, the polymer will respond, i.e., relax, to achieve a new state which is closer to equilibrium with the new environment or situation.

Relaxation refers to a time-dependent change. A good example of such time-dependent character is "silly putty," an uncrosslinked silicone material. Given sufficient time it creeps and flows, but throw it hard against a rigid surface (a very short-time experiment) and you observe a highly elastic rebound.

One can also speak of a *transition,* which is a temperature-dependent change.[14] For example, a crystalline material consisting of low-molecular weight molecules will generally exhibit a phase change at a precise temperature in the transition from solid to liquid. We say that the material has melted. This is an equilibrium process under ideal conditions and is independent of time. Many thermodynamic properties, such as density and enthalpy, show discontinuities when heated through their melting points or boiling points. Glass, on the other hand, shows no such discontinuity. Glass, depending on composition, will go from a rigid, brittle material to a soft, viscous material on heating to the vicinity of 1000°C. Glassy or amorphous polymers do precisely the same thing at lower temperatures. As a rigid, amorphous polymer is heated, a temperature range is reached where the mechanical properties, such as tensile strength or modulus or elasticity, change dramatically. This temperature range is said to be the glass transition temperature. "The glass transition is a relaxation phenomenon and represents the point at which the relaxation times of the polymer chain are of the same order of magnitude as the time required to perform the measurement." [Reference (43), p. 180.] "Molecular motion in a polymer sample is promoted by its thermal energy. It is opposed by the cohesive force between structural segments both along the chain and between neighboring chains." The molecular motions each have a characteristic frequency "... determined by the temperature and moment of inertia of the participating groups. The onset of these molecular motions causes changes in the macroscopic physical properties of the polymer, giving rise to transition or relaxation phenomena." [Reference (14), pp. 241–242.]

A discontinuity in properties is generally evident upon heating a thin polymer sample at a reasonable rate (~10°C/min). Heating at a very slow rate, close to equilibrium conditions, makes the discontinuity difficult to detect; under very slow heating conditions it is difficult to speak of a specific glass transition temperature. Transitions can generally be observed by differential thermal analysis (DTA) or differential scanning calorimetry (DSC) as a change in the heat capacity, C_p, with temperature[15] (Table 2). One can view the temperature-dependent transitions as occurring because as temperature increases there is more and more thermal energy in the polymer structure which can activate motions and processes. In addition, increasing temperature leads to increased volume due to thermal expansion, providing a greater volume in the bulk material which makes motions easier to achieve. It is as if increasing temperature provides additional "elbow room" for the polymer molecules in which to move.[14]

Another way of looking at the situation is to consider that molecular motions have a characteristic frequency. A natural frequency is determined by the temperature as well as the moments of inertia of the participating segments. Micro-Brownian motion of large segments of the polymer chain become possible above the glass transition temperature. Parts of the chain, perhaps of the order of 10 monomer units,[1,2,14-17] can move in a sort of cooperative fashion, as indicated in Figure 1, which shows chains, loops, and coils in motion under a particular set of conditions.[2] The time and temperature characteristics of these motions are, of course, directly interrelated. At a high temperature a polymer segment may be able to move or respond to a stimulus which is applied for only a very short period of time. At lower temperatures the polymer does not have the capacity to respond to the stimulus unless it is held for a longer period of time. Thus, if we wait long enough, motions and responses to environments can occur, in principle, at any temperature.

TABLE 2
Fundamental Differences between the Glass Transition and Local Mode Transitions[a]

Property	Structural transition (main chain segmental motions), T_α, T_g, etc.)	Local mode transition (such as side chain rotation), T_β
V vs. T	Discontinuous at T_g	Little or no change at T_β
C_p vs. T	Discontinuous at T_g	Little or no change at T_β
Arrhenius equation	Not obeyed at T_g	Obeyed at T_β
Sample history	Sensitive to thermal history and processing	Insensitive to thermal history and processing

[a] Collected from References (1) and (14).

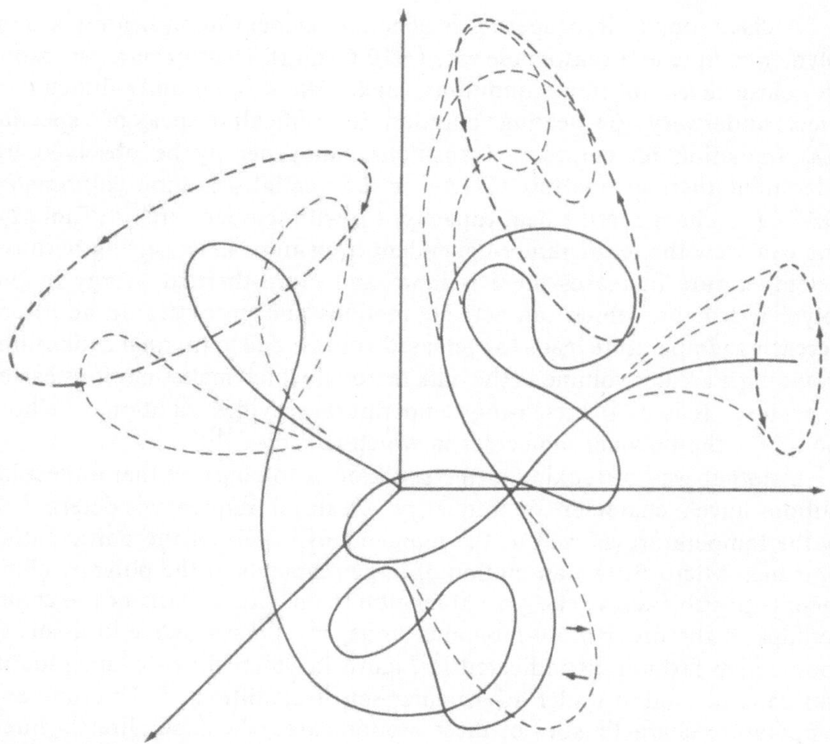

FIGURE 1. Schematic view of the motion of chain ends, loops, and segments of a polymer molecule in a glass transition zone. [Reprinted from Reference (2), p. 203, by permission. Copyright 1971 by Prentice Hall, Inc., Englewood Cliffs, New Jersey.]

In addition to the relatively large cooperative motions of the main chain, there are a variety of other motions present in synthetic polymers. For example, the rotation of a side chain about a carbon–carbon bond, sometimes called the beta relaxation or beta transition, is particularly easy to see in acrylate and methacrylate systems, which have ester-linked side chains (Figure 2). In the case of the methacrylates, these motions are activated in the vicinity of room temperature, whereas the main chain glass transition temperature for poly(methyl methacrylate) is in the vicinity of 130°C. Some of the differences between side chain rotations (local mode transitions) and main chain large-scale motions (structural or glass transitions) were given in Table 2.

Elastomeric materials have glass transitions considerably below room temperature. For example, the glass transition of poly(dimethylsiloxane) is around −130°C. Thus at room temperature and at 37°C this polymer is nearly 170° above its glass transition temperature. The polymer segments

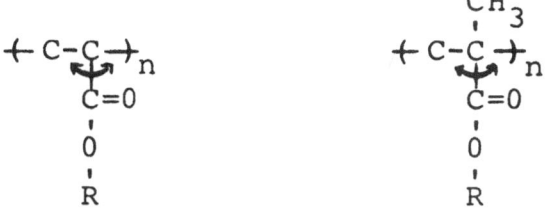

FIGURE 2. Structure of polyacrylates (left) and polymethacrylates (right). R denotes an alkyl or other substituent. The arrow indicates side chain rotation about the carbon-carbon bond, which is responsible for the beta relaxation in these polymers.

are in motion; it is a highly flexible, open structure which, of course, strongly influences its elasticity characteristics.

The polyurethane materials commonly used for cardiac assist devices and total artificial hearts are two-phase block copolymer systems with one of the blocks, the so-called soft segment, having a glass transition considerably below room temperature, thus providing the elasticity for the material. The other block, the so-called hard segment, normally has a transition temperature considerably above room temperature, which provides rigid pinning points and is largely responsible for the strength of these elastomers. (See Chapter 3.)

3. PROBING POLYMER MOTIONS AND TRANSITIONS[1,14–17]

3.1. BULK METHODS

There are two fundamental ways to probe polymer motions and relaxation phenomena. The response of some thermodynamic or physical property can be measured as a function of temperature or one can look at a resonance response to externally applied oscillations, be they mechanical, electrical, or other. There are many experimental methods by which to probe motions, relaxations, and transitions in polymers. We will briefly mention several which are widely used (Table 3).

Differential scanning calorimetry (DSC) consists in measuring the heat taken up (endothermic) or given off (exothermic) by a polymer sample as a function of temperature. The glass transition temperature is manifested as a change in the slope of the heat capacity vs. temperature curve. Side chain and other relaxation phenomena in polymers are very difficult to detect by this technique. The glass transition causes a discontinuity in specific heat, C_p. Thermal methods are not generally sensitive to local mode transitions (Table 2).

TABLE 3

Most Common Experimental Methods to Probe Polymer Transitions and Relaxations[a]

Method	Principle	Information derived	References
DSC/DTA	Measurement of enthalpy or heat capacity ($\Delta H/\Delta T$) as function of T	Specific heat or heat capacity in a temperature range. Enthalpy of a transition (exothermic or endothermic). Insensitive to local mode transitions.	14, 15, 20
Mechanical relaxation	Measure of mechanical "loss" at a certain frequency as a function of temperature (see Figure 3)	"Loss" t–T plot, from which activation energy and t–T dependence of relaxation can be obtained. Most sensitive to those motions which affect mechanical response	14, 21, 22
Dielectric relaxation	Measure of dielectric "loss" at a certain frequency as a function of temperature (see Figure 3, 4)	"Loss" t–T plot, from which activation energy and t–T dependence of relaxation can be obtained. Most sensitive to those motions which affect dielectric response	1, 9

[a] Abbreviations: DTA = differential thermal analysis; DSC = differential scanning calorimetry; t = time; T = temperature.

Dynamic methods generally utilize oscillatory stress or electric fields which can lead to an energy transfer and a loss which can be detected. The theoretical background for this is discussed in detail in most polymer textbooks and is presented briefly in Table 4 and Figure 3. The methods basically fall into two categories. One is to apply a fixed mechanical oscillation to the sample and measure a phase lag in the oscillation transmit-

TABLE 4
Comparison between Mechanical and Dielectric Relaxation[a,b]

Comparison	Mechanical	Dielectric
Basic relation (no loss)	$\gamma = J\sigma$ or $\sigma = G\gamma$: ideal spring	$D = \varepsilon E$: ideal capacitor
Basic relation (loss)	$\sigma = \eta\, d\gamma/dt$: viscous flow (viscous loss)	$V = RI = R\, dq/dt$; Ohm's law (resistive loss)
Real system	f (ideal, elastic, no-loss features and viscous, dissipative lossy features). Real system is modeled as series, parallel, or combination arrangements of the no-loss and loss relationships	
Static case	Hold stress constant, measure strain → creep; hold strain constant, measure stress → stress relaxation; → relaxation time at a particular T	Hold charge constant, follow voltage; hold voltage constant, follow charge or capacitance; → relaxation time at a particular T
Dynamic case	Applied cyclic stress (sinusoidal): $\sigma = \sigma_0\, e^{i\omega t}$ produces a cyclic material response: $J^* = J' + iJ''$ and $G^* = G' + iG''$ as a function of ω	Applied cyclic field (sinusoidal): $\varepsilon^* = \varepsilon_0\, e^{i\omega t}$ produces a cyclic dielectric response: $\varepsilon^* = \varepsilon' - i\varepsilon''$ as a function of ω
Loss	Imaginary component or out-of-phase component represents a "loss" due to material response to the stress or field, which is a function of frequency, ω	
Loss modulus or loss tangent	$\tan\delta_m \equiv J''/J' = G''/G'$, a function of ω	$\tan\delta_e \equiv \varepsilon''/\varepsilon'$, a function of ω
Experimental data	Plot J'', J', $\tan\delta_m$ or G'', G', $\tan\delta_m$ as function of frequency for various temperatures (Figure 6a)	Plot ε'', ε', $\tan\delta_e$ as function of frequency for various temperatures (Figure 6b)

[a] References (1) and (14).
[b] Definitions: G = modulus of elasticity = $1/J$; γ = strain; σ = stress; J = compliance = $1/G$; G = elastic modulus; E = electric field (analogous to stress); D = electric displacement (analogous to strain); ε = dielectric permittivity (analogous to compliance); η = viscosity; t = time; q = charge; I = current; V = voltage; Superscript * = imaginary quantity; Superscript ′ = real part; Superscript ″ = imaginary part; subscript m = mechanical; subscript e = electrical or dielectric; tan δ = loss tangent = loss component/no-loss component = out-of-phase part/in-phase part; ω = angular frequency; δ = phase or lag angle.

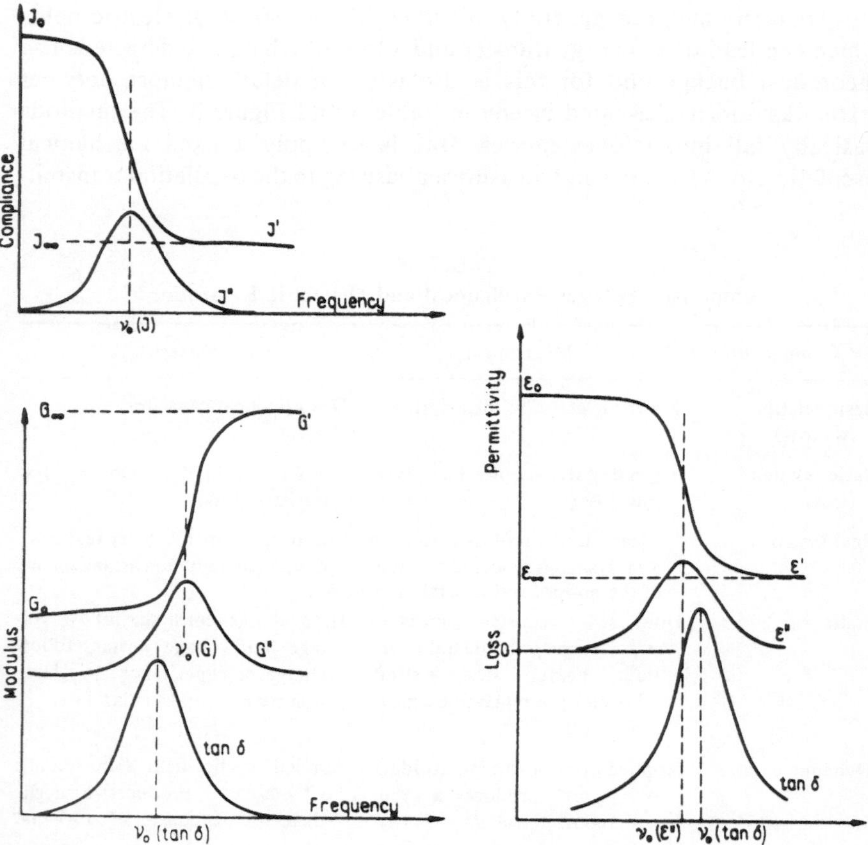

FIGURE 3. (*Right*) Illustration of the dielectric response as a function of frequency for a material exhibiting a single relaxation time. The real and imaginary parts of the dielectric permittivity are plotted as a function of frequency. The ratio of the loss to the no-loss permittivity (the loss tangent) is also shown. (*Upper left*) An analogous plot for a mechanical relaxation experiment is displayed wherein the real and imaginary parts of the compliance are plotted against frequency. Normally, the reciprocal of the compliance, the elastic modulus, is used (*lower left*), which also shows the loss tangent plot. Although these plots are against frequency, because of the time-temperature superposition relationships in polymers the same general results would be obtained if a single frequency experiment was plotted against temperature. See Figure 4. [Reprinted from Reference (1), pp. 56–57, by permission of Adam Hilger, Ltd.]

ted through the sample (Figure 3). The other very common method is dielectric relaxation in which the sample is placed between two electrodes forming a capacitor; the capacitance is measured as a function of ac frequency (Figure 3). This method requires that the polymers have a different dielectric constant for each of the motions to be examined.

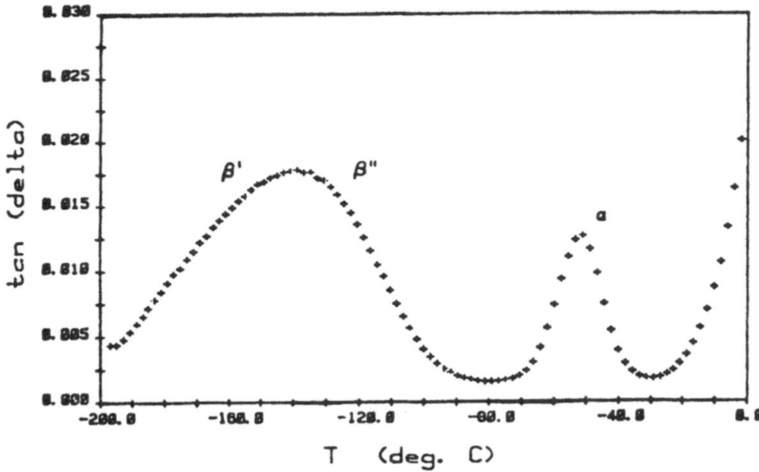

FIGURE 4. The dielectric loss tangent of a fluorinated polyphosphazene material, PNF, containing trifluoroethoxy and fluoroalkoxy side chain relaxations, denoted as β' and β'' and glass transition denoted as α. [Reprinted by permission from Reference (9), p. 801; copyright 1982 by J. Wiley and Sons.]

Although one might expect that apolar polymers could not be measured, the presence of various defects and related structures does produce local dipole moments which can be monitored. Dielectric spectroscopy provides information on those motions which most couple with the dielectric phenomena, whereas mechanical oscillations provide information on those motions which have a greater impact on the mechanical properties. Basic theory, experimental methods, and data interpretation are well covered in the literature.[1,14]

Figure 4 shows a typical dielectric relaxation spectrum at a fixed frequency where the polymer is PNF-200 fluorophosphazene elastomer.[9] The glass transition of this elastomer is clearly evident at about −50°C. There are two different side chains in this polymer, both of which show beta or side chain relaxations which in this spectrum are merged together at around −140°C. Figure 5 presents a dielectric loss and a mechanical loss spectrum of poly(methyl methacrylate) (PMMA) as a function of temperature. The dielectric loss (top) shows a relatively small peak corresponding to the glass transition at about 130°C. The larger peak, corresponding to the beta side chain relaxation at around 30°C, demonstrates that the side chain has a larger dielectric effect in the dielectric loss experiment for this polymer. The mechanical loss spectrum (bottom) shows just the opposite, i.e., a fairly large glass transition peak, demonstrating the effect which the large segmental motions at the glass transition have on mechanical properties, and a relatively small side chain peak, demonstrating that in this

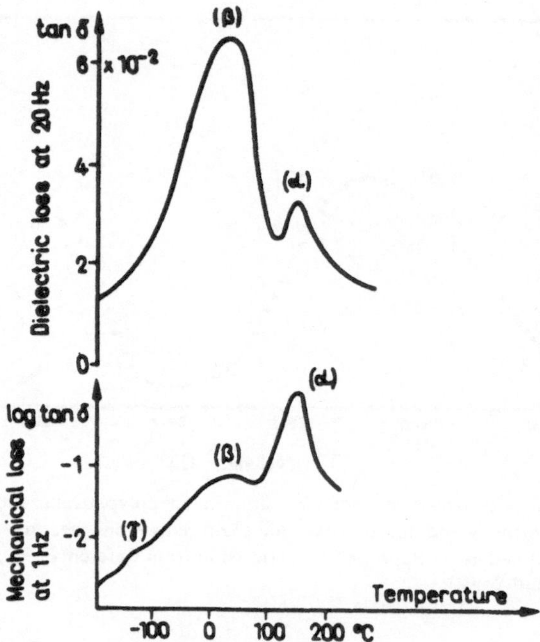

FIGURE 5. Comparison of dielectric and mechanical relaxation spectra of amorphous poly(methyl methacrylate) as a function of temperature. Note the intensity of the glass transition peak in the mechanical loss spectrum, but the higher intensity of the β relaxation in the dielectric spectrum. [Reprinted by permission from Reference (1), p. 78; copyright 1977 by Adam Hilger, Ltd.]

polymer the side chain motions do not have as large an effect on the mechanical properties.[1]

Figure 6 shows a rigid unplasticized poly(vinyl chloride) [PVC] for which the dielectric permittivity has been measured as a function of temperature at different frequencies. At very low frequencies, essentially static conditions, the glass transition is about 80°C. As one goes to higher and higher frequencies it is more difficult for the polymer to relax and move in response to the higher frequencies, and therefore, higher temperatures are required to achieve this resonance. As one goes to relatively high frequencies, the glass transition can move as high as 150°C and beyond. This demonstrates the time–temperature response and correlation in polymeric materials.

Relaxation times can be calculated from the dielectric or mechanical loss data using relatively simple models[1,14] By plotting log (relaxation time) vs. $1/T$ (°K) and applying the Arrhenius equation, the activation energy for the transition can be calculated.[1,14] Figure 7a shows the various transitions of PMMA, plotted as relaxation time vs. $1/T$. The activation

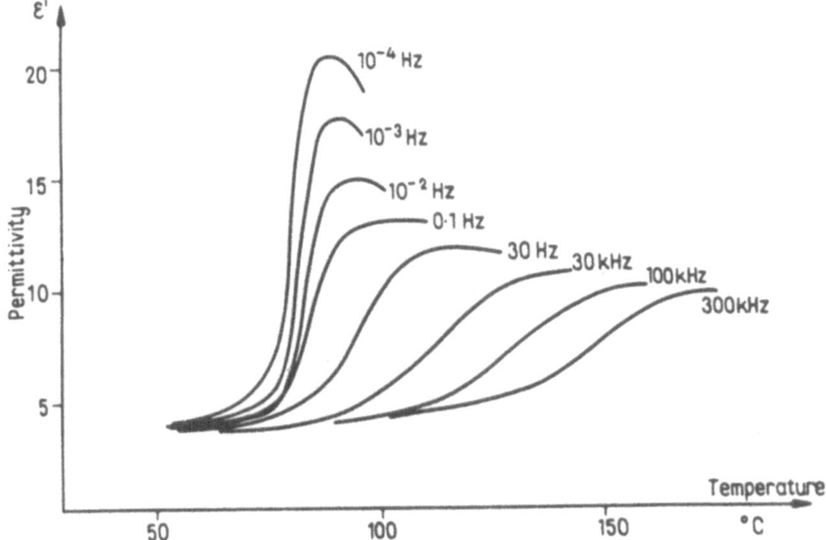

FIGURE 6. Dielectric permittivity vs. temperature of unplasticized poly(vinyl chloride), showing the effect of frequency (time) on relaxation temperature. [Reprinted by permission from Reference (1), p. 86, copyright 1977 by Adam Hilger, Ltd.]

energy for beta or side chain rotation is about 20 Kcal/mol. The beta transition is activated around room temperature. Figure 7b shows poly(methyl acrylate); note the decreased transition temperatures for both the glass transition (α) and the beta (β) relaxation as compared to PMMA. The Arrhenius equation only applies to sub T_g transitions. In the range of T_g and above, other t-T relationships must be used.[14] Many such relaxation time-temperature maps are available in Reference (1).

Other methods of probing polymer motions and relaxations include fluorescence,[42] nuclear magnetic resonance spectroscopy,[23] electron spin resonance,[23] and inverse gas chromatography (IGC).[24] Only the latter will be discussed here because of its importance in probing the surface properties of polymers.

3.2. INVERSE GAS CHROMATOGRAPHY (IGC)

The inverse GC method uses a chromatographic column on which the normal packing material has been coated with a thin film of the polymer of interest. One then performs a GC experiment in which a probe solute is injected into the carrier stream and the retention volume of that probe molecule is recorded at different temperatures. The retention volume, V, is then corrected to a standard set of conditions, and the specific (per gram

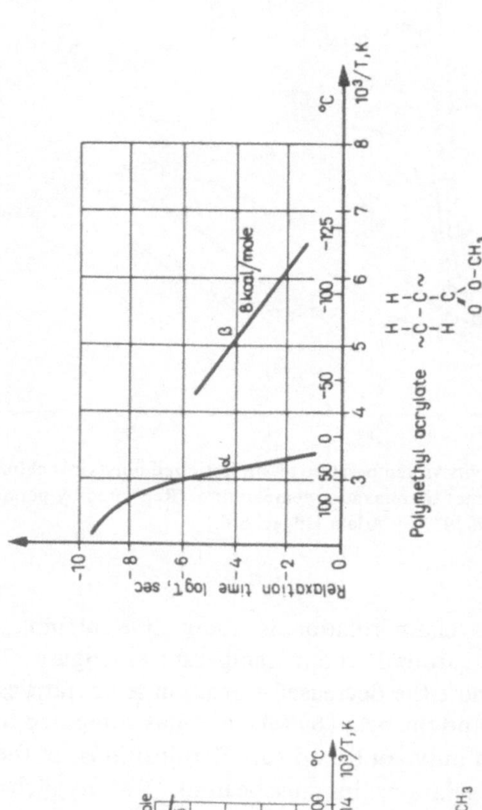

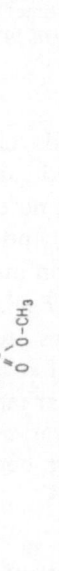

FIGURE 7. A relaxation time–temperature map for poly(methyl methacrylate) (*Left*) and poly(methyl acrylate) (*Right*). The slopes of the lines yield the Arrhenius activation energy for the relaxation. [Reprinted by permission from Reference (1), pp. 399, 400; copyright 1977 by Adam Hilger, Ltd.]

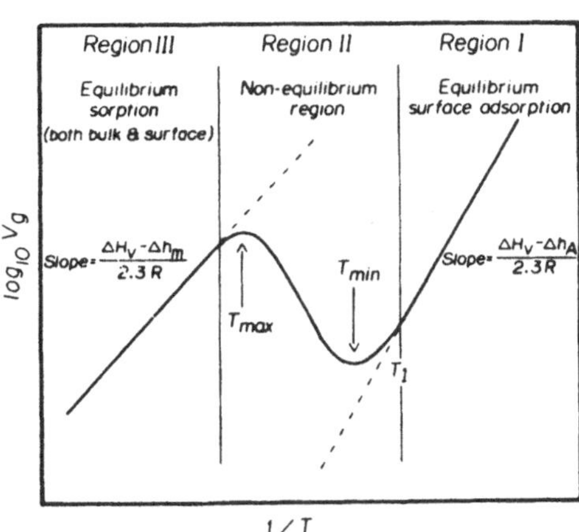

FIGURE 8. Plot of the log of the retention volume, V_g, against $1/T$ for an inverse gas chromatography experiment using an ideal amorphous polymer material. See text for discussion. [Reprinted by permission from Reference (25), p. 884; copyright 1975 by the American Chemical Society.]

or area of column packing material) retention volume, V_g, is plotted against $1/T$. Standard GC texts show that $V_g \propto \exp(-\Delta H/RT)$ or $[\ln V_g/(1/T)] = -\Delta H/R$, where ΔH is the molar enthalpy of the retention process. Assuming no changes in the support with temperature and an adsorption only process, ΔH would be the adsorption enthalpy. Figure 8 is an idealized plot showing a glass transition effect.[(24,25)] In Region III of the plot, which represents the situation at elevated temperatures, the probe molecules are both adsorbing and absorbing into the polymer and the retention, though a complex process, is dominated by absorption effects. As the temperature decreases ($1/T$ increases) the retention volume decreases (Region II) because now the probe molecules find it progressively more difficult to absorb into the bulk polymer due to the decreased motion of the chains. A point is reached (the glass transition region) where the polymer chains are relatively frozen and immobile, at least in terms of bulk segmental motions, and the probe molecules cannot penetrate. As the temperature continues to decrease (Region I) the separation process is primarily due to surface adsorption. The adsorption constant increases as temperature decreases, and the retention volume also increases. The T_g is generally defined as that temperature where the Region I slope deviates from linearity.[(25)] Figure 9 shows inverse GC data for PVC with different levels of dioctylphthalate plasticizer, demonstrating the well-known effect that T_g decreases with increasing plasticizer content. Figure 10 presents inverse GC data on atactic polystyrene as a function of degree of loading of the chromatographic support.

As one goes to extremely thin films, the technique becomes progressively less sensitive to T_g; the T_g indeed appears to increase with very thin films (Figure 10), perhaps suggesting some influence of the substrate on the transition properties of the thin polymer film adjacent to it.

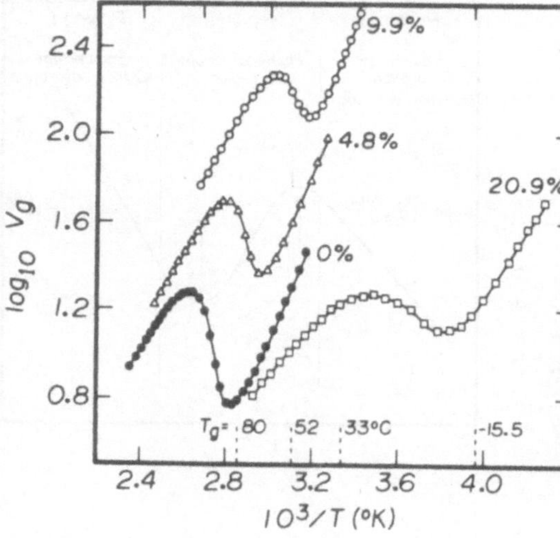

FIGURE 9. Retention diagrams for *n*-decane and *n*-pentane (□) on poly(vinyl chloride) plasticized with dioctylphthalate. Note the change in the glass transition with plasticizer loading. [Reprinted from Reference (24), p. 107, by permission.)

Thus, a gas chromatographic experiment can directly measure the glass transition temperature of polymers. This technique is relatively new; as sensitivities improve and applications increase, one may expect in certain cases to sense sub-glass transitions as well.

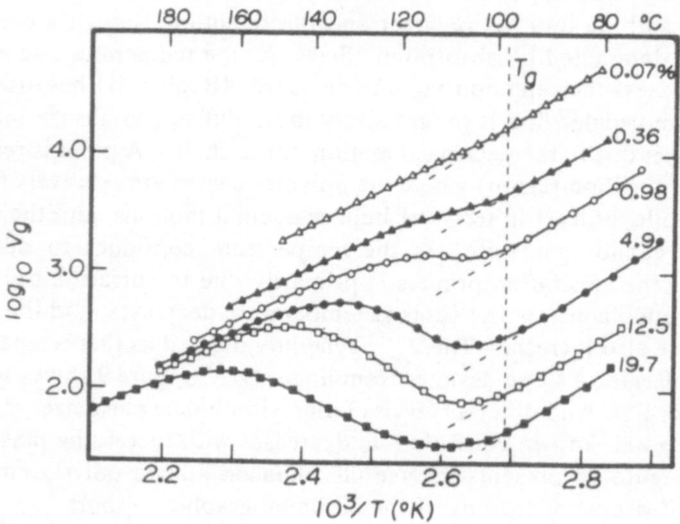

FIGURE 10. Demonstration of the effect of polymer loading, which relates to polymer coating thickness, on the inverse GC behavior of polystyrene using *n*-hexadecane as the gas probe. Note the apparent shift in the transition temperature at very low loadings. [Reprinted from Reference (25), p. 884, by permission; copyright 1975 by the American Chemical Society.]

4. POLYMER SURFACE MOTIONS

One would intuitively expect that polymer molecules in the vicinity of the surface or interface would also exhibit motions and relaxations, although they should be different from the motions observed in the bulk due to the different interfacial environment. A number of techniques are available with which to probe this hypothesis.

One is inverse gas chromatography, just discussed. A second approach is to use classical measures of polymer transitions and relaxations in highly filled polymers, where a significant proportion of the total polymer molecules are adjacent or close to a solid interface. A third method is to directly measure the wetting properties and surface energetics of polymers as a function of time and temperature. These will be discussed in order.

4.1. INVERSE GC

Kessaissia *et al.*[26] have shown that transitions can be observed with short alkyl chains chemically attached to silica supports. They used argon, nitrogen, and methane to probe alkyl-derivatized silica. The log V_g–$1/T$ plots were smooth for underivatized silica (Figure 11a). Alkyl chains grafted

(a)

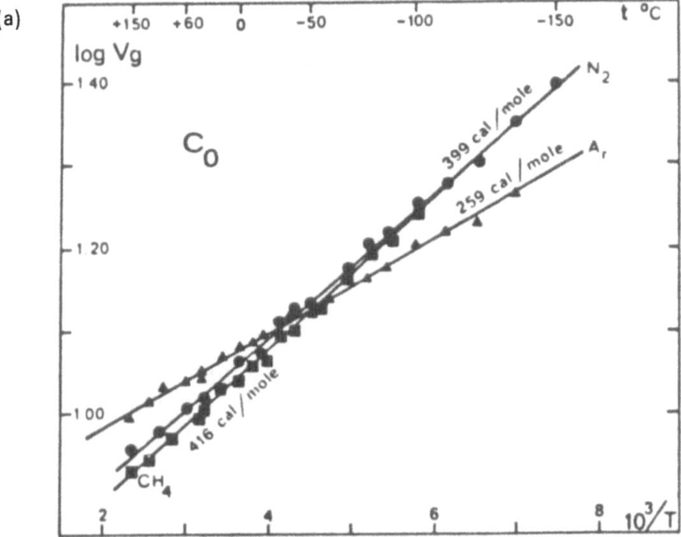

FIGURE 11. (a) Inverse GC data for unmodified silica using N_2, Ar, and CH_4 probe gases. Note the apparent linearity of the log V_g vs. $1/T$ plot for all three probes. (b) Results of the same experiment as (a) but for a C_3-esterified silica. Note now the abrupt changes in the plot at about −20 and −75°C. (c) Same as (b) but for a C_6 esterified silica. Again, note the different thermal transitions present for the three probe gases. (d) Plot of the observed transition temperatures against number of carbons for the different modified silica materials. [Reprinted from Reference (26), pp. 257-263; copyright 1981 by Academic Press, Inc.]

(b)

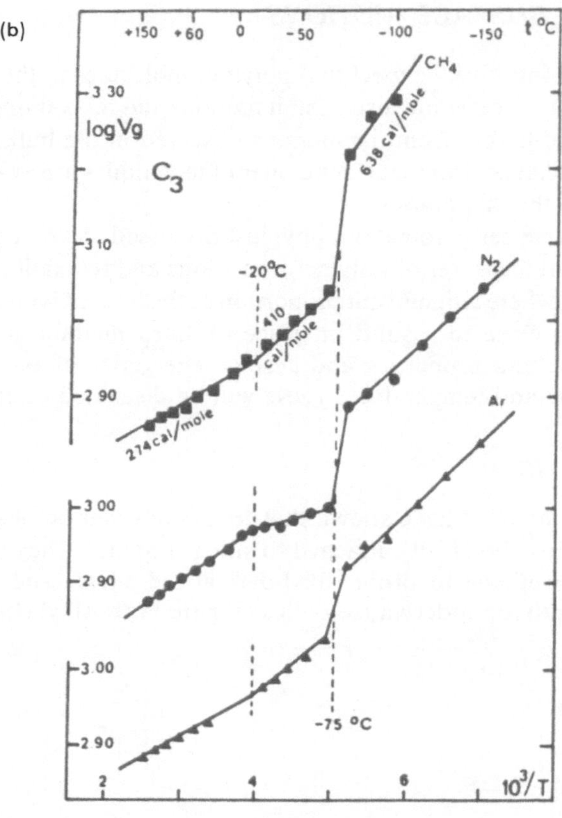

FIGURE 11 (continued)

to the polymer at a density of two alkyl chains per 100 Å^2 showed several transitions (Figure 11b, c). A plot of the transition temperatures as a function of the alkyl chain length showed common transitions at −20 and −80°C, and an intermediate transition which varied with chain length (Figure 11d).[26] The significance of this with respect to the surface properties of alkyl-grafted silicas is not immediately apparent, but it does serve to demonstrate the sensitivity of this technique to what must be relatively local, short-range motions and relaxations. This study utilized an instrument modified for work at low temperatures.[26]

Schreiber and coworkers[27] have shown that inverse GC measurements of polymer films prepared from different solvents show different retention times. The retention times for PMMA at room temperature were a function of the nature of the solvent from which the film was prepared, whereas no such effect was noted for polystyrene. They suggest that the different casting solvents provide chain conformations in solution which result in different

(c)

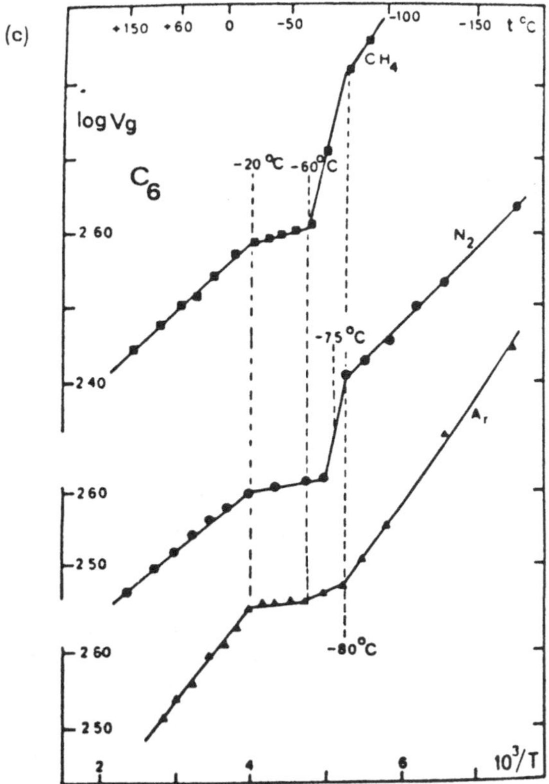

(d)

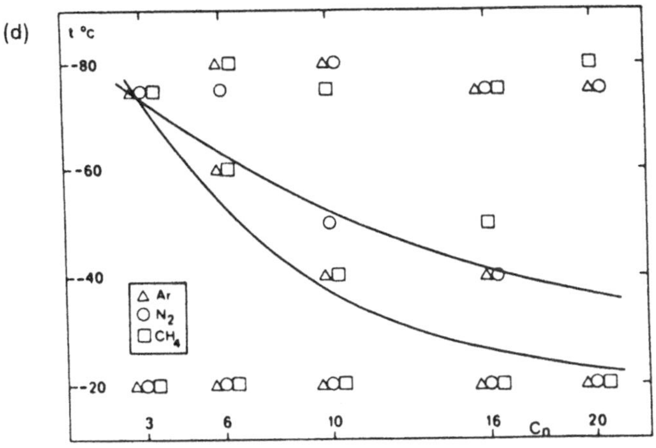

FIGURE 11 (continued)

surface conformations in the solid state. They further note "... that for any polymer only a single equilibrium surface structure can obtain; when a nonequilibrium surface condition is produced, slow but significant time-dependent variations in film properties are to be expected as the equilibrium condition is sought and attained."[(27)] This study suggests that relatively subtle changes in the surface properties result in different retention times as detected by inverse GC.

4.2. FILLED SYSTEMS

The study of highly filled polymer systems allows one to measure the relaxations and transitions by classical methods, that is by mechanical and dielectric spectroscopies and by thermal methods. Although the literature is still somewhat controversial, it is expected that the fact that the polymer is at a rigid interface, such as a silica filler, must, in principle, constrain its motions and decrease its allowable degrees of freedom, influencing the glass transition and other transition temperatures. Lipatov emphasized the non-specific, rigid surface effect;[(28)] a direct interaction effect is treated by Howard and Shanks[(29)] and by Yim *et al.*[(30)] A correlation was observed between the shift in T_g and the polymer–filler interaction energy.[(30)]

Measurements of transitions in highly filled polymers are highly sensitive to the preparation and thermal history of the polymers.[(31)] In fact, it is important to note that since most solid polymers are nonequilibrium structures, subtle changes in thermal history and preparation conditions may dramatically affect the internal structure and, therefore, the transitions observed. This is already noted in the case of the casting solvent effect studies by Schreiber and Croucher.[(27)]

4.3. CONTACT ANGLE METHODS

Another approach to the measurement of polymer surface relaxation or motions is to prepare the surface under one set of conditions and examine it as a function of time and temperature in another environment. This has been done by Pennings and Bosman.[(32)] They prepared polymer films against high-energy substrates and then measured the surface energetics as a function of time and temperature (Figure 12) by standard contact angle, critical surface tension, techniques (see Chapter 7).

A poly(vinyl chloride–vinyl acetate) copolymer was compression molded against gold foil at 403°K. The vinyl acetate segments preferentially adsorb on the gold surface. Surface tension, γ, deduced from contact angles, ranges from a maximum of 49 to a minimum of 39 mN/m. A plot of $\Delta\gamma$ vs. time shows a temperature-dependent relaxation process (see Figures 12, 13, and caption), which yields an Arrhenius activation energy of 43 kJ/mole.

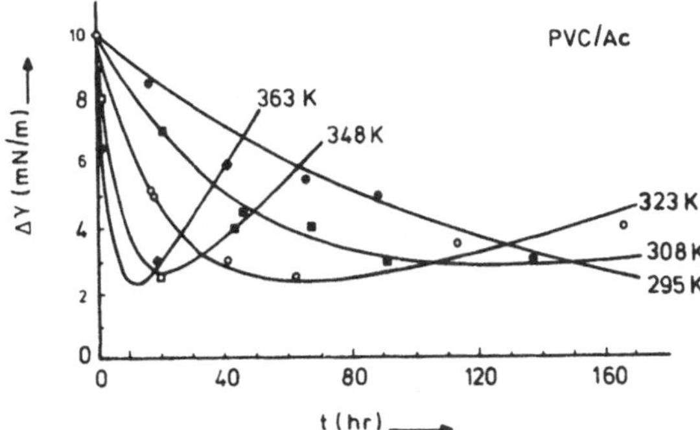

FIGURE 12. Change in surface tension of a poly(vinyl chloride-vinyl acetate) copolymer as a function of time and temperature in vacuum after being compression molded against gold foil at 403 K. The surface tension immediately after molding was 49, the equilibrium value was 39 nN/m. 87% vinyl chloride; MW = 22,000. The upward trend at higher temperatures is due to polymer degradation. Note the decrease in surface tension with time. [Reprinted by permission from Reference (32); copyright 1979 by Springer-Verlag.]

The beta relaxation energy is 40 kJ/mole for pure PVAc and 63 for PVC, suggesting that the relaxation process is largely due to motions of the vinyl acetate side chains.[32]

A similar experiment has been performed by Carre and Schreiber[33] by casting PMMA onto high-energy mercury and low-energy polytetrafluoroethylene (PTFE) surfaces and then measuring the surface energy as a function of time by critical surface tension methods. The critical surface tension of PMMA produced against air or PTFE were identical (~42 mN/m), whereas against Hg it was ~47 mN/m and decayed with time.

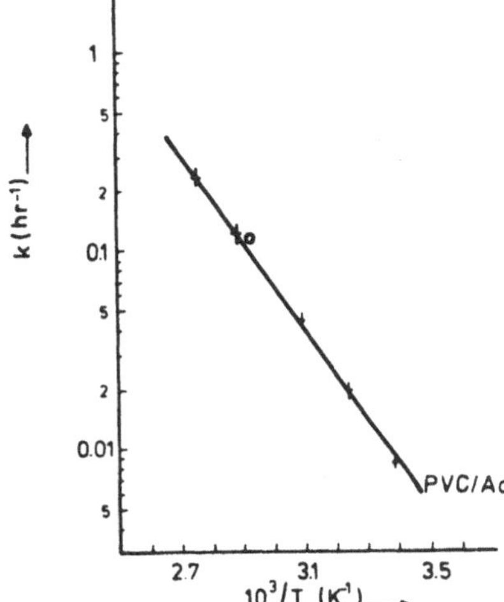

FIGURE 13. The relaxation rate initial slope vs. $1/T$ for the data of Figure 12, producing an activation energy of 43 kJ/mole. [Reprinted from Reference (32), pp. 720–724, by permission; copyright 1979 by Springer-Verlag.]

Holly and Refojo[5] suggested that the water contact angle hysteresis observed on hydrophilic, hydrated, polymethacrylates was due to reorientation of the hydrophilic side chains at the polymer–air and polymer–water interfaces. Yasuda *et al.*[34] have shown that water contact angle hysteresis can be related to reorientation of hydrophilic side chains. Contact angle hysteresis measurements of many polymer systems suggest that the surface regions of a polymer can reorient in response to air or water environments. These effects are discussed in Chapter 7 in some detail.

Studies by Okawa[38] using a model system of methacrylates and acrylates of varying T_g and T_β show that the receding water contact angle is a function of water contact time. If one plots the water hydration time to achieve a constant and minimum water receding angle as a function of the glass transition temperature of the polymer, a crude correlation (Figure 14) is obtained. The surface appears to be restructuring in response to the aqueous environment, although the restructuring is kinetically slow and related to the transition temperatures, as expected. Part of this effect may be due to water absorbing into and solubilizing the surface region of the

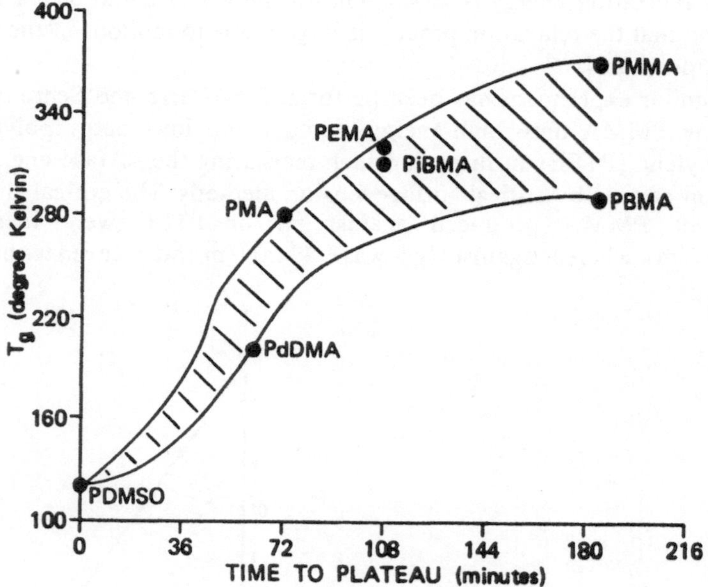

FIGURE 14. Plot of the glass transition temperature of a series of acrylate and methacrylate polymers as a function of the water hydration time required to achieve a constant and minimum water receding contact angle. Data obtained by the Wilhelmy plate contact angle procedure.[38] PBMA equals poly(butyl methacrylate), PdDMA equals poly(dodecyl methacrylate), PDMSO equals poly(dimethyl siloxane), PEMA equals poly(ethyl methacrylate), PiBMA equals poly(isobutyl methacrylate), PMA equals poly(methyl acrylate), PMMA equals poly(methyl methacrylate).

polymers. This water penetration effect is commonly invoked as a mechanism to explain contact angle hysteresis, particularly with small, polar, highly penetrating liquids such as water (see Chapter 7). The absorbed water, in addition to conferring its own hydrophilicity to the surface, would probably plasticize the surface region, thereby augmenting transitions and reorientations.

4.4. OTHER METHODS

Monomolecular film studies of polymers, including pressure–area isotherms and surface potential determinations of the spread monolayers,[35] clearly indicate that such polymer monolayers are highly oriented at the solution–air surface. Surface and interfacial tension studies of dimethylsiloxane oligomers at the siloxane–water surface[36,37] suggest significant interfacial orientation and indeed a modest surface activity for these materials. These effects are particularly pronounced in aqueous systems in which the high surface tension of water provides a high interfacial energy driving force for polymer surfaces to orient such as to minimize the interfacial tension.

There are other studies documenting surface motions and relaxations of polymers. Ohara[39] has demonstrated a correlation between frictional electrification of polymers and molecular motions and relaxations. Klein and Luckham[40] have measured the interaction force between two adsorbed poly(ethylene oxide) layers, demonstrating incomplete relaxation of the adsorbed polymer layers at short experiment times. Fluorescence methods have also been used to probe motions and relaxations.[42]

Baszkin *et al.*[41] showed that a hydrophilic, surface-modified polyethylene becomes hydrophobic when heated in an inert atmosphere, attributed to a surface tension-driven "burial" of the hydrophilic groups when the temperature is high enough to permit sufficient molecular mobility.

The surface modification of polymers in air by the diffusion of low-energy species to the surface is now well-known.[48] The controlled modification of polymer studies by diffusion of functional groups from the bulk to the surface has been proposed.[47] Careful studies of block copolymers show that the low-energy phase generally dominates the surface in air or vacuum (minimizing the surface free energy), but the high-energy phase dominates the interface underwater (minimizing the interfacial free energy).[46]

Direct measurements have now been made relating surface mobility of poly(ethylene oxide) side chains to protein and platelet interactions.[45]

Surface relaxation and reorientation effects have also been observed by X-ray photoelectron spectroscopy.[49]

Many other studies, including inverse GC, contact angle, filler, and monomolecular film studies, all attest to the fact that polymer surfaces can and do respond to their environments. A further discussion of these points with respect to water contact angle hysteresis is presented in Chapter 7.

5. CONCLUSIONS

Given sufficient mobility, polymer surfaces will reorient or restructure in response to their local micro-environment so as to minimize their interfacial free energy with the surrounding phase.

The interfacial free energy at a polymer–water interface is a sufficient driving force to cause restructuring of the polymer surface and orientation of the dipolar and other groups which can directly interact with water towards the aqueous phase. These processes are time–temperature dependent, and correspond to the relaxation characteristics of the polymer; thus, long equilibration times with water may be required before the effect is maximally manifested.

Even relatively rigid polymers, such as poly(methyl methacrylate), reorient at the polymer–water surface, due to relaxation mechanisms in the surface region which may occur at lower temperatures than in the bulk, perhaps due to surface-induced water plasticization of the interfacial region, and due to segmental side chain motions which are activated at or near room temperature.

In systems containing hydrophilic phases of submicroscopic dimensions, such as common diblock and triblock copolymers, given sufficient mobility the hydrophilic phase will dominate the interface in water, the hydrophobic phase will dominate in air.

It is suggested that the adsorption of biological and other macromolecules at a polymer-water interface will result in considerable restructuring of the polymer surface in response to the local microenvironment of the adsorbed macromolecule as well as the local water and other solution components.

We propose that it is necessary to characterize the surface properties directly at the solid–water interface as well as the more commonly and classically performed solid–air or vapor interface when searching for correlations between the surface properties of polymers and their biological behavior. Further, the polymer-water interfacial properties may need to be characterized as a function of hydration time or after suitable water equilibration.

The surface motions of polymers are probably different from the bulk motions. Few data are available; experimental methods with which to probe surface motions are becoming available and have been discussed.

Polymer surface restructuring effects in response to a surrounding liquid phase are probably most pronounced in aqueous systems due to the unique hydrogen bonding and acid–base characteristics of water.

Finally, these effects are not readily detectable by classical advancing contact angle measurements, including determinations of critical surface tension, nor by X-ray photoelectron spectroscopy or other analysis techniques which primarily probe the solid-vacuum or solid-air interface.

6. SUMMARY

A brief review of polymer transitions and relaxation phenomena has been presented, together with a summary of hypotheses and speculations regarding the correlation between polymer surface motions and biological interactions. Methods of probing polymer transitions and relaxations have been discussed, with particular emphasis placed on techniques useful for studying surface motions. Surface motions and relaxations have been demonstrated by inverse gas chromatography and contact angle data. The basic conclusion is that polymer surface motions do occur, resulting in relaxation and re-equilibration of polymer surfaces in response to different environments. Polymer surfaces are highly sensitive to processing and fabrication conditions and relax or re-equilibrate in the use environment. This effect is particularly dramatic in the case of an aqueous environment in which the interaction of the polymer with water provides a strong driving force for reducing interfacial tension by reorientation of polar surface groups to optimally interact with the aqueous phase. In attempting to draw correlations between biological interactions and polymer surface properties, it is important to be aware of the fact that the surface underwater may be, and generally is, very different from the surface in air or in other characterization environments.

ACKNOWLEDGMENTS

This work was partially supported by NIH grants HL18519 and HL26469. Discussions with R. Ward, M. Owen, R. N. King, A. Okawa, and M. Reichert are appreciated.

REFERENCES

1. P. Hedvig, *Dielectric Spectroscopy of Polymers*, Wiley, New York (1977).
2. D. J. Williams, *Polymer Science and Engineering*, Prentice-Hall, Inc., Englewood Cliffs, New Jersey (1971).
3. J. D. Andrade, D. E. Gregonis, and L. M. Smith, in: *Physicochemical Aspects of Polymer Surfaces* (K. L. Mittal, ed.), pp. 911-922, Plenum Press, New York (1983).
4. E. Nyilas and R. S. Ward, Jr., Development of blood compatible elastomers. V., *J. Biomed. Materials Res. Symp.* **8**, 69-84 (1977); also *Proc. 23rd Ann. Conf. Eng. Med. Biol.* **12**, 147-148 (1970).

5. F. J. Holly and M. F. Refojo, Wettability of hydrogels, *J. Biomed. Materials Res.* **9**, 315-326 (1975).
6. E. W. Merrill, Behavior of blood at surfaces, *Ann. N.Y. Acad. Sci.* **283**, 6-16 (1977).
7. S. A. Barenberg, J. S. Schultz, J. M. Anderson, and P. H. Geil, Hemocompatibility: macromolecular motions and order of the polymer interface, *Trans. Am. Soc. Artificial Internal Organs* **25**, 159-162 (1979).
8. S. A. Barenberg, J. M. Anderson, and K. A. Mauritz, Thrombogenesis: an epitaxial phenomena, *J. Biomed. Materials Res.* **15**, 231-245 (1981).
9. W. M. Reichert, R. E. Filisko, and S. A. Barenberg, Polyphosphazenes: effect of molecular motions on thrombogenesis, *J: Biomed. Materials Res.* **16**, 301-312 (1982).
10. Y. Sakurai, T. Akaike, K. Kataoka, and T. Okano, in: *Biomedical Polymers* (E. P. Goldberg and A. Nakajima, eds.), pp. 335-379, Academic Press, New York (1980).
11. D. Brier-Russell, E. W. Salzman, J. Lindon, R. Handin, E. C. Merrill, A. K. Dincer, and J.-S. Wu, Interaction of blood with model surfaces, *J. Colloid Interface Sci.* **81**, 311-318 (1981).
12. D. L. Coleman, *In Vitro Blood-Materials Interactions: A Multi-Test Approach*, PhD. Thesis, University of Utah, August, 1980.
13. A. M. North, in: *Molecular Behaviour and the Development of Polymeric Materials* (A. Ledwith and A. M. North, eds.), pp. 368-403, Chapman and Hall, London (1974).
14. C. D. Armeniades and E. Baer, in: *Introduction to Polymer Science and Technology* (H. S. Kaufman, ed.), pp. 239-299, John Wiley and Sons, New York (1977).
15. M. J. Richardson and N. G. Savill, What information will DSC give on glassy polymers?, *Brit. Polym. J.* **11**, 123-129 (1979).
16. R. A. Pethrick, Molecular motion in semi-flexible macromolecules, *Sci. Prog. Oxf.* **6**, 571-592 (1980).
17. J. M. G. Cowie, Relaxation processes in the glassy state: molecular aspects, *J. Macromol. Sci. Phys.* **B18**, 569-623 (1980).
18. J. Brandrup and E. H. Immergut, eds. *Polymer Handbook*, 2nd Edition, John Wiley and Sons, New York (1975).
19. L. R. Brostrom, D. L. Coleman, D. E. Gregonis, and J. D. Andrade, Thermal analysis of polymethacrylates, *Makromol. Chem., Rapid Comm.* **1**, 341-343 (1980).
20. E. A. Turi, *Thermal Characterization of Polymeric Materials*, Omnitherm Corp., Arlington Heights, Illinois (1982).
21. J. K. Gillham, S. J. Standicki, and Y. Hazony, Low-frequency thermomechanical spectrometry of polymeric materials: tactic poly(methyl methacrylates), *J. Appl. Polym. Sci.* **21**, 401-424 (1977).
22. N. G. McCrum, B. E. Read, and G. Williams, *Anelastic and Dielectric Effects in Polymeric Solids*, John Wiley and Sons, New York (1967).
23. A. E. Woodward and F. A. Bovey, eds., *Polymer Characterization by ESR and NMR*, Am. Chem. Soc. Symp. Series **142** (1980).
24. J.-M. Braun and J. E. Guillet, Study of polymers by inverse gas chromatography, *Adv. Polym. Sci.* **21**, 107-145 (1976).
25. J.-M. Braun and J. E. Guillet, Studies of polystyrene in the region of the glass transition temperature by inverse gas chromatography, *Macromolecules* **8**, 882-888 (1975).
26. Z. Kessaissia, E. Papirer, and J.-B. Donnet, Molecular transitions of alkyl chains grafted onto silicas observed by gas chromatography, *J. Colloid Interface Sci.* **79**, 257-263 (1981).
27. H. P. Schreiber and M. D. Croucher, Surface characteristics of solvent-cast polymers, *J. Appl. Polym. Sci.* **25**, 1961-1968 (1980).
28. Y. S. Lipatov and L. M. Sergeeva, *Adsorption of Polymers*, Chapter 4, John Wiley and Sons, New York (1974).
29. G. J. Howard and R. A. Shanks, Influence of filler particles on the mobility of polymer molecules, *J. Macromol. Sci., Chem.* **A17**, 287-295 (1982).

30. A. Yim, R. S. Chahal, and L. E. St. Pierre, Effect of polymer-filler interaction energy on the Tg of filled polymers, *J. Colloid Interface Sci.* **43**, 583-590 (1973).
31. P. Peyser and W. D. Bascom, Effect of filler and cooling rate on the glass transition of polymers, *J. Macromol. Sci-Phys.* **B13**, 597-610 (1977).
32. J. F. M. Pennings and B. Bosman, Relaxation of the surface energy of solid polymers, *Colloid Polym. Sci.* **257**, 720-724 (1979).
33. A. Carre and H. P. Schreiber, Solvent history effects and multi-valued surface properties of PMMA coatings, *Proc. FATIPEC Congress*, Belgium, May, 1982.
34. H. Yasuda, A. K. Sharma, and T. Yasuda, Effect of orientation and mobility of polymer molecules at surfaces on contact angle and its hysteresis, *J. Polym. Sci., Polym. Physics* **19**, 1285-1291 (1981).
35. N. Beredijick, in: *Newer Methods of Polymer Characterization* (B. Ke, ed.), Interscience, New York (1964).
36. H. W. Fox, P. W. Taylor, W. A. Zisman, Polyorganosiloxanes: surface active properties, *Ind. Eng. Chem.* **39**, 1401-1409 (1947).
37. M. J. Owen, The surface activity of silicones, *Ind. Eng. Chem., Prod. Res. Develop.* **19**, 97-103 (1980).
38. A. Okawa, B.Sc. Thesis, Department of Materials Science, University of Utah, June, 1983.
39. K. Ohara, Relationship between frictional electrification and molecular motion of polymers, *J. Electrostatics* **9**, 107-115 (1980).
40. J. Klein and P. Luckham, Forces between two adsorbed PEO layers immersed in a good aqueous solvent, *Nature* **300**, 429-430 (1982).
41. A. Baszkin, N. Nishino, and L. Ter-Minassian-Saraga, Solid-liquid adhesion of oxidized polyethylene films, *J. Colloid Interface Sci.* **54**, 317-322 (1976).
42. H. Morawetz, Fluorescence studies of conformational mobility, *Pure Appl. Chem.* **52**, 277-284 (1980).
43. K. C. Rusch, Time-temperature superposition and relaxation behavior in polymer glasses, *J. Macromol. Sci. Phys.* **B12**, 179-204 (1968).
44. L.-H. Lee, Surface wettability and glass temperatures, *J. Appl. Polym. Sci.* **12**, 719-730 (1968).
45. S. Nagaoka, Y. Mori, H. Takiuchi, K. Yokota, H. Tanzawa, and S. Nichiumi, Interaction between blood components and hydrogels with poly(oxyethylene) chain, *Polymer Preprints* **24**, 67-68 (1983).
46. T. Matsuda and T. Akutsu, Blood/materials interactions of hydrophobic and hydrophilic segmented polyurethanes, *Organic Coatings and Applied Polymer Science Preprints* **48**, 647-648 (1983).
47. T. J. McCarthy, Polymer surface modification by diffusion of functional groups, *Organic Coatings and Applied Polymer Science Preprints* **48**, 520-522 (1983).
48. R. S. Ward, Jr., Development of thermoplastics, *Organic Coatings and Plastics Preprints* **42**, 227-228 (1980).
49. D. S. Everhart and C. N. Reilley, Functional group mobility, *Surface and Interface Analysis* **3**, 126-133 (1981).

3

Model Polymers for Probing Surface and Interfacial Phenomena

D. E. Gregonis and J. D. Andrade

1. INTRODUCTION

This chapter will review the use of model polymer systems for studying surface and interfacial properties. It is not intended to be all inclusive in regard to the overall chemistry of the systems, or even to review each and every polymer which has been investigated. This chapter is meant to provide a condensed overview of the bulk and surface characteristics of selected systems. Generally these model polymers are systematically prepared in order to change the bulk composition of polymeric material. The bulk composition is then extrapolated to the interface. It must be emphasized that this extrapolation is often not direct. The tendency is for the polymer to minimize its interfacial energies; thus, a polymer cast against a clean (high energy) glass surface may exhibit different surface properties than the same polymer's air-exposed surface. These effects are more pronounced in block copolymers which may have large domains of different surface properties. A polymer with high glass transition temperature may retain its surface energies for extended periods of time or until annealing. Polymers with their glass transition below room temperature may reorient quite rapidly upon exposure to different environments and may have different groups exposed in an air environment as compared to an aqueous environment. (See Chapter 2.)

Plasticization of a polymer generally decreases the glass transition temperature. A hydrogel, such as poly(hydroxyethyl methacrylate), can have a high glass transition temperature (115°C)[1] in the dry state but becomes a highly flexible polymer when swollen with water. Even polymers

---

D. E. Gregonis and J. D. Andrade • Department of Bioengineering, University of Utah, Salt Lake City, Utah 84112.

that are not highly swollen in the aqueous environment may experience enough plasticization of the polymer chains at the interface to have different thermal motions as compared to the polymer chains found in the bulk.

What must be emphasized is that extrapolation from bulk polymer measurements to surface properties is not direct. The ability to measure the top ten angstroms of the polymer interface are limited; most of the available techniques measure the dry, nonhydrated polymer surface. However, until surface measurement techniques become more advanced, knowledge of the bulk polymer system is still important for investigating surface trends.

2. STEREOREGULAR POLYMERS

2.1. PURPOSE (STEREOCHEMISTRY)

All addition polymers which have unsymmetrical groups off the main chain exhibit asymmetrical centers along the polymer chain. The orientation of these groups in relation to adjacent neighbors is referred to as stereoregularity. Polymers such as polyethylene, polyisobutylene, and poly(vinylidene chloride) have symmetric pendant groups, and thus do not exhibit stereoisomerism. Although several forms of stereoisomerism exist,[(2)] only the methacrylate polymer system, which can exist in isotactic, heterotactic, and syndiotactic configuration, will be covered here. Even though it seems reasonable that the stereochemistry of the polymer may influence its biological interactions, no systematic study of these polymeric systems has been done to substantiate this hypothesis.

One of the most studied systems, the methacrylates, has a methyl group and a carboxylic acid or derivatized carboxylic acid group attached to every other carbon atom along the main chain. Thus, if the polymer chain is

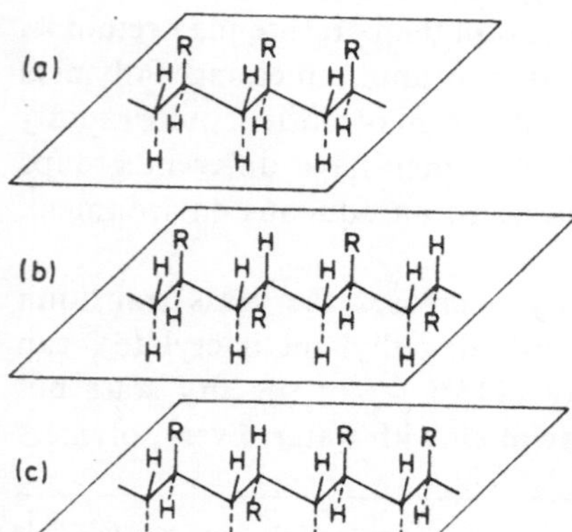

FIGURE 1. Zig-zag planar conformation of the main chain of vinyl polymers having different types of stereoregularity: (a) isotactic chain; (b) syndiotactic chain; (c) atactic chain [reprinted with permission from Reference (76)].

stretched, in the extreme cases of tacticity the same group is always on the same side (isotactic) or on alternate sides (syndiotactic) of the polymer chain. Random orientation of the groups is referred to as atactic (Figure 1).

The stereoregularity of the polymer usually has a strong influence on bulk characteristics. For example, the glass transition (T_g) for poly(methylmethacrylate) is 41.5°C for isotactic, 125.6°C for syndiotactic, and 104°C for atactic polymer.[3] Tactic polymers generally show detectable changes in properties such as density, solubility, rate and extent of solubility, crystallinity, thermal and mechanical transitions, and spectroscopic characteristics such as infrared and nuclear magnetic resonance (NMR) spectra and X-ray diffraction patterns. Although many of these methods have been used to quantitate stereoregularity, the NMR technique is most sensitive for exact measurements in most polymer systems.[4]

2.2. CHEMISTRY

The synthesis of methyl and other alkyl methacrylates has been widely studied. The stereoregularity can be varied by both radical and anionic initiation methods. Under identical reaction conditions, the stereochemistry of the resulting polymer can be strongly influenced by the size of the ester substituent.

In general, for poly(methyl methacrylate) and other small ester methacrylates, the tacticity of the polymer polymerized free radically produces a 60% syndio, 40% hetero, 0% isotactic configuration.[5] Decreasing the temperature of polymerization results in a greater degree of syndiotacticity (Table 1). For example, poly(hydroxyethyl methacrylate) polymerized by radical means at −40°C produced a 84% syndio, 16% hetero, and <1% isotactic configuration.[6]

TABLE 1
Effect of Temperature on Tacticity in Radical Polymerization of Methyl Methacrylate[a]

Temperature (°C)	Fraction of syndiotactic placement(s)
−40	0.86
60	0.76
100	0.73
150	0.67
250	0.64

[a] From Reference (74).

TABLE 2
Effect of Solvent on Tacticity of Poly(methyl methacrylate)[a]

Polymerization system	Tacticity (triad analysis) Iso	Hetero	Syndio
Radical, bulk polymerization at 60°C	0.08	0.33	0.59
n-C_4H_9Li at −78°C			
% Tetrahydrofuran in toluene:			
0	0.78	0.16	0.06
2.5	0.30	0.31	0.39
5	0.24	0.34	0.42
10	0.13	0.35	0.52

[a] From Reference (74).

The anionic polymerization technique provides much more versatility in preparation of these tactic polymers. The polymerization is much more sensitive to reaction conditions, e.g., temperature, solvent composition, order and possible rate of addition of reactants.[(7)] One of the more common anionic initiators is n-butyl lithium. When polymerization takes place in an apolar solvent such as toluene, highly isotactic product results; but upon addition of the polar solvent tetrahydrofuran to the reaction, a syndiotactic polymer results, as shown in Table 2. The polar solvent acts to increase the distance between the gegen ion from the propagating anion, negating its stereochemical influence on the incoming monomer. This effect is also noted in ether-containing monomers, such as methoxyethyl methacrylate. When anionically polymerized in an apolar solvent, the formation of syndiotactic polymer results due to ether solvation of the gegen ion.[(8)]

Stereoregular poly(hydroxyethyl methacrylate) hydrogels have been prepared using the above concepts. The hydroxyl group was protected as the benzoate ester during anionic polymerization, and then removed by selective hydrolysis. The isotactic polymer exhibits a more strongly temperature-dependent aqueous swelling profile than the syndiotactic polymer.[(6)]

2.3. BULK PROPERTIES

As mentioned above, the bulk properties are generally different for polymers of different stereoregularity. It has been proposed that isotactic poly(methyl methacrylate) exists as a double helix[(9)]; however, this recently has been questioned.[(10)] The relation between the tacticity of poly(methyl methacrylate) and glass transition temperature is shown in Table 3.

TABLE 3
Glass Transition as a Function of Tacticity of Poly(methyl methacrylate)[a]

Glass transition temperature T(°C)	Tacticity (triad analysis)		
	Iso	Hetero	Syndio
41.5	0.95	0.05	
54.3	0.73	0.16	0.11
61.6	0.62	0.20	0.18
104.0	0.06	0.37	0.56
114.2	0.10	0.31	0.59
120.0	0.10	0.20	0.70
125.6	0.09	0.36	0.64

[a] From Reference (3).

2.4. SURFACE CHARACTERIZATION

The surface wettability of tactic polymers as measured by critical surface tension contact angle methods exhibits no difference between isotactic and atactic poly(methyl methacrylate) ($\gamma_c = 36$ dyn/cm); however, isotactic poly(α-chloroethyl methacrylate) and poly(phenyl methacrylate) show lower γ_c by about 4 dyn/cm than those of the atactic polymer.[(11)] It is postulated that this effect is due to specific conformation requirements of the bulkier ester group which would minimize hydrogen bonding interactions between the methacrylate carbonyl and wetting liquids. The smaller methyl ester substituent would minimize this effect.

Methacrylate polymers and copolymers may show time-dependent interfacial properties in contact with water. This effect has been discussed in some detail in Chapter 2 and will not be repeated here. It is important to point out, however, that surface properties may and generally are time, temperature, and environment dependent.

3. HYDROXYETHYL METHACRYLATE COPOLYMERS

3.1. PURPOSE (SYSTEMIC HYDRATION)

The methacrylate system offers the versatility of providing a wide range of polymer properties by varying the side chain ester functionality. Many of these monomers are commercially available or can be conveniently prepared in the laboratory. A system which has been widely studied[(12)] is the poly(methyl methacrylate-hydroxyethyl methacrylate) (MMA-HEMA) copolymer system. Poly(methyl methacrylate) is a hard, rigid glassy polymer that absorbs approximately 1-2% water when hydrated. Poly(hydroxyethyl methacrylate) in the dry state is also a hard, glassy polymer; it becomes a hydrogel by absorbing 40% by weight of water when hydrated. Other hydrophobic comonomers have been substituted for MMA in this system, for example, ethyl methacrylate[13] and methoxyethyl methacrylate, resulting in relatively similar properties to those of the copolymers.

3.2. CHEMISTRY

The polymers of MMA and HEMA are polymerized generally by radical initiators. For biomedical applications, azobisisobutyronitrile (AIBN) or AIBN derivatives are generally preferred due to the similarity of the initiating group to the methacrylate backbone, and the absence of charge. The chemistry of the hydroxyethyl methacrylate system has recently been reviewed.[12,14]

The reactivity ratios of HEMA to MMA, ethyl methacrylate (EMA), and *n*-butyl methacrylate (BMA) have been reported by Varma *et al.*[15,16] for bulk polymerizations; they report a higher reactivity of the HEMA radical relative to the HEMA monomer and a higher reactivity for the MMA and other hydrophobic methacrylate radicals relative to the HEMA monomer.

The monomer reactivity of HEMA with hydrophobic methacrylates at 60°C using the Fineman–Ross method is:

r_1HEMA	r_2MMA	r_1HEMA	r_2EMA	r_1HEMA	r_2BMA
1.054	0.296	1.87	0.55	2.08	0.45

This indicates that the initial polymer produced is higher in HEMA composition in comparison to the monomer feed. However, other reports[17] indicate the reactivity ratios of HEMA and MMA to be r_1(HEMA), 0.66, and r_2(MMA), 0.86, indicating this copolymer tends toward an alternating system.

Soluble polymers of the methacrylate system are obtained by polymerization at high solvent dilution, typically one part monomer to ten parts solvent. Polymers prepared by this method at 60°C are ~60% syndio, 40% hetero, and <1% isotactic. The soluble polymers are generally used for preparation of solvent cast films. It should be mentioned that uncrosslinked PHEMA, although it swells in water, will not dissolve due to the hydrophobic nature of the methacrylate backbone.

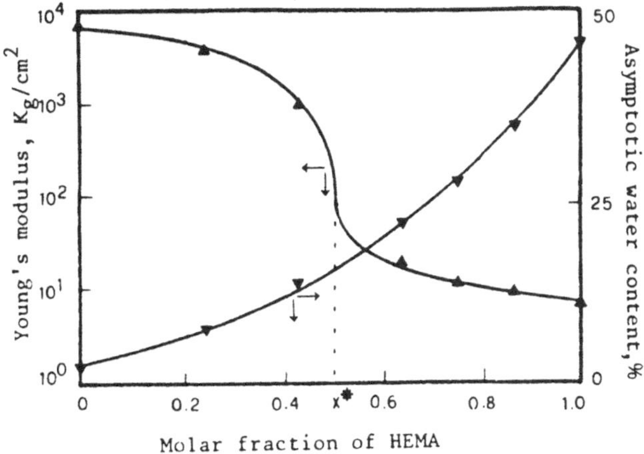

FIGURE 2. Water content and modulus of elasticity as a function of molar fraction of HEMA in HEMA-MMA copolymers [reprinted with permission from Reference (18)].

3.3. BULK PROPERTIES

The aqueous swelling of HEMA and MMA or methoxyethyl methacrylate (MEMA) copolymers results in a linear relationship when volume percentages of HEMA and MEMA are copolymerized.[(12)] The water fraction of MEMA homopolymer is 0.035% water and pHEMA homopolymer is 0.40% water. When molar fractions of MMA and HEMA are polymerized, the water content again increases with increasing percentage of HEMA; however, this does not follow exactly a linear relationship. Young's modulus undergoes a sharp decrease at near equal molar ratio.[(18)] This is the transition region where the hydrated copolymer changes from a glass to a plasticized rubbery polymer as shown in Figure 2.

3.4. SURFACE CHARACTERISTICS

The interface between a hydrogel and water is more diffuse when compared to a hydrophobic polymer-water interface.[(19)] It has been shown that contact angle methods can lead to local dimensional changes of the highly deformable gel surface.[(20)] (See Chapter 7.)

Captive underwater air and octane droplet data have been reported for the HEMA-MMA and HEMA-EMA copolymers.[(21)] This study utilized the Hamilton procedure[(22,23)] for calculation of dispersion and polar components at the interface and the surface-water interfacial free energies as shown in Figure 3. The surface-water interfacial free energies approach zero at values as low as 30% water suggesting that the hydrophilic phase dominates the underwater interface.

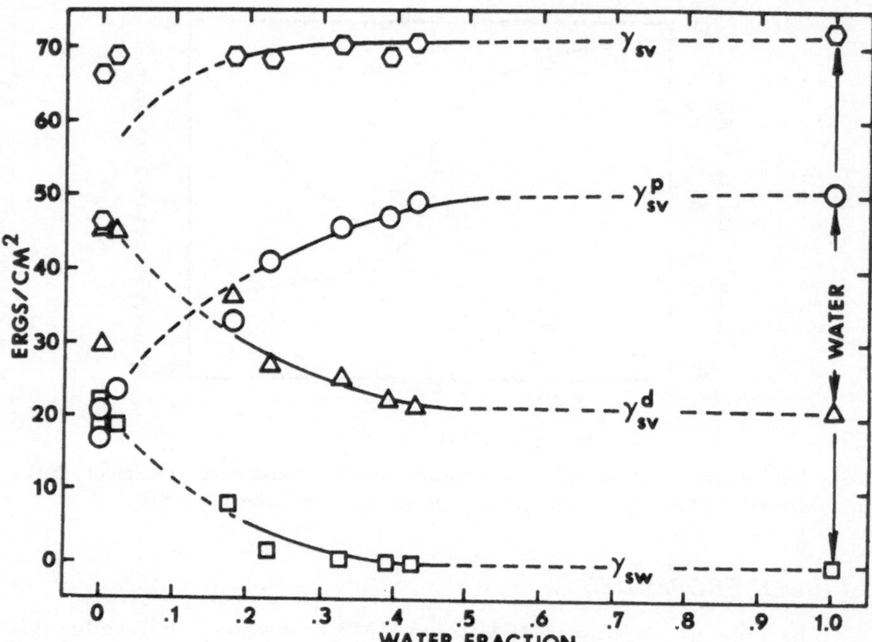

FIGURE 3. γ_{SV}^{d}, γ_{SV}^{p}, γ_{SV}, and γ_{SW} as a function of water fraction for methacrylate polymers of various bulk water contents. The data at water fraction = 1.0 are for pure water [reprinted with permission from Reference (21)].

The aqueous wetting character of HEMA–MMA copolymers has also been reported as a function of the mole ratio of HEMA as in Figure 4.[24] In this study, the wetting characteristics are measured both by the Wilhelmy plate procedure and the captive underwater air bubble procedure. The dynamic advancing and receding Wilhelmy angles are recorded as the surface is immersed and removed from the water liquid at a rate of 40 mm/min. To obtain the static Wilhelmy plate angles, the surface was stopped for 15 seconds at several points in both the advancing and receding mode, before measurements were made. The receding contact angle changes created at higher amounts of hydroxyethyl methacrylate in the copolymer. The advancing contact angle shows less change over the copolymer range. The underwater captive air bubble closely approximates the receding contact angle measurements of the Wilhelmy plate.

Analysis of ethyl methacrylate–HEMA copolymers by X-ray photoelectron spectroscopy (ESCA) have been reported on solvent cast and radiation grafted surfaces.[13]

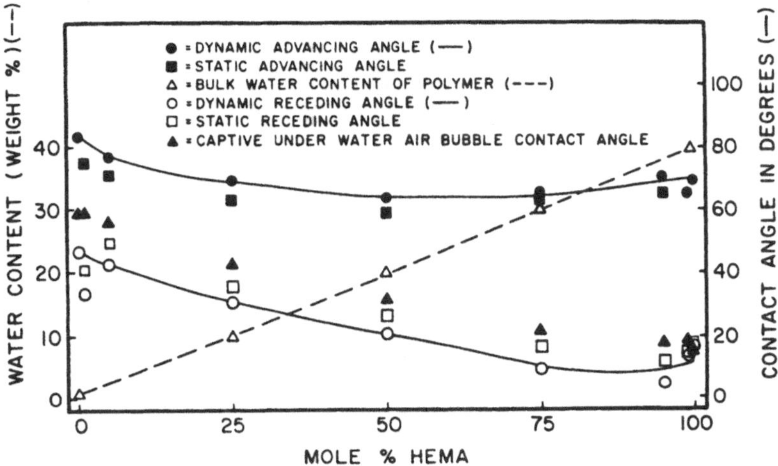

FIGURE 4. Contact angle measurements at 24-hour hydration of methyl methacrylate [MMA] and hydroxyethyl methacrylate [HEMA] copolymers using both captive air bubble and Wilhelmy plate techniques. Also shown are equilibrium water content measurements of the copolymers [reprinted with permission from Reference (24)].

4. BLOCK COPOLYMERS

Mixtures of two homopolymers are usually not compatible; that is they phase separate into macroscopic domains consisting of essentially pure homopolymers.[25] If these two polymeric materials were covalently coupled, they would still be phase separated, but the phase domains will be limited to the molecular weight of the polymeric chains and the relative molecular weight ratios of the two polymers. The polymeric blocks can be arranged in various structures, i.e., A—B, A—B—A, and $(A—B)_n$ blocks, so-called diblocks, triblocks, and multiblock polymeric systems.

4.1. PURPOSE (MORPHOLOGY)

Block copolymer systems permit the control of morphology for the study of protein interactions and cellular adhesion. For example, if one investigates the general detail of A—B diblocks or A—B—A triblock copolymers, and changes the volume fraction of the blocks, one finds a systematic order of the phases as shown in Figure 5.

If the morphology of these polymers extends to the interface, the nature of protein adsorption is expected to change compared to that on a homogeneous surface. When a protein molecule diffuses to the vicinity of

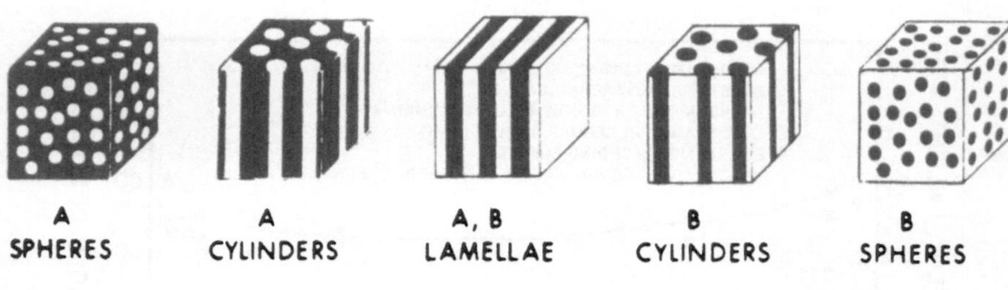

FIGURE 5. Schematic illustration of various phase structures composed of components A and B [reprinted with permission from Reference (77)].

the polymer-solution interface, it may adsorb on either phase A, B, or even the A—B interface bridging the two phases. Table 4 shows the techniques which are useful for the differentiation of block copolymers versus homopolymer blends. (See Chapters 10 and 13.)

Several block copolymer materials are of commercial importance. These include the styrene-butadiene or styrene-isoprene A—B and A—B—A block systems produced by Shell Chemical Company (Kratons) and Phillips Petroleum Company (some grades of Solprenes). Several $(A—B)_x$ materials are commercially available including the polyether-polyester blocks produced by DuPont Co. (Hytrels) and the polyurethanes, which may be poly(ester-urethane) blocks, poly(ether-urethane) blocks, poly(ether-urethane-urea) blocks, etc. Makers of these polyurethanes include Upjohn Corp. (Pellethanes), DuPont Co. (Lycra), Mobay Co. (Texins), and B. F. Goodrich (Estanes). Because of the biomedical importance of the urethanes, they are included in a separate listing in this chapter.

4.2. STYRENE–HYDROXYLATED BUTADIENE–STYRENE TRIBLOCKS

Styrene-butadiene-styrene triblock materials are commercially available from Shell Chemical Co. (Kraton) and from Phillips Petroleum Co. (Solprene) in a wide variety of molecular weight and composition ratios. Several groups have used these materials to prepare hydrophilic-hydrophobic segments by oxidation of the butadiene block either by peracid followed by hydrolysis[26,27] or by hydroboration procedures.[28,29] The peracid method leads primarily to dihydroxylation of the double bond as shown in Figure 6, whereas the hydroboration leads to anti-Markovnikov monohydroxylation of the double bond as shown in Figure 7. Due to the nature of

TABLE 4
Techniques for Block Copolymer Characterization[a]

Block copolymer versus homopolymer blend

1. Solubility characteristics
   a. Solid phase extraction
   b. Solution fractionation
2. Film clarity
3. Solution compatibility
4. Molecular weight distribution
   a. Density gradient ultracentrifugation
   b. Gel permeation chromatography
5. Rheological characteristics

Block copolymer versus random copolymer

1. Proton magnetic resonance
2. Infrared spectroscopy
3. Dynamic mechanical behavior
4. Differential scanning calorimetry
5. Electron microscopy
6. Small-angle X-ray scattering
7. Mechanical properties
8. Rheological characteristics
9. Crystallinity characteristics
10. Solution light scattering
11. Thermomechanical analysis

Molecular structure

1. Osmometry
2. Solution light scattering
3. Ultracentrifugation
4. Gel permeation chromatography
5. Solution viscometry
6. Oligomer analysis
7. Selective degradation

Architecture and purity

1. Elastic recovery
2. Rheological characteristics
3. Gel permeation chromatography
4. Density gradient ultracentrifugation

Supermolecular structure

1. Dynamic mechanical behavior
2. Differential scanning calorimetry
3. Rheological characteristics
4. Electron microscopy and scanning electron microscopy
5. Wide-angle X-ray scattering
6. Small-angle X-ray scattering
7. Birefringence
8. Small-angle light scattering

[a] From reference (43).

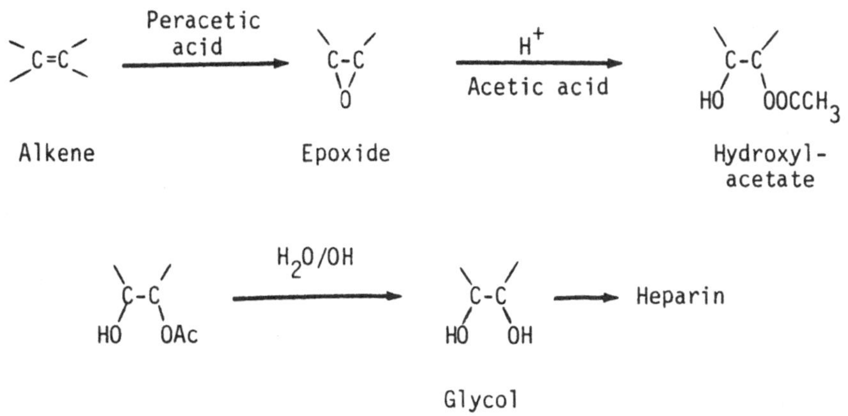

>C=C< —(Peracetic acid)→ Epoxide (C-C bridged by O) —(H^+, Acetic acid)→ Hydroxyl-acetate (HO–C–C–$OOCCH_3$)

Alkene | Epoxide | Hydroxyl-acetate

HO–C–C–OAc —(H_2O/OH)→ HO–C–C–OH → Heparin

Glycol

FIGURE 6. Hydroxylation reaction scheme for polybutadienes [reprinted with permission from Reference (6)].

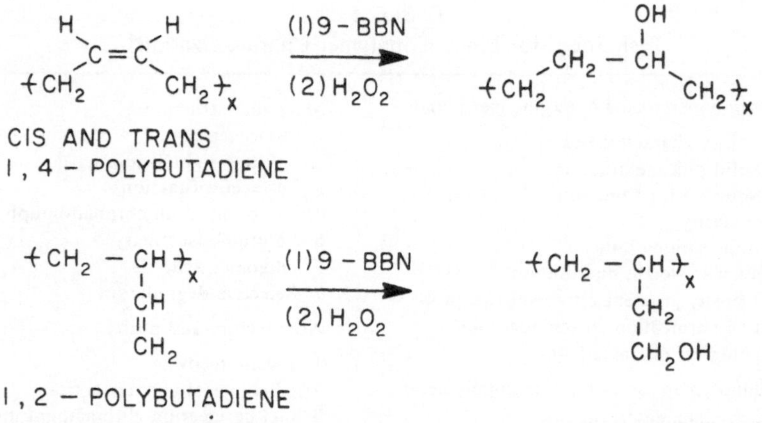

$$\text{+}CH_2\text{–}CH{=}CH\text{–}CH_2\text{+}_x \xrightarrow[(2)\,H_2O_2]{(1)\,9\text{-}BBN} \text{+}CH_2\text{–}CH_2\text{–}CH(OH)\text{–}CH_2\text{+}_x$$

CIS AND TRANS
1, 4 - POLYBUTADIENE

$$\text{+}CH_2\text{–}CH(CH{=}CH_2)\text{+}_x \xrightarrow[(2)\,H_2O_2]{(1)\,9\text{-}BBN} \text{+}CH_2\text{–}CH(CH_2CH_2OH)\text{+}_x$$

1, 2 – POLYBUTADIENE

FIGURE 7. Hydroxylation of olefin-containing polymers by hydroboration procedures (9-BBN = 9-borobicyclo[3.3.1]nonane) (reprinted with permission from Reference (24)].

the peracid method, which proceeds through a reactive epoxide intermediate, crosslinked polymer inevitably results; the hydroboration procedure leads to uncrosslinked polymer.

The butadiene segments of the S—B—S triblock polymers of different styrene–butadiene ratios were hydroborated and the bulk water content and aqueous wetting angles were measured[24] (Figure 8). The receding contact angle shows the most change and becomes highly wetting at or above 20%

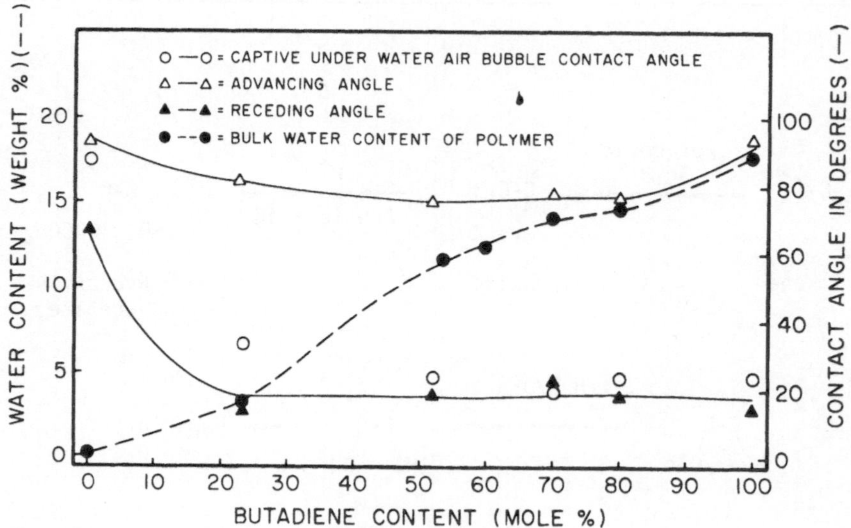

FIGURE 8. Contact angle measurements at 24-hour hydration and equilibrium water content of styrene–hydroxylated butadiene–styrene triblock copolymers [reprinted with permission from Reference (24)].

hydroxylated butadiene content. The advancing contact angle, however, changes very little over the entire copolymer composition.

4.3. HEMA–STYRENE–HEMA TRIBLOCKS

Preparation of these A—B—A block polymers with the blocks having different wetting character has been reported.[27,28] The (A) blocks consist of hydroxyethyl methacrylate segments; the (B) block consists of polystyrene. This material was prepared from an amino-terminated poly(hydroxyethyl methacrylate) and a *bis*-isocyanate-terminated styrene in order to couple the blocks by urea functionalities as shown in Figure 9.

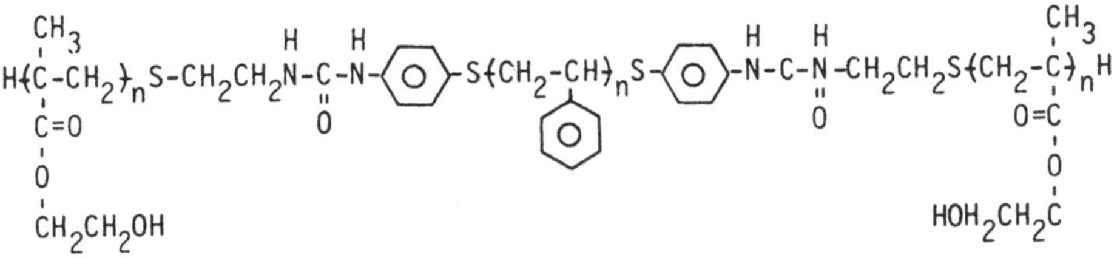

$$H{+}C(CH_3)(C{=}O{-}O{-}CH_2CH_2OH){-}CH_2{+}_n S{-}CH_2CH_2NH{-}C(=O){-}NH{-}C_6H_4{-}S{+}CH_2{-}CH(C_6H_5){+}_n S{-}C_6H_4{-}NH{-}C(=O){-}NH{-}CH_2CH_2S{+}CH_2{-}C(CH_3)(C(=O){-}O{-}CH_2CH_2OH){+}_n H$$

FIGURE 9. Chemical structure of HEMA-Styrene ABA-type block copolymers [reprinted with permission from Reference (78)].

The wetting characteristic using a water droplet on the dry polymeric surface is shown in Figure 10 for both the A—B—A triblock system along with the HEMA-styrene random copolymer wetting characteristics. This contact

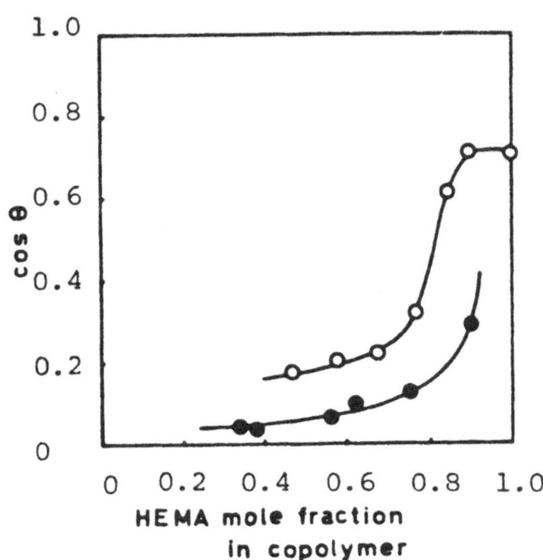

FIGURE 10. Relation between wettability and copolymer composition: (●) HEMA-Styrene ABA-type block copolymer system; (○) HEMA-Styrene co-oligomer system [reprinted with permission from Reference (31)].

angle measurement gives results approximating the advancing contact angle measurement of the Wilhelmy plate procedure. The contact angle becomes more wetting above 0.8 mole fraction HEMA in copolymer.

5. ALKYL METHACRYLATES

5.1. PURPOSE (SYSTEMATIC CHANGE IN T_g)

Polymer science is aware of the importance of polymer main chain and side chain segmental mobility on the physical and mechanical properties of polymers. It is very reasonable that such characteristics would significantly influence the interfacial interactions of solute molecules with the polymer surface. (See Chapter 2.) One can think of this in terms of an entropic argument: if a protein molecule were to adsorb on a highly mobile polymer surface, it would influence the polymer surface locally and the mobility of the polymer would be constrained. Therefore, from the entropic point of view, adsorption would not be favorable assuming all other factors are constant.[32]

The mobility of polymer molecules is expressed via a very commonly measured bulk property of polymers, the glass transition temperature (T_g). The T_g of a polymer is that temperature at which large segments of the molecule or entire polymer molecules can move within the solid. At temperatures considerably below the glass transition temperature, the mobility is constrained or does not occur, and as a result the polymer is relatively brittle and not deformable. If the polymer is crosslinked to a three-dimensional network, then above the glass transition temperature the network will generally have elastic character. Thus, all of our common elastomers and rubbers are three-dimensional polymer networks whose glass transition temperature is considerably below room temperature. Merrill[32] has suggested that such a surface is in constant motion and would be relatively unfavorable for protein adsorption. It is difficult to vary T_g in most systems without varying the chemistry of the system. The family of methacrylate polymers with different length alkyl ester side chains provides a means of being able to vary T_g without greatly altering surface wetting characteristics.

5.2. CHEMISTRY

The chemistry of the methacrylates has been reviewed.[33,34] The methacrylates can be polymerized either by radical or anionic means as shown opposite:

$$CH_2{=}\overset{CH_3}{\overset{|}{C}}{-}\underset{\underset{O-R}{|}}{C}{=}O \xrightarrow[\text{or anion }(-)]{\text{radical}} \left(CH_2{-}\underset{\underset{\underset{O-R}{|}}{C=O}}{\overset{\overset{CH_3}{|}}{C}}\right)_x$$

R = alkyl

Many of the alkyl methacrylate monomers and polymers are commercially available, although preparation of specific alkyl methacrylate monomers is easily accomplished.[35] For highly pure polymers it is usually advisable to purify the monomer and polymerize it in the lab rather than to obtain the commercial polymer. The polymer is most useable in the uncrosslinked form, because it can be solubilized and solvent cast. Uncrosslinked polymer can be prepared by polymerizing the monomer at relatively high dilution (1 to 10) in a good solvent such as toluene or tetrahydrofuran (for these *n*-alkyl methacrylates) using a radical initiator such as AIBN (azobisisobutyronitrile). The polymer is isolated by precipitation in a non-solvent such as methanol and then purified by redissolving and reprecipitation.

5.3. BULK CHARACTERISTICS

The glass transition temperatures of the *n*-alkyl methacrylates are shown in Table 5. The brittle point temperatures of the acrylates and methacrylates are plotted as a function of the alkyl ester functionality in Figure 11.

The T_g of the methacrylates reaches a minimum with the *n*-octyl ester and then increases with temperature due to interactions of the longer alkyl ester side chains. The accurate measurement of the polymer transitions requires the complete removal of all solvent or other additives which can function as plasticizers. This is done by heating the sample in vacuum above its glass transition temperature.

5.4. SURFACE CHARACTERISTICS

The critical surface tension of the polymers will range from 35 ergs/cm^2 for poly(methyl methacrylate) to around 24 ergs/cm^2 for the long chain length ester methacrylates. Air and octane underwater contact angles[36] are shown for various chain length methacrylate esters in Table 6. The solid-water interfacial energies are determined from these values along with the dispersion and polar components from these surfaces. The dispersion component of the surface shows little change over the entire range (from methyl

TABLE 5
Glass Transition Temperatures (T_g)of Alkyl Methacrylate Polymers

$$\left(CH_2 - \underset{\underset{\underset{O-R}{|}}{\overset{|}{C=O}}}{\overset{\overset{CH_3}{|}}{C}} \right)$$

R =	Name	T_g(°C)
$-CH_3$	methyl methacrylate (MMA)	105
$-CH_2CH_3$	ethyl methacrylate	47, 65
$-(CH_2)_2CH_3$	*n*-propyl methacrylate	33
$-(CH_2)_3CH_3$	*n*-butyl methacrylate (BMA)	17
$-C-(CH_3)_3$	*t*-butyl methacrylate	107
$-(CH_2)_5CH_3$	*n*-hexyl methacrylate (HMA)	−5
$-(CH_2)_7CH_3$	*n*-octyl methacrylate (OMA)	−70
$-(CH_2)_{11}CH_3$	*n*-dodecyl methacrylate (DDMA)	−65
$-(CH_2)_{17}CH_3)$	*n*-octadecyl methacrylate (ODMA)	<+60

[a] From Reference (75).

to *n*-octadecyl) of *n*-alkyl methacrylates. The polar force component shows a slight decrease with increasing alkyl ester chain length. (See also Chapter 2.)

6. DERIVATIZED AGAROSE

6.1. PURPOSE (MODIFICATION FOR SYSTEMIC HYDROPHOBICITY)

The adsorption of plasma proteins on various surfaces can lead to activation of the blood coagulation, complement, and fibrinolytic systems. The adsorbed plasma proteins may also influence the deposition of platelets and other blood cellular components at a surface. Although investigators have initially investigated hydrogels for minimizing interactions with blood components, it is now known that hydrogels such as poly(hydroxyethyl methacrylate) show fairly strong interactions with certain blood proteins.

There are, however, certain hydrogels which are demonstrated to show little protein adsorption. These are hydrogels such as crosslinked dextran, crosslinked polyacrylamide, and agarose gels used by biochemists for gel permeation chromatography of proteins.[37] The gel permeation process relies upon size separation of molecules and minimal or non-interaction of

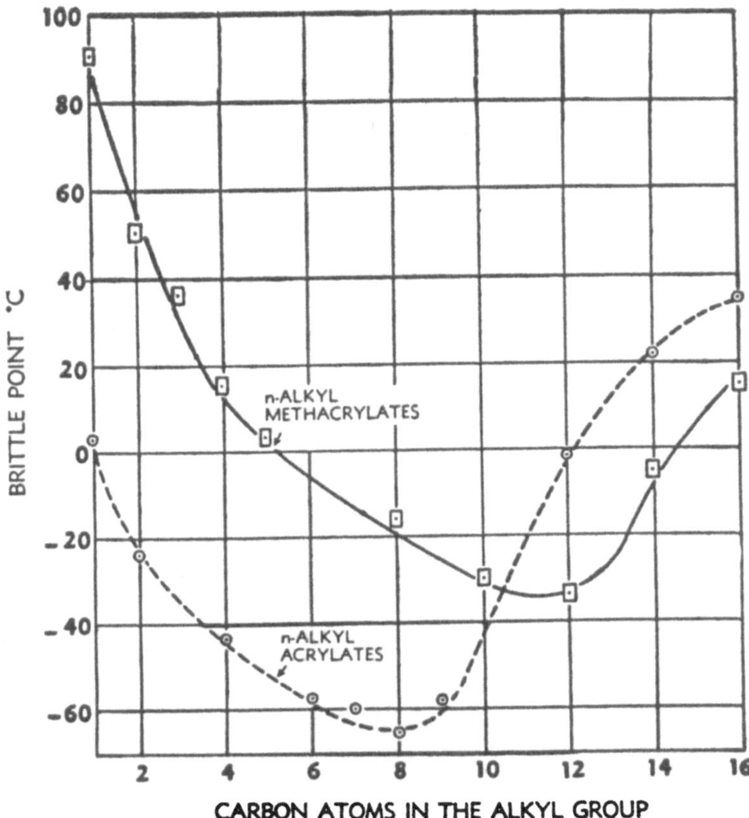

FIGURE 11. Brittle points of polymeric *n*-alkyl acrylates and methacrylates [reprinted with permission from Reference (75)].

TABLE 6
Surface Characterization of Alkyl Methacrylates[a]

Methacrylate ester polymer	T_g (°K)	Water contact angle (degrees)		Surface energetics (ergs/cm^2)			
		θ(air)	θ(octane)	γ_{SW}	γ^P	γ^d	γ^p/γ^d
methyl	332/368	62 ± 1.1	87 ± 2.4	19.2	17.6	35.4	0.50
n-butyl	310/312	72 ± 3.2	103[b]	29.5	11.8	40.0	0.30
n-hexyl	265/257	76	110[b]	32.9	9.9	40.4	0.25
n-octyl	192/175	73	105[b]	29.9	11.2	38.6	0.29
n-dodecyl	NA	82	118[b]	36.2	7.7	38.5	0.20
n-octadecyl	185/179	90	130[b]	41.1	4.9	36.1	0.14

[a] From Reference (36).
[b] Estimated values.

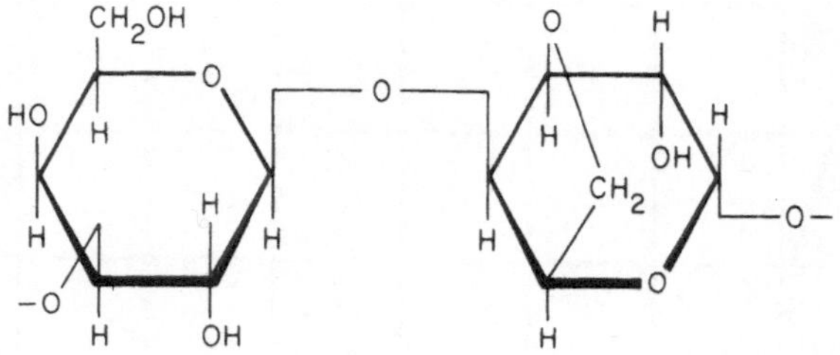

FIGURE 12. Repeating subunit of agarose [reprinted with permission from Reference (24)].

the protein with the stationary chromatography support is a requirement for the gel permeation process.

The agarose material is a polysaccharide containing a repeating 1,3-linked β-D-galactopyranose and 1,4-linked 3,6-anhydro-α-L-galactopyranose structure (Figure 12). All the other gel chromatographic supports require chemical crosslinked structures to maintain their gel properties: agarose, however, gels by self-aggregation due to helical chain interactions and by hydrogen bonding and hydrophobic interactions. Agarose is soluble in hot water (>80–90°C) but forms a porous, mechanically strong network upon cooling to 30 to 40°C.

A modification of the agarose gel network to cause selective protein interactions is a technique now known as hydrophobic chromatography.[38–40] This is accomplished by various procedures to covalently bond aryl or various chain length alkyl groups to the agarose network. All of these previous procedures have resulted in the formation or require the use of chemically crosslinked agarose chromatography supports. Thus, the resulting derivatized agarose materials can no longer be dissolved for solvent casting of films.

6.2. CHEMISTRY

We developed a procedure to overcome the crosslink formation upon alkyl derivatization.[24] To accomplish this, the agarose beads were exchanged with dry acetone. It has previously been shown that this acetone exchange does not disrupt the agarose structure.[41] The agarose can now be reacted with alkyl and aryl chloroformates to covalently bond the hydrophobic group to the agarose molecule via a carbonate linkage. This linkage is shown to be quite stable in aqueous environments from pH 3 to 8. The derivatized agarose has been stored at 5°C in distilled water for over six months without detectable alkyl group hydrolysis. The degree of alkyl group coupling to the agarose can be measured using tritiated alkyl alcohols

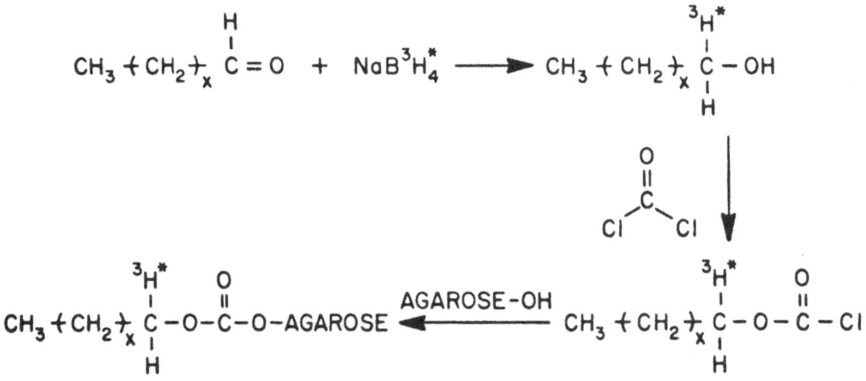

$$CH_3 \leftarrow CH_2 \rightarrow_x CH{=}O + NaB^3H_4^* \longrightarrow CH_3 \leftarrow CH_2 \rightarrow_x C(^3H^*)(H){-}OH$$

$$\xrightarrow{Cl{-}CO{-}Cl} CH_3 \leftarrow CH_2 \rightarrow_x C(^3H^*)(H){-}O{-}C(=O){-}Cl \xrightarrow{AGAROSE{-}OH} CH_3 \leftarrow CH_2 \rightarrow_x C(^3H^*)(H){-}O{-}C(=O){-}O{-}AGAROSE$$

FIGURE 13. Preparation of α-titriated alcohols and covalent bonding to agarose (reprinted with permission from Reference (24)].

which are readily prepared by the reduction of alkyl aldehydes with tritium-labelled borohydride. This procedure is outlined in Figure 13.

The agarose and derivatized agarose materials are soluble in dimethyl sulfoxide which is used for solvent casting of surfaces. The derivatized agarose is quantitated in terms of μmoles alkyl group per ml of packed gel for aqueous suspensions of the beads or moles alkyl group per mole anhydrodisaccharide repeat unit for dried agarose powders. The latter terminology is a more exact representation of coupling at high derivatization since the gel structure starts to collapse in water due to the increased internal hydrophobic forces above a value of ~70 μmole/ml packed gel.

6.3. BULK CHARACTERISTICS

Spectroscopic measurements such as transmission infrared and 300 MHz ^{1}H-nuclear magnetic resonance spectroscopy confirm the alkyl coupling to the agarose structure and can be used as quantitative procedures to determine the degree of coupling. For example, in infrared analysis the ratio of the hydroxyl absorption at 3400 cm^{-1} (2.9 μm) to the carbonate absorption at 1750 cm^{-1} (5.7 μm) correlates with the percentage of alkyl group coupling. The carbonate functional group is attached to the agarose structure by the alkyl chloroformate coupling reaction.

Proton nuclear magnetic resonance (NMR) has been used by us and others to quantitate the amount of alkyl group coupling. The agarose samples are dried and then dissolved in a solvent consisting of 25% D_2SO_4 using 0.1 g agarose per 0.5 ml. Typical NMR spectra of the underivatized and derivatized agarose are shown in Figure 14. The area of the signals as shown in spectrum C of Figure 14 are ratioed according to the following

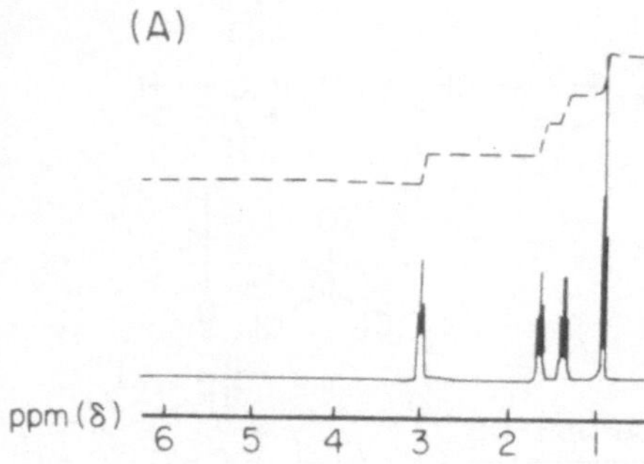

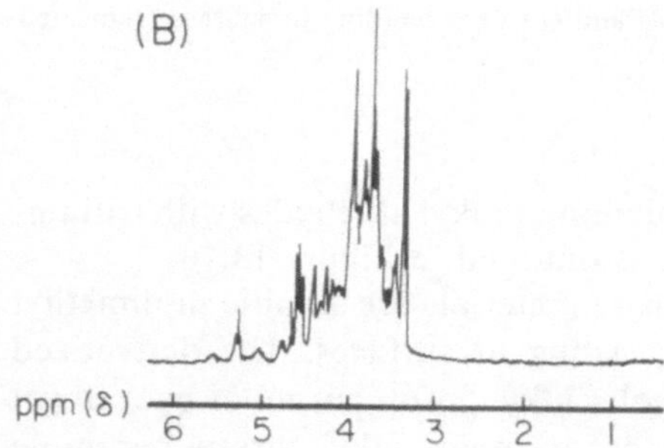

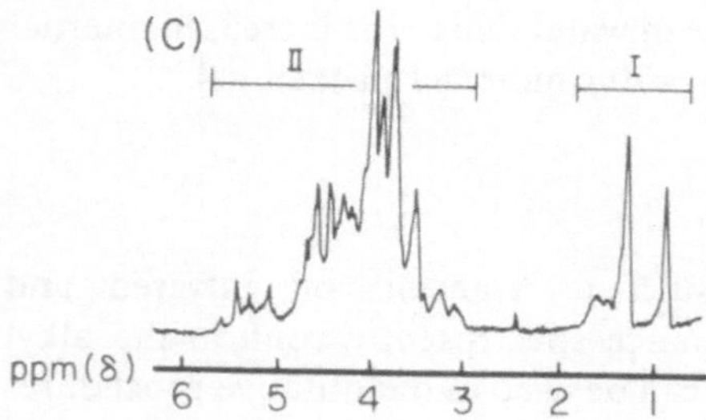

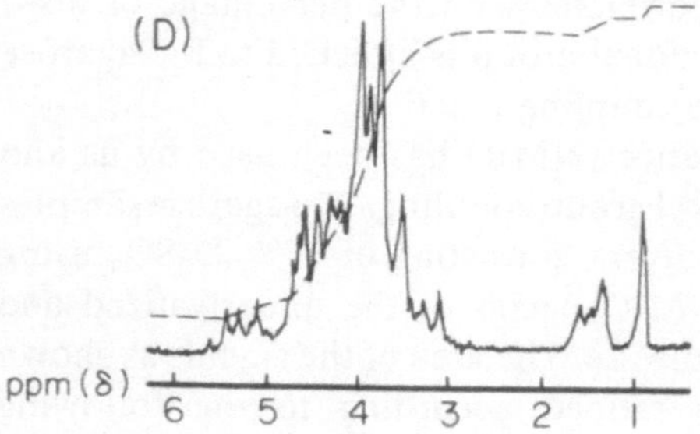

FIGURE 14. 300 MHz ^{1}H-NMR analysis of agarose and alkyl-derivatized agarose (spectra taken in 25% D_2SO_4 in D_2O): (A) *n*-butylamine standard; (B) Sepharose 4B-200; (C) *n*-hexyl agarose; (D) *n*-butyl agarose.

TABLE 7
300 MHz ^{1}H-NMR Analysis of *n*-Alkyl Agarose Derivatives

Derivative	1st analysis[a]	2nd analysis[a]
Hexyl agarose 1	18.7	17.1
Hexyl agarose 2	28.3	28.1
Butyl 1	11.1	10.4
Butyl 2	17.2	18.3
Butyl 3	30.0	30.9
Butyl 4	12.7	13.1
Butyl 5	21.1	16.8
Butyl 6	29.6	29.4

[a] Reported in μmole alkyl group/ml packed gel.

equation[42]:

$$\frac{\text{moles alkyl group}}{\text{moles anhydrodisaccharide}} = \frac{I/Z}{\left(II - \frac{2I}{Z}\right)\Big/7}$$

where $Z = 2n - 2$ for $NH_2{-}(CH_2)_n{-}CH_3$ or $HO{-}(CH_2)_n{-}CH_3$ and I and II correspond to areas shown in Figure 14C.

The reproducibility of the technique appears to be quite good. Duplicate analyses of derivatized agarose by the NMR method are shown in Table 7. The NMR results have also been compared with radioisotopic quantitation methods (the radiolabeled samples were obtained from Dr. Herbert Jennissen, Ludwig-Maximilians-Universitat Munchen, Munich, West Germany).

As seen in Table 8, good correlations are observed for the hexyl agarose samples; however, not quite as good results were observed for the butyl samples. We haven't been able to determine if this discrepancy is due to the radiolabeled samples or the NMR quantitation method.

6.4. SURFACE CHARACTERISTICS

The surface characterization of the agarose and derivatized agarose films were done primarily by two methods, ESCA (Chapter 5) and advancing/receding water contact angles (Chapter 7). The ESCA technique measures the top 50 ± 20 Å of the surface of the polymer, whereas the contact angle method (Wilhelmy plate) measures the top 10 Å for most surfaces. The contact angle method as applied to hydrogels is not very well

TABLE 8
Comparison of NMR Method and Radioisotope Method for Analysis of Alkyl-Derivatized Agarose

	Derivatization: NMR method[a]	Derivatization: radioisotope method[b]
Hexyl agarose 1	12.2	14.1 ± 0.3
Hexyl agarose 2	17.1, 18.7	20.7 ± 1.0
Hexyl agarose 3	28.1, 28.3	28.3 ± 4.6
Butyl agarose 1	10.4, 11.1	21.3 ± 0.8
Butyl agarose 2	18.3, 17.2	34.1 ± 2.5
Butyl agarose 3	30.9, 30.0	45.9 ± 1.3

[a] Average of 2 determinations.
[b] Average ± standard deviation for 3 determinations.

modeled due to the diffuse water–bulk gel interface and the problem of contact angle-induced surface deformation (Chapter 7).

The ESCA studies used a Hewlett–Packard 5950B X-ray photoelectron spectrometer utilizing Al $K\alpha$ radiation (1487 eV) and a vacuum of 10^{-8} to 10^{-9} torr at ambient temperature. Water contact angles utilized a Wilhelmy plate apparatus built in our labs. The agarose samples were solvent cast and dried on glass plates from 0.2-μ-filtered dimethyl sulfoxide solutions.

The contact angles of butyl and dodecyl agarose after one hour of hydration are plotted versus the degree of derivatization in Figure 15. The underivatized agarose is, of course, very hydrophilic and exhibits essentially no hysteresis and essentially a zero contact angle. After relatively low degrees of derivatization (a value of 3×10^{-3} moles alkyl group per gram underivatized agarose corresponds to one alkyl group per each anhydrodisaccharide unit repeat of agarose), the advancing and receding contact angles remain constant. It is interesting to note that the advancing and receding contact angles of the butyl agarose always show less hysteresis than those of the longer alkyl chain dodecyl agarose surfaces.

ESCA analysis was done on underivatized agarose samples and *n*-butyl, *n*-hexyl, *n*-octyl, and *n*-dodecyl agarose coupled at approximately 40 μmole/ml packed gel. These samples were prepared on 1-cm^2 glass plates by solvent casting from dimethyl sulfoxide. The solvent was evaporated under nitrogen atmosphere at 60°C overnight. A survey scan for sulfur showed that all the solvent was removed by this process.

The carbon ESCA peaks are shown in Table 9. It is observed that the underivatized agarose, which should only contain ether-like carbons, contains a substantial amount of alkyl carbon contaminants. This may not be unexpected since high-energy surfaces are known to pick up a hydrocarbon layer in normal laboratory environments.

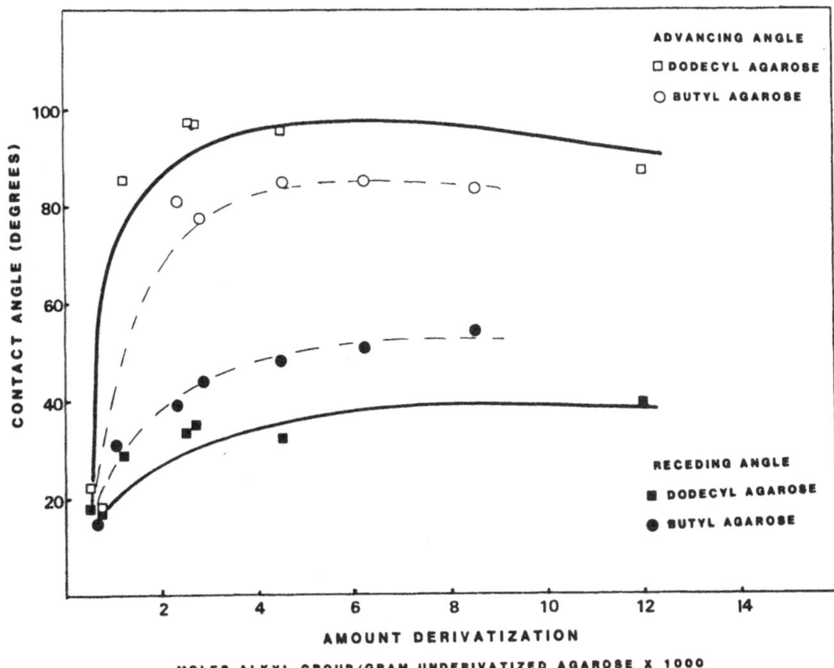

FIGURE 15. Contact angle changes with alkyl derivatization for *n*-butyl and *n*-dodecyl agarose.

Note that the relative percentage of ether carbon decreases as the length of the alkyl chain increases and the alkyl carbon peak area increases as the alkyl chain length increases. The ratio of the ether carbon peak to the carbonate carbon peak remains relatively constant, which would be expected since they are derivatized to essentially the same degree.

TABLE 9
ESCA Analysis of Derivatized Agarose Surfaces

	Alkyl carbon (%)	Carbonate carbon (%)	Ether carbon (%)	Ether carbon: carbonate carbon ratio
Agarose	33.9	0	55.3	—
n-Butyl agarose	31.3	6.8	49.8	7.3
n-Hexyl agarose	36.6	5.0	43.6	8.7
n-Octyl agarose	47.2	6.2	39.1	6.3
n-Dodecyl agarose	52.5	4.1	35.6	8.7

7. POLYURETHANES

7.1. PURPOSE (PRACTICAL BIOMEDICAL ELASTOMERS)

Polyurethanes have been studied fairly extensively as biomedical polymers. They can be prepared with a wide range of mechanical properties, from flexible elastomers to rigid plastics. The poly(ether urethanes) exhibit greater hydrolytic stability than the poly(ester urethanes) and, thus, find greater use in biomedical applications. The elastomeric urethanes are particularly suited for flexing membranes as used in artificial heart and heart assist devices due to their adequate blood compatibility, good elastomeric properties, and ability to undergo repeated flexing without failure.

7.2. CHEMISTRY

Polyurethane chemistry involves primarily the chemistry of the isocyanate group. This will be discussed here for background purposes, but has been reviewed extensively in the literature.[43,44] Isocyanates react with alcohols by route (*a*) to form urethane structures and with amines by route (*b*) to form urea structures.

$$\underset{\text{isocyanate + alcohol}}{\mathrm{R{-}N{=}C{=}O + HO{-}R'}} \xrightarrow{a} \underset{\text{urethane}}{\mathrm{R{-}\overset{\displaystyle H}{\overset{|}{N}}{-}\overset{\displaystyle O}{\overset{\|}{C}}{-}O{-}R'}}$$

$$\underset{\text{isocyanate + amine}}{\mathrm{R{-}N{=}C{=}O + H_2N{-}R'}} \xrightarrow{b} \underset{\text{urea}}{\mathrm{R{-}\overset{\displaystyle H}{\overset{|}{N}}{-}\overset{\displaystyle O}{\overset{\|}{C}}{-}\overset{\displaystyle H}{\overset{|}{N}}{-}R'}}$$

Both urethane and urea structures are found in various biomedical urethanes. The basic polyurethane structures of interest here are prepared by the reaction of three components: diisocyanates, polyols, and chain extenders. The chain extenders may be either a diol or a diamine. Polyols are linear polyethers terminated at both ends by hydroxyl groups. The molecular weights of the polyols which exhibit suitable mechanical properties in polyurethane formulations are in the range of 500 to 5000 Daltons. By using different diisocyanates, polyols, and chain extenders, a large number of possible polymer structures can be formed. A common feature of all these polymers is the phase morphology of the material due to the incompatibility of the segments and aggregation of the hard segments due to hydrogen bonding. The hard segments refer to the diisocyanate and the chain extender moiety in the polymer; the soft segment consists of the polyol portion of the polymer. In general, the higher the molecular weight of the polyol used in the polymer, the softer, more elastic the resulting polyurethane.

Most of the biomedical polymers use similar components. The diisocyanate most used is methylene-4,4′-diphenyl diisocyanate, commonly referred to as MDI. The polyol used most commonly is poly(tetramethylene glycol), PTMG, of molecular weight from 500 to 3000. The chain extender can be a diol such as 1,4-butanediol or a diamine such as ethylenediamine. The diamine chain extender gives generally better mechanical properties and greater phase separation of the segments due to the stronger hydrogen bonding nature of the urea as compared to the urethane unit.[45] Most injection moldable (thermoplastic) urethanes are diol chain extended which produces weaker hydrogen bonding interactions in the hard segment, allowing it to melt below the degradation temperature.

The addition and stoichiometry of the components can be combined in a variety of methods. For small batch biomedical polymers, the polyurethanes are generally prepared by a two-step process. One mole of polyol is "end capped" with two moles of diisocyanate in the first stage. This is followed by the addition of one mole of chain extender in the second stage as shown in Figure 16.

It is noted that the diamine chain extender results in a poly(ether urethane urea) polymeric structure. Other combinations of this stoichiometry may be used. For example, a 3:2 ratio of diisocyanate to polyol results in a dimeric, end capped polyol under the proper conditions which is then polymerized using one mole of chain extender.

For larger-volume production of polyurethanes, it is more economical to mix the polyol and chain extender together and combine with the diisocyanate in absence of solvent to form the polyurethane. The chain extender in this case is a diol to provide equal reactivity with the polyol. The segments of the blocks are not quite as well defined in this case resulting in a statistical linking of the components.

Other components used in urethane polymers are diisocyanates such as 2,4-toluene diisocyanate (TDI) and methylene-*bis*(4-dicyclohexyl isocyanate), a hydrogenated form of MDI. The hydrogenated MDI diisocyanate yields polymers with better photooxidative stability. Poly(propylene glycol) and poly(ethylene glycol) have been used as polyol components for the polyurethanes. The poly(ethylene glycol)-containing polyurethane becomes more hydrophilic with increasing molecular weight of the polyol and can be prepared as a hydrogel formulation.

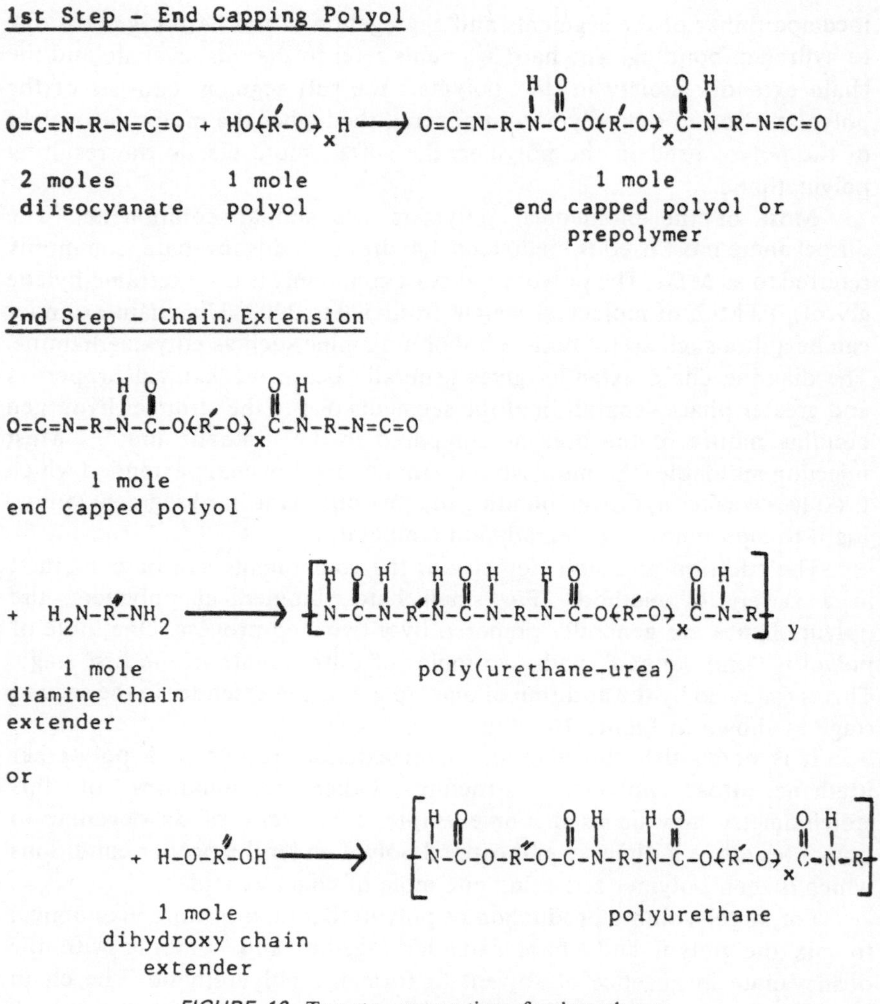

1st Step - End Capping Polyol

$$O{=}C{=}N{-}R{-}N{=}C{=}O + HO(R'{-}O)_xH \longrightarrow O{=}C{=}N{-}R{-}\overset{H}{\underset{|}{N}}{-}\overset{O}{\underset{\|}{C}}{-}O(R'{-}O)_x\overset{O}{\underset{\|}{C}}{-}\overset{H}{\underset{|}{N}}{-}R{-}N{=}C{=}O$$

2 moles diisocyanate — 1 mole polyol — 1 mole end capped polyol or prepolymer

2nd Step - Chain Extension

$$O{=}C{=}N{-}R{-}\overset{H}{\underset{|}{N}}{-}\overset{O}{\underset{\|}{C}}{-}O(R'{-}O)_x\overset{O}{\underset{\|}{C}}{-}\overset{H}{\underset{|}{N}}{-}R{-}N{=}C{=}O$$

1 mole
end capped polyol

$$+ \quad H_2N{-}R''{-}NH_2 \longrightarrow \left[\overset{H}{N}{-}\overset{O}{C}{-}\overset{H}{N}{-}R''{-}\overset{H}{N}{-}\overset{O}{C}{-}\overset{H}{N}{-}R{-}\overset{H}{N}{-}\overset{O}{C}{-}O(R'{-}O)_x\overset{O}{C}{-}\overset{H}{N}{-}R\right]_y$$

1 mole diamine chain extender — poly(urethane-urea)

or

$$+ \quad H{-}O{-}R''{-}OH \longrightarrow \left[\overset{H}{N}{-}\overset{O}{C}{-}O{-}R''{-}O{-}\overset{O}{C}{-}\overset{H}{N}{-}R{-}\overset{H}{N}{-}\overset{O}{C}{-}O(R'{-}O)_x\overset{O}{C}{-}\overset{H}{N}{-}R\right]$$

1 mole dihydroxy chain extender — polyurethane

FIGURE 16. Two-step preparation of polyurethanes.

7.3. BULK PROPERTIES

Considerable work has been reported on structure-property relationships of segmented polyurethanes.[46-48] Much of this work has investigated the domain morphology of these materials, which is quite complex and dependent upon solvent casting temperatures and on annealing conditions. Usually the diamine chain extended polymers have better phase separation than the diol extended polymer. Phase separation is also dependent upon the molecular weight of the polyol. The morphology has been studied by wide- and small-angle X-ray diffraction,[49] infrared spectroscopy,[50] infrared dichroism,[51] low-angle light scattering,[46,52] and stress birefringence.[53]

7.4. SURFACE CHARACTERISTICS

The surface characteristics of the polyurethanes have been studied by internal reflection infrared spectroscopy[54] and by ESCA techniques.[55,56] Both of these techniques show changes between the surface and the bulk structure of the polyurethanes. The infrared data show differences between the molecular and secondary (hydrogen) bonding nature of the surface. The surface nature of polyurethanes has been studied by Ratner,[57,58] Cooper,[59] Merrill,[55] and Knutson and Lyman.[60] Surface structure depends on the preparation conditions, including casting substrate, solvent removal rate, and the environment in which the surface characterization is made, i.e., air, water, etc. The surface properties are also, of course, highly dependent on the nature of the soft segment, i.e., PEG, PPG, or PTMG, as well as its molecular weight. There is inconsistency among some of the various studies in the literature. The surface nature of biomedical polyurethanes is not yet fully understood.

8. POLY(α-AMINO ACIDS)

8.1. PURPOSE (NATURALLY OCCURRING REPEAT UNITS)

Polymers of a wide variety of characteristics can be prepared from over 20 commonly occurring amino acids. They can be made as hydrogels or very hydrophobic polymers and be either biodegradable or biostable. Biodegradable poly(amino acids) have been proposed for use in sutures,[61] sustained release drug systems,[62] and artificial skin substitutes.[63] The more stable poly(amino acids) have been suggested for use in hemodialysis and blood oxygenator membranes.[64]

8.2. CHEMISTRY

The most useful reaction for the preparation of synthetic poly(α-amino acids) involves n-carboxy anhydrides (Leuchs anhydrides) prepared by reaction of α amino acids with phosgene.[65,66]

$$\mathrm{R{-}\overset{\displaystyle H}{\underset{\displaystyle C{=}O{-}OH}{C}}{-}NH_2} + \mathrm{Cl{-}\overset{\displaystyle O}{\overset{\|}{C}}{-}Cl} \longrightarrow \mathrm{R{-}\overset{\displaystyle H}{\underset{\displaystyle O{=}C{-}O}{C}}{-}\overset{\displaystyle H}{\underset{\displaystyle C{=}O}{N}}} + \mathrm{2HCl}$$

With amino acids containing other reactive groups, such as glutamic acid and lysine, the groups must be blocked prior to the *n*-carboxy anhydride formation. The *n*-carboxy anhydrides react with nucleophiles such as water or amines to yield high-molecular weight poly(amino acids), liberating carbon dioxide in the process.

$$\mathrm{R{-}\overset{H}{\underset{|}{\overset{|}{C}}}{-}\overset{H}{\underset{|}{\overset{|}{N}}}}\ (\text{ring: } C(=O){-}O{-}C(=O)) + R_3N \longrightarrow H{-}\!\left(\underset{H}{N}{-}\underset{R}{\overset{H}{C}}{-}\overset{O}{\overset{\|}{C}}\right)_x\!{-}OH$$

The polymerization initiation mechanism may differ depending upon the type of nucleophile used[67]; the exact reaction mechanism is still not completely understood.

Co-polymers of the amino acids are also prepared from the *n*-carboxy anhydride intermediates. For the preparation of random copolymers, the *n*-carboxy anhydrides of two different amino acids are both mixed in the same reaction vessel before initiation. The usual procedure is to polymerize the *n*-carboxy anhydrides to near 100% conversion, isolate the copolymer, and measure the average amino acid composition. Depending upon the particular amino acid *n*-carboxy anhydrides used, the reactivity of the comonomers may be close to unity, leading to an essentially random copolymer, or be widely different resulting in a very heterogeneous polymer composition. The solvent used for polymerization can also exert an effect on the comonomer reactivities.

Block copolymers have also been prepared from amino acid *n*-carboxy anhydrides by polymerizing *n*-carboxy anhydride monomer almost to completion, and then adding a second monomer. The size of the blocks are controlled by the stoichiometry of the initiator and *n*-carboxy anhydride-amino acid ratios.

The hydrophobicity of the poly(α-amino acids) can be controlled by the amount and type of amino acid incorporated into the polymer. Hydrogels or water-soluble poly(α-amino acids) are prepared by using glutamic acid or aspartic acid and derivatives such as hydroxyethyl and hydroxypropyl glutamine, prepared from nucleophilic displacement of poly(α-benzyl glutamate) with the corresponding hydroxyalkyl amine.[68]

The crosslinking of poly(α-amino acids) has been accomplished by two methods. One method uses a hexamethylene diisocyanate poly(oxyethylene glycol) to transesterify with poly(L-glutamic acid).[69] Alternatively the formation of hydrogels can be accomplished by the displacement of poly(γ-benzyl glutamate) with ω-hydroxyalkyl amines with α,ω-diaminoalkanes amino added to function as crosslinker.[70,71]

TABLE 10
Water Contact Angles on Copolypeptide Films[a] (from Ref. 73)

Material[b,c]	Casting solvent	Conformation	Contact angle (average) (degrees)
GL 1:1	$CHCl_3$	α-helix	80
GA 4:1	TFA	α + (random)	71
GA 1:1	$CHCl_3$ + Tr TFA	α-helix	58
GL 4:1	$CHCl_3$	α-helix	63.5
ZL 4:1	$CHCl_3$	α-helix	53.5
ZL 1:1	$CHCl_3$	α-helix	60
ZL 1:4	$CHCl_3$ + Tr TFA	α-helix	63
G(OH)V 4:1	TFA	α + (random)	37
G(OH)V 1:1	TFA	α + (random)	38
G(OH)V 1:4	TFA	β-sheet	47
G(OH)L 1:1	TFA	α-helix	40

[a] From Reference (73).
[b] G = γ-benzyl-L-glutamate; L = L-leucine; A = L-alanine; Z = ε-carbobenzoxy-L-lysine; V = L-valine; G(OH) = L-glutamic acid.
[c] Note that conformation of the G(OH) polypeptides may change slightly on hydration. Anhydrous film conformations are listed.

8.3. SURFACE ANALYSIS

The surface analysis and characterization of poly(amino acid) films has been limited to optical microscopy, scanning electron microscopy,[(72)] and some contact angle analysis.[(73)]

The contact angle measurements were advancing water droplets measured in air after ten minutes of contact (Table 10). As expected, the more hydrophilic copolypeptides which contain L-glutamine acid result in lower wetting angles.

9. SUMMARY/CONCLUSIONS

We have briefly reviewed a series of different polymer systems which have been used and can be used as model materials for probing biosurface phenomena. Although a wide variety of other polymer systems exist and can potentially be used as models, the ones reported here allow some study of stereochemistry, hydrophilicity, morphology, side chain and main chain mobility, hydrophobicity, and related properties. In addition, the hydroxyethyl methacrylate copolymer systems and the poly(α-amino acid) copolymer systems can also be used to probe surface charge phenomena, although this was not emphasized here.

It is clear that by application of suitable model systems and by careful quality control and surface property evaluation, coupled with careful biomedical interaction measurements such as *in vitro* blood interactions, one can indeed obtain significant correlations between the surface properties of materials and their biological behavior.[36]

ACKNOWLEDGMENTS

The agarose work presented in this chapter was supported in part by NIH grant HL 26569.

REFERENCES

1. Y. K. Sung, D. E. Gregonis, G. A. Russell, and J. D. Andrade, Effect of water and tacticity on the glass transition temperature of poly(2-hydroxyethyl methacrylate), *Polymer* **19**, 1362–1363 (1978).
2. A. D. Jenkins, Stereochemical definitions and notations relating to polymers, *Pure Appl. Chem.* **51**, 1101–1121 (1979).
3. J. Brandrup and E. H. Immergut, *Polymer Handbook*, 2nd ed., Wiley, New York (1975).
4. F. A. Bovey, *High Resolution NMR of Macromolecules*, Academic Press, New York (1972).
5. F. A. Bovey and G. V. D. Tiers, Polymer NSR spectroscopy, *J. Polym. Sci.* **44**, 173–182 (1960).
6. D. E. Gregonis, G. A. Russell, J. D. Andrade, and A. C. deVisser, Preparations and properties of stereoregular poly(hydroxyethyl methacrylate) polymers and hydrogels, *Polymer* **19**, 1279–1284 (1978).
7. H. Yuki and K. Hatada, in: *Advances in Polymer Science* (H.-J. Cantow, G. Dall'Asta, K. Dušek, J. D. Ferry, H. Fujita, M. Gordon, W. Kern, G. Natta, S. Okamura, C. G. Overberger, T. Saegusa, G. V. Schulz, W. P. Slichter and J. K. Stille, eds.), Vol. 31, pp. 1-45, Springer-Verlag, New York (1979).
8. P. Vlcek, D. Doskocilova, and J. Trekoval, Anionic copolymerization of methacrylates, *J. Polym. Sci. Symp.* **42**, 231–238 (1973).
9. H. Kusangi, H. Tadokoro, and Y. Chatani, Double strand helix of isotactic poly(methyl methacrylate), *Macromolecules* **9**, 531–532 (1976).
10. R. Lovell and A. H. Windel, Structure of noncrystalline isotactic poly(methyl methacrylates). Evidence against double helices, *Macromolecules* **14**, 211–212 (1981).
11. M. Toyama and T. Ito, Studies of surface wettability of stereospecific poly(methacrylic acid esters), *J. Colloid Interface Sci.* **49**, 139–142 (1974).
12. D. E. Gregonis, C. M. Chen, and J. D. Andrade, in: *Hydrogels for Medical and Related Applications*, (J. D. Andrade, ed.), Am. Chem. Soc. Symp. Ser. **31**, 88–104 (1976).
13. B. D. Ratner and A. S. Hoffman, in: *Adhesion and Adsorption of Polymers* (L. H. Lee, ed.), Part B, pp. 691–706, Plenum, New York (1980).
14. O. Wichterle, Hydrogels, *Encyclopedia of Polymer Science* **15**, 273-291 (1971).
15. M. S. Choudhary and I. K. Varma, Copolymerization of 2-hydroxyethyl methacrylate with alkyl methacrylates, *Eur. Polym. J.* **15**, 957-959 (1979).
16. I. K. Varma and S. Patnaik, Copolymerization of 2-hydroxyethyl methacrylate with alkyl acrylates, *Eur. Polym. J.* **12**, 259-261 (1976).
17. T. Okano, J. Aoyagi, and I. Shinohara, The wettability and composition of 2-hydroxyethyl methacrylate copolymers. *Nippon Kagaku Kaishi* **1976(1)** 161–170 (1976).

18. C. Migliaresi, L. Nicodemo, and L. Nicolais, 2-Hydroxyethyl methacrylate/methyl methacrylate copolymers for biomedical use, *Society for Biomaterials Abstracts*, 8th Annual Meeting, p. 123 (1982).
19. A. Silberberg, in: *Hydrogels for Medical and Related Applications* (J. D. Andrade, ed.), Am. Chem. Soc. Symp. Ser. **31**, 198-205 (1976).
20. J. D. Andrade, R. N. King, D. E. Gregonis, and D. L. Coleman, Surface characterization of poly(hydroxyethyl methacrylate) and related polymers, *J. Polym. Sci. Symp.* **66**, 313-336 (1979).
21. J. D. Andrade, S. M. Ma, R. N. King, and D. E. Gregonis, Contact angles at the solid-water interface, *J. Colloid Interface Sci.* **72**, 488-494 (1979).
22. W. C. Hamilton, A technique for the characterization of hydrophilic solid surfaces, *J. Colloid Interface Sci.* **40**, 219-222 (1972).
23. W. C. Hamilton, Measurement of the polar force contribution to adhesive bonding, *J. Colloid Interface Sci.* **47**, 672-675 (1974).
23. W. C. Hamilton, Measurement of the polar force contribution to adhesive bonding, *J. Colloid Interface Sci.* **47**, 672-675 (1974).
24. D. E. Gregonis, R. Hsu, D. E. Buerger, L. M. Smith, and J. D. Andrade, in: *Macromolecular Solutions* (R. B. Seymour and G. A. Stahl, eds.), pp. 120-133, Pergamon, New York (1982).
25. D. R. Paul and S. Newman, *Polymer Blends*, Vol. 1, Academic Press, New York (1978).
26. M. F. Sefton and E. W. Merrill, Surface hydroxylation of styrene-butadiene-styrene block copolymers for biomaterials, *J. Biomed. Materials Res.* **10**, 33-45 (1976).
27. M. F. Sefton and E. W. Merrill, Infrared spectroscopic analysis of complex polymer systems, *J. Appl. Polym. Sci.* **20**, 157-168 (1976).
28. H. C. Brown, *Organic Synthesis via Boranes*, J. Wiley, New York (1975).
29. C. P. Pinazzi, A. Menil, J. C. Rabadeax, and A. Pleurdeau, Polyisoprene and polybutadiene derivatives of potential biomedical interest, *J. Polym. Sci. Symp.* **52**, 1-7 (1975).
30. T. Okano, M. Ikemi, and I. Shinohara, The viscosity behavior of ABA-type block copolymer composed of 2-hydroxyethyl methacrylate and styrene in organic solvent mixture, *Polym. J.* **10**, 477-484 (1978).
31. T. Okano, M. Katayama, and I. Shinohara, The influence of hydrophilic and hydrophobic domains on water wettability of 2-hydroxyethyl methacrylate-styrene copolymers, *J. Appl. Polym. Sci.* **22**, 369-377 (1978).
32. E. W. Merrill, Properties of materials affecting the behavior of blood at their interfaces, *Ann. N.Y. Acad. Sci.* **283**, 6-16 (1977).
33. L. S. Luskin in: *Encyclopedia of Industrial Chemical Analysis* (F. D. Snell and C. L. Hilton, eds.), Vol. 4, pp. 181-218, Interscience, New York (1967).
34. R. S. Corley, in: *Monomers* (E. R. S. Blout and H. Mark, eds.), Interscience, New York (1951).
35. C. E. Rehberg and C. H. Fisher, Preparation and properties of *n*-alkyl acrylates, *J. Am. Chem. Soc.* **66**, 1203-1207 (1944).
36. D. L. Coleman, *In Vitro Blood-Materials Interactions: A Multi-Test Approach*, Ph.D. Thesis, Department of Pharmaceutics, University of Utah, August, 1980.
37. H. Determan, *Gel Chromatography*, Springer-Verlag, New York (1968).
38. S. Shaltiel, Hydrophobic chromatography, *Methods Enzymol.* **34**, 126-140 (1974).
39. J. L. Ochoa, Hydrophobic (interaction) chromatography, *Biochimie* **60**, 1-15 (1978).
40. B. H. J. Hofstee and N. F. Otillio, Non-ionic adsorption chromatography of proteins, *J. Chromatogr.* **159**, 57-70 (1978).
41. T. C. J. Gribnau, C. A. G. van Ekelen, C. Stumm, and G. I. Tesser, Microscopic observations on agarose beads, *J. Chromatogr.* **132**, 519-524 (1977).
42. J. Rosengren, S. Pahlman, M. Glad, and S. Hjerten, Hydrophobic interaction chromatography on noncharged sepharose derivatives, *Biochem. Biophys. Acta* **412**, 51-61 (1975).

43. A. Noshay and J. E. McGrath, *Block Copolymers, Overview and Critical Survey*, Academic Press, New York (1977).
44. D. C. Allport and W. H. James, *Block Copolymers*, Wiley, New York (1973).
45. G. L. Wilkes, T. S. Dziemianowicz, Z. H. Ophir, E. Artz, and R. Wildnauer, Thermally induced time dependence of mechanical properties in biomedical grade polyurethanes, *J. Biomed. Materials Res.* **13**, 189–206 (1979).
46. K. A. Pigott, in: *Kirk Othmer Encycl. Chem. Technol.* (H. Mark, ed.), Vol. 21, pp. 56–106, Interscience, New York (1970).
47. J. M. Buist and H. Gudgeon, *Adv. Polyurethane Tech.* Wiley, New York (1968).
48. C. S. Schollenberger and K. J. Dinberg, Thermoplastic urethane molecular weight–property relations, *J. Elastoplast.* **5**, 222–251 (1973).
49. C. S. P. Sung, C. B. Hu, and C. S. Wu, Properties of segmented poly(urethane ureas) based on 2,4-toluene diisocyanate, *Macromolecules* **13**, 111–116 (1980).
50. C. S. P. Sung and C. B. Hu, Orientation studies of segmented polyether poly(urethane urea) elastomers by infrared dichroism, *Macromolecules* **14**, 212–215 (1981).
51. C. S. P. Sung and N. S. Schnedier, Infrared studies of hydrogen bonding in toluene diisocyanate based polyurethanes, *Macromolecules* **8**, 68–73 (1975).
52. S. L. Samuels and G. L. Wilkes, Anisotropic superstructure in segmented polyurethanes as measured by photographic light scattering, *Polym. Lett.* **9**, 761–766 (1971).
53. R. W. Seymour, G. M. Estes, D. S. Huh, and S. L. Cooper, Rheo-optical studies of polyurethane block polymers, *J. Polym. Sci.* **10**, 1521–1527 (1972).
54. K. Knutson and D. J. Lyman, Morphology of block copolyurethanes. II. FTIR and ESCA techniques for studying surface morphology, *Organic Coatings and Plastics Preprints* **42**, 621–627 (1980).
55. V. S. da Costa, D. Brier-Russell, D. W. Salzman, and E. W. Merrill, ESCA studies of polyurethanes: blood activation in relation to surface composition, *J. Colloid Interface Sci.* **80**, 445–452 (1981).
56. B. D. Ratner, ESCA and SEM studies on polyurethanes for biomedical applications, *Polymer Preprints* **21**, 152–153 (1980).
57. B. D. Ratner in: *Photon, Electron, and Ion Probes of Polymer Structure and Properties* (D. W. Dwight, T. J. Fabish, and H. R. Thomas, eds.), Am. Chem. Soc. Symp. Ser. **162**, 371–382 (1981).
58. B. D. Ratner in: *Physicochemical Aspects of Polymer Surfaces* (K. L. Mittal, ed.), Vol. 2, pp. 969–983, Plenum, New York (1983).
59. M. D. Lelah, L. K. Lambrecht, B. R. Young, and S. L. Cooper, Physiochemical characterization and *in vivo* blood tolerability of cast and extruded biomer, *J. Biomed. Materials Res.* **17**, 1–22 (1981).
60. K. Knutson and D. J. Lyman in: *Biomaterials: Interfacial Phenomena and Applications* (S. L. Cooper and N. A. Peppas, eds.), Advan. Chem. Ser. **199**, 109–132 (1982).
61. T. Miyamae, S. Mori, and Y. Takeda, Poly-L-glutamic acid surgical sutures, U.S. Patent 3,371,069 (1968).
62. R. V. Peterson, C. G. Anderson, S. M. Fang, D. E. Gregonis, S. W. Kim, J. Feijen, J. M. Anderson, and S. Mitra in: *Controlled Release of Bioactive Materials* (R. Baker, ed.), pp. 45–60, Academic Press, New York (1980).
63. C. W. Hall, M. Spira, F. Gerow, L. Adams, E. Martin, and S. B. Hardy. Evaluation of artificial skin models: presentation of three clinical cases, *Trans. Am. Soc. Art. Int. Org.* **16**, 12–16 (1970).
64. E. Klein, P. D. May, J. K. Smith, and N. Leger, Permeability of synthetic polypeptide membranes, *Biopolymers* **10**, 647–655 (1971).
65. F. Fuchs, Uber N-carbonsaure-anhydride, *Chem. Ber.* **55**, 2943 (1922).
66. A. C. Farthing, Synthetic polypeptides, *J. Chem. Soc.*, 3213–3217 (1950).

67. H. R. Kricheldorf, Mechanisms of NCA polymerization, *Makromol. Chem.* **178**, 905–939 (1977).
68. N. Lupu-Lotan, A. Yaron, A. Berger, and M. Sela, Conformational changes in the nonionizable water soluble synthetic polypeptide poly-N5(3-hydroxypropyl) L-glutamine, *Biopolymers* **3**, 625–655 (1965).
69. T. Tanaka, T. Mori, K. Ogawa, and R. Tanaka, Heterogeneous network polymers. IX. Poly(L-glutamic acid) crosslinked with polyether diisocyanates, *Polymer J.* **11**, 731–736 (1979).
70. T. Sugie, J. M. Anderson, and P. A. Hiltner, Structure and deformation of a crosslinked poly(α-amino acid), *Polymer Preprints* **20**, 439–441 (1979).
71. T. Sugie and A. Hiltner, Structure and deformation of a crosslinked poly(α-amino acid), *J. Macromol. Sci., Phys.* **B17**, 769–785 (1980).
72. D. D. Solomon, D. H. Cowan, J. M. Anderson, and A. G. Walton, Platelet interaction with synthetic copolypeptide films, *J. Biomed. Materials Res.* **13**, 765–782 (1980).
73. D. D. Solomon, D. H. Cowan, and A. G. Walton in: *Colloid and Interface Science*, Vol. 5 (M. Kerker, ed.), pp. 1–21, Academic Press, New York (1976).
74. G. Odian, *Principles of Polymerization*, McGraw-Hill, New York (1970).
75. E. H. Riddle, *Monomeric Acrylic Esters*, Reinhold, New York (1954).
76. P. Pino and R. Mulhaupt, Stereospecific polymerization of propylene: an outlook 25 years after its discovery, *Angew Chem., Int. Ed. Engl.* **19**, 857–875 (1980).
77. S. L. Aggarwal, *Block Copolymers*, Plenum, New York (1970).
78. T. Okano, M. Shimada, I. Shinohara, K. Kataoka, T. Akaike, and Y. Sakurai, in: *Biomaterials 1980*, (G. D. Winters, D. F. Gibbons, and H. Plenk, Jr., eds.), pp. 445–450, John Wiley, New York (1982).

4

Polymeric Oriented Monolayers and Multilayers as Model Surfaces

Bernd Hupfer and Helmut Ringsdorf

In the beginning, there must have been a membrane.[1]

1. INTRODUCTION

1.1. BIOLOGICAL MEMBRANES

All living cells are surrounded by a lipid bilayer membrane in which a variety of proteins (e.g., enzymes) are embedded (fluid mosaic model; Figure 1). Phospholipids and cholesterol represent the major part of the lipids of a biomembrane. Figure 2 illustrates the structure of some typical amphiphilic membrane components with hydrophobic alkyl chains and hydrophilic head groups. The amount of protein in biological membranes varies between 40 and 60%[3]; however, in highly specialized membranes values between 20% (myelin sheath of nerve axons; electrical isolator) and 75% (mitochondrial inner membrane; enzyme system of the respiratory chain) may occur. Furthermore, the incorporation of proteins in a membrane and in particular as reticulum on its inside (spectrin of erythrocyte membranes[4]) increases its stability.

The basic functions of biological membranes can be divided into four groups:

1. Organization: energetically most favorable spatial arrangements of chemical reactions, i.e., enzymes (mitochondrial inner membrane, endoplasmatic reticulum).

Bernd Hupfer and Helmut Ringsdorf • Institut fur Organische Chemie, Johannes Gutenberg-Universität, J.-J.-Becher-Weg 18–20, D-6500 Mainz, Federal Republic of Germany.

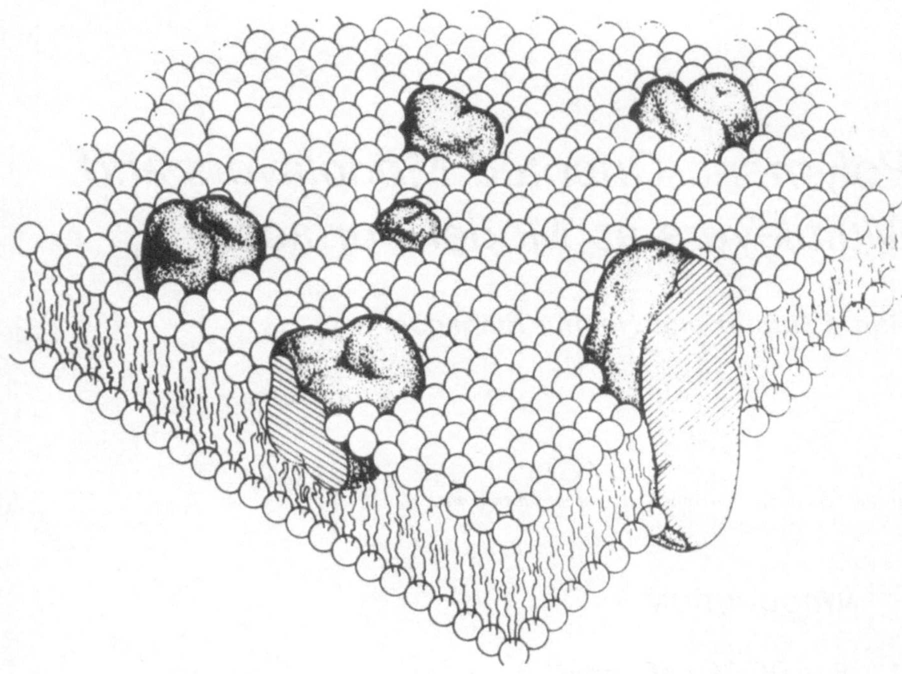

FIGURE 1. Fluid moasic model of a cell membrane.[2]

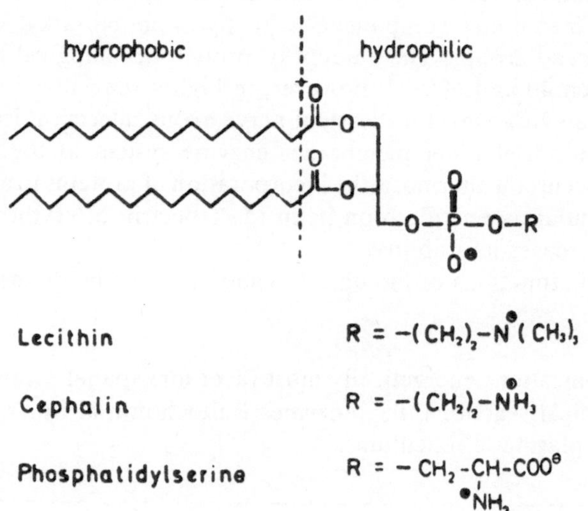

FIGURE 2. Structures of the most common biomembrane phospholipids.[5]

2. Interaction: transport of substances, i.e., of substrates and metabolic intermediates of end products.
3. Information: cell recognition and adhesion, contact inhibition, immunologic reactions (glycoproteins, glycolipids).
4. Compartmentation: formation of closed "spheres," in which biochemical reactions may take place under optimal conditions without being disturbed by the environment.

The functions which are understood in detail have resulted from the study of highly specialized membranes. These are membranes which incorporate only a few molecular components and, because of their simplicity, are more easily understood.

1.2. MODELS FOR BIOMEMBRANES

For studying specific processes going on in a complex natural membrane, model systems, which are based on the spontaneous self-organization of lipid molecules when brought in contact with an aqueous medium, can be useful. The three most frequently used model membrane systems are monolayers at the gas/water interface,[6] Biomolecular Lipid Membranes (BLM),[7] and vesicles (liposomes)[8,9] (Figure 3). BLM and liposomes consist—as do membranes in nature—of lipid bilayers separating two aqueous compartments. They enable the measurement of, e.g., membrane resistance, capacity, and permeability. The monolayer system, however, representing only one half of the naturally occurring bilayer, allows one to vary the packing density of the spread lipids by applying different surface pressures. Moreover, it can be used for surface potential, surface viscosity, and protein adsorption measurements.[10]

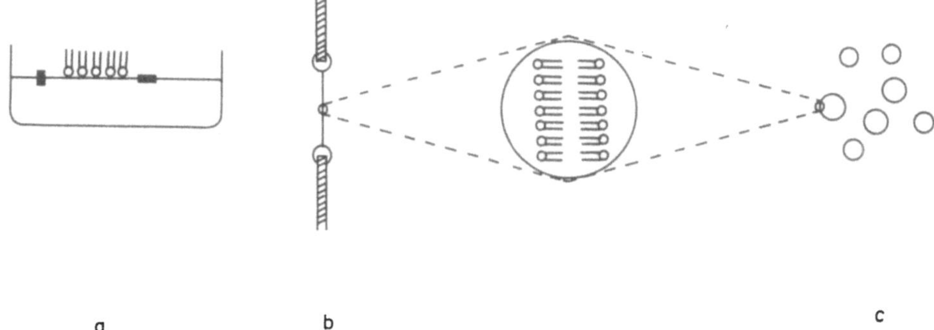

FIGURE 3. Orientation of amphiphilic compounds in model membranes: (a) monolayer; (b) Biomecular Lipid Membrane (BLM); (c) vesicle (liposome). Between (b) and (c) a cross section through the BLM or liposome wall is shown.[5]

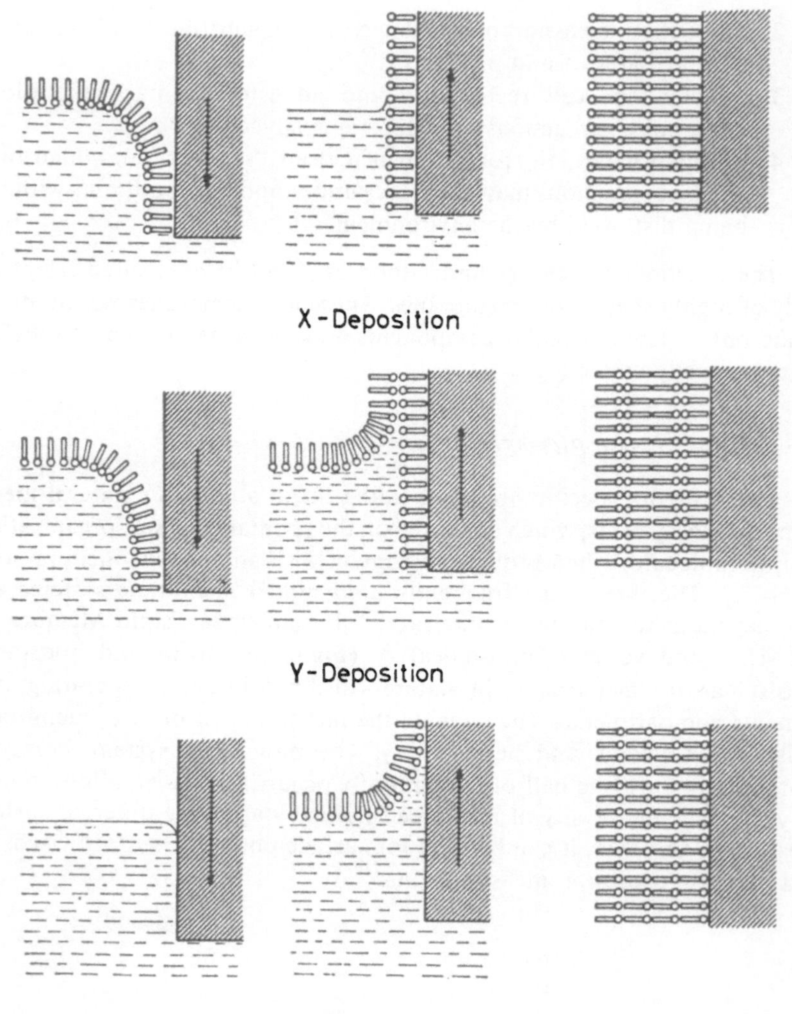

FIGURE 4. Types of monolayer deposition and resulting monolayer system if no rearrangement occurs. Each lipid molecule is represented by a circle (hydrophilic head group) and a bar (long hydrohpobic chain).[10]

In addition, there is the possibility of building up multilayers on solid supports by the Langmuir–Blodgett technique[10–12] (Figure 4). These films generated by slowly dipping the support through a monolayer perpendicular to the water surface, have a defined thickness which is only controlled by the dimensions of the lipids. Since the architecture of these layers can be specifically designed, they are particularly suitable for assembling systems in which photoinitiated charge separation processes can be studied (conversion of solar energy into electrically or chemically stored energy).

1.3. CHARACTERIZATION OF MONO- AND MULTILAYERS

Monomolecular films of lipids on a water surface can be prepared by spreading. A lipid solution is dropped on the surface of the water. The solvent evaporates and the lipid molecules left behind on the surface of the water can be packed by decreasing the surface area by a movable barrier (Figure 5). The resulting surface pressure is recorded by a measuring barrier on the opposite side of the trough. The decreasing area per molecule is in general accompanied by several phase changes. At large areas the film will form a gaseous phase, which is converted to a liquid expanded (liquid crystalline) phase and finally to a condensed (crystalline) phase. The latter can be characterized by a dense packing of the head groups (head packing) or of the alkyl chains (chain packing). Further decrease of the surface area results in the collapse of the monolayer by formation of undefined bi- and multilayers.

Well-defined multilayers are obtained by transfer of a (usually condensed) monolayer kept under constant surface pressure onto a solid substrate (glass, quartz, metals, polymers) (Figure 4). Whether deposition takes place only on immersion (X-type), on both dipping and withdrawing (Y-type), or on withdrawal only (Z-type) is dependent on the nature of the substrate (hydrophilic or hydrophobic), temperature, and dipping and with-

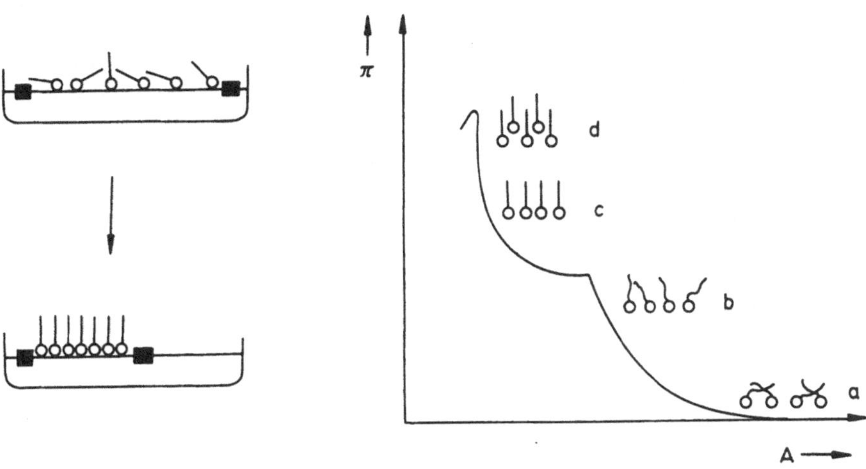

FIGURE 5. Principle of monolayer characterization via surface pressure (π)-area (A) isotherms: (a) gaseous phase, (b) liquid expanded phase, (c) condensed phase (head packing), (d) condensed phase (chain packing).[5]

drawing speed, as well as on the applied surface pressure. Additional methods for characterizing mono- and multilayers (surface potential, surface viscosity, optical investigations, X-ray diffraction, photoelectron spectroscopy, electron microscopy, electron diffraction, electrical properties) have already been extensively reviewed.[6,10,13]

1.4. STABILITY OF CONVENTIONAL MODEL MEMBRANE SYSTEMS

Despite the great usefulness of conventional model membrane systems (Figure 3) they all have one major shortcoming: their limited long-term stability. Liposomes on prolonged standing tend to fuse.[14] In addition, when administered as drug carriers *in vivo*,[15] they are subject to lipid exchange and removal processes, uncontrolled leakage of entrapped drugs, and vesicle-vesicle and vesicle-cell fusion.

Depending on the lipids and the conditions used for their preparation, BLM are only stable for minutes or hours and finally break apart like soap bubbles. Monolayers if kept under constant surface pressure are subject to film collapse or loss of material by solution or evaporation.[6] In LB-multilayers reorientation (overturning) of the system may take place during or after the coating procedure.[10] In addition, during storage of multilayer systems at room temperature and with exposure to the air, crystallization has been observed by microscopic and X-ray studies.[10] If exposed to a high vacuum, desorption of monolayer material may occur.

It would, therefore, be desirable to have model systems available that do not exhibit these disadvantages. Mother nature stabilizes cell membranes by means of integral and especially peripheral proteins. However, the idea of using membrane proteins for model membrane stabilization does not yet seem to be realizable. A simpler approach is the attempt to incorporate the lipids into the stabilization process. Khorana and Chakatrabarti[16] synthesized phospholipids bearing photoreactive groups and linked them to one another or to membrane proteins by UV-irradiation. A better approach is a complete polymerization or—in the case of bifunctional monomers—crosslinking of the membrane. As a matter of fact, polymerized monolayers have already been known for a long time, but only recently has this concept been transferred to BLM and vesicles.

2. POLYMERIZED MODEL MEMBRANES

2.1. POLYMERIZABLE LIPIDS

There are different possibilities for the localization of the polymerizable group in the lipid molecules. A polyreaction in the hydrophobic part of the

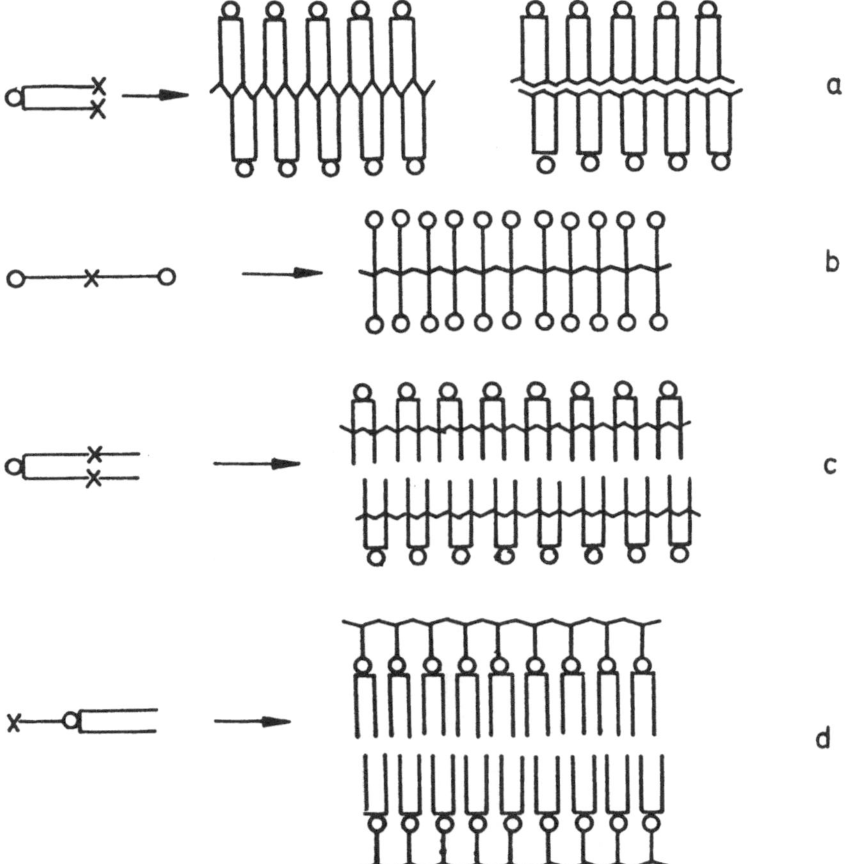

FIGURE 6. Possible preparation of polymeric model membranes (X = polymerizable group). (a)–(c): Polymerization preserving head group properties. (d) Polymerization preserving chain mobility.[5]

monomers (Figure 6a–c) impacts the ordered to fluid phase transition of the hydrocarbon chains, whereas a polymerization in the hydrophilic part of the molecules (Figure 6d) changes the properties of the head groups. All polymeric lipid systems will show an increase in viscosity and a decrease of lateral mobility of the molecules.

Some examples of the variety of amphiphiles carrying polymerizable groups known so far are listed in Table 1. Besides the very simple carbonic acid (**3, 4**) and ammonium structures (**1, 7, 13**), lipids with naturally occurring head groups (**2, 5, 8**) were also synthesized. As polyreactive units diacetylenic and butadienic, as well as acryloylic and methacryloylic moieties were used.

TABLE 1
Model Membrane Forming Polymerizable Amphiphiles

Type[a]	Number	Formula[b]	Reference
(a)	(1)	$R{-}NH{-}(CH_2)_{10}{-}CO{-}O{-}(CH_2)_2$ and $R{-}NH{-}(CH_2)_{10}{-}CO{-}O{-}(CH_2)_2$ both bonded to $\overset{+}{N}(CH_3)_2$ Br^-	(17)
(b)	(2)	$H_2O_3PO{-}(CH_2)_9{-}C{\equiv}C{-}C{\equiv}C{-}(CH_2)_9{-}OPO_3H_2$	(18)
(c)	(3)	$H_3C{-}(CH_2)_n{-}C{\equiv}C{-}C{\equiv}C{-}(CH_2)_8{-}COOH$	(19)
(c)	(4)	$H_3C{-}(CH_2)_{12}{-}CH{=}CH{-}CH{=}CH{-}COOH$	(20)
(c)	(5)	$H_3C{-}(CH_2)_{12}{-}C{\equiv}C{-}C{\equiv}C{-}(CH_2)_9{-}O{-}P(=O)(O^-){-}O(CH_2)_2{-}\overset{+}{N}H_3$	(21)
		$H_3C{-}(CH_2)_{12}{-}C{\equiv}C{-}C{\equiv}C{-}(CH_2)_8{-}CO{-}O$ and $H_3C{-}(CH_2)_{12}{-}C{\equiv}C{-}C{\equiv}C{-}(CH_2)_8{-}CO{-}O$ both bonded to X	
(c)	(6)	$X{:}{-}(CH_2)_2{-}O{-}(CH_2)_2{-}O{-}(CH_2)_2{-}O{-}(CH_2)_2{-}$	(19)-(22)
(c)	(7)	$X{:}{-}(CH_2)_2{-}\overset{+}{N}(CH_3)_2{-}(CH_2)_2$ Br^-	(19)-(22)
(c)	(8)	$X{:}{-}CH_2{-}CH{-}CH_2{-}O{-}PO(O^-){-}O{-}(CH_2)_2{-}\overset{+}{N}(CH_3)_3$	(22), (23)

(c)	**(9)**	$H_3C-(CH_2)_{12}-CH=CH-CH=CH-CO-O-CH_2$ $H_3C-(CH_2)_{12}-CH=CH-CH=CH-CO-O-CH$ $CH_2-O-P(=O)(O^-)-O(CH_2)_2-\overset{+}{N}(CH_3)_3$	(17), (23)
(c)	**(10)**	$H_3C-(CH_2)_{12}-C\equiv C-C\equiv C-(CH_2)_8-CO-O-(CH_2)_2$ $H_3C-(CH_2)_{12}-C\equiv C-C\equiv C-(CH_2)_8-CO-O-(CH_2)_2$ $\overset{+}{N}$, $(CH_2)_2-SO_3^-$, H	(24)
(d)	**(11)**	$H_3C-(CH_2)_{17}-OR$	(25)
(d)	**(12)**	$H_3C-(CH_2)_{17}-O-CH_2$ $H_3C-(CH_2)_{17}-O-CH$ CH_2-O-R	(17)
(d)	**(13)**	$H_3C-(CH_2)_{17}$, CH_3 $\overset{+}{N}$ Br^- $H_3C-(CH_2)_{17}$, $(CH_2)_3-NH-R$	(26)

[a] Cf. Figure 6.
[b] $R = CH_2=C(CH_3)-CO$.

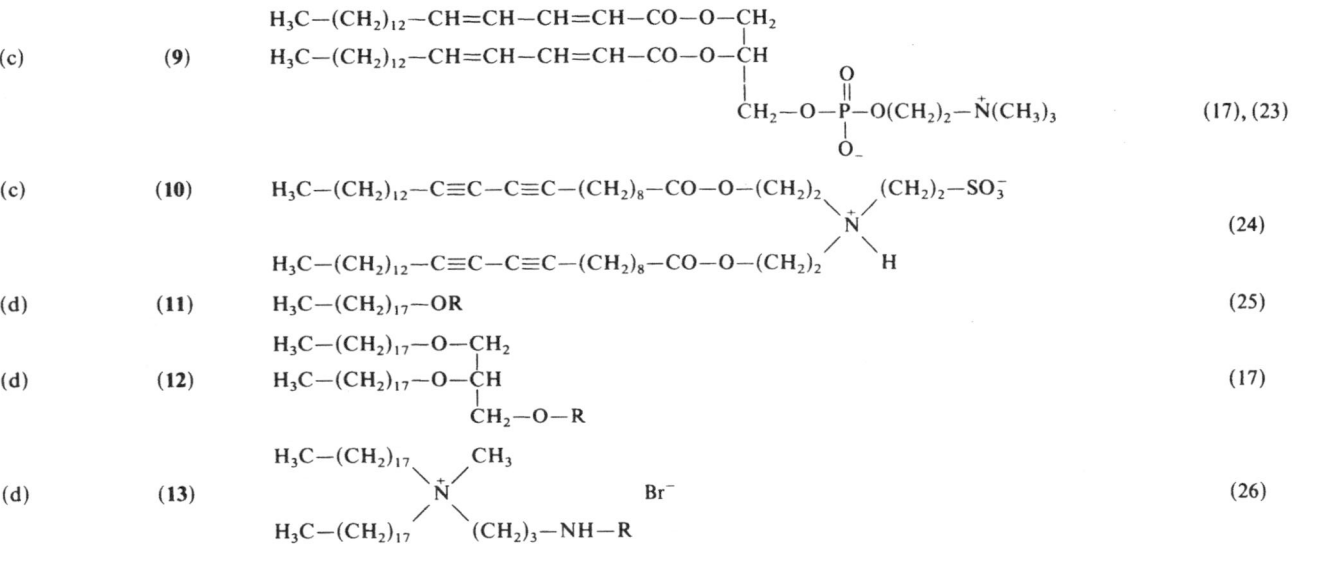

The high interest in this field is documented by the reports of several other laboratories[27-32] which also describe the synthesis and polymerization in model membrane systems of a number of novel amphiphiles.

The formation of monolayers at the gas/water interface provides the oldest and simplest of the membrane models in Figure 3. It is especially suited for studying polymerization reactions. Therefore, in the following it will be described in more detail.

2.2. POLYMERIZED MONOLAYERS

2.2.1. Monolayers with One Component

2.2.1.1. Spreading Behavior of Monomers. Influence of subphase temperature, pH, and molecular structure of amphiphiles can easily be studied by recording surface pressure/area isotherms (Figure 5). The effect of chain length and structure of polymerizable and natural lecithins is illustrated in Figure 7. At 30°C distearoyllecithin is still fully in the condensed state, whereas butadiene lecithin (**9**), which carries the same number of carbon atoms per alkyl chain, is already completely in the expanded state.[23] Although diacetylene lecithin (**8**) bears 26 carbon atoms per chain, it forms

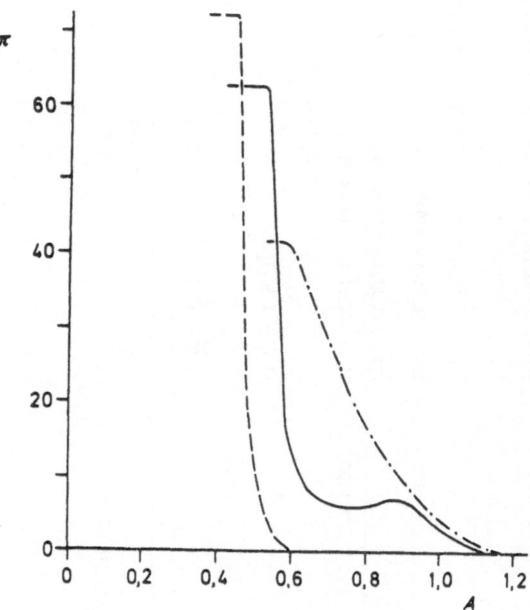

FIGURE 7. Surface pressure–area isotherms of polymerizable and natural lecithins at 30°C.[23] Key: (-----) distearoyllecithin; (——) diacetylene lecithin **8**; (—·—) butadiene lecithin **9**. π: Surface pressure in $mN \cdot m^{-1}$, A: surface area in nm^2/molecule.

both an expanded and a condensed phase at 30°C. The reason for these marked differences is the disturbance of the packing of the hydrophobic side chains by the double and triple bonds of the polymerizable lipids. At 2°C, however, all three lecithins are in the condensed state. Johnston *et al.*[27] reported the spreading behavior of two homologues of **8** containing 23 and 25 carbon atoms per chain. These compounds exhibit expanded phases even at subphase temperatures as low as 7°C which is certainly due to impurities.

On the other hand, changes of subphase pH have no significant influence on the shape of the surface pressure/area isotherms of lecithins, since they do not contain movable protons. This is different from the case of cephalins (e.g., **5**) which occupy considerably larger areas at high pH's and smaller areas at low pH's. This effect is thought to be due to head group conformation changes.[33]

A more drastic influence of subphase pH can be observed for the sulfolipid analogue **10** (Figure 8). At pH values from 2 to 5.5., **10** forms *two* condensed phases, whereas at pH 12, where the nitrogen is deprotonated, there is only one phase present.[24] This behavior has considerable influence on the monolayer polymerization properties of **10**, which is discussed in section 2.2.1.2.

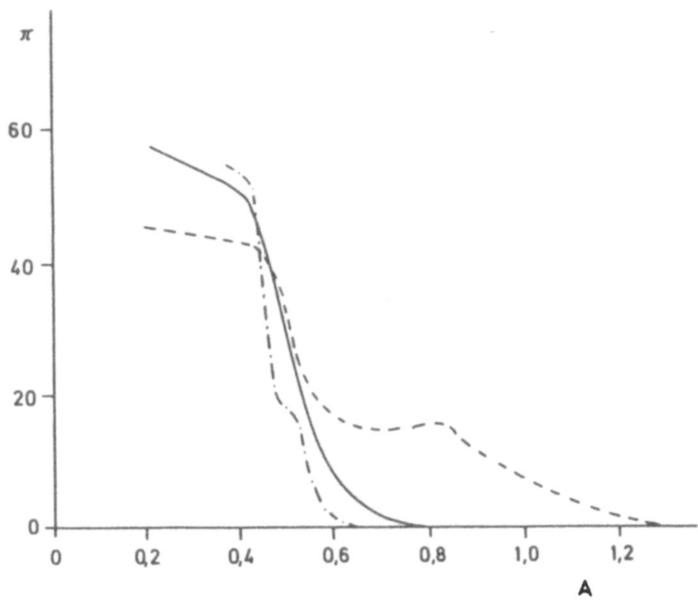

FIGURE 8. Surface pressure-area isotherms of sulfolipid **10** at different *p*H's.[24] Key: (—·—) *p*H 5.5, 41°C; (——) *p*H 12, 10°C; (- - -) *p*H 12, 21°C. π: surface pressure in $mN \cdot m^{-1}$, *A*: surface area in nm^2/molecule.

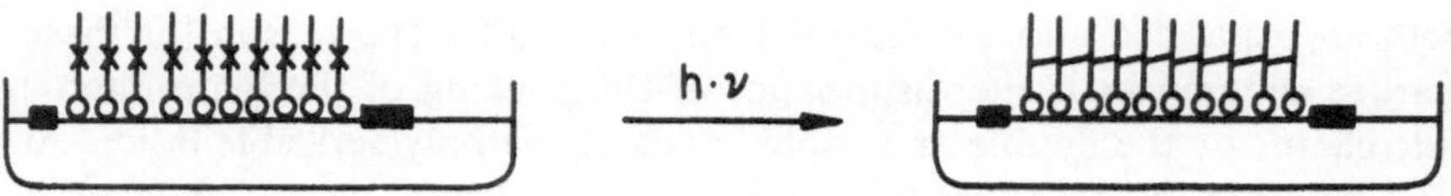

FIGURE 9. Formation of polymeric monolayers from polymerizable lipids (X = polymerizable group).

2.2.1.2. Polymerization behavior. A complete polymerization of a monolayer was first reported in 1935,[34] when Gee described the spontaneous polyreaction of the maleic anhydride-β-elaeostearin complex at the gas/water interface. Since then a variety of papers on this issue have been published, where mainly simply structured lipids bearing polyreactive groups (e.g., **11**) were polymerized by high-energy irradiation (Figure 9). These films were characterized by a number of physical methods (Table 2) without checking on their usefulness as models for biomembranes.

A first indication of the improved stability of polymerized monolayers was given by Day and Lando,[55] who could form bilayers by sandwiching two polymeric monolayers of a diacetylene carbonic acid. These bilayers can cover holes larger than 0.5 mm in diameter and are completely self-supporting. In air they are stable for periods of up to one year.

Since the extinction coefficient of the final fully conjugated diacetylene polymer **14** is very large ($\varepsilon > 10^4$ liter·cm^{-1}·mol^{-1}), the course of the polyreaction can be followed spectroscopically even in a single monolayer.[19,54]

CH_3–$(CH_2)_{12}$–C≡C–C≡C–$(CH_2)_9$–OPO_3H_2 (two monomer chains) $\xrightarrow{h\nu}$ polymer: =C[$(CH_2)_{12}CH_3$]–C≡C–C[$(CH_2)_9OPO_3H_2$]= (repeating)

2 **14**

It turns out[22] that single chain and symmetric double chain diacetylenic lipids show the same transitions from colorless monomers to a blue and finally a red polymer as had been known for a long time for the

TABLE 2

Monomers, Initiators, and Methods Used for Monolayer Polymerization and Characterization

Monomer	Initiator	Methods	Reference
Vinyl stearate	UV, peroxides, electron beam	SAD[a], surface potential, surface viscosity, ATR-IR	(36)-(41)
Octadecyl methacrylate (**10**)	UV, peroxides, electron beam	SAD[a], IR, mass spectrometry	(25), (35), (42)-(45)
Octadecylacrylate	UV, electron beam	SAD[a], ATR-IR, mass spectrometry	(39), (46)-(48)
Octadecylacrylamide	UV, peroxides	SAD[a], mass spectrometry	(43)
Divinyl esters	UV	SAD[a]	(42)
Vinyl isobutyl ether	BF_3, benzoyl peroxide	SAD[a]	(49)
Glycerol esters	UV	SAD[a], surface viscosity	(50)
Butadiene	H_2O_2	surface viscosity, surface pressure	(51)
Oleic acid/elaidic acid	UV	SAD[a]	(52)
Diacetylene carbonic acids	UV	SAD[a], UV, optical microscopy, electron microscopy, electron diffraction, contact angle, ellipsometry, ESCA, streaming potential	(19), (53)-(58), (62)
Butadienic lipids	UV	SAD[a]	(20)

[a] Surface pressure/area diagram.

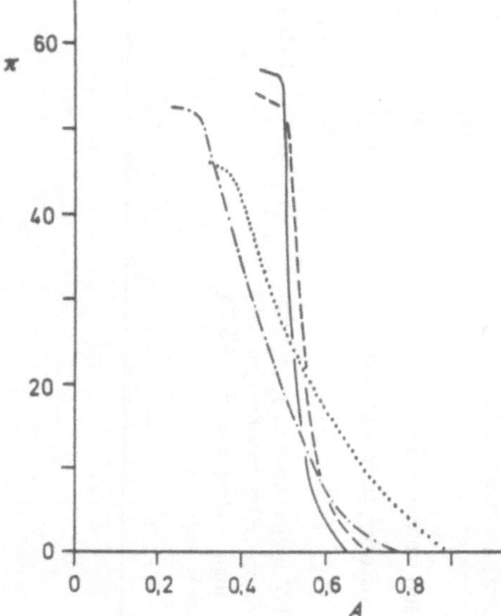

FIGURE 10. Surface pressure-area isotherms of polymeric lecithins at 20°C. Key (——) poly-**8**, (—·—·—) poly-**9**. Monomers, (- - -) **8**, (· · · ·) **9** shown for comparison.

topochemical polymerization of diacetylenes in the solid state.[59] Unsymmetric double chain amphiphiles (e.g., **8**), where the diyne moieties are at different distances from the hydrophilic head group, instantly form an orange-red polymer without going through the blue intermediate.[33]

In all cases the resulting polymer films occupy smaller areas than the corresponding monomers and are less compressible. In addition, their collapse usually occurs at higher surface pressure which indicates their improved stability. This is shown in Figure 10 for the phospholipids **8** and **9**.[23]

In contrast to all other monomers, films of diacetylenic lipids such as **3**, **5**, **7**, **8**, and **10** are only polymerizable if they are in the condensed state. This means that in monolayers also the polyreaction is topochemically controlled. This is especially well documented by the polymerization behavior of the sulfolipid **10** (cf. Figure 8). At 41°C and pH 2 to pH 5.5 **10** is only polymerizable in the "first" condensed phase, i.e., at surface pressures up to 15 mN/m, whereas a polyreaction does not take place in the "second" condensed phase at surface pressures from 20 to 50 mN/m.[24] Apparently in the second phase either the packing of the diyne groups is too tight to permit a topochemical polymerization or a vertical shift of the molecules at the gas/water interface causes a transition from head to chain packing and prevents the formation of polymer.

In contrast, methacryloyl (**1**, **11–13**) and butadiene (**4**, **9**) monomers are polymerizable in both the condensed and the expanded states, the rate of the polyreaction being somewhat lower in the fluid phase. Butadienes polymerize to form a 1,4-*trans*-polybutadiene backbone **15**.[20]

$$2\ CH_3(CH_2)_{12}CH{=}CH{-}CH{=}CH{-}COOH \xrightarrow{h\nu} \cdots[CH(C_{12}H_{24}CH_3){-}CH{=}CH{-}CH(COOH)]_n\cdots$$

15

Polycondensation reactions at the gas/water interface leading to highly oriented polyamides and polypeptides were recently described by two groups.[60,61] These polyreactions do not require an initiator and catalyst. In the case of long-chain esters of long-chain α-amino acids[61] stable polymerized model membranes are formed that are biodegradable.

2.2.2. *Mixed Monolayers*

Since the lipids of biomembranes have a wide distribution of head groups and alkyl chains, model membranes consisting of only one component do not represent very well the situation present in natural systems. In addition, as pointed out earlier, polymeric membranes exhibit a decreased mobility of the lipid alkyl chains which can completely suppress the ordered–fluid phase transition. If one wants to build up fairly stable model membranes that are still flexible one has to create mixed systems from polymerizable and natural lipids. Hereby, the polymerization behavior will greatly depend on the miscibility of the components. The question of whether an intimately mixed monolayer is formed at all can be answered by recording surface pressure/area isotherms.

2.2.2.1. Spreading Behavior. In Figure 11 surface pressure/area isotherms of monolayers of the cationic lipid **7** with fully saturated distearoylphosphatidylcholine (DSPC) are illustrated. While the lecithin is completely in the condensed state at 20°C (collapse at 69 mN/m), **7** exhibits an expanded phase at surface pressures up to 4 mN/m and collapses already at 42 mN/m.[19] All mixtures only show one collapse point, which continuously increases from 42 mN/m (pure **7**) to 69 mN/m (pure DSPC). The plot of the mean area per molecule versus the mole fraction reveals a considerable deviation from the additivity rule[63] (Figure 12). The deviation

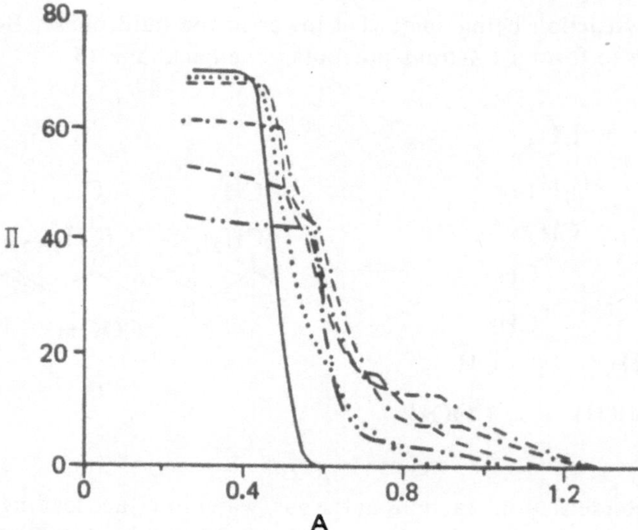

FIGURE 11. Surface pressure–area isotherms of mixtures of ammonium lipid **7** and distearoyllecithin at 20°C. Mole fraction of **7**: 0.0 (1); 0.2 (2); 0.4 (3); 0.61 (4); 0.8 (5); 1.0 (6). π: surface pressure in $mN \cdot m^{-1}$, A: area in nm^2/molecule.

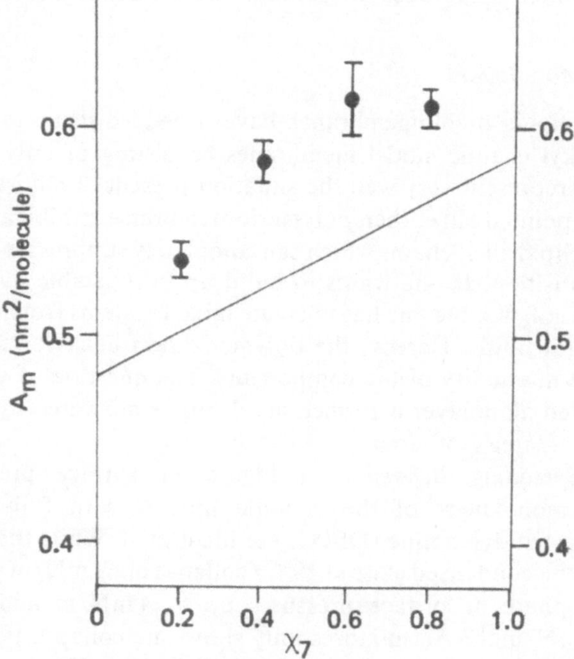

FIGURE 12. Mean molecular area (A_m) as a function of the composition for lipid **7** with distearoyllecithin at a surface pressure of $30\,mN \cdot m^{-1}$. (——) expected for immiscibility or ideal miscibility.[63]

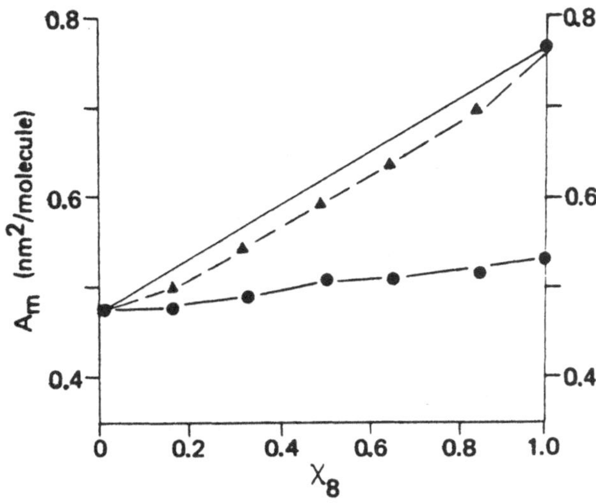

FIGURE 13. Mean molecular area (A_m) as a function of the composition for phospholipid **8** with distearoyllecithin at surface pressures of 30 $mN \cdot m^{-1}$ at 20°C (●) and 40°C (▲).

is indicative of the interaction of the molecules in the monolayer and indicates miscibility. In the present case, however, the miscibility cannot be complete because even mixtures with only 20 mole% of **7** are still photopolymerizable (see below).

The diacetylene phosphocholine **8** is immiscible with DSPC at temperatures where either phospholipid is in the condensed state. This is not only proved by the very small deviation from additivity (Figure 13), but also by the fact that photopolymerization also occurs if the mole fraction of **8** is smaller than 0.25. At temperatures above the phase transition of **8**, however, the negative deviation from the additivity rule indicates miscibility (Figure 13, 40°C). The same is true for mixture of the neutral lipid **6** with DSPC and dioleoylphosphatidylcholine, respectively.[64]

2.2.2.2. Polymerization Behavior. If the mole fraction of nonpolymerizable lipid exceeds 0.75, in the case of complete miscibility—assuming a hexagonal packing of the molecules—the topochemical diacetylene polymerization can take place at an only very reduced rate since then each polymerizable lipid is encompassed by nonpolymerizable amphiphiles. The rate of polymerization is then controlled by the time constant of the lateral movement of the lipids in the film, which is small in a condensed monolayer. On the other hand, if each component forms its own "patches," an unchanged polyreaction rate should be expected at all compositions.

In fact, the rate of polymerization of **7**/DSPC mixed films is smaller than that of pure **7**. This was determined by monitoring the monolayer

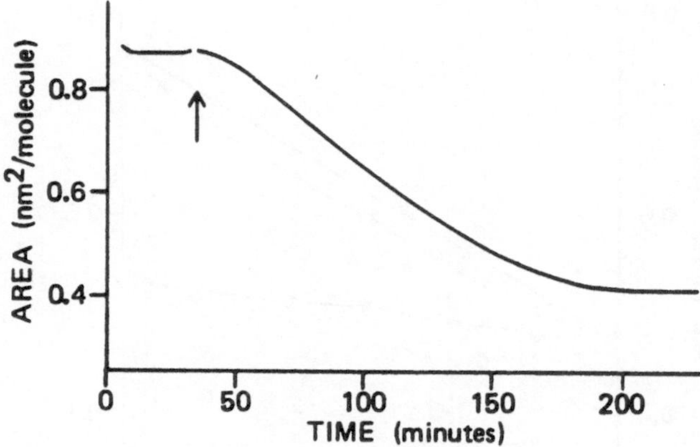

FIGURE 14. Area decrease as a function of time of a 1:1 mixed monolayer of lipid **6** and dilauroyllecithin caused by action of phospholipase A_2 (23°C).

absorption spectra.[19] In the case of **6**/DSPC mixtures, due to immiscibility at all mole fractions, the polymerization rate is equal to that of pure **6**.

2.2.2.3. Polymeric Mixed Monolayers as Model Surfaces. Since it is possible to create mixed or demixed partially polymerized monolayers only by making the right choice of the hydrophilic head groups, it was interesting to investigate how the mixing behavior affects an enzyme that can only react with the unpolymerized part of the membrane. Phospholipase A_2, which is able to selectively split off the fatty acid in the β-position of L-glycerophospholipids, was chosen for this purpose. This reaction results in the formation of a lysophospholipid and a fatty acid. If the chain lengths of the products are small enough, they will leave the monolayer and dissolve in the subphase. For this reason dilauroyllecithin (DLPC) was used as the substrate. If the mixed films are held under constant surface pressure the reaction can directly be monitored via the decrease in area. The resulting area versus time plots all look very similar to that in Figure 14.

As expected for 1:1 mixtures the surface area diminishes by half, corresponding to complete hydrolysis and removal of DLPC. In some experiments lower and higher values than 50% area decrease were observed. This may, on one hand, be due to incomplete removal of DLPC from the monolayer, and, on the other hand, be caused by partial solubilization of polymerizable lipid by lauric acid and α-lauroyl lyophosphatidylcholine present in the subphase.

A more quantitative interpretation of the curves can be given by determining the initial rate of hydrolysis from the initial slopes of the exponential plots. The results are summarized in Table 3. The hydrolysis rates of the monomeric mixed monolayers are lower by a factor of 0.5 for

TABLE 3
Hydrolysis of Dilauroyl Phosphatidylcholine (DLPC) Containing 1:1 Mixed Monolayers by Phospholipase A_2 at $10\,mN \cdot m^{-1}$ (23°C) Followed by Surface Area Decrease

Monolayer	Initial surface area (nm^2/molecule)	Final surface area (nm^2/molecule)	Hydrolysis rate (pm^2/mol min)
DLPC	0.80	0.00	38
(**7**)/DLPC, monomeric	0.94	0.43	0.8
(**7**)/DLPC, polymeric	0.86	0.45	0.7
(**6**)/DLPC, monomeric	0.87	0.40	14.9
(**6**)/DLPC, polymeric	0.81	0.37	8.9

the neutral lipid **6** and by a factor of 0.02 for the cationic **7**, which can be attributed to the phase behavior discussed above in detail: apparently, the enzyme can much more easily hydrolyse the DLPC in the patches present in demixed **6**/DLPC monolayers than that in almost homogeneously mixed films from **7** and DLPC. No significant differences of the hydrolysis rates are observed for the polymerized mixed films.

By this method it is possible to "drill" defined holes into a partially polymerized membrane, which is of significance in the specific opening of polymerized liposomes (cf. Section 2.5.).

2.3. POLYMERIC MULTILAYERS

As pointed out in section 1.2., lipid monolayers can be transferred to solid substrates by the Langmuir-Blodgett technique leading to multilayers of defined compositions and thickness. By using lipids bearing polymerizable groups such layers can be stabilized by polymerization under retention of the layer structure. Not all compounds that form monolayers, however, are able to build up multilayers. Table 4 represents a survey of the investigations conducted on polymerized multilayers. With one exception[57] no attempts have been made to characterize the surface of these thin polymer films.

2.4. POLYMERIC BLACK LIPID MEMBRANES

So far, only one example of the polymerization of a BLM has been reported.[26] The acrylic ammonium compound **13** forms bilayer membranes on the hole in a Teflon foil that are stable enough to survive the UV-initiated polymerization. Although the membrane undergoes no visible optical changes during that process, its capacitance increases while its resistance

TABLE 4
Monomers, Initiators, and Methods Used for LB-Multilayer Polymerization and Characterization

Monomer	Initiator	Methods	References
Maleic acid Octadecyl ester	UV	UV, mass spectrometry	(65)
Fumaric acid Octadecyl ester	UV	UV, IR, X-ray diffraction, electron diffraction	(65), (66)
Fumaric acid-N-octadecylamide	UV	UV, mass spectrometry	(65)
Maleic acid-N-octadecylamide	UV	UV, mass spectrometry	(65)
Vinyl stearate	UV, electron beam	ATR-IR, X-ray diffraction, electron diffraction, contact angle	(69)–(71), (81)
ω-Tricosenoic acid	Electron beam	SEM	(73)
Amino acid esters	none	IR	(60)
Octadecylacrylate	γ, electron beam	IR, X-ray diffraction	(74), (80)
Octadecyl methacrylate	γ	IR, X-ray diffraction	(74)
Octadecylacrylamide	γ	IR, electron diffraction, X-ray diffraction	(72)
Diacetylene carbonic acids	UV	UV, electron microscopy, electron diffraction, ATR-IR, SAXS[a]	(62), (67), (68), (75)

[a] Small-angle X-ray scattering.

decreases drastically. This is explained by a polymerization-induced contraction of some parts of the membrane which creates holes through which the surrounding electrolyte solution can penetrate. In contrast to this very slow membrane collapse, the rupture of a typical monomeric BLM after its induction is complete within a fraction of a second. Means to overcome this problem could be the polymerization of mixed BLM from natural and polymerizable lipids, where the natural lipids could fill the created holes. Since Black Lipid Membranes usually can only exist if the lipids are in the fluid state, no polymeric BLM can be formed from diacetylenic monomers.

2.5. POLYMERIC VESICLES (LIPOSOMES)

Kunitake in 1977 showed that not only naturally occurring phospholipids but also a great variety of simpler structured amphiphiles are able to form spherical bilayer assemblies if dispersed in water.[76,77] It did not take long until several investigators transferred this concept to lipids carrying polymerizable groups.[17,18,21,22,24,27,29-32,61] The resulting polymerized vesicles showed the desired enhanced stability and decreased leakage rate of entrapped substances (Figure 15). As monomers all amphiphiles listed in Table 1 may be used. Additional lipids carrying the polymerizable group in the head group, amidst, or at the end of the alkyl chains can be found in the literature cited above. So far almost 100 compounds capable of forming polymeric bilayers have been synthesized.

In some cases (e.g., **2**) the addition of cholesterol is mandatory for vesicle formation.[18] The resulting single-walled liposomes are promising systems for further research because *unsymmetric* α,ω dipolar amphiphiles

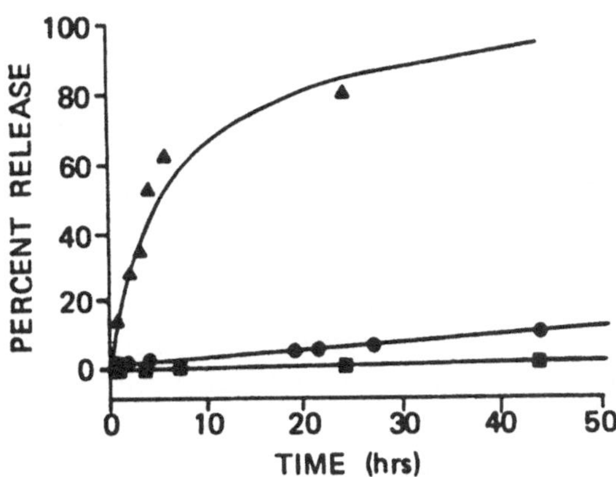

FIGURE 15. Release of entrapped 6-carboxyfluorescein from liposomes of monomeric (▲) and polymeric (■) **9**. For comparison: dipalmitoylphosphatidylcholine vesicles (●).[5]

with two head groups varying in size and chemical properties should arrange into "inside–outside vesicles" with distinct inner and outer membrane surfaces. Reactive groups such as electron acceptor and donor residues localized on different membrane boundaries of such a stable polymeric liposome may be useful tools in a study of transmembrane interactions and may provide new approaches to charge transfer processes involved in solar energy conversion.

Since glycoproteins and glycolipids are responsible for processes such as cell recognition and immune response, as a first model reaction liposomes from polymerized glycolipids have been reacted with lectins and shown to be agglutinated by these proteins.[78] Further aspects of this issue are the incorporation of lectins in the membrane of polymeric vesicles and the reaction of these vesicles with sugar-carrying liposomes as a first model reaction for cell–cell interaction occurring in nature.[5] This approach would not be feasible with unpolymerized liposomes since, if they come in contact with living tissue, they are either destroyed or taken up by fusion or endocytosis. The possible use of polymeric liposomes as antitumor agents on a cellular level has been extensively described recently.[5]

Although DSC measurements indicate that membranes from diacetylene monomers lose their ability to undergo the ordered–fluid phase transition after polymerization, it is possible to incorporate membrane proteins such as the F_0F_1-ATP synthetase complex into this rigid bilayer.[79] The enzyme incorporated in polymeric vesicles even exhibits a twofold increase in activity compared to monomeric vesicles. This is explained by the incorporation of the enzyme in "monomeric domains" which are necessary for linking the *planar* polymerized parts of a *spherical* liposome.

In general, the build-up of "mixed polymerized liposomes"[64] provides the following interesting possibilities:

1. Liposomes from polymerized and natural lipids possess an enhanced stability compared to unpolymerized vesicles and a higher mobility of the membrane than fully polymerized liposomes. By using nonpolymerizable lipids carrying chemically labile groups, the unpolymerized areas could be selectively destroyed and entrapped substances be selectively released (cf. section 2.2.2.3.).
2. The incorporation of glycolipids, glycoproteins, or antibodies makes them useful as models for cell–cell recognition.
3. By combination of polymerized lipids, natural lipids, membrane proteins, and lipids or proteins carrying cell-recognizing groups, a stable cell model could be constructed.

First approaches and methods to achieve these goals are discussed in detail in Reference (5).

3. CONCLUSIONS

Polymerized model membrane systems have received considerable interest during the last years. While however, initially major emphasis was laid upon creation and characterization of polymeric mono- and multilayers, nowadays systems such as Black Lipid Membranes and liposomes get the most attention, because they are better models for biological membranes.

Because of their controlled thickness, chemical resistance, possibilities for variation of polar groups, and long-term stability, polymeric mono- and multilayers could be utilized for many applications such as photoresists in microlithography,[73] one-dimensional semiconductors or photoconductors (diacetylenes only), or for studying adsorption processes.

Polymeric BLM as yet have been investigated only out of academic interest. Their spherical analogues, the vesicles (liposomes), are of special interest because they have the ability to entrap and very slowly release substances. This could make them useful as drug carriers as well as promoters of the separation of charged photoproducts. Their possible use as antitumor agents on the cellular level[5] is a completely new perspective that currently is only beginning to get beyond the stage of pure speculation.

REFERENCES

1. G. Weissmann, in: *Cell Membranes—Biochemistry, Cell Biology and Pathology* (G. Weissmann and R. Clairborne, eds.), pp. 257-266, HP Publishing, New York (1975).
2. S. J. Singer and G. L. Nicolson, The fluid mosaic model of the structure of cell membranes, *Science* **175**, 720-731 (1972).
3. A. L. Lehninger, *Biochemistry*, p. 210, Worth, New York (1970).
4. S. J. Singer, in: *Cell Membranes—Biochemistry, Cell Biology and Pathology* (G. Weissmann and R. Claribrone, eds.), pp. 35-44, HP Publishing, New York (1975).
5. L. Gros, H. Ringsdorf, and H. Schupp, Polymeric antitumor agents on a molecular and on a cellular level?, *Angew. Chem., Int. Ed. Engl.* **20**, 305-325 (1981).
6. G. L. Gaines, *Insoluble Monolayers at Liquid-Gas Interfaces*, Interscience, New York (1966).
7. H. T. Tien, *Bimolecular Lipid Membranes, Theory and Practice*, Marcel Dekker, New York (1974).
8. R. E. Pagano and J. N. Weinstein, Interaction of liposomes with mammalian cells, *Ann. Rev. Biophys. Bioeng.* **7**, 435-468 (1978).
9. F. Szoka and D. Papahadjopoulos, Comparative properties and methods of preparation of lipid vesicles (liposomes) *Ann. Rev. Biophys. Bioeng.* **9**, 467-508 (1980).
10. H. Kuhn, D. Moebius, and H. Buecher, in: *Physical Methods of Chemistry* (A. Weissberger and B. W. Rossiter, eds.), Vol. I, part 3B, pp. 577-702, Wiley, New York (1972).
11. I. Langmuir, The mechanism of the surface phenomena of flotation, *Trans. Faraday Soc.* **15**, 62-74 (1920).
12. K. B. Blodgett, Films built by depositing successive monomolecular layers on a solid surface, *J. Am. Chem. Soc.* **57**, 1007-1022 (1935).

13. K. Kuhn and D. Moebius, Systems of monomolecular layers—Assembling and physico-chemical properties, *Angew. Chem., Int. Ed. Engl.* **10**, 620–637 (1971).
14. J. H. Fendler, Surfactant vesicles as membrane mimetic agents: Characterization and utilization, *Acc. Chem. Res.* **13**, 7–13 (1980).
15. G. Gregoriadis and A. C. Allison, *Liposomes in Biological Systems*, Wiley, New York (1980).
16. H. G. Khorana and P. Chakakrabarti, A new approach to the study of phospholipid-protein interaction in biological membranes. Synthesis of fatty acids and phospholipids containing photosensitive groups, *Biochemistry* **14**, 5021–5033 (1975).
17. A. Akimoto, K. Dorn, L. Gros, H. Ringsdorf, and H. Schupp, Polymer model membranes, *Angew. Chem., Int. Ed. Engl.* **20**, 90–91 (1981).
18. H. Bader and H. Risgsdorf, Liposomes from α,ω-dipolar amphiphiles with a polymerizable diyne moiety in the hydrophobic chain, *J. Polym. Sci., Polym. Chem. Ed.* **20**, 1623–1628 (1982).
19. D. Day, H. H. Hub, and H. Ringsdorf, Polymerization of mono- and bi-functional diacetylene derivatives in monolayers at the gas–water interface, *Isr. J. Chem.* **18**, 325–329 (1979).
20. H. Ringsdorf and H. Schupp, Polymerization of substituted butadienes at the gas–water interface, *J. Macromol. Sci., Chem.* **A15**, 1015–1026 (1981).
21. H. H. Hub, B. Hupfer, H. Koch, and H. Ringsdorf, Polymerization of lipid and lysolipid like diacetylenes in monolayers and liposomes, *J. Macromol. Sci., Chem.* **A15**, 701–715 (1981).
22. H. H. Hub, B. Hupfer, H. Koch, and H. Ringsdorf, Polymerizable phospholipid analogs—New stable biomembrane and cell models *Angew. Chem., Int. Ed. Engl.* **19**, 938–940 (1980).
23. B. Hupfer, H. Ringsdorf, and H. Schupp, Polymeric phospholipid monolayers, *Makromol. Chem.* **182**, 247–253 (1981).
24. H. Koch and H. Ringsdorf, Topochemical polymerization of a diacetylene sulfolipid analog in monolayers and liposomes, *Makromol. Chem.* **182**, 255–259 (1981).
25. D. Naegele and H. Ringsdorf, Polymerization in ordered systems: Polymerization of octadecylmethacrylate in monolayers at the gas/water interface, *J. Polym. Sci., Polym. Chem. Ed.* **15**, 2821–2834 (1977).
26. R. Benz, W. Press, and H. Ringsdorf, Black Lipid Membranes from polymerizable lipids, *Angew. Chem., Int. Ed. Engl.* **21**, 368–369 (1982).
27. D. S. Johnston, S. Sanghera, M. Pons, and D. Chapman, Phospholipid polymers—Synthesis and spectral characteristics, *Biochim. Biophys. Acta* **602**, 57–69 (1980).
28. D. S. Johnston, S. Sanghera, A. Manjon-Rubio, and D. Chapman, The formation of polymeric model biomembranes from diacetylenic fatty acids and phospholipids, *Biochim. Biophys. Acta* **602**, 213–216 (1980).
29. E. Lopez, D. F. O'Brien, and T. H. Whitesides, Structural effects on the photopolymerization of bilayer membranes, *J. Am. Chem. Soc.* **104**, 305–307 (1982).
30. P. Tundo, D. J. Kippenberger, P. L. Klahn, N. E. Prieto, T.-C. Jao, and J. H. Fendler, Functionally polymerized surfactant vesicles, *J. Am. Chem. Soc.* **104**, 456–461 (1982).
31. S. L. Regen, A. Singh, G. Oehme, and M. Singh, Polymerized phosphatidylcholine vesicles. Synthesis and characterization, *J. Am. Chem. Soc.* **104**, 791–795 (1982).
32. T. Kunitake, N. Nakashima, K. Takarabe, M. Nagai, A. Tsuge, and H. Yanagi, Vesicles of polymeric bilayer and monolayer membranes, *J. Am. Chem. Soc.* **103**, 5945–5947 (1981).
33. B. Hupfer and H. Ringsdorf, Spreading and polymerization behavior of diacetylenic phospholipids at the gas/water interface, *Chem. Physics Lipids* **33**, 263–282 (1983).
34. G. Gee, Reactions in the monolayers of drying oils. II. Polymerization of the oxidized forms of the maleic anhydride compound of β-elaeostearin, *Proc. Roy. Soc.* **A153**, 129–141 (1935).
35. C. H. Bamford and G. C. Eastwood, Solid-phase addition polymerization, *Quart. Rev.* **23**, 271–299 (1969).

36. S. A. Letts, T. Fort, Jr., and J. B. Lando, Polymerization of oriented monolayers of vinyl stearate, *J. Colloid Interface Sci.* **56**, 64–75 (1976).
37. K. Fukuda, Polymerization of vinyl stearate in emulsion and in monomolecular film, *Sci. Rep. Saitama University* **AIII**, 143–152 (1959); *Chem. Abstr.* **54**, 12648h (1960).
38. H. Z. Friedlaender, Oriented polymeric films, U.S. Patent 3,031,721, May 1, 1962; *Chem. Abstr.* **57**, 14008b (1962).
39. M. Hatada, M. Nishii, and K. Hirota, Radiation-induced polymerization of vinyl monomers at gas/water interfaces, *Japan Atomic Energy Res. Inst. Repts.* **5028**, 1–11 (1973); *Chem. Abstr.* **80**, 3828p, 27492a (1974).
40. M. Nishii, M. Hatada, and K. Hirota, Radiation-induced polymerization of vinyl stearate monomer layer at gas/water interface, *Japan Atomic Energy Res. Inst. Repts.* **5029**, 18–25 (1974); *Chem. Abstr.* **81**, 91995q (1974).
41. J. B. Lando and T. Fort, Jr., in: *Polymerization of Organized Systems* (H. G. Elias, ed.), Midland Macromolecular Monographs, Vol. 3, pp. 63–78, Gordon & Breach, New York (1977).
42. N. Beredjick and W. J. Burlant, Polymerization of monolayers of vinyl and divinyl monomers, *J. Polym. Sci., Part A-1*, 2807–2818 (1970).
43. R. Ackermann, O. Inacker, and H. Ringsdorf, Polyreactions in oriented systems. 1. Polymerization of acrylic and methacrylic compounds in monomolecular layers, *Kolloid-Z., Z. Polym.* **249**, 1118–1126 (1971).
44. M. Hatada and M. Nishii, Polymerization induced by electron beam. Irradiation of octadecyl methacrylate in the form of a multilayer or monolayer, *J. Polym. Sci., Polym. Chem. Ed.* **15**, 927–935 (1977).
45. A. Dubault, C. Casagrande, and M. Veyssie, Two dimensional polymerization processes in mono- and diacrylic esters, *J. Phys. Chem.* **79**, 2254–2259 (1975).
46. M. Hatada, M. Nishii, and K. Hirota, Radiation-induced polymerization of monomolecular films of octadecylacrylate at the gas/water interface, *J. Colloid Interface Sci.* **45**, 502–505 (1973).
47. M. Hatada, M. Nishii, and K. Hirota, Radiation-induced polymerization of octadecyl acrylate at a nitrogen–water interface under constant surface pressure conditions, *Macromolecules* **8**, 19–22 (1975).
48. M. Hatada, M. Nishii, and K. Hirota, Radiation-induced polymerization of octadecyl acrylate at nitrogen–water interface. Surface balance for constant pressure operation and analyses of irradiated film substances, *Japan Atomic Energy Res. Inst. Repts.* **5029**, 8–17 (1974); *Chem. Abstr.* **81**, 92035p (1974).
49. G. Scheibe and H. Schuller, About the polymerization of monomolecular films of vinyl isobutylether, *Z. Elekrochem.* **59**, 861–862 (1955).
50. A. Dubault, M. Veyssie, L. Liebert, and L. Strzelecki, Crosslinked polymerization of monolayers of 1-*n*-octadecyloxy-2,3-bis(acryloyloxy)propane, *Nature (London)* **245**, 94–95 (1973).
51. G. Gee, C. B. Davies, and H. W. Melville, The catalyzed polymerization of butadiene at liquid-gas interface, *Trans. Faraday Soc.* **35**, 1298–1312 (1939).
52. C. Golian, J. G. Hawke, J. Green, and J. M. Gebicki, Photocontraction of unsaturated monolayers at the air/liquid interface, *Experientia* **31**, 34–35 (1975).
53. D. Day and H. Ringsdorf, Polymerization of diacetylene carbonic acid monolayers at the gas/water interface, *J. Polym. Sci., Polym Lett. Ed.* **16**, 205–210 (1978).
54. D. R. Day and H. Ringsdorf, The monolayer polymerization of 10,12-nonacosadiynoic acid studied by a spectroscopic technique, *Makromol. Chem.* **180**, 1059–1063 (1979).
55. D. Day and J. B. Lando, Morphology of crystalline diacetylene monolayers polymerized at the gas–water interface, *Macromolecules* **13**, 1478–1483 (1980).
56. D. Day and J. B. Lando, Structure determination of a poly(diacetylene)monolayer, *Macromolecules* **13**, 1483–1487 (1980).

57. H. Schupp, B. Hupfer, R. A. Van Wagenen, J. D. Andrade, and H. Ringsdorf, Surface characterization of functional poly(diacetylene) and poly(butadiene) mono- and multilayers, *Colloid Polym. Sci.* **260**, 262–267 (1982).
58. B. Hupfer, H. Schupp, J. D. Andrade, and H. Ringsdorf, Photoelectron mean free paths in poly(diacetylene) mono- and multilayers, *J. Electron Spectrosc. Relat. Phenom.* **23**, 103–107 (1981).
59. G. Wegner, Topochemical polymerization of monomers with conjugated triple bonds, *Makromol. Chem.* **154**, 35–48 (1972).
60. K. Fukuda, Y. Shibasaki, and H. Kakahara, Polycondensation of long-chain esters of α-amino acids in monolayers at air/water interface and in multilayers on solid surface, *J. Macromol. Sci., Chem.* **A15**, 999–1014 (1981).
61. T. Folda, L. Gros, and H. Ringsdorf, Formation of oriented polypeptides and polyamides in monolayers and liposomes, *Makromol. Chem. Rapid Commun.* **3**, 167–174 (1982).
62. B. Tieke and G. Lieser, Influences of the structure of long-chain diynoic acids on their polymerization properties in Langmuir–Blodgett multilayers, *J. Colloid Interface Sci.* **88**, 471–486 (1982).
63. G. L. Gaines, *Insoluble Monolayers at Liquid-Gas Interfaces*, pp. 281–300, Interscience, New York (1966).
64. R. Bueschl, B, Hupfer, and H. Ringsdorf, Partially polymerized mixed monolayers and liposomes, *Makromol. Chem. Rapid Commun.* **3**, 589–596 (1982).
65. R. Ackermann, D. Naegele, and H. Ringsdorf, Polyreactions in oriented media. 4 Photoreactions of fumaric and maleic acid derivatives in multilayers, *Makromol. Chem.* **175**, 699–700 (1974).
66. D. Naegele, J. B. Lando, and H. Ringsdorf, Polymerization of cadmium octadecylfumarate in multilayers, *Macromolecules* **10**, 1339–1344 (1977).
67. B. Tieke, G. Lieser, and G. Wegner, Polymerization of diacetylenes in multilayers, *J. Polym. Sci., Polym. Chem. Ed.* **17**, 1631–1644 (1979).
68. G. Lieser, B. Tieke, and G. Wegner, Structure, phase transitions and polymerizability of multilayers of some diacetylene monocarboxylic acids, *Thin Solid Films* **68**, 77–90 (1980).
69. A. Cemel, T. Fort, Jr. and J. B. Lando, Polymerization of vinyl stearate multilayers, *J. Polym. Sci., Part A-1*, 2061–2083 (1972).
70. M. Putermann, T. Fort, Jr., and J. B. Lando, The polymerization and structure of mixed multilayers of ethyl and vinyl stearate, *J. Colloid Interface Sci.* **47**, 705–718 (1974).
71. V. Enkelmann and J. B. Lando, Polymerization of ordered tail-to-tail vinyl stearate bilayers, *J. Polym. Sci., Polym. Chem. Ed.* **15**, 1843–1854 (1977).
72. A. Banerjie and J. B. Lando, Radiation-induced solid state polymerization of oriented ultrathin films of octadecylacrylamide, *Thin Solid Films* **68**, 67–75 (1980).
73. A. Barraud, C. Rosilio, and A. Ruaudel-Teixier, Polymerized monomolecular layers: a new class of ultrathin resins for microlithography, *Thin Solid Films* **68**, 91–98 (1980).
74. K. Fukuda and T. Shiozawa, Conditions for formation and structural characterization of X-type and Y-type multilayers of long-chain esters, *Thin Solid Films* **68**, 55–66 (1980).
75. B. Tieke and G. Lieser, Polymerization of diacetylenes in mixed multilayers, *J. Colloid Interface Sci.* **83**, 230–239 (1981).
76. T. Kunitake and Y. Okahata, A totally synthetic bilayer membrane, *J. Am. Chem. Soc.* **99**, 3860–3861 (1977).
77. T. Kunitake, Chemistry of synthetic bilayer membranes, *J. Macromol. Sci., Chem.* **A13**, 587–602 (1979).
78. H. Bader, H. Ringsdorf, and J. Skura, Liposomes from polymerizable glycolipids, *Angew. Chem., Int. Ed. Engl.* **20**, 91–92 (1981).
79. N. Wagner, K. Dose, H. Koch, and H. Ringsdorf, Incorporation of ATP synthetase into long-term stable liposomes from a polymerizable sulfolipid, *FEBS Lett.* **132**, 313–318 (1981).

80. M. Hatada, M. Nishii and K. Hirota, Radiation-induced polymerization of octadecyl acrylate multilayers by electron beam irradiation, *Japan Atomic Energy Res. Inst. Repts.* **5030**, 26–32 (1975); *Chem. Abstr.* **84**, 5557f (1976).
81. M. Nishii and M. Hatada, Polymerization of vinyl stearate multilayers by electron beam irradiation, *Japan Atomic Energy Res. Inst. Repts.* **5030**, 33–37 (1975); *Chem. Abstr.* **84**, 44718d (1976).

5

X-ray Photoelectron Spectroscopy (XPS)

Joseph D. Andrade

1. INTRODUCTION

X-ray photoelectron spectroscopy (XPS) is generally regarded as an important and key technique for the surface characterization and analysis of biomedical polymers.[1] This technique, also called ESCA (Electron Spectroscopy for Chemical Analysis), provides a total elemental analysis, except for hydrogen and helium, of the top 10–200 Å (depending on the sample and instrumental conditions) of any solid surface which is vacuum stable or can be made vacuum stable by cooling. Chemical bonding information is also provided. Of all the presently available instrumental techniques for surface analysis, XPS is generally regarded as being the most quantitative, the most readily interpretable, and the most informative with regard to chemical information. For these reasons it has been highly recommended and used by biomedical researchers for the analysis of medical polymers. The basic advantages and disadvantages of the technique are given in Table 1. Although the method requires relatively sophisticated instrumentation, many universities, industrial R and D groups, and commercial service laboratories provide access to their instruments on a collaborative or fee for service basis.

This chapter will review the basic principles of XPS. It is assumed that the reader is primarily interested in applying the technique to the routine surface analysis of biomedical polymers. It is further assumed that the reader is not a practicing X-ray photoelectron spectroscopist and is not directly involved with XPS instrumentation. The main objective of the chapter is to provide the information necessary to intelligently determine

---

Joseph D. Andrade • Departments of Bioengineering and Materials Science, College of Engineering, University of Utah, Salt Lake City, Utah 84112.

TABLE 1
Advantages and Disadvantages of XPS for Polymer Surface Analysis

Advantages:
Nondestructive
Surface sensitive (10–200 Å)
Elemental sensitivity (parts per 1000)
All elements (except H and He)
Quantitative
Chemical bonding information
Interpretation and theory straightforward
High information content

Disadvantages:
Large analysis area (several mm^2)[a]
Expensive ($200,000–$500,000/instrument, $50–$500/sample)
High vacuum (10^{-8} to 10^{-11} torr)
Slow (1/2 to 8 hours/sample)
Charging and energy referencing can be a problem
Low resolution (~0.1–1.0 eV)

[a] A new state-of-the-art commercial instrument has an analysis area of ~150 μm in diameter (Surface Science Labs, Palo Alto, CA).

XPS analysis strategies for optimum problem solving and characterization purposes, to prepare samples for routine XPS analysis, and to interpret and analyze the resulting spectra. Emphasis will be placed on polymer surfaces; those principles and details not normally useful for polymer surface analysis will only be briefly mentioned. Although the technique is applicable to gases and liquids, in this chapter we will concentrate only on solids. Many of the sources listed in Table 2 contain information on application to other states of matter. Some subjects which are highly important for polymer surface analysis, but may not be particularly important for surface analysis of other solids, will be covered in some detail, including variable angle methods and models, surface reactions to improve functional group detection, and charging and electrical surface property considerations.

XPS is a relatively simple and straightforward technique; the information content in a typical XPS spectrum is enormous. There are various levels or hierarchies of spectral interpretation, including (1) simple elemental analysis; (2) detailed considerations of chemical shifts and chemical bonding nature in the surface region; and (3) various loss or relaxation structures which provide further information on the chemical nature of the surface. There are also many potential artifacts in an XPS experiment, often related to the preparation of samples. The interpretation becomes progressively more difficult as one goes to more complex surfaces, such as multicomponent and multiphase materials.

Because of the high information content in the spectrum the reader will have to consult a number of texts and review papers and must examine a large number of spectra to obtain practical experience with spectral interpretation and analysis for a wide range of sample types. This is best accomplished by thoroughly studying reviews and papers containing a large number of specific examples. Table 2 presents a summary of those texts, monographs, and review articles which are generally regarded as basic to X-ray photoelectron spectroscopy.

Those names primarily associated with polymer surface analysis by XPS include Clark,[12,13,21] whose group has provided a significant proportion of the total available literature on XPS application to polymers; Briggs,[8] Thomas,[21,22] and Dilks,[14,15] students of Clark and alumni of the Durham group; Ratner,[16] who has pioneered the applications of the technique to biomedical polymers; and Millard,[23] who pioneered many applications of XPS to biological systems. A good historical account of the development of XPS is given in the Nobel lecture of K. Siegbahn,[2] who shared the Nobel prize in physics in 1982, primarily for his development of X-ray photoelectron spectroscopy.

A number of other instrumental surface analysis techniques are available and are widely used, in particular Auger electron spectroscopy (AES)[7] and secondary ion mass spectroscopy (SIMS).[17] The Auger technique is actually much older in practical application than XPS. Because AES utilizes an electron beam, it is generally highly damaging to organic polymer surfaces. AES is a form of electron spectroscopy and thus is complementary to XPS; various aspects of the Auger process will be discussed in this chapter. SIMS is now beginning to be applied to polymer surfaces,[17] but is not yet a routine technique nor as easy to interpret as XPS. SIMS is under extensive development and may prove useful for routine polymer surface analysis in the future (see Chapter 13).

There are a number of key journals and serials which are largely devoted to XPS. Some of these were noted briefly in Chapter 1. They include: *Journal of Electron Spectroscopy and Related Phenomena; Surface and Interface Analysis*; and the series *Electron Spectroscopy; Theory, Techniques, and Applications*, edited by C. R. Brundle and A. D. Baker.

2. BASIC PRINCIPLES OF XPS

2.1. THE PHOTOELECTRIC EFFECT

The basic principle of XPS is the photoelectric effect, the phenomenon for which Einstein received his Nobel prize. Figure 1 presents schematically the interaction of a photon with an atomic orbital electron. Three possibilities are evident:

TABLE 2
Basic Reference Sources on XPS for Surface Analysis

Author(s)	Year	Subject	Comments	Reference
Briggs, ed.	1977	Handbook XPS, UPS	A collection of chapters and appendices on all aspects of XPS—a useful handbook; chapters by Briggs and Clark particularly useful for polymer work.	8
Brundle and Baker, eds.	1977–1983	All areas	A review series with chapters on specific areas of XPS. The chapters by Dilks (Vol. 4), Briggs (Vol. 3), Fadley (Vol. 2), and Heilbronner and Maier (Vol. 1) are especially helpful to XPS polymer surface analysis.	11
Carlson	1975	Photoelectron spectroscopy	Good theoretical treatment of basics of XPS, UPS, and Auger electron spectroscopy; basic source of information (now somewhat dated) on chemical shifts of the elements. A key text.	7
Clark and Feast	1975, 1977	Polymers	The basic reviews on XPS of polymers.	12, 13
Dilks	1981	Polymers	Important reviews on polymer applications.	14, 15
Fadley	1978	Basic concepts	Outstanding, complete, in-depth source on all aspects of XPS.	3
Ghosh	1983	Photoelectron spectroscopy	A recent monograph—good text.	100
Millard	1982	Fibers and polymers	Review of industrial applications.	23
Ratner	1983	Medical polymers	Review on XPS of medical polymers.	16
Siegbahn *et al.*	1967	XPS—general	The first ESCA book—the "bible" of XPS—emphasizes solids; now somewhat dated but excellent treatment of principles and concepts—presentation of binding energies and chemical shifts. Many biomedical examples.	5
Siegbahn *et al.*	1969	XPS—gases	The second ESCA book—gas phase XPS. Best source of spectra and chemical shift data of simple molecules.	6
Siegbahn and Karlsson	1985	XPS—general	Basic modern monograph on all apects of XPS.	4
Wagner *et al.*	1979	Handbook of XPS	Good general introduction to XPS. A complete collection of XPS spectra of pure elements and simple compounds. An essential reference.	9
Windawi and Ho	1982	Applied XPS	Collection of chapters on XPS applications.	10

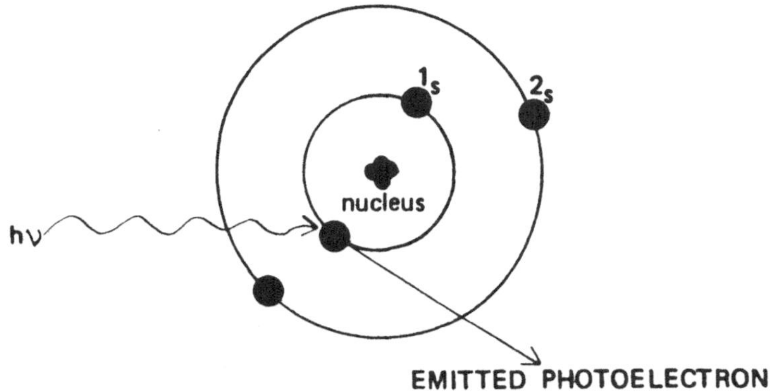

FIGURE 1. A schematic view of the interaction of an X-ray photon of energy $h\nu$ with an atomic orbital electron. The photon is shown to interact with a $1s$ electron, which is photoemitted into the vacuum with kinetic energy, $E_k = h\nu - E_b$.

1. The photon may traverse through the atom without significantly interacting with either the orbital electrons or the nucleus.
2. The photon may be scattered by the atomic orbital electron resulting in a partial loss of photon energy. This process is called Compton scattering.
3. The photon may interact with the atomic orbital electron such that there is total and complete transfer of the photon's energy to the electron. This is the basic process in XPS.

Given that the photon energy is greater than the binding energy of the electron in the atom, the electron is then ejected from the atom with a kinetic energy approximately equal to the difference between the photon energy and the binding energy. Therefore, the basic equation for XPS is:

$$E_b \simeq h\nu - E_k \tag{1}$$

where E_b is the electron binding energy, E_k is the electron kinetic energy, measured by the instrument, and $h\nu$ is the photon energy (h is Planck's constant and ν is the X-ray frequency). All energies are usually expressed in electron volts (eV). Measuring the kinetic energy allows one to calculate the binding energy. Knowing the binding energy, we can identify the atom (Table 3).

2.2. BASIC EXPERIMENT

We have just described a gas phase experiment. Figure 2 presents a schematic version of the experiment for a solid sample. The solid sample

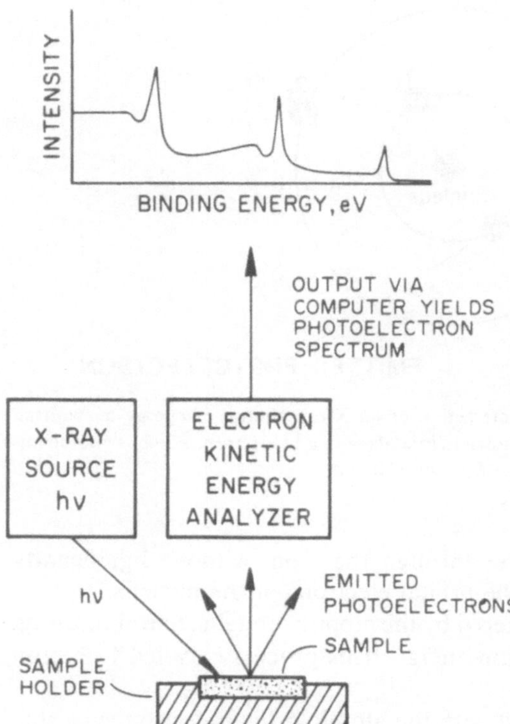

FIGURE 2. Schematic diagram of an X-ray photoelectron spectroscopy experiment. An X-ray source, usually aluminum or magnesium $K\alpha$, provides X-ray photons of energy $h\nu$ that hit the sample. Electrons are then emitted because of the photoelectric effect. The kinetic energy of the electrons is easily measured in the kinetic energy analyzer to about 0.1 eV. The result is a spectrum of photoelectron intensity as a function of binding energy, as shown at the top of the figure. The sample is usually mounted in a suitable holder which is in electrical contact with the spectrometer ground. [Reprinted by permission from Reference (99), copyright 1981 by Canon Comm., Inc., Santa Monica, Calif.]

is mounted in a suitable sample holder and placed in the high-vacuum environment of a typical instrument. An X-ray source emitting monochromatic X-ray photons, usually magnesium or aluminum Kα, is directed towards the sample. The X-ray photons statistically interact with the atomic and molecular orbital electrons in the sample. Some fraction of the photoelectrons produced by this process is directed up and out of the sample and analyzed by the analyzer. The analyzer basically measures the number of electrons of different kinetic energies. The information is generally processed by a computer to produce a spectrum of photoelectron intensity as a function of binding energy. The binding energy position of each of the key peaks allows elemental identification to be made. Quantitation is achieved by measuring the area under each peak and applying appropriate correction and sensitivity factors, to be discussed later.

Table 3 presents the photoelectron binding energies, taken primarily from the tabulation of Siegbahn[(5)] and compiled and arranged by R. N. King.[(8)] This table shows that, for every element in the periodic table, there exists at least one and generally a series of photoelectron lines within the energy analysis range of conventional X-ray sources. This table will be discussed in detail later.

XPS is surface sensitive. Although the X-ray photons penetrate many microns deep into the sample, the electrons generated at that depth simply cannot make it out into the vacuum to be detected. Relatively low kinetic energy electrons (0 to 1500 eV), the type generally excited by XPS, cannot travel very large distances in matter. Therefore, the sampling depth in XPS is generally quoted in the range of 20–200 angstroms, depending on equipment geometry and sample type. Only those electrons which emerge from the sample without any loss in energy comprise the main photoelectron peaks (Table 3).

2.3. INSTRUMENT

Figure 3 shows the components of one type of XPS instrument. A typical XPS instrument consists of an X-ray source, sample mounting assembly, an analyzer, a detector, and data processing and display. We will first discuss the X-ray source.

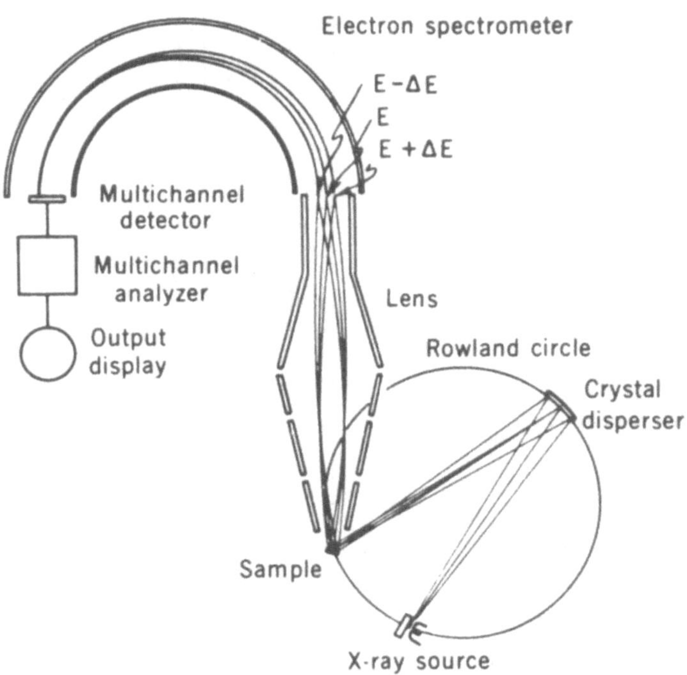

FIGURE 3. Schematic of an XPS instrument showing a pre-retardation type of electron kinetic energy analyzer and employing an X-ray monochromator. (Reprinted from Hewlett-Packard 5950A Manual).

TABLE 3

Photoelectron binding energies (large numbers) and Scofield photoionization cross sections (smaller numbers) for aluminum $K\alpha_{1,2}$ radiation, 1487 eV. Compiled by R. N. King[18] with data taken from Siegbahn[5] and Scofield.[27][a]

	$1S_{1/2}$	$2S_{1/2}$	$2P_{1/2}$	$2P_{3/2}$	$3S_{1/2}$	$3P_{1/2}$	$3P_{3/2}$	$3D_{3/2}$	$3D_{5/2}$
H	14 0002								
He	25 0082								
Li	55 0568								
Be	111 1947								
B	188 486			5 0002					
C	284 1 00			7 0015					
N	399 1 80			9 0065					
O	532 2 93	24 1405		7 0193					
F	686 4 43	31 210		9 0478					
Ne	867 6 30	45 296		18 103					
Na	1072 8 52	63 422		31 1941	1 0064				
Mg		89 575		52 3335	2 0285				
Al		118 753	74 1811	73 356	1 0535				
Si		149 955	100 276	99 541	8 0808		3 014		
P		189 1 18	136 403	135 789	16 1116		10 0368		
S		229 1 43	165 567	164 1 11	16 1465		8 0774		
Cl		270 1 69	202 775	200 1 51	18 1852		7 1433		
Ar		320 1 97	247 1 03	245 2 01	25 227		12 2418		
K		377 2 27	297 1 35	294 2 62	34 286		18 3619		
Ca		438 2 59	350 1 72	347 3 35	44 351		26 507		5
Sc		500 2 91	407 2 17	402 4 21	54 411		32 650		7 0042
Ti		564 3 24	461 2 69	455 5 22	59 473		34 813		3 0136
V		628 3 57	520 3 29	513 6 37	66 538		38 996		2 0309
Cr		695 3 91	584 3 98	575 7 69	74 596		43 1 173		2 0651
Mn		769 4 23	652 4 74	641 9 17	84 674		49 1 423		4 1046
Fe		846 4 57	723 5 60	710 10 82	95 745		56 1 669		6 1711
Co		926 4 88	794 6 54	779 12 62	101 818		60 1 930		3 2664
Ni		1008 5 16	872 7 57	855 14 61	112 892		68 2 217		4 3979
Cu		1096 5 46	951 8 66	931 16 73	120 957		74 2 478		2 589
Zn		1194 5 76	1044 9 80	1021 18 92	137 1 04		87 2 828		9 81

	$2P_{1/2}$	$2P_{3/2}$	$3S_{1/2}$	$3P_{1/2}$	$3P_{3/2}$	$3D_{3/2}$	$3D_{5/2}$	$4S_{1/2}$	$4P_{1/2}$	$4P_{3/2}$
Ga	1143 11 09	1116 21 40	158 1 13	107 1.10	103 2 11		18 1 085			1 018
Ge	1249 12 52	1217 24 15	181 1 23	129 1 24	122 2.39		29 1 42			3 058
As			204 1 32	147 1 39	141 2 68		41 1 82			3 121
Se			232 1 43	168 1 55	162 2 98		57 2 29			6 210
Br			257 1 53	189 1 72	182 3 31	70 1 16	69 1 68	27 1863		5 328
Kr			289 1 64	223 1 89	214 3.65		89 3 48	24 213		11 476
Rb			322 1 75	248 2 07	239 4 00	112 1 72	111 2 49	30 251	15 214	14 411
Sr			358 1 86	280 2 25	269 4 37	135 2 06	133 2 99	38 291		20 775
Y			395 1 98	313 2 44	301 4 75	160 2 44	158 3 54	46 329		26 091
Zr			431 2 10	345 2 64	331 5 14	183 2 87	180 4 17	52 367		29 1 05
Nb			469 2 22	379 2 84	363 5 53	208 3 35	205 4 86	58 402		34 1 17
Mo			505 2 34	410 3 04	393 5 94	230 3 88	227 5 62	62 440		35 1 31
Tc			544 2 45	445 3 23	425 6 36	257 4 46	253 6 47	68 479		39 1 45
Ru			585 2 57	483 3 44	461 6 78	284 5 10	279 7 39	75 519		43 1 59
Rh			627 2 70	521 3 64	496 7 21	312 5 80	307 8 39	81 560		48 1 75
Pd			670 2 81	559 3 83	531 7 63	340 6 56	335 9 48	86 598		51 1 88
Ag			717 2 93	602 4 03	571 8 06	373 7 38	367 10 66	95 644	62 700	56 1 36
Cd			770 3 04	651 4 22	617 8 50	411 8 27	404 11 95	108 692		67 2 25
In			826 3 16	702 4 40	664 8 93	451 9 22	443 13 32	122 742		77 2.45
Sn			884 3.26	757 4.58	715 9 35	494 10 25	485 14.80	137 794		89 2 67
Sb			944 3 36	812 4 76	766 9.77	537 11.35	528 16.39	152 848		99 2 88
Te			1006 3 46	870 4.92	819 10.21	582 12.52	572 18.06	168 903		110 3 11
I			1072 3 53	931 5.06	875 10.62	631 13.77	620 19.87	186 959		123 3 34
Xe			1145 3 62	999 5 20	937 10.99	685 15.10	672 21.79	208 1.02		147 3 58
Cs			1217 3 73	1065 5.29	998 11.38	740 16.46	726 23.76	231 1.08	172 1 27	162 2.56
Ba				1137 5.42	1063 11 71	796 17.92	781 25.84	253 1.13	192 1.34	180 2.73
La				1205 5 55	1124 12.11	849 19.50	832 28.12	271 1.19	206 1.42	192 2.91
Ce					1186 12.53	902 21.12	884 30.50	290 1.24	224 1.47	208 3.03
Pr					1243 12.94	951 22.72	931 32.85	305 1.28	237 1.53	218 3.17
Nd						1000 24.27	978 35.29	316 1.33	244 1.59	225 3.31

[a] Note that binding energies are generally given for the pure or metallic form of each element where available. Binding energies in different tabulations are often 1–2 eV apart; thus all of the energies in the table should be taken as approximate. Other tabulations are available for

4D3/2	4D5/2	4F's	5S1/2	5P1/2	5P3/2
3 031					
3 085					
4 198					
2 316					
2 470					
2 667					
3 908					
1 1.24					
3 1.55					
9 1.89				2	
16 2.28				1 0195	
24 2.70			1 0922	1 058	
32 3.14			7 1085	2 1145	
40 3.63			12 1251	2 189	
50 4.13			14 1421	3 2828	
63 4.68			18 1596	7 3961	
79 2.15	77 3.10		23 1843	13 1697	12 332
93 2.40	90 3.46		40 210	17 202	15 400
99 6.52			33 234	15 688	
111 6.93		1 1389	38 230	20 660	
114 7.48		2 2545	38 .238	23 .685	
118 8.03		2 4068	38 .247	22 .708	

	3D3/2	3D5/2	4S1/2	4P1/2	4P3/2	4D3/2	4D5/2	4F5/2	4F7/2	5S1/2	5P1/2	5P3/2	5D3/2	5D5/2	6S1/2	6P's
Pm	1052 26.08	1027 37.65	331 1.38	255 1.64	237 3.45	121 8.59		4 .604		38 .254	22 .730					
Sm	1107 27.96	1081 40.37	347 1.42	267 1.70	249 3.59	130 9.16		7 .851		39 .261	22 .750					
Eu	1161 29.91	1131 43.24	360 1.46	284 1.75	257 3.72	134 9.73		0 1.155		32 .268	22 .770					
Gd	1218 31.98	1186 46.23	376 1.51	289 1.80	271 3.88	141 10.40		0 1.434		36 .288	21 .847					
Tb		1242 49.42	398 1.54	311 1.84	286 3.99	148 10.87		3 1.967		40 .281	26 .804					
Dy			416 1.58	332 1.88	293 4.12	154 11.43		4 2.49		63 .287	26 .821					
Ho			436 1.61	343 1.91	306 4.24	161 12.00		4 3.10		51 .293	20 .836					
Er			449 1.64	366 1.95	320 4.37	177 5.15	168 7.41	4 3.82		60 .298	29 .849					
Tm			472 1.67	386 1.98	337 4.48	180 13.12		5 4.64		53 .303	32 .864					
Yb			487 1.70	396 2.00	343 4.60	197 5.61	184 8.07	6 5.58		53 .308	23 .876					
Lu			506 1.73	410 2.03	359 4.74	205 5.87	195 8.45	7 6.50		57 .326	28 .949		5 .0593			
Hf			538 1.76	437 2.06	380 4.88	224 6.13	214 8.84	19 3.32	18 4.20	65 .344	38 .325	31 .699	7 .1526			
Ta			566 1.79	465 2.08	405 5.02	242 6.40	230 9.24	27 3.80	25 4.82	71 .363	45 .346	37 .754	6 .2778			
W			595 1.81	492 2.10	426 5.16	259 6.68	246 9.65	37 4.32	34 5.48	77 .383	47 .367	37 .811	6 .4344			
Re			625 1.84	518 2.12	445 5.30	274 6.95	260 10.06	47 4.88	45 6.20	83 .402	46 .387	35 .869	4 .624			
Os			655 1.86	547 2.13	469 5.45	290 7.23	273 10.48	52 5.48	50 6.96	84 .422	58 .408	46 .928	0 .847			
Ir			690 1.88	577 2.14	495 5.59	312 7.51	295 10.90	63 6.12	60 7.78	96 .438	63 .422	51 .967	4 1.238			
Pt			724 1.90	608 2.14	519 5.74	331 7.78	314 11.32	74 6.81	70 8.65	102 .459	66 .444	51 1.04	2 1.477			
Au			759 1.92	644 2.14	546 5.89	352 8.06	334 11.74	87 7.54	83 9.58	108 .479	72 .463	54 1.10	3 1.808			
Hg			800 1.94	677 2.14	571 6.04	379 8.33	360 12.17	103 8.32	99 10.57	120 .500	81 .484	58 1.17	7 2.079			
Tl			846 1.95	722 2.13	609 6.19	407 8.60	386 12.60	122 9.14	118 11.62	137 .520	100 .505	76 1.25	16 .991	13 1.39		
Pb			894 1.96	764 2.12	645 6.33	435 8.87	413 13.02	143 10.01	138 12.73	148 .542	105 .526	86 1.33	22 1.11	20 1.58	3 .0742	1 .0439
Bi			939 1.96	806 2.10	679 6.48	464 9.14	440 13.44	163 10.93	158 13.90	160 .563	117 .546	93 1.41	27 1.24	25 1.76	8 .0840	3 .0841
Po			995 1.97	851 2.07	705 6.62	500 9.40	473 13.87	184 27.04		177 .584	132 .566	104 1.50	31 3.31		12 .0937	5 .1356
At			1042 1.96	886 2.04	740 6.77	533 9.65	507 14.29	210 29.36		195 .605	148 .584	115 1.58	40 3.63		18 .1033	8 .1892
Rn			1097 1.95	929 2.00	768 6.92	567 9.90	541 14.70	238 31.81		214 .625	164 .602	127 1.67	48 3.95		26 .1129	11 .2719
Fr			1153 1.95	980 1.97	810 7.07	603 10.16	577 15.11	268 34.36		234 .645	182 .618	140 1.77	58 4.28		34 .1257	15 .3366
Ra			1208 1.95	1058 1.91	879 7.20	636 10.40	603 15.53	299 37.04		254 .665	200 .633	153 1.86	68 4.61		44 .1383	19 .3959
Ac				1080 1.86	890 7.33	675 10.61	639 15.93	319 39.83		272 .684	215 .647	167 1.95	80 4.9*			
Th				1168 1.80	968 7.46	714 10.82	677 16.31	344 18.81	335 23.94	290 .702	229 .660	182 2.05	95 2.15	88 3.15	60 .1625	49,43 .133 .366

2.3.1. X-ray Source

Monochromatic X-ray photons are produced by an X-ray gun or source. Generally, a heated filament is used to produce electrons which are then accelerated to an appropriate target over a potential of up to 20 kV. Interaction of the high-energy electrons with the target results in the emission of characteristic target X-rays (Figure 4). The targets are generally aluminum or magnesium, because these materials provide a stable, reliable beam of Kα X-rays dominated by the K$\alpha_{1,2}$, which is nearly monochromatic. The X-rays are then directed towards the sample, usually through a thin metal "window" to separate the X-ray chamber from the sample analysis chamber. The window serves to prevent scattered electrons from entering the analysis region and filters the high-energy white X-radiation which is produced.[91] Although the X-ray emission is dominated by the K$\alpha_{1,2}$, a variety of other characteristic X-rays are also produced, such as K$\alpha_{3,4}$, Kβ, and others.

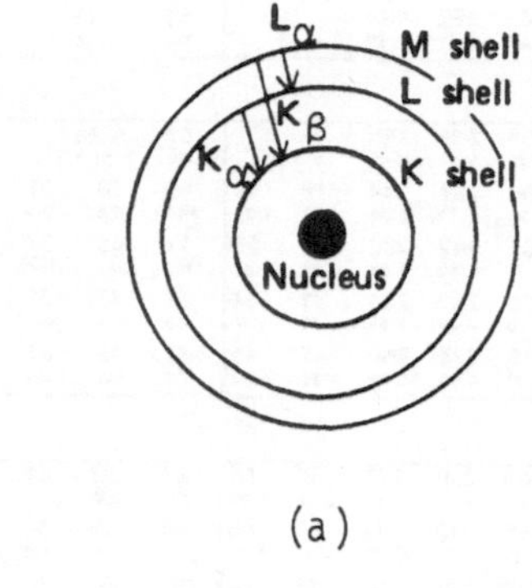

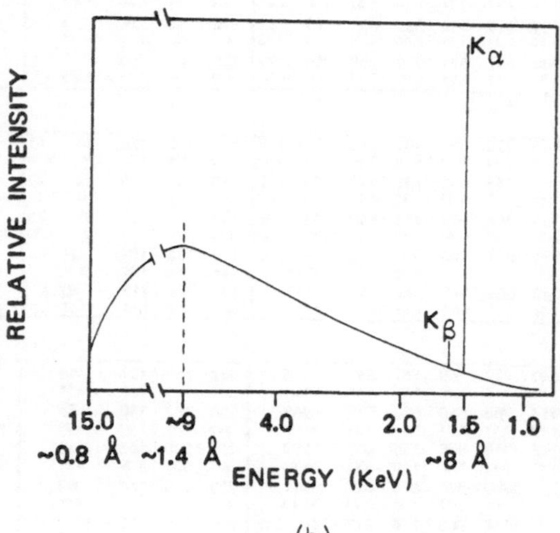

FIGURE 4. (a) The atomic orbital nomenclature for X-ray emission. (b) Approximate energy distribution of unmonochromatized Al X-rays, showing the strong characteristic $K\alpha$ and $K\beta$ peaks and the high-energy white (Bremsstrahlung) radiation. The use of an X-ray filter results in a significant decrease in the white radiation. [Adapted from Hewlett-Packard 5950A ESCA Manual and References (18) and (96), pp. 6-17].

TABLE 4
X-Ray Satellite Lines Present in Mg and Al-Excited Photoelectron Spectra

	$\alpha_{1,2}$	α_3	α_4	α_5	α_6	β
Mg:						
Change in energy (eV)	0	8.7	10.3	17.3	19.7	48.4
~C-1*s* (eV)	285	276	275	268	265	237
~0-1*s* (eV)	533	524	523	516	313	485
Relative intensity	100	8	4	0.5	0.5	0.5
Al:						
Change in energy (eV)	0	9.5	11.5	19.5	21.5	70
~C-1*s* (eV)	285	276	274	266	264	215
~0-1*s* (eV)	533	524	522	514	512	463
Relative intensity	100	6	3	0.4	0.3	0.6

These other X-rays are often called satellite X-rays (Table 4); a complete discussion is available in References (3), (89), and (91). In addition to the characteristic X-ray production, there is a considerable amount of high-energy broad-band X-ray production, commonly called white radiation, which leads to a high X-ray background in the instrument (Figure 4). To remove this component, as well as the characteristic X-rays not directly useful in XPS analysis (K$\alpha_{3,4}$, Kβ, etc.), some manufacturers have utilized an X-ray monochromator. By diffracting the aluminum Kα X-rays from spherically bent quartz crystals, a narrow truly monochromatic beam of aluminum K$\alpha_{1,2}$ photons is produced (Figure 5). This has the advantages of higher resolution, because the spread of energies in the X-ray beam is decreased; simpler spectra, because all of the features due to satellite X-rays have been removed; and lower backgrounds, because the high-energy white radiation is totally removed. Discussions of monochromatized X-ray sources are available.[89–91]

Instruments with X-ray monochromators have generally been the most useful for rigorous chemical shift studies of polymers. In addition, the elimination of the high-energy white radiation significantly reduces the total X-ray flux, minimizing radiation damage to the sample. In an instrument not equipped with a monochromator, significant radiation damage effects can occur with certain polymers and with biological samples. A severe price is paid for utilizing a monochromator not only in terms of expense and complexity, but also in terms of signal. X-ray diffraction is a very inefficient process, and the overall monochromatic X-ray flux to the sample is reduced by several orders of magnitude. In order to achieve suitable signal levels,

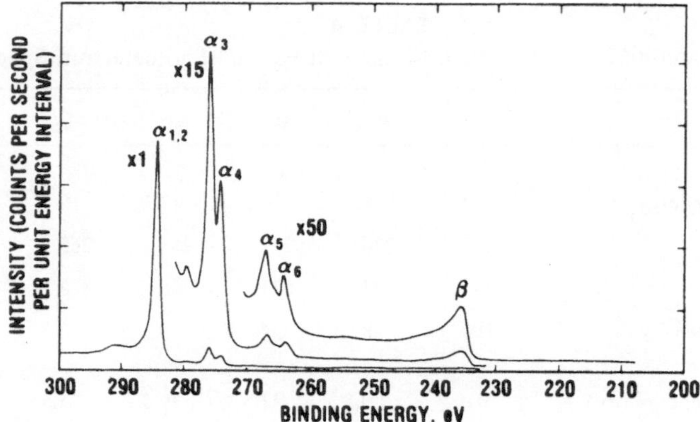

FIGURE 5. The effect of the X-ray satellites (Table 4) produced by unmonochromatized sources on the XPS spectrum. Mg X-ray satellites and the C-1*s* graphite region are shown. [Reprinted from Reference (9), p. 13, by permission.]

this decrease must be compensated for by a much more efficient detection system.

Some instruments are now being fitted with titanium and chromium X-ray sources (Table 5), which, although they have higher line widths and therefore decreased resolution, have considerably greater photon energies and, therefore, can eject photoelectrons from deeper lying core shells, as well as produce Auger electrons which are often useful for chemical state analysis.

In addition to X-ray photoelectron spectroscopy, ultraviolet photoelectron spectroscopy (UPS) is widely used. Helium UV lines have extremely narrow line widths and are used as photon sources in UPS (Table 5); very high-resolution studies are possible. This is only practical, of course, for molecular orbital and valence band studies, because the energies involved

TABLE 5
Photon Sources Commonly Used in X-Ray (XPS) and Ultraviolet Photoelectron Spectroscopy (UPS)

Photon source	$h\nu$ (eV)	Line width (eV)	Type
Mg $K\alpha$	1254	0.70	XPS
Al $K\alpha$	1487	0.85	XPS
Ti $K\alpha$	4510	2.00	XPS
Cr $K\alpha$	5417	2.10	XPS
He I	21.2	0.002	UPS
He II	40.8	0.002	UPS

are relatively low (Table 5). In addition, many researchers are now using synchrotron radiation which, in essence, provides a tuneable source from the UV into the X-ray region.[19]

2.3.2. Sample

The X-ray beam is then impinged upon the sample, mounted in a suitable holder (Figure 2) and electrically connected to the rest of the instrument and to electrical ground. Practically any sample which is vacuum stable can be analyzed by the XPS technique. Films, foils, and conventional solid samples are particularly easy to mount and to analyze. Fibers, textiles, meshes, and weaves can also be easily mounted and analyzed. Powders often present some difficulty, but a mounting can usually be devised. Double-stick tape of low volatility is often used for routine sample mounting and analysis. If the sample is volatile, for example, the plasticizer in soft poly(vinyl chloride), the sample can be cooled to −50°C or even lower and then analyzed under cryogenic conditions.

Liquids can be condensed on the probe surface and analyzed as condensed solids. Ratner *et al.*,[24] have performed an elegant freeze-etch study of hydrated polymer surfaces by freezing the hydrated polymer directly on the probe tip and then allowing the ice to etch or sublime off in the high vacuum, thereby measuring the surface characteristics of a frozen, hydrated polymer surface. See also Chapter 10.

Electrically conducting samples are electrically connected to the holder and the probe so that it is at ground potential. The interaction between the X-ray photons and the sample results in photoelectron production. Those photoelectrons produced in the near-surface region of the sample and traveling outward from the surface may emerge with little or no energy loss, continuing on to the analyzer with a kinetic energy given by Eq. 1. The problem is to measure the kinetic energy of the emitted photoelectrons.

2.3.3. Analyzer/Detector

The kinetic energy of the photoelectron is usually measured with a hemispherical analyzer coupled with an entrance device consisting of electron lenses. Ideally, the analyzer should be capable of resolving 0.1 eV for electrons with kinetic energies of up to 1000 eV with high efficiency. Various designs of electron kinetic energy analyzers have been used and are discussed in several references.[3,5,7,89-91,100] Two common approaches are to use a hemispherical electrostatic analyzer or a cylindrical mirror electrostatic analyzer. In these approaches the electrons are spatially dispersed based on their kinetic energy, usually by pre-retardation of the electron kinetic energy at the entrance to the analyzer. Because the electrons entering the analyzer all have kinetic energies in a narrow range, the analyzer operates in a constant resolution mode. The actual kinetic energy scanning is done by ramping voltages in the pre-retardation section. This method permits electrons of relatively constant kinetic energy to enter the hemispherical analyzer, where a highly precise energy analysis is performed under constant resolution conditions.

Electrons which match the tuning conditions of the spectrometer and pass through the pre-retardation section and through the hemispherical analyzer are then directed to a channeltron electron multiplier detector. The result is that a single electron impacting the detector results in a pulse of electrons leaving the detector with multiplications of 10^6 to 10^8 being common. The multiplied electron pulses are then usually directed to a phosphorescent screen to produce light pulses which are then counted by a photomultiplier tube or suitable light detection system. In some instruments, particularly those incorporating X-ray monochromators, multichannel detection systems are utilized to partially compensate for the decreased intensity which would otherwise result. In this case the phosphorescent screen can be coupled to a TV camera or other multichannel detection device to obtain a spatially dispersed multichannel signal.

2.3.4. Instrument Sources

An instrument manufactured by the Hewlett-Packard Corporation until 1976 utilized an aluminum $K\alpha$ quartz crystal monochromator, a channeltron electron multiplier, and a vidicon TV camera multichannel detection system to produce very high resolution spectra with very low sample radiation damage. In addition, this instrument was designed so that a portion of the energy dispersion inherent in the X-ray source was geometrically compensated via the electron optics design.[89-91] This dispersion compensation approach has produced the highest instrumental resolution so far attainable by XPS. A modern updated version of this instrument is now in commercial production by Surface Science Laboratories in Palo Alto, California.[20] This instrument, in addition to using monochromatic aluminum $K\alpha$ radiation and optical multichannel detection, features the use of a 150 μm diameter X-ray beam, thus permitting routine micro area XPS measurements for the first time (see also Chapter 13).

Other manufacturers of XPS instruments include: Vacuum Generators, Inc.[20] (VG), Leybold-Heraeus,[20] Perkin-Elmer Corp.,[20] and Kratos.[20] Many other XPS instruments are available and in wide use throughout the world. Instruments no longer manufactured include Varian, McPherson, and DuPont, as well as Hewlett-Packard. These instruments have been reviewed in Reference (7) and their operating characteristics are available.

It is important for the investigator to be aware of limitations, peculiarities, and characteristics of the instrument being used in order to fully interpret the spectra. This will become evident later.

Most of the examples presented in this chapter were obtained with a Hewlett-Packard 5950B X-ray photoelectron spectrometer.

2.4. CHARGING AND ENERGY REFERENCING

The electrical characteristics of the sample are important in interpreting the X-ray photoelectron spectrum. As photoelectrons are constantly being emitted from the sample surface, the sample will charge positively if it is an insulator or if it is not connected to a source of electrons. Conducting and semiconducting samples are mounted in direct electrical contact with the probe, which in turn is in contact with the spectrometer ground circuit which provides a source of electrons. The surface charge which results for an insulator or for any sample not in electrical contact with the spectrometer is an electrical potential barrier which the electron must overcome in making its way to the vacuum and to the analyzer. Crossing such a potential barrier requires energy; the electron, therefore, emerges from the sample with a lower kinetic energy and therefore appears on the XPS spectrum as a higher binding energy. This results in photoelectron peaks from insulating samples being shifted to higher binding energies.

If the charging process is uniform, the spectrum simply shifts a potential equal to the surface potential of the sample. However, for a variety of reasons, the charging is generally not uniform. For example, the X-ray beam generally has an intensity distribution on the surface and, therefore, the number of photoelectrons being emitted from the fringes of the beam is less than in the center of the beam, producing a charge distribution across the sample surface. In addition, the sample surface may not be completely homogeneous, and different regions may emit photoelectrons at different rates, thereby leading to differential charging. Also, depending on how the sample is mounted, if a portion of the illuminated surface is in contact with metal which is attached to the spectrometer, then some of the charge on the surface adjacent to the metal will drain to the metal, and again one has a charge distribution across part of the sample surface. Thus as charging develops, not only do the peaks move up-stream in binding energy, but because of the differential charging, the peaks often tend to broaden.

In a conventional nonmonochromatized XPS instrument, this process is self-limiting, largely because the high-energy white background radiation present in the sample analysis chamber results in photoelectron emission, not only from the sample but also from the walls of the chamber, providing an electron atmosphere in the vacuum. As the sample surface begins to develop a positive potential, these electrons are attracted to the surface and partially neutralize that charge. This results in a steady-state charging of up to about 4 eV, depending on the particular instrument, the sample, and the way it is mounted. It can and has been shown that this shift is uniform throughout the entire binding energy range of the XPS spectrum.

The charging shift varies from sample to sample and is a characteristic of the particular instrument and the sample, including sample type, dimensions, thickness, and method of mounting. The most common way of calibrating or compensating for this charging effect is through the use of an internal reference. Take a known photoelectron line in the sample and place it at a known binding energy. Commonly this is the carbon 1*s* photoelectron line, which is normally taken at 285.0 eV. If, for example, it appears at 287.0, and we *know* that the sample contains hydrocarbon at the surface, the entire spectrum is simply shifted 2 eV. This procedure works reasonably well for most polymers and other insulators. Most samples, unless they have been cleaned under high-vacuum conditions, contain a layer of adsorbed organic material with a significant hydrocarbon component, which provides a C-1*s* line useful for energy referencing. Although there has been considerable discussion and some controversy in the literature over the best way of charge referencing and charge calibration of XPS spectra, referencing to a known line in the sample is still the most widely used and probably the most effective.

In the case of an instrument utilizing an X-ray monochromator, the situation is considerably different. The high-energy white X-radiaton is not present, the electron gas in the analysis chamber is not present, and there is no mechanism for charge compensation. If one analyzes an insulating polymer, for example, in the Hewlett–Packard instrument, the carbon 1*s* line can appear at much higher than 285 eV; charging shifts of 40 or more eV are not uncommon. In order to compensate for this, such instruments include an electron flood gun to bathe the surface of the sample with a gas of low-energy electrons. As the sample begins to charge, electrons produced by the flood gun are directed and attracted towards the sample surface, flooding the surface with low-energy electrons which provide charge compensation. One would expect that the electron flux from the flood gun should be adjusted to just compensate the charging, that is to drive the carbon 1*s* line to exactly 285.0 eV. Because of heterogeneous charging and other considerations, it is more optimum to overdrive, that is to place the surface at a net negative surface potential, generally 6 to 7 eV negative such that the carbon 1*s* appears at about 278 eV.[101] This results in sharper, more well-defined peaks and better separation of chemical bonding features (Figure 6). This probably works because the process results in a surface of constant potential, minimizing heterogeneous surface charge effects. Now the entire spectrum is shifted *down* in binding energy by up to 7 eV. Again, one takes a reference line, for example, the carbon 1*s*, places it at 285 eV and thereby shifts the entire spectrum back up 7 eV. This method works remarkably well. Figure 6 gives some examples for an acrylic polymer sample. Figure 6a is a wide scan for poly(methyl methacrylate) under optimum flood gun conditions. Figure 6b shows the C-1*s* and O-1*s* regions under different flood gun conditions.

The added feature of having control over the charging characteristics of the sample by use of an electron flood gun provides a totally new source of information from the XPS spectrum. For example, if one has insulating domains in a conductive matrix, one can easily detect such domains by looking at the XPS spectrum with the flood gun on and then off.[32]

Generally all spectra recorded in the literature are charge referenced to some appropriate line (usually the C-1*s* at 285.0 eV) and the binding energy axis adjusted accordingly.

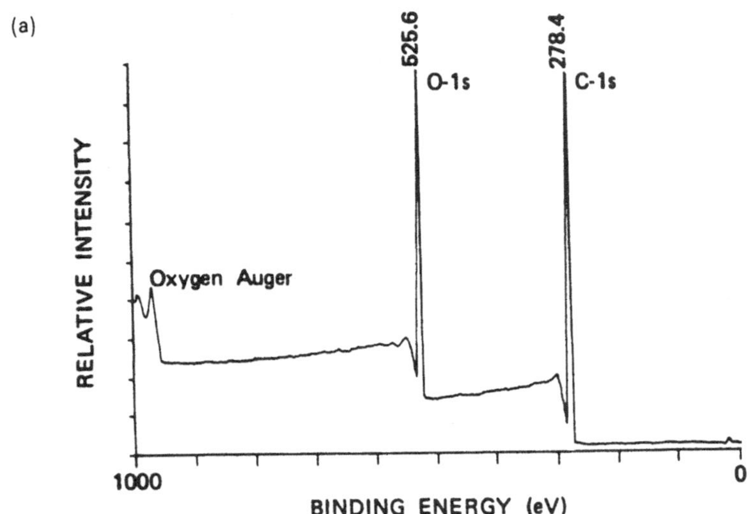

FIGURE 6. A poly(methyl methacrylate) thin film sample. (a) A wide scan under optimum flood gun conditions; (b) *Top*: normal C-1*s* and 0-1*s* scans with the flood gun operating under normal charge compensation conditions. Note the sharpness of the peaks and the fact that their energy position is 6 eV lower than normally tabulated due to overcompensation by the flood gun. *Center*: the same sample spectrum recorded with the flood gun off. Note the fact that the peaks have broadened considerably and have been shifted upward about 35 eV due to sample charging. *Bottom*: flood gun conditions adjusted so that the C-1*s* alkyl line is about 284 eV. Although this spectrum appears comparable to that normally recorded with non-monochromatic instruments, note the poor resolution of the C-1*s* and the 0-1*s* lines compared with those at the top of the figure. These spectra were taken on a monochromatic Hewlett-Packard 5950B instrument.

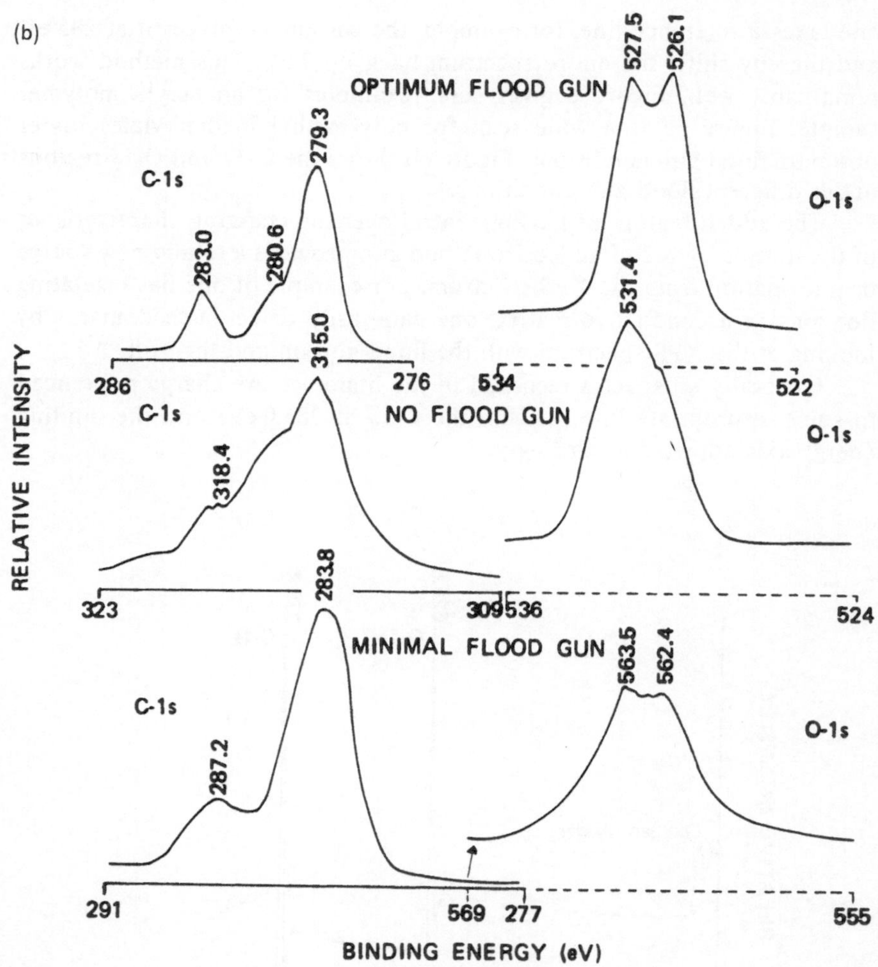

FIGURE 6 (continued)

2.5. SPECTRA AND SPECTRAL FEATURES

Refer to Table 3 often as you read this section.

Figure 6a showed a typical wide scan spectrum of an acrylic sample. Only 3 major features are present—the C-1*s* line at about 285.0 (278.4) eV, the O-1*s* at about 532.2 (525.6) eV, and the oxygen Auger line (Table 9) at about 977 (970) eV. Auger lines will be discussed later. The values in parentheses are due to charging effects discussed above. The area under each peak is used to determine elemental ratios (Section 3). Note the "stair step" appearance (right to left) of the spectrum.

A significant proportion of the electrons which are detected have lost some energy in interactions with the sample. Because those electrons have a lower kinetic energy, they appear at a higher binding energy and result in a high binding energy tail or background. Therefore, all typical photoelectron spectra have a stair-step appearance (Figures 2 and 6). The high binding energy tail adjacent to every intense photoelectron peak consists primarily of those photoelectrons which have lost some energy but have been detected by the instrument.

Figure 7 is a wide scan (0 to 1000 eV binding energy range) of gold, commonly used as a standard. Referring to Table 3, note the richness of spectral features. All of the atomic orbitals of gold, within the energy range of the instrument, are represented. Gold is a conductor, so we expect no charging shifts if the sample is properly mounted. Reading from right to left, one sees the 5*d*, 5*p*, 4*f*, 5*s*, 4*d*, 4*p*, and 4*s* orbitals of gold. Note the stair-step appearance and the growing background as one goes to higher binding energy. A C-1*s* peak is also observed at ~284 eV.

Table 6 is useful in learning atomic orbital nomenclature. The spectral lines or bands observed in X-ray photoelectron spectra are identified in terms of the principal quantum number, *n*, with values of 1, 2, 3, 4, 5, or 6 and the angular momentum quantum number, ℓ, with values of 0, 1, 2, 3, commonly called *s*, *p*, *d*, or *f*, respectively.

A characteristic feature of an XPS spectrum is the spin-orbit splitting observed for *p*, *d*, and *f* orbitals (Table 3) due to spin-orbit interactions.[7,26,79] An electron can have two different magnetic spin states, basically spin-up or spin-down. The orbital motion of the electron also

TABLE 6
Atomic Orbital Nomenclature[a]

Principal quantum number:	1, 2, 3, 4, 5 *K*, *L*, *M*, *N*, *O*
Angular momentum quantum number (l):	0, 1, 2, 3 *s*, *p*, *d*, *f*
Subscript:	$l + 1/2$ or $l - 1/2$
Roman numerals:	I, II, III, IV, V, VI, VII $s^{1/2}, p^{1/2}, p^{3/2}, d^{3/2}, d^{5/2}, f^{5/2}, f^{7/2}$
Example:	$4d^{5/2}$ principal quantum number; orbital type ($l = 2$); spin-orbit splitting ($l + 1/2$)

[a] References (7), (26), and (79).

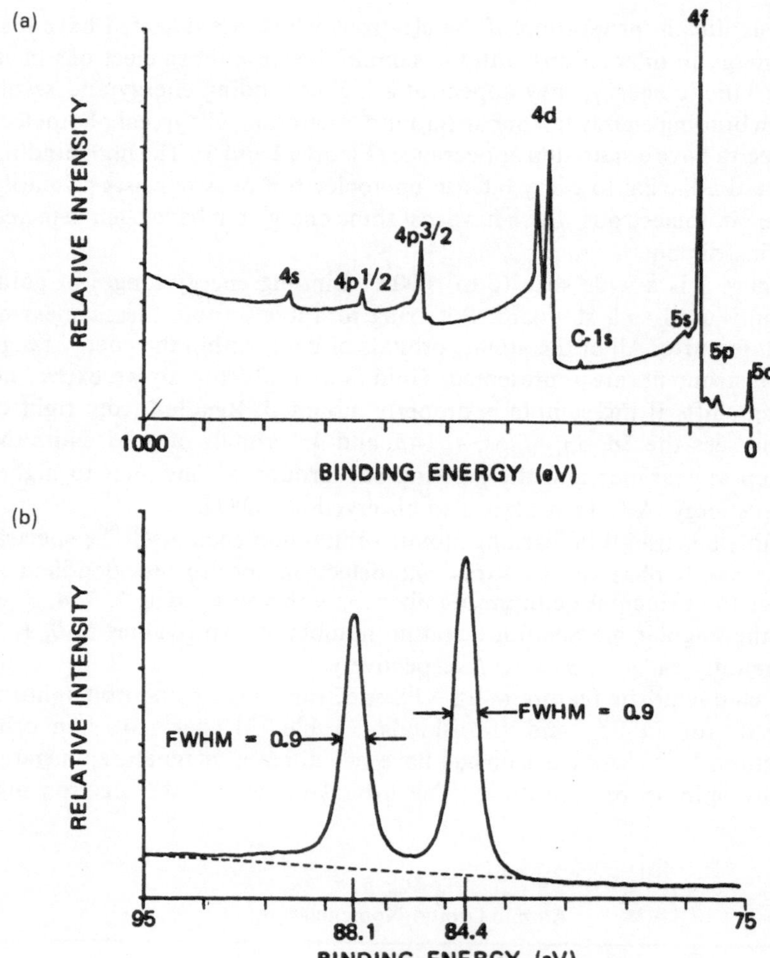

FIGURE 7. (a) A 0–1000 eV wide scan of gold, argon ion etched prior to analysis. Going from right to left on the spectrum, we are moving from the valence level of gold into progressively deeper atomic orbitals to the 4*s* orbital (~761 eV). The deeper orbitals are not within the analysis range of the instrument and photon source used. Note also that the inner core levels tend to broaden and ride on a high background, thereby compromising their usefulness for analytical purposes. A trace of carbon is observed as noted. (b) The gold 4*f* spin orbit doublet, commonly used for energy calibration of XPS instruments. The main features of the peak are noted, including energy position, peak height, full width at half maximum (FWHM), and general peak shape.

produces a magnetic field. These two magnetic fields can couple (said to be spin–orbit coupling or a spin–orbit interaction). The coupling results in two different interaction energies, depending upon whether the electron is spin-up or spin-down. The coupling is also a function of the nuclear charge because that is what is largely responsible for the orbital magnetic field or angular momentum. The coupling is negligible in the case of very small atoms but begins to be substantial as the nuclear charge increases. Coupling is not observed in *s* orbitals because they are symmetrical. The spin orbit effect is observable in the spectra and designated by the subscripts 1/2 and 3/2 for *p* orbitals, 3/2 and 5/2 for the *d*, and 5/2 and 7/2 for *f* orbitals, with intensity ratios of the peaks of about 1:2, 2:3, and 3:4, respectively.

The spin-orbit split begins to become evident in high-resolution XPS spectra of silicon. Beginning with chlorine and higher mass atoms, the spin-orbit split in the *p* orbitals is well separated and easily resolved. As one goes to very high mass, the spin-orbit split may be as great as 100 eV (see gold 4*p*, Figure 7a). The spin-orbit doublets are particularly diagnostic and useful for elemental analysis in XPS. Figure 7b shows the gold 4*f* spin-orbit doublet at high resolution. Note the shape of the peaks, the background, the area, energy position, and peak width at half maximum (called FWHM; full width at half maximum). These features are used to analyze XPS peaks. The gold 4*f* doublet is commonly used to calibrate and check the performance of XPS instruments; values of 83.8 eV for the Au $4f^{7/2}$ and 932.4 eV for Cu $2p^{3/2}$ lines are often assumed.[9]

A brief and basic discussion of photoelectron spectroscopy is now available in modern physical chemistry textbooks.[25,26] Such texts also discuss the nature of atomic orbitals and of molecular orbitals.

3. ELEMENTAL ANALYSIS

3.1. QUALITATIVE AND SEMIQUANTITATIVE ANALYSIS

A wide scan (0 to 1000 eV) spectrum is usually used for qualitative elemental analysis. For example, Figure 8 is a wide scan of an unknown elastomer. The data workup is given in Table 7. By reference to Tables 3 and 8 the key elements can be readily identified. Table 8 is a useful version of Tables 3 and 9, and gives the major XPS and Auger peaks of the elements in increasing binding energy. It is important to use *all* available lines to verify an elemental identification.

The areas under the key peaks are measured and tabulated (Table 8). Different elements and different orbitals have different probabilities for interaction with X-ray photons. This results in different sensitivity factors and thus different peak areas. In Table 3 there is an entry (in small type) beneath every binding energy value. This value is a sensitivity factor for the particular element and orbital—it has been calculated and tabulated by Scofield.[27] Such factors are called photoionization cross sections. The cross sections in Table 3 are for interaction with 1487-eV Al Kα photons. Values

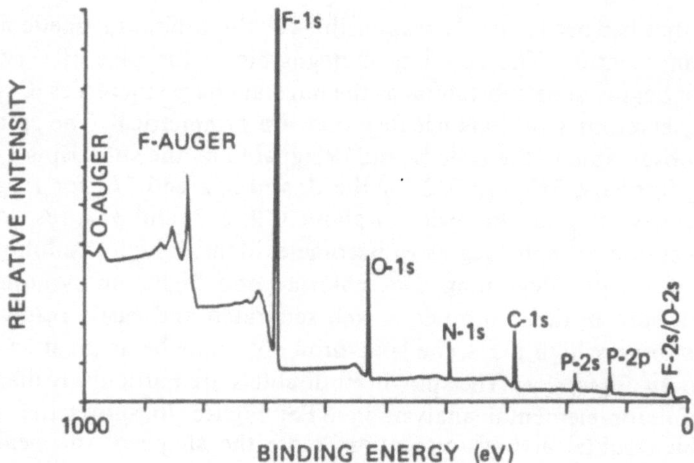

FIGURE 8. Wide scan spectrum of a polyphosphazene elastomer (PNF 200; material supplied courtesy of Firestone, Inc.). The major elemental lines have been identified. Note the presence of carbon, oxygen, nitrogen, phosphorus, and fluorine photoelectron peaks as well as the fluorine and oxygen Auger peaks.

are also available for Mg Kα.[27] The photoionization cross sections are also listed in Table 8.

The simplest approach to rough quantitation is to simply divide the peak area by the Scofield factor. The normalized values are then used to calculate atomic ratios or percentages. All data in XPS are in terms of *numbers* of atoms—concentrations are in atomic percents, not weight percents or weight ratios. It is useful to ratio only *s* peaks when possible, as this eliminates the need to consider orbital asymmetry effects (Section 4).

TABLE 7
Semiquantitative Elemental Analysis of Figure 8[a]

Binding energy (referred to C-1*s* at 285.0 eV)	Element and line identification	Peak area (counts, 10 scans)	Scofield section (Table 3)	Normalized area	% Atomic
285.0	C-1*s*	87,423	1.00	87,423	32.1
685.6	F-1*s*	549,500	4.43	124,041	45.5
131.8	P-2*p*	21,608	1.19	18,158	6.7
395.0	N-1*s*	25,809	1.80	14,338	5.2
531.0	O-1*s*	84,269	2.93	28,761	10.5
		Total normalized "area" =		272,721	100%

[a] Data used was obtained from 20-eV wide high resolution scans (not shown in Figure 8).

TABLE 8
Data of Tables 3 and 9 in Order of Increasing Binding Energy[a]

Element		Orbital	Scofield cross section
0	Eu	$4f$'s	1.155
0	Gd	$4f$'s	1.434
0	Os	$5d$'s	.847
1	Na	$3s^{1/2}$	.0064
1	Al	$3s^{1/2}$	.0535
1	Ga	$4p$'s	.018
1	Pd	$4d$'s	1.24
1	In	$5p$'s	.0195
1	Sn	$5s^{1/2}$	.0922
1	Sn	$5p$'s	.058
1	Ce	$4f$'s	.1389
1	Pb	$6p$'s	.0439
2	Mg	$3s^{1/2}$	.0285
2	V	$3d$'s	.0309
2	Cr	$3d$'s	.0651
2	Cu	$3d$'s	.589
2	Mo	$4d$'s	.316
2	Tc	$4d$'s	.470
2	Ru	$4d$'s	.667
2	Cd	$5p$'s	*b*
2	Sb	$5p$'s	.1145
2	Te	$5p$'s	.189
2	Pt	$5d$'s	1.477
2	Pr	$4f$'s	.2545
2	Nd	$4f$'s	.4068
3	Si	$3p$'s	.014
3	Ti	$3d$'s	.0136
3	Co	$3d$'s	.2664
3	Ge	$4p$'s	.058
3	As	$4p$'s	.121
3	Y	$4d$'s	.031
3	Zr	$4d$'s	.085
3	Rh	$4d$'s	.908
3	I	$5p$'s	.2828
3	Ag	$4d$'s	1.55
3	Tb	$4f$'s	1.967
3	Au	$5d$'s	1.808
3	Pb	$6s^{1/2}$	.0742
3	Bi	$6p$'s	.0841
4	Mn	$3d$'s	.1046
4	Ni	$3d$'s	.3979
4	Nb	$4d$'s	.198
4	Pm	$4f$'s	.604
4	Dy	$4f$'s	2.49
4	Ho	$4f$'s	3.10
4	Er	$4f$'s	3.82
4	Re	$5d$'s	.624
4	Ir	$5d$'s	1.238
5	B	$2p$'s	.0002
5	Ca	$3d$'s	*
5	Br	$4p$'s	.328
5	Lu	$5d$'s	.0593
5	Tm	$4f$'s	4.64
5	Po	$6p$'s	.1356
6	Fe	$3d$'s	.1711
6	Se	$4p$'s	.210
6	Yb	$4f$'s	5.58
6	Ta	$5d$'s	.2778
6	W	$5d$'s	.4344
7	C	$2p$'s	.0015
7	O	$2p$'s	.0193
7	Cl	$3p$'s	.1433
7	Sc	$3d$'s	.0042
7	Sb	$5s^{1/2}$	.1085
7	Xe	$5p$'s	.3961
7	Sm	$4f$'s	.851
7	Lu	$4f$'s	6.50
7	Hf	$5d$'s	.1526
7	Hg	$5d$'s	2.079
8	Si	$3s^{1/2}$	.0808
8	S	$3p$'s	.0774
8	Bi	$6s^{1/2}$	.0840
8	At	$6p$'s	.1892
9	N	$2p$'s	.0065
9	F	$2p$'s	.0478
9	Zn	$3d$'s	.81
9	Cd	$4d$'s	1.89
10	P	$3p$'s	.0368
11	Kr	$4p$'s	.476
11	Rn	$6p$'s	.2719
12	Ar	$3p$'s	.2418
12	Te	$5s^{1/2}$	.1251

continued

TABLE 8 (*continued*)

Element		Orbital	Scofield cross section	Element		Orbital	Scofield cross section
12	Cs	$5p^{3/2}$	.332	24	O	$2s^{1/2}$	.1405
12	Po	$6s^{1/2}$	.0937	24	Kr	$4s^{1/2}$	.213
				24	Sn	$4d$'s	2.70
13	Cs	$5p^{1/2}$	.1697				
13	Tl	$5d^{5/2}$	1.39	25	He	$1s^{1/2}$	.0082
				25	Ar	$3s^{1/2}$	.227
14	Rb	$4p^{3/2}$	.411	25	Bi	$5d^{5/2}$	1.76
14	I	$5s^{1/2}$	.1421	25	Ta	$4f^{7/2}$	4.82
14	H	$1s^{1/2}$	.0002				
				26	Ca	$3p$'s	.507
15	Rb	$4p^{1/2}$	.214	26	Y	$4p$'s	.091
15	Ba	$5p^{3/2}$	.400	26	Tb	$5p$'s	.804
15	Fr	$6p$'s	.3366	26	Dy	$5p$'s	.821
15	La	$5p$'s	.688	26	Rn	$6s$'s	.1129
16	S	$3s^{1/2}$	.1465	27	Br	$4s^{1/2}$	.1863
16	P	$3s^{1/2}$	.1116	27	Bi	$5d^{3/2}$	1.24
16	In	$4d$'s	2.28	27	Ta	$4f^{5/2}$	3.80
16	Tl	$5d^{3/2}$	.991				
				28	Lu	$5p$'s	.949
17	Ba	$5p^{1/2}$	.202				
				29	Ge	$3d^{3/2}$	1.42
18	Ne	$2p$'s	.103	29	Zr	$4p$'s	1.05
18	Cl	$3s^{1/2}$	.1852	29	Er	$5p$'s	.849
18	K	$3p$'s	.3619				
18	Xe	$5s^{1/2}$	.1596	30	Rb	$4s^{1/2}$	.251
18	At	$6s^{1/2}$	.1033				
18	Ga	$3d$'s	1.085	31	F	$2s^{1/2}$	.210
18	Hf	$4f^{7/2}$	4.20	31	Na	$2p$'s	.1941
				31	Hf	$5p^{3/2}$	.699
19	Ra	$6p$'s	.3959	31	Po	$5d$'s	3.31
19	Hf	$4f^{5/2}$	3.32				
				32	Sc	$3p$'s	.650
20	Sr	$4p$'s	.775	32	Sb	$4d$'s	3.14
20	Ce	$5p$'s	.660	32	Eu	$5s^{1/2}$	.268
20	Ho	$5p$'s	.836	32	Tm	$5p$'s	.864
20	Pb	$5d^{5/2}$	1.58				
				33	La	$5s^{1/2}$	.234
21	Gd	$5p$'s	.847				
				34	K	$3s^{1/2}$	.286
22	Nd	$5p$'s	.708	34	Ti	$3p$'s	.813
22	Pm	$5p$'s	.703	34	Nb	$4p$'s	1.17
22	Sm	$5p$'s	.750	34	Fr	$6s^{1/2}$	.1257
22	Eu	$5p$'s	.770	34	W	$4f^{7/2}$	5.48
22	Pb	$5d^{3/2}$	1.11				
				35	Mo	$4p$'s	1.31
23	Cs	$5s^{1/2}$	.1843	35	Re	$5p^{3/2}$	.869
23	Pr	$5p$'s	.685				
23	Yb	$5p$'s	.876	36	Gd	$5s^{1/2}$	.288

TABLE 8 (*continued*)

Element	Orbital	Scofield cross section
37 Ta	$5p^{3/2}$	.754
37 W	$5p^{3/2}$	.811
37 W	$4f^{5/2}$	4.32
38 V	$3p$'s	.996
38 Sr	$4s^{1/2}$	.291
38 Ce	$5s^{1/2}$	.230
38 Pr	$5s^{1/2}$	.238
38 Nd	$5s^{1/2}$	.247
38 Pm	$5s^{1/2}$	.254
38 Hf	$5p^{1/2}$	.325
39 Tc	$4p$'s	1.45
39 Sm	$5s^{1/2}$	.261
40 Te	$4d$'s	3.63
40 Ba	$5s^{1/2}$	.210
40 Tb	$5s^{1/2}$	.281
40 At	$5d$'s	3.63
41 As	$3d$'s	1.82
43 Cr	$3p$'s	1.173
43 Ru	$4p$'s	1.59
43 Th	$6p$'s	.366
44 Ca	$3s^{1/2}$	.351
44 Ra	$6s^{1/2}$	.1383
45 Ne	$2s^{1/2}$	.296
45 Re	$4f^{7/2}$	6.20
45 Ta	$5p^{1/2}$	.346
46 Y	$4s^{1/2}$	.329
46 Re	$5p^{1/2}$	.387
46 Os	$5p^{3/2}$	.928
47 Re	$4f^{5/2}$	4.88
47 W	$5p^{1/2}$	.367
48 Rh	$4p$'s	1.75
48 Rn	$5s^{3/2}$	3.95
49 Th	$6p$'s	.133
49 Mn	$3p$'s	1.423
50 I	$4d$'s	4.13
50 Os	$4f^{7/2}$	6.96
51 Pd	$4p$'s	1.88
51 Ho	$5s^{1/2}$	.293
51 Ir	$5p^{3/2}$	.967
51 Pt	$5p^{3/2}$	1.04
52 Mg	$2p$'s	.3335
52 Zr	$4s^{1/2}$	.367
52 Os	$4f^{5/2}$	5.48
53 Tm	$5s^{1/2}$	.308
53 Yb	$5s^{1/2}$	.308
54 Sc	$3s^{1/2}$	.411
54 Au	$5p^{3/2}$	1.10
55 Li	$1s^{1/2}$	.0568
56 Fe	$3p$'s	1.669
56 Ag	$4p^{3/2}$	1.36
57 Se	$3d$'s	2.29
57 Lu	$5s^{1/2}$	.326
58 Nb	$4s^{1/2}$	.402
58 Os	$5p^{1/2}$	.408
58 Hg	$5p^{3/2}$	1.17
58 Fr	$5d$'s	4.28
59 Ti	$3s^{1/2}$	.473
60 Co	$3p$'s	1.930
60 Er	$5s^{1/2}$	.298
60 Th	$6s^{1/2}$	.1625
60 Ir	$4f^{7/2}$	7.78
62 Mo	$4s^{1/2}$	.440
62 Ag	$4p^{1/2}$	.700
63 Xe	$4d$'s	4.68
63 Dy	$5s^{1/2}$	.287
63 Ir	$5p^{1/2}$	.422
63 Na	$2s^{1/2}$	.422
63 Ir	$4f^{5/2}$	6.12
65 Hf	$5s^{1/2}$	.344
66 V	$3s^{1/2}$	.538
66 Pt	$5p^{1/2}$	.444
67 Cd	$4p$'s	2.25
68 Ni	$3p$'s	2.217
68 Tc	$4s^{1/2}$	.479
68 Ra	$5d$'s	4.61
69 Br	$3d^{5/2}$	1.68
70 Br	$3d^{3/2}$	1.16
70 Pt	$4f^{7/2}$	8.65

continued

TABLE 8 (*continued*)

Element		Orbital	Scofield cross section
71	Ta	$5s^{1/2}$	.363
72	Au	$5p^{1/2}$	.463
73	Al	$2p^{3/2}$	.356
74	Al	$2p^{1/2}$	.1811
74	Cr	$3s^{1/2}$	.596
74	Cu	$3p$'s	2.478
74	Pt	$4f^{5/2}$	6.81
75	Ru	$4s^{1/2}$	.519
76	Tl	$5d^{3/2}$	1.25
77	In	$4p$'s	2.45
77	Cs	$4d^{5/2}$	3.10
77	W	$5s^{1/2}$	.383
79	Cs	$4d^{3/2}$	2.15
80	Ac	$5d$'s	4.96
81	Rh	$4s^{1/2}$	.560
81	Hg	$5p^{1/2}$	.484
83	Re	$5s^{1/2}$	.402
83	Au	$4f^{7/2}$	9.58
84	Mn	$3s^{1/2}$	.674
84	Os	$5s^{1/2}$	.422
86	Pd	$4s^{1/2}$	.598
86	Pb	$5p^{3/2}$	1.33
87	Zn	$3p$'s	2.828
87	Au	$4f^{5/2}$	7.54
88	Th	$5d^{5/2}$	3.15
89	Mg	$2s^{1/2}$	.575
89	Kr	$3d$'s	3.48
89	Sn	$4p$'s	2.67
90	Ba	$4d^{5/2}$	3.46
93	Ba	$4d^{3/2}$	2.40
93	Bi	$5p^{3/2}$	1.41
95	Fe	$3s^{1/2}$	.745
95	Ag	$4s^{1/2}$	.644
95	Th	$5d^{3/2}$	2.15
96	Ir	$5s^{1/2}$	.438
99	Si	$2p^{3/2}$	.541
99	Sb	$4p$'s	2.88
99	La	$4d$'s	6.52
99	Hg	$4f^{7/2}$	10.57
100	Si	$2p^{1/2}$	.276
100	Tl	$5p^{1/2}$	.505
101	Co	$3s^{1/2}$	.818
102	Pt	$5s^{1/2}$	.459
103	Ga	$3p^{3/2}$	2.11
103	Hg	$4f^{5/2}$	8.32
104	Po	$5p^{3/2}$	1.50
105	Pb	$5p^{1/2}$	.526
107	Ga	$3p^{1/2}$	1.10
108	Au	$5s^{1/2}$	.479
108	Cd	$4s^{1/2}$	.692
110	Te	$4p$'s	3.11
111	Rb	$3d^{5/2}$	2.49
111	Be	$1s^{1/2}$	.1947
111	Ce	$4d$'s	6.93
112	Ni	$3s^{1/2}$	.892
112	Rb	$3d^{3/2}$	1.72
114	Pr	$4d$'s	7.48
115	At	$5p^{3/2}$	1.58
117	Bi	$5p^{1/2}$	.546
118	Nd	$4d$'s	8.03
118	Al	$2s^{1/2}$	.753
118	Tl	$4f^{7/2}$	11.62
120	Cu	$3s^{1/2}$	.957
120	Hg	$5s^{1/2}$	.500
121	Pm	$4d$'s	8.59
122	Ge	$3p^{3/2}$	2.39
122	In	$4s^{1/2}$	.742
122	Tl	$4f^{5/2}$	9.14
123	I	$4p$'s	3.34

TABLE 8 (*continued*)

Element		Orbital	Scofield cross section	Element		Orbital	Scofield cross section
127	Rn	$5p^{3/2}$	1.67	162	Se	$3p^{3/2}$	2.98
				162	Cs	$4p^{3/2}$	2.56
129	Ge	$3p^{1/2}$	1.24	163	Bi	$4f^{5/2}$	10.93
130	Sm	$4d$'s	9.16	164	S	$2p^{3/2}$	1.11
132	Po	$5p^{1/2}$	.566	164	Rn	$5p^{1/2}$	.602
133	Sr	$3d^{5/2}$	2.99	165	S	$2p^{1/2}$	.567
134	Eu	$4d$'s	9.73	167	Ac	$5p^{3/2}$	1.95
135	P	$2p^{3/2}$	.789	168	Se	$3p^{1/2}$	1.55
135	Sr	$3d^{3/2}$	2.06	168	Te	$4s^{1/2}$	.903
136	P	$2p^{1/2}$	.403	168	Er	$4d^{5/2}$	7.41
137	Zn	$3s^{1/2}$	1.04	172	Cs	$4p^{1/2}$	1.27
137	Sn	$4s^{1/2}$	.794	177	Er	$4d^{3/2}$	5.15
137	Tl	$5s^{1/2}$	.520	177	Po	$5s^{1/2}$	.584
138	Pb	$4f^{7/2}$	12.73	180	Zr	$3d^{5/2}$	4.17
140	Fr	$5p^{3/2}$	1.77	180	Ba	$4p^{3/2}$	2.73
				180	Tm	$4d$'s	13.12
141	As	$3p^{3/2}$	2.68				
141	Gd	$4d$'s	10.40	181	Ge	$3s^{1/2}$	1.23
142	Se	Auger		182	Br	$3p^{3/2}$	3.31
143	Pb	$4f^{5/2}$	10.01	182	Fr	$5p^{1/2}$	.618
				182	Th	$5p^{3/2}$	2.05
147	As	$3p^{1/2}$	1.39	183	Zr	$3d^{3/2}$	2.87
147	Xe	$4p$'s	3.58				
148	At	$5p^{1/2}$	.584	184	Po	$4f$'s	27.04
148	Pb	$5s^{1/2}$	.542	184	Yb	$4d^{5/2}$	8.07
148	Tb	$4d$'s	10.87	184	Se	Auger	
149	Si	$2s^{1/2}$	.955	186	I	$4s^{1/2}$	.959
152	Sb	$4s^{1/2}$	.848	188	B	$1s^{1/2}$	.486
153	Ra	$5p^{3/2}$	1.86	189	P	$2s^{1/2}$	1.18
154	Dy	$4d$'s	11.43	189	Br	$3p^{1/2}$	1.72
158	Ga	$3s^{1/2}$	1.13	192	Ba	$4p^{1/2}$	1.34
158	Y	$3d^{5/2}$	3.54	192	La	$4p^{3/2}$	2.91
158	Bi	$4f^{7/2}$	13.90				
160	Y	$3d^{3/2}$	2.44	195	Lu	$4d^{5/2}$	8.45
160	Bi	$5s^{1/2}$	.563	195	At	$5s^{1/2}$	.605
161	Ho	$4d$'s	12.00	197	Yb	$4d^{3/2}$	5.61

continued

TABLE 8 (*continued*)

Element		Orbital	Scofield cross section	Element		Orbital	Scofield cross section
200	Cl	$2p^{3/2}$	1.51	242	Ta	$4d^{3/2}$	6.40
200	Ra	$5p^{1/2}$	.633	244	Nd	$4p^{1/2}$	1.59
202	Cl	$2p^{1/2}$	.775	245	Ar	$2p^{1/2}$	2.01
204	As	$3s^{1/2}$	1.32	246	W	$4d^{5/2}$	9.65
205	Nb	$3d^{5/2}$	4.86	247	Ar	$2p^{1/2}$	1.03
205	Lu	$4d^{3/2}$	5.87	248	Rb	$3p^{1/2}$	2.07
206	La	$4p^{1/2}$	1.42	249	Sm	$4p^{3/2}$	3.59
208	Nb	$3d^{3/2}$	3.35	253	Tc	$3d^{5/2}$	6.47
208	Xe	$4s^{1/2}$	1.02	253	Ba	$4s^{1/2}$	1.13
208	Ce	$4p^{3/2}$	3.03	254	Ra	$5s^{1/2}$	.665
210	At	$4f$'s	29.36	255	Pm	$4p^{1/2}$	1.64
214	Kr	$3p^{3/2}$	3.65	257	Br	$3s^{1/2}$	1.53
214	Hf	$4d^{5/2}$	8.84	257	Tc	$3d^{3/2}$	4.46
214	Rn	$5s^{1/2}$	.625	257	Eu	$4p^{3/2}$	3.72
215	Ac	$5p^{1/2}$	.647	259	W	$4d^{3/2}$	6.68
218	Pr	$4p^{3/2}$	3.17	259	Se	Auger	
223	Kr	$3p^{1/2}$	1.89	260	Re	$4d^{5/2}$	10.06
224	Ce	$4p^{1/2}$	1.47	266	As	Auger	
224	Hf	$4d^{3/2}$	6.13	267	Sm	$4p^{1/2}$	1.70
225	Nd	$4p^{3/2}$	3.31	268	Fr	$4f$'s	34.36
227	Mo	$3d^{5/2}$	5.62	269	Sr	$3p^{3/2}$	4.37
229	S	$2s^{1/2}$	1.43	270	Cl	$2s^{1/2}$	1.69
229	Th	$5p^{1/2}$	.660	271	La	$4s^{1/2}$	1.19
230	As	Auger		271	Gd	$4p^{3/2}$	3.88
230	Mo	$3d^{3/2}$	3.88	272	Ac	$5s^{1/2}$	.684
230	Ta	$4d^{5/2}$	9.24	273	Os	$4d^{5/2}$	10.48
231	Cs	$4s^{1/2}$	1.08	274	Re	$4d^{3/2}$	6.95
232	Se	$3s^{1/2}$	1.43	279	Ru	$3d^{5/2}$	7.39
234	Fr	$5s^{1/2}$	.645	280	Sr	$3p^{1/2}$	2.25
237	Pr	$4p^{1/2}$	1.53	284	C	$1s^{1/2}$	1.00
237	Pm	$4p^{3/2}$	3.45	284	Ru	$3d^{3/2}$	5.10
238	Rn	$4f$'s	31.81	284	Eu	$4p^{1/2}$	1.75
239	Rb	$3p^{3/2}$	4.00				

TABLE 8 (*continued*)

Element		Orbital	Scofield cross section	Element		Orbital	Scofield cross section
286	Tb	$4p^{3/2}$	3.99	332	Dy	$4p^{1/2}$	1.88
288	Se	Auger		334	Au	$4d^{5/2}$	11.74
289	Kr	$3s^{1/2}$	1.64	335	Pd	$3d^{5/2}$	9.48
289	Gd	$4p^{1/2}$	1.80	335	Th	$4f^{7/2}$	23.94
290	Ce	$4s^{1/2}$	1.24	337	Tm	$4p^{3/2}$	4.48
290	Os	$4d^{3/2}$	7.23	339	As	Auger	
290	Th	$5s^{1/2}$	.702	340	Pd	$4d^{3/2}$	6.56
293	Dy	$4p^{3/2}$	4.12	343	Yb	$4p^{3/2}$	4.60
294	K	$2p^{3/2}$	2.62	343	Ho	$4p^{1/2}$	1.91
295	Ir	$4d^{5/2}$	10.90	344	Th	$4f^{5/2}$	18.81
297	K	$2p^{3/2}$	1.35	345	Zr	$3p^{1/2}$	2.64
299	Ra	$4f$'s	37.04	347	Ca	$2p^{3/2}$	3.35
301	Y	$3p^{3/2}$	4.75	347	Sm	$4s^{1/2}$	1.42
301	Se	Auger		350	Ge	Auger	
305	Pr	$4s^{1/2}$	1.28	350	Ca	$2p^{1/2}$	1.72
306	Ho	$4p^{3/2}$	4.24	352	Au	$4d^{3/2}$	8.06
307	Rh	$3d^{5/2}$	8.39	353	Mg	Auger	
307	Mg	Auger		358	Sr	$3s^{1/2}$	1.86
311	Tb	$4p^{1/2}$	1.84	359	Lu	$4p^{3/2}$	4.74
312	Ir	$4d^{3/2}$	7.51	360	Eu	$4s^{1/2}$	1.46
312	Rh	$3d^{3/2}$	5.80	360	Hg	$4d^{5/2}$	12.17
313	Y	$3p^{1/2}$	2.44	363	Nb	$3p^{3/2}$	5.53
314	Pt	$4d^{5/2}$	11.32	364	As	Auger	
316	Nd	$4s^{1/2}$	1.33	366	Er	$4p^{1/2}$	1.95
318	Ge	Auger		367	Ag	$3d^{5/2}$	10.66
319	Ac	$4f$'s	39.83	373	Ag	$3d^{3/2}$	7.38
320	Ar	$2s^{1/2}$	1.97	376	Gd	$4s^{1/2}$	1.51
320	Er	$4p^{3/2}$	4.37	376	As	Auger	
322	Rb	$3s^{1/2}$	1.75	377	K	$2s^{1/2}$	2.27
331	Zr	$3p^{3/2}$	5.14	379	Nb	$3p^{1/2}$	2.84
331	Pm	$4s^{1/2}$	1.38	379	Hg	$4d^{3/2}$	8.33
331	Pt	$4d^{3/2}$	7.78	380	Hf	$4p^{3/2}$	4.88

continued

TABLE 8 (*continued*)

Element		Orbital	Scofield cross section	Element		Orbital	Scofield cross section
386	Tm	$4p^{1/2}$	1.98	449	Er	$4s^{1/2}$	1.64
386	Tl	$4d^{5/2}$	12.60	450	Ge	Auger	
393	Mo	$3p^{3/2}$	5.94	451	In	$3d^{3/2}$	9.22
395	Y	$3s^{1/2}$	1.98	455	Ti	$2p^{3/2}$	5.22
396	Yb	$4p^{1/2}$	2.00	461	Ti	$2p^{1/2}$	2.69
396	Ga	Auger		461	Ru	$3p^{3/2}$	6.78
398	Tb	$4s^{1/2}$	1.54	464	Bi	$4d^{3/2}$	9.14
399	N	$1s^{1/2}$	1.80	465	Ta	$4p^{1/2}$	2.08
402	Sc	$2p^{3/2}$	4.21	469	Nb	$3s^{1/2}$	2.22
404	Cd	$3d^{5/2}$	11.95	469	Os	$4p^{3/2}$	5.45
405	Ta	$4p^{3/2}$	5.02	472	Tm	$4s^{1/2}$	1.67
407	Tl	$4d^{3/2}$	8.60	473	Po	$4d^{5/2}$	13.87
407	Sc	$2p^{1/2}$	2.17	476	Zn	Auger	
410	Mo	$3p^{1/2}$	3.04	483	Ru	$3p^{1/2}$	3.44
410	Lu	$4p^{1/2}$	2.03	485	Sn	$3d^{5/2}$	14.80
411	Cd	$3d^{3/2}$	8.27	487	Yb	$4s^{1/2}$	1.70
413	Pb	$4d^{5/2}$	13.02	489	Ga	Auger	
416	Dy	$4s^{1/2}$	1.58	492	W	$4p^{1/2}$	2.10
419	Ge	Auger		494	Sn	$3d^{3/2}$	10.25
423	Ga	Auger		495	Ir	$4p^{3/2}$	5.59
425	Tc	$3p^{3/2}$	6.36	496	Rh	$3p^{3/2}$	7.21
426	W	$4p^{1/2}$	5.16	497	Na	Auger	
431	Zr	$3s^{1/2}$	2.10	499	Zn	Auger	
435	Pb	$4d^{3/2}$	8.87	500	Sc	$2s^{1/2}$	2.91
436	Ho	$4s^{1/2}$	1.61	500	Po	$4d^{3/2}$	9.40
437	Hf	$4p^{1/2}$	2.06	505	Mo	$3s^{1/2}$	2.34
438	Ca	$2s^{1/2}$	2.59	506	Lu	$4s^{1/2}$	1.73
440	Ge	Auger		507	At	$4d^{5/2}$	14.29
440	Bi	$4d^{5/2}$	13.44	507	Ga	Auger	
443	In	$3d^{5/2}$	13.32	513	V	$2p^{3/2}$	6.37
445	Tc	$3p^{1/2}$	3.23				
445	Re	$4p^{3/2}$	5.30				

TABLE 8 (*continued*)

Element		Orbital	Scofield cross section	Element		Orbital	Scofield cross section
518	Ga	Auger		584	Cr	$2p^{1/2}$	3.98
518	Re	$4p^{1/2}$	2.12	585	Ru	$3s^{1/2}$	2.57
519	Pt	$4p^{3/2}$	5.74	587	Zn	Auger	
520	V	$2p^{1/2}$	3.29	595	W	$4s^{1/2}$	1.81
521	Rh	$3p^{1/2}$	3.64	602	Ag	$3p^{1/2}$	4.03
528	Sb	$3d^{5/2}$	16.39	603	Fr	$4d^{3/2}$	10.16
				603	Ra	$4d^{5/2}$	15.53
531	Pd	$3p^{3/2}$	7.63	608	Pt	$4p^{1/2}$	2.14
532	O	$1s^{1/2}$	2.93	609	Tl	$4p^{3/2}$	6.19
533	At	$4d^{3/2}$	9.65	617	Cd	$3p^{3/2}$	8.50
536	Na	Auger		619	Ni	Auger	
537	Sb	$3d^{3/2}$	11.35	620	I	$3d^{5/2}$	19.87
538	Hf	$4s^{1/2}$	1.76	623	Ni	Auger	
541	Rn	$4d^{5/2}$	14.70	623	Cu	Auger	
543	Cu	Auger		625	Re	$4s^{1/2}$	1.84
544	Tc	$3s^{1/2}$	2.45	627	Rh	$3s^{1/2}$	2.70
546	Au	$4p^{3/2}$	5.89	628	V	$2s^{1/2}$	3.57
547	Os	$4p^{1/2}$	2.13	631	I	$3d^{3/2}$	13.77
559	Pd	$3p^{1/2}$	3.83	635	Cu	Auger	
563	Zn	Auger		636	Ra	$4d^{3/2}$	10.40
564	Ti	$2s^{1/2}$	3.24	637	Ni	Auger	
564	Cu	Auger		639	Ac	$4d^{5/2}$	15.93
566	Ta	$4s^{1/2}$	1.79	641	Mn	$2p^{3/2}$	9.17
567	Rn	$4d^{3/2}$	9.90	643	Cu	Auger	
571	Ag	$3p^{3/2}$	8.06	644	Au	$4p^{1/2}$	2.14
571	Hg	$4p^{3/2}$	6.04	645	Pb	$4p^{3/2}$	6.33
572	Te	$3d^{5/2}$	18.06	651	Cd	$3p^{1/2}$	4.22
575	Cr	$2p^{3/2}$	7.69	652	Mn	$2p^{1/2}$	4.74
577	Ir	$4p^{1/2}$	2.14	655	Os	$4s^{1/2}$	1.86
577	Fr	$4d^{5/2}$	15.11	664	In	$3p^{3/2}$	8.93
577	Zn	Auger					
582	Te	$3d^{3/2}$	12.52	670	Pd	$3s^{1/2}$	2.81

continued

TABLE 8 (*continued*)

Element	Orbital	Scofield cross section
671 Sm	Auger	
672 Xe	$3d^{5/2}$	21.79
675 Ac	$4d^{3/2}$	10.61
677 Hg	$4p^{1/2}$	2.14
677 Th	$4d^{5/2}$	16.31
679 Bi	$4p^{3/2}$	6.48
685 Xe	$3d^{3/2}$	15.10
686 F	$1s^{1/2}$	4.43
690 Ir	$4s^{1/2}$	1.88
695 Cr	$2s^{1/2}$	3.91
701 Ni	Auger	
702 In	$3p^{1/2}$	4.40
705 Po	$4p^{3/2}$	6.62
708 Co	Auger	
708 Ni	Auger	
710 Fe	$2p^{3/2}$	10.82
710 Cu	Auger	
714 Th	$4d^{3/2}$	10.82
715 Sn	$3p^{3/2}$	9.35
717 Ag	$3s^{1/2}$	2.93
722 Tl	$4p^{1/2}$	2.13
723 Fe	$2p^{1/2}$	5.60
724 Pt	$4s^{1/2}$	1.90
726 Cs	$3d^{5/2}$	23.76
740 Cs	$3d^{3/2}$	16.46
740 At	$4p^{3/2}$	6.77
751 Nd	Auger	
757 Sn	$3p^{1/2}$	4.58
759 Au	$4s^{1/2}$	1.92
763 Ni	Auger	
764 Pb	$4p^{1/2}$	2.12
766 Sb	$3p^{3/2}$	9.77
768 Rn	$4p^{3/2}$	6.92
769 Mn	$2s^{1/2}$	4.23
770 Cd	$3s^{1/2}$	3.04
773 Ni	Auger	
779 Co	$2p^{3/2}$	12.62
780 Fe	Auger	
781 Ba	$3d^{5/2}$	25.84
794 Co	$2p^{1/2}$	6.54
796 Ba	$3d^{3/2}$	17.92
800 Hg	$4s^{1/2}$	1.94
806 Bi	$4p^{1/2}$	2.10
810 Fr	$4p^{3/2}$	7.07
812 Sb	$3p^{1/2}$	4.76
819 Te	$3p^{3/2}$	10.21
823 Ce	Auger	
826 In	$3s^{1/2}$	3.16
830 F	Auger	
832 La	$3d^{5/}2$	28.12
841 Fe	Auger	
846 Fe	$2s^{1/2}$	4.57
846 Tl	$4s^{1/2}$	1.95
849 La	$3d^{3/2}$	19.50
851 Po	$4p^{1/2}$	2.07
853 Mn	Auger	
855 Ni	$2p^{3/2}$	14.61
857 F	Auger	
867 Ne	$1s^{1/2}$	6.30
870 Te	$3p^{1/2}$	4.92

TABLE 8 (*continued*)

Element		Orbital	Scofield cross section	Element		Orbital	Scofield cross section
872	Ni	$2p^{1/2}$	7.57	998	Cs	$3p^{3/2}$	11.38
875	I	$3p^{3/2}$	10.62	999	Xe	$3p^{1/2}$	5.20
879	Ra	$4p^{3/2}$	7.20	1000	Nd	$3d^{3/2}$	24.27
884	Sn	$3s^{1/2}$	3.26	1006	Te	$3s^{1/2}$	3.46
884	Ce	$3d^{5/2}$	30.50	1008	Ni	$2s^{1/2}$	5.16
886	At	$4p^{1/2}$	2.04	1015	V	Auger	
889	Ba	Auger		1021	Zn	$2p^{3/2}$	18.92
890	Ac	$4p^{3/2}$	7.33	1027	Pm	$3d^{5/2}$	37.65
894	Pb	$4s^{1/2}$	1.96	1028	V	Auger	
902	Ce	$3d^{3/2}$	21.12	1036	Sb	Auger	
902	Mn	Auger		1042	At	$4s^{1/2}$	1.96
903	Fe	Auger		1044	Zn	$2p^{1/2}$	9.80
903	Ba	Auger		1052	Pm	$3d^{3/2}$	26.08
926	Co	$2s^{1/2}$	4.88	1052	Sn	Auger	
929	Rn	$4p^{1/2}$	2.00	1055	V	Auger	
931	Cu	$2p^{3/2}$	16.73	1058	Ra	$4p^{1/2}$	1.91
931	I	$3p^{1/2}$	5.06	1059	Sn	Auger	
931	Pr	$3d^{5/2}$	32.85	1063	Ba	$3p^{3/2}$	11.71
937	Xe	$3p^{3/2}$	10.99	1065	Cs	$3p^{1/2}$	5.29
939	Bi	$4s^{1/2}$	1.96	1068	Ti	Auger	
944	Sb	$3s^{1/2}$	3.36	1072	Na	$1s^{1/2}$	8.52
951	Cu	$2p^{1/2}$	8.66	1072	I	$3s^{1/2}$	3.53
951	Pr	$3d^{3/2}$	22.72	1076	In	Auger	
958	Mn	Auger		1080	Ac	$4p^{1/2}$	1.86
968	Th	$4p^{3/2}$	7.46	1081	Sm	$3d^{5/2}$	40.37
970	I	Auger		1083	In	Auger	
974	O	Auger		1096	Cu	$1s^{1/2}$	5.46
978	Nd	$3d^{5/2}$	35.29	1097	Rn	$4s^{1/2}$	1.95
979	I	Auger		1104	Ti	Auger	
980	Fr	$4p^{1/2}$	1.97	1106	Cd	Auger	
995	Po	$4s^{1/2}$	1.97				
997	O	Auger					

continued

TABLE 8 (*continued*)

Element		Orbital	Scofield cross section	Element		Orbital	Scofield cross section
1107	Sm	$3d^{3/2}$	27.96	1186	Ce	$3p^{3/2}$	12.53
1110	N	Auger		1186	Gd	$3d^{5/2}$	46.23
1112	Cd	Auger		1186	Rh	Auger	
1116	Ga	$2p^{3/2}$	21.40	1194	Zn	$2s^{1/2}$	5.76
1124	La	$3p^{3/2}$	12.11	1194	Ca	Auger	
1127	Ag	Auger		1205	La	$3p^{1/2}$	5.55
1131	Eu	$3d^{5/2}$	43.24	1208	Ra	$4s^{1/2}$	1.95
1132	Ag	Auger		1211	Ru	Auger	
1137	Ba	$3p^{1/2}$	5.42	1216	C	Auger	
1143	Ga	$2p^{1/2}$	11.09	1217	Cs	$3s^{1/2}$	3.73
1145	Xe	$3s^{1/2}$	3.62	1217	Ge	$2p^{3/2}$	24.15
1153	Fr	$4s^{1/2}$	1.95	1218	Gd	$3d^{3/2}$	31.98
1154	Pd	Auger		1235	K	Auger	
1161	Eu	$3d^{3/2}$	29.91	1242	Tb	$3d^{5/2}$	49.42
1168	Th	$4p^{1/2}$	1.80	1243	Pr	$3p^{3/2}$	12.94
				1249	Ge	$2p^{1/2}$	12.52

[a] The most intense line for each particular element is underlined.
[b] No cross section data available.

This simplistic approach to quantitation gives rough semiquantitative results. This method ignores such factors as instrument sensitivity, sampling depths, etc.

3.2. QUANTITATIVE ELEMENTAL ANALYSES[3,28–33]

3.2.1. Basics

Quantitative analysis of XPS data can be routinely performed. The intensity of an X-ray photoelectron peak is related to a number of factors, including the intensity of the X-rays, the probability for X-ray/atomic orbital interaction (the photoionization cross section), the probability that the photoelectron will actually be emitted into the vacuum, the probability that once emitted into the vacuum it is actually collected and detected by the spectrometer, and of course the concentration of the element present in the sample.

A simplified form of the basic expression for quantitative XPS is:

$$N_{i,k} = I_0 \rho_i \sigma_{i,k} \lambda_{i,k} T_{i,k} \tag{2}$$

$N_{i,k}$ is the experimentally determined peak intensity (area) for the kth shell of atom of type i in the sample; I_0 is the X-ray flux incident on the sample; ρ_i is the volume density of element i in the surface volume examined by XPS—this is the quantitative information normally desired in the experiment; $\sigma_{i,k}$ is the differential photoionization cross section for the kth shell of atom i; $\lambda_{i,k}$ is the mean free path for the kth electron of atom i in the sample of interest; and $T_{i,k}$ is the instrument transmission or throughput function at the kinetic energy of the electrons from the kth shell of atom i. σ, λ, and T are, of course, functions of kinetic energy and thus of binding energy. More rigorous and complete forms of the equation, specifically considering all relevant parameters, are available in the literature.[3,31] A number of quantitative treatments based on Eq. (2) or variants of it have been described in the literature. Perhaps the first was discussed by Carter *et al.*[33] We will examine each of the parameters in Eq. (2).

3.2.2. X-ray Flux, I_0

Generally XPS quantitation has been limited to ratioing elements found in a particular sample. Therefore, most quantitation schemes normally assume that the X-ray flux (I_0) is a constant. It is often necessary, however, to compare the spectra of samples obtained at different times on the same instrument or even on different instruments. Given proper allowances for instrument variations and for the values of the variables in Eq. (2), absolute quantitative comparisons can be made.[29]

Although the X-ray flux in most instruments can be assumed to be constant during a limited observation time, such as several hours, it is certainly not constant on a longer time scale of the order of weeks to months. X-ray intensity is usually maximum shortly after the X-ray anode has been cleaned or refurbished and then decays with time by some function which is related to the design characteristics of the anode, the operating parameters selected by the user, and the state of the vacuum in the chamber containing the anode.

Our lab performs a routine set of instrument diagnostics once each week, using a set of standard samples (gold, copper, elemental carbon, and polytetrafluoroethylene). In the event that spectra from different samples obtained at widely varying times on an instrument may need to be compared, it is useful for laboratories to plot the intensity decay of their X-ray source.

For the remainder of this paper we will assume we are interested in quantitative XPS of samples whose spectra were obtained at approximately the same time; therefore, we assume that the X-ray flux, I_0, is constant.

3.2.3. Differential Photoionization Cross Section, $\sigma_{i,k}$

$\sigma_{i,k}$ is the differential photoionization cross section and consists of two parameters [Eq. (3)]; the total photoionization cross section, calculated and tabulated by Scofield,[27] and an angular asymmetry factor which has been discussed in detail by Reilman *et al.*[34]:

$$\sigma_{i,k}(h\nu, \phi) = \sigma_{i,k}(h\nu) \cdot F_{i,k}(\beta, \phi) \tag{3}$$

The first term of Eq. (3) is the Scofield *total* photoionization cross section, which is a function of X-ray photon energy, $h\nu$. The second term is given below and consists of an angular term and a parameter, β, which is a characteristic of atomic orbital and atomic number.[34]

$$F_{i,k}(\beta, \phi) = 1 + \beta/2\,(3/2 \sin^2 \phi - 1) \tag{4}$$

ϕ in Eq. (4) is the angle between the X-ray photons and the emitted photoelectrons. This normally is a constant in any particular instrument but varies among different instrument types.[29]

By use of the Scofield tabulations and the Reilman *et al.*[34] tabulations for β and knowing ϕ for the particular instrument, σ can be calculated for any particular element and orbital. The ϕ value for the Hewlett-Packard instrument is 73°. Values for other instruments are available.[29]

3.2.4. Instrument Transmission or Throughput Function, $T_{i,k}$

Instrument transmission functions have been considered by Seah[29] and others[33,67] with reference to XPS quantitation. Many instruments with a pretardation analyzer have an $(E_{\text{kin}})^{-1}$ transmission function. The transmission function for a particular instrument is usually available from the manufacturer or can be found in the literature [see Reference (67)].

Using the Scofield cross sections,[27] the asymmetry parameter β,[34] and the instrument transmission function, one can calculate an overall sensitivity factor for each element and orbital for each instrument used. Such information is now tabulated for the Hewlett-Packard instrument[82] and is becoming available from manufacturers of currently sold instruments. It is important to point out, however, that one cannot directly use these sensitivity factors for quantitation because the remaining term in Eq. (2), the mean free path (λ), has not yet been considered.

3.2.5. Mean Free Path, $\lambda_{i,k}$*

Photoelectron mean free paths and their kinetic energy dependencies have been reviewed recently by Seah and Dench[35] and by Wagner *et al.*[36] An earlier theoretical study by Penn[37] developed mean free path functions for free electron elements. We have previously presented experimental data for mean free paths in barium stearate monolayers for kinetic energies from approximately 270 to 1500 eV.[38] There is considerable controversy in the literature regarding experimental values for mean free paths and in the theoretical development of such values.[39,40] Wagner *et al.*[36] have shown that most of the available experimental data can be interpreted as the power function:

$$\lambda_{i,k} = CE_{\text{kin}}^{m} \tag{5}$$

where C and m are functions of the solid material. They have shown that for kinetic energies greater than 300 eV, m is approximately 0.75 for inorganic solids and is in the range of 0.7 to 1.0 for organics. A theoretical model recently developed by Ashley[69-72] provides very reasonable mean free paths for organic polymers (see Section 7.2 for a detailed discussion).

Given appropriate values for σ [Eq. (3)] and for the instrument transmission function, coupled with reasonable estimates of mean free path dependencies on kinetic energy for the sample of interest, one can perform a rigorous quantitative elemental analysis.

As the kinetic energy dependence of λ is $\sim E_k^{1.0}$ for organics, and the transmission function of some instruments is $\sim(E_k)^{-1}$, the two functions roughly cancel. It has been shown that these two parameters approximately cancel over the effective binding energy range of 0–600 eV for the Hewlett-Packard instrument.[38] This fact provides some justification for the semi-quantitative elemental analysis procedure discussed earlier. This is a very rough cancellation, however, and for rigorous quantitation it cannot be assumed. Therefore, rigorous XPS quantitation requires that one have mean free path values for the particular sample.

3.2.6. Measured Intensities, $N_{i,k}$

The problem now is how to obtain the intensity of the photoelectron peak. For organic polymers with relatively weak loss processes, this is relatively straightforward as the peaks are generally symmetrical and baselines are relatively straight; determining the peak area is rather simple. Of course, all satellites and related structures due to relaxation effects

* Discussed in more detail in Section 7.2.

(discussed in Section 4.0) must be included in the area of the photoelectron peak. The high-energy tails due primarily to inelastic loss processes are, of course, accounted for through the mean free path function and should not be included in the area of the main peaks. In some systems it is difficult to separate the relaxation and the inelastic loss contributions to the peak; in these systems it is difficult to take relevant areas of high accuracy. This problem has been discussed for the case of metallic systems by Wertheim and Hufner.[42] The problem of the appropriate baseline function to apply for determining the area of the peak has been discussed.[29,43,44] Our work and that of others with organic polymers suggests that a sloping straight-line baseline is usually adequate for such materials.[43-45]

The total photoionization cross section, tabulated by Scofield,[27] varies by approximately a factor of 20 across the periodic table and the various orbitals. Scofield's values are only good for electrons whose kinetic energies are greater than about 500 eV. Thus for spectra derived from Al Kα instruments, lines with binding energies greater than 1000 eV should not be used (750 eV for Mg Kα instruments).

Variation in the asymmetry parameter is also a function of atomic number and orbital. Thus s shells are often used for quantitation as they have a constant asymmetry factor, which cancels out during ratioing. Ignoring the asymmetry parameter term can lead to errors in the differential photoionization cross section term as large as 10–20%, depending on which orbitals are being ratioed. The term in the quantitation expression with the greatest uncertainty, however, is the photoelectron mean free path, λ.

Generally quantitation means ratioing the elements present in the sample. It is not possible to determine an absolute surface concentration due to uncertainties in I_0, sample area illuminated, mean free path values, etc. By ratioing, such common terms as X-ray intensity cancel out. If the elements present have peaks of fairly similar binding energy, these should be used for quantitation because then the mean free path and instrument throughput functions, which are functions of kinetic energy, will tend to cancel.

We have assumed that the sample is absolutely homogeneous for this quantitation discussion. The sample is not homogeneous if it contains overlayers or shows compositional segregation at the surface; in such cases the analysis is more complex (see Section 7).

In general if one does as rigorous a quantitation as is possible, considering all the terms discussed above, uses samples which are homogeneous and reproducible, and obtains their spectra under nearly identical conditions, quantitative XPS on an elemental ratio basis is good to better than ±5%. If, however, one uses only the Scofield total photoionization cross section as a sensitivity factor, the quantitation is probably no better than ±10% and perhaps even ±20%.

Modern computer-based XPS instruments can be expected to have built into their software the differential photoionization cross sections, the instrument throughput function, and possibly some appropriate mean free path functions for standard instrument operating conditions.

3.2.7. Quantitative Analysis Summary

1. Find ϕ for your particular instrument.
2. Find the instrument response function, T. Is T proportional to $(E_k)^{-1}$? This information can usually be obtained from the instrument manufacturer or supplier.
3. Using β, σ, and ϕ, calculate a cross section or sensitivity factor for each element and orbital.
4. Determine or estimate λ for your particular samples (see Section 7).
5. Obtain spectra under nearly identical conditions.
6. Measure peak areas including satellites, chemically shifted peaks, etc., but do not include extrinsic loss backgrounds.
7. Ratio the peaks present.

4. CHEMICAL BONDING

4.1. CHEMICAL SHIFTS

High-resolution spectra of any particular photoelectron line are easily obtained, generally using high-resolution (20 eV wide) scans. Figures 6a and 7b showed high resolution scans of the C-1*s*, O-1*s*, and Au-4*f* regions, respectively. Such scans usually have much better signal:noise ratios than survey scans (1000 eV wide scans), because one spends the time accumulating data only over a narrow energy region. Normally we use a wide or survey scan to identify the elements and then high-resolution scans to obtain peak positions, chemical states, and peak areas for quantitation.

Referring to Figure 6a, the C-1*s* peak consists of three major features. The major carbon constituent in poly(methyl methacrylate) is alkyl carbon, the major peak at the right in Figure 6. There are two additional features to the left of the main peak. These are due to the carbon atoms bonded to oxygen. If one examines the oxygen 1*s* region, Figure 6a, we see that the oxygen peak consists of two components, which represent the double- and single-bonded oxygens.

It is clear from many studies including the pioneering studies by Siegbahn *et al.*[(5)] that there is a chemical shift effect in XPS. Even though the C-1*s* and O-1*s* core electrons are not directly involved in chemical bonding, the valence electrons' bonding environment affects the overall

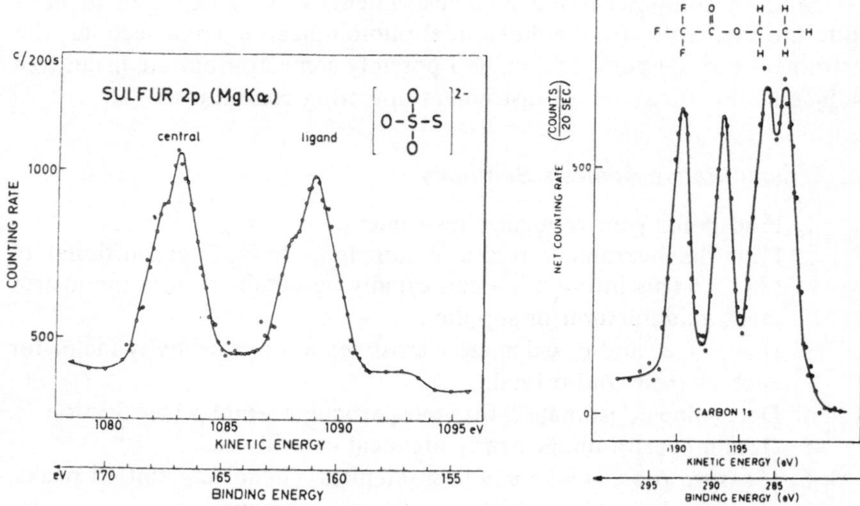

FIGURE 9. Demonstrations of the chemical shift effect originally presented by Siegbahn.[5] *Left*: sulfur region of sodium thiosulfate, showing the large chemical shift in the sulfur lines due to sulfur atoms present in the compound in two greatly different binding environments. *Right*: carbon 1*s* region of ethyl trifluoroacetate. Here we see carbons in alkyl, ether, ester-like, and fluorinated environments, with very significant chemical shifts. [Reprinted from Reference (5), pp. 15, 17.]

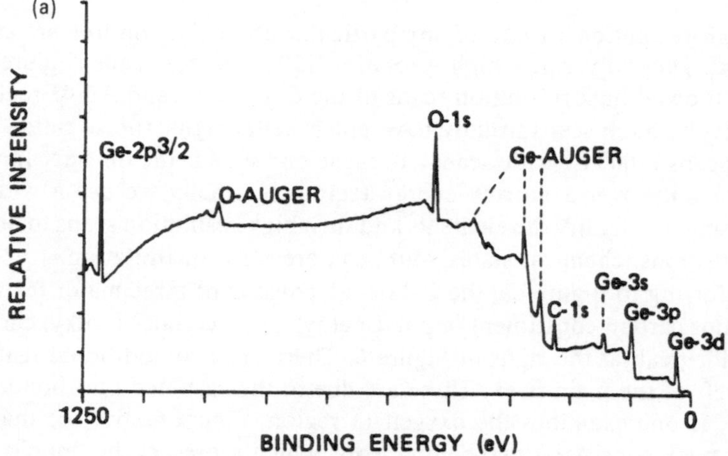

FIGURE 10. (a) Wide scan of clean, oxidized germanium showing the 3*d* through 2*p* orbitals of the element (note the presence of the *LMM* Auger lines). (b) High resolution scans of each of the orbitals demonstrates an oxide-shifted component with the same binding energy shift (about 3.5 eV) for each of the orbitals, confirming that all of the core levels are equally influenced by the valence electron environment. Note also the fact that the ratio of metal to oxide component in each of the peaks changes with binding energy due to the influence of kinetic energy on mean free paths, as discussed in the text.

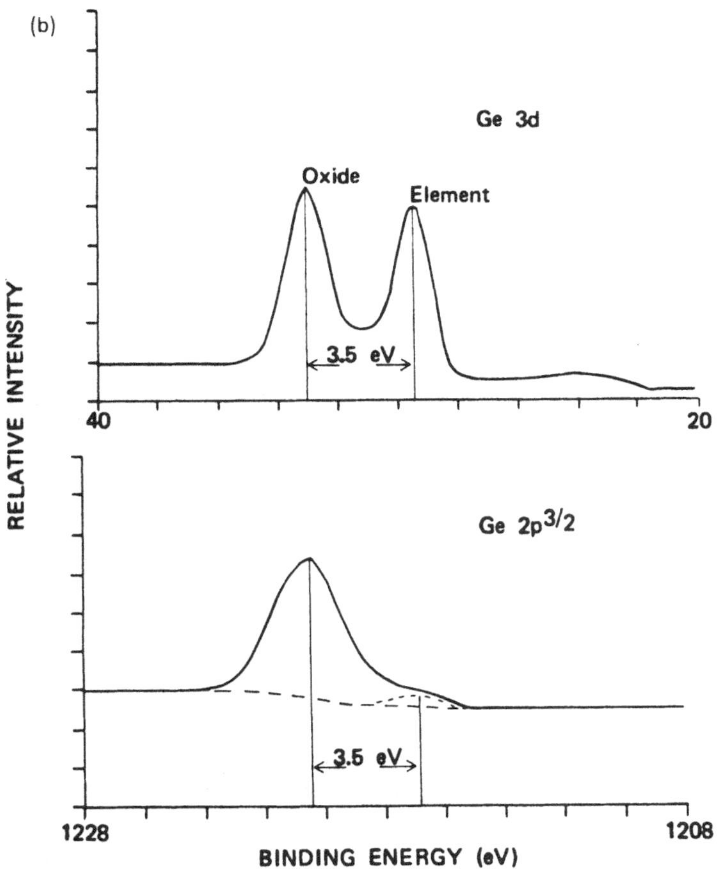

FIGURE 10 (continued)

electrostatic interaction in the atom. Changes in the valence environment result in a change in the binding energy of all of the inner core electrons. Although this chemical shift effect is seen for all atoms in the periodic table, it is particularly pronounced for those elements of particular interest to polymer chemistry and biochemistry, i.e., carbon, oxygen, sulfur, nitrogen, and phosphorus. Thus XPS is of particular utility for examining surface chemistry and bonding of polymers and biologicals.

Figure 9 presents two very classical spectra which are commonly used to demonstrate the chemical shift, both taken from the work of Siegbahn[(5)]—the spectrum of sodium thiosulfate, in which the sulfur is present in two very different environments, and that of ethyl trifluoroacetate, in which the carbon is present in a variety of bonding environments. The chemical shift effect is also readily observed with many pure metal samples containing a thin metal oxide film, such as germanium (Figure 10), silicon, or aluminum (Figure 11).

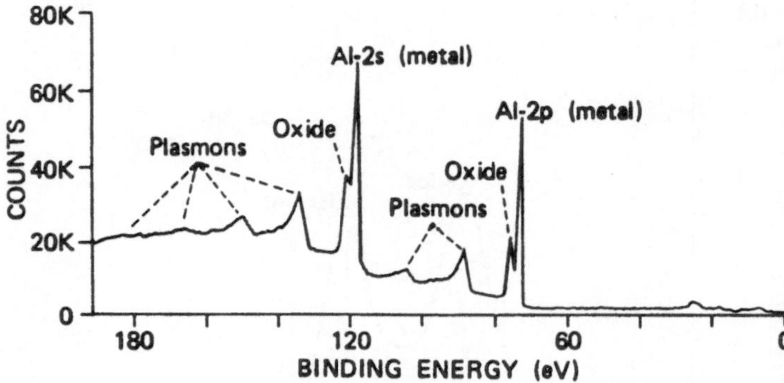

FIGURE 11. A spectrum of aluminum showing the chemically shifted oxide peaks and bulk plasmon inelastic loss features [from Reference (3), p. 5.]

A qualitative explanation for the chemical shift effect is that atoms bound to highly electronegative species, such as carbon bound to oxygen, result in the carbon electrons being attracted to the oxygen atom due to the electronegativity difference. Therefore there is less negative charge to interact with the carbon nucleus, resulting in less electrostatic shielding of the carbon 1*s* electron, meaning the carbon 1*s* electron is more strongly attracted to the nucleus, and thus a higher carbon 1*s* electron binding energy is observed. Conversely an electropositive or electron-donating ligand would contribute electron density to the carbon, resulting in increased shielding or increased competition for the nuclear charge and therefore a decreased 1*s* electron binding energy. Although this explanation is very simplistic, it appears to be qualitatively correct. More rigorous theoretical models of the chemical shift effect are readily available and have been successful in predicting binding energies and chemical shifts in polymers and many other species.[3-8]

It is generally found that the core electron binding energy is influenced by the local chemical bonding environment and increases with oxidation state. The shifts are relatively large and can be comparable to those of chemical reaction energies. A 1-eV shift corresponds to about 23 kcal/mole. Figure 12 compares the shifts for a polyester, a poly(ethylene oxide), and an anthraquinone. The carbon-oxygen shifts due to single bonded oxygen, carbonyl oxygen, and the ester oxygens are clearly evident.

By studying a range of model compounds of known structure and bonding types, one can develop a correlation chart for each element analogous to the chemical shift correlation charts developed for infrared, Raman, and NMR spectroscopy. Reference (9) is the best source of such correlation charts; see also References (5)-(7). The best source of information for chemical shifts of polymers is the extensive set of data obtained

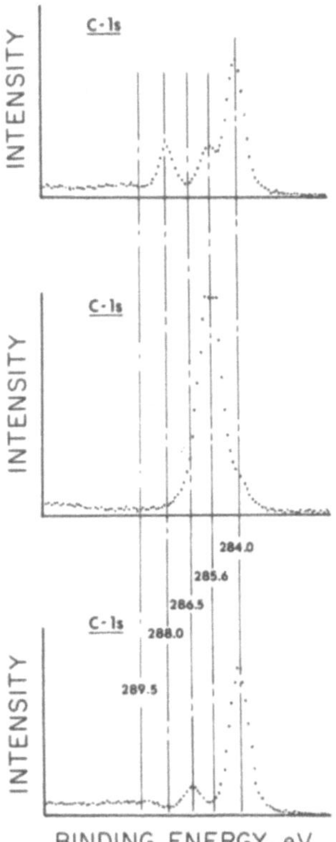

FIGURE 12. Carbon 1*s* regions of a series of carbon-oxygen materials. *Top*: poly(ethylene terephthalate) (Mylar polyester). *Center*: poly(ethylene oxide), a pure polyether material. *Bottom*: anthraquinone. Note the ether, quinone carbonyl, and ester chemical shifts of 1.6, 2.5, and 4.0 eV, respectively. [Reprinted from Reference (97), p. 63, by permission.]

by Clark and Thomas(21,22) which has been presented nicely by Dilks(14) (Figure 13). Figure 14 is a brief correlation chart taken from our own work. A particularly useful set of group chemical shifts is given by Carlson.(7)

The germanium spectrum (Figure 10a) shows the expected 2*p*, 3*s*, 3*p*, and 3*d* peaks. Oxygen and carbon are also present. If one examines each of the major elemental lines under high resolution, one finds that each of the lines are split. There is a 3.5-eV higher energy feature due to those germanium atoms bonded to oxygen in the GeO_2 oxide overlayer. The chemical shift is identical for the 2*p*, 3*s*, 3*p*, and 3*d* orbitals, suggesting that all of the inner core levels feel the same influence of the valence electrons. This can be rationalized via an electrostatic sphere model.

The fact that germanium has both high and low binding energy electrons, the 2*p* and 3*d*, leads to an interesting result. Note in Figure 10b that the intensities of the metal and metal oxide peaks are different for the 2*p* and 3*d* regions. The reason for this has to do with the mean free path of the photoelectrons. The low binding energy or high kinetic energy 3*d* electrons travel farther in matter than the high binding energy, low kinetic

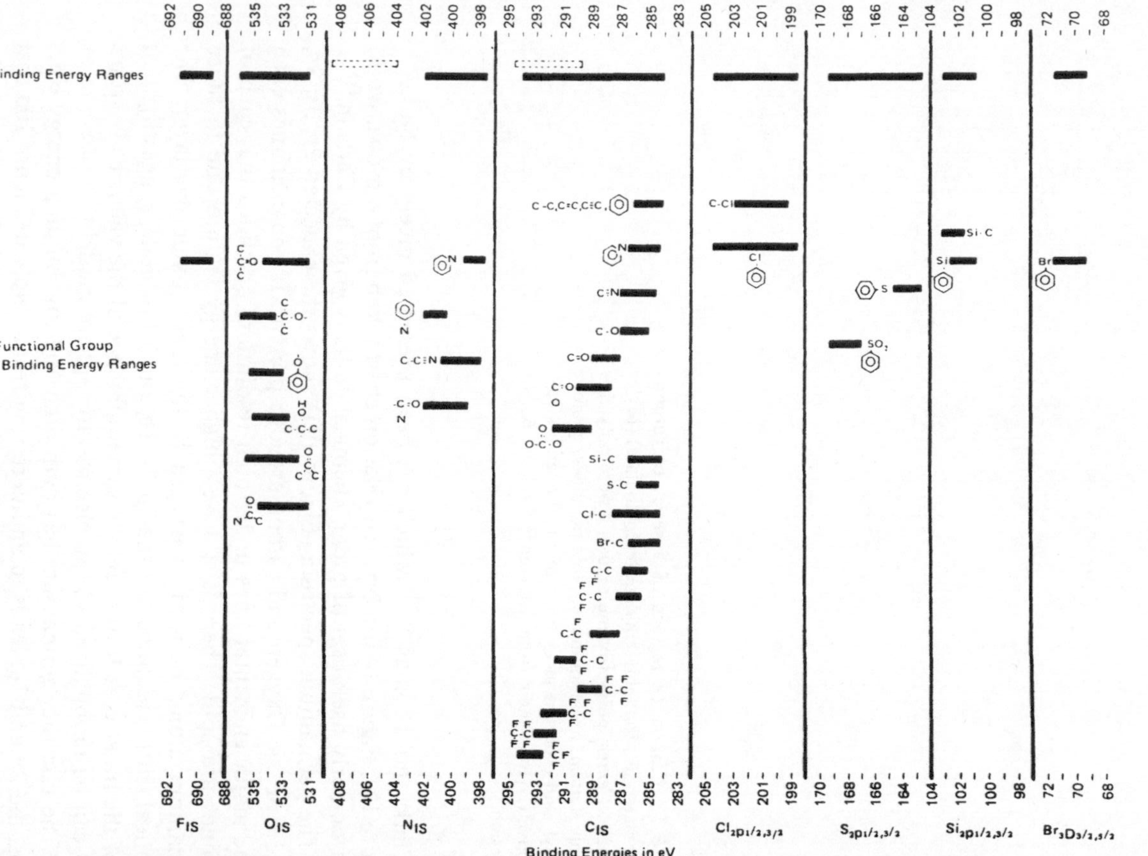

FIGURE 13. Summary of chemical shifts found in polymers. The black bar shows the range of binding energies found in a variety of polymers for the functional group given. The range is in part due to the effect of other substitutents on the binding energy of the group in question. These charts are based on the method of presentation in Reference (9). Data taken from References (21), (22), and (14) and from unpublished data.

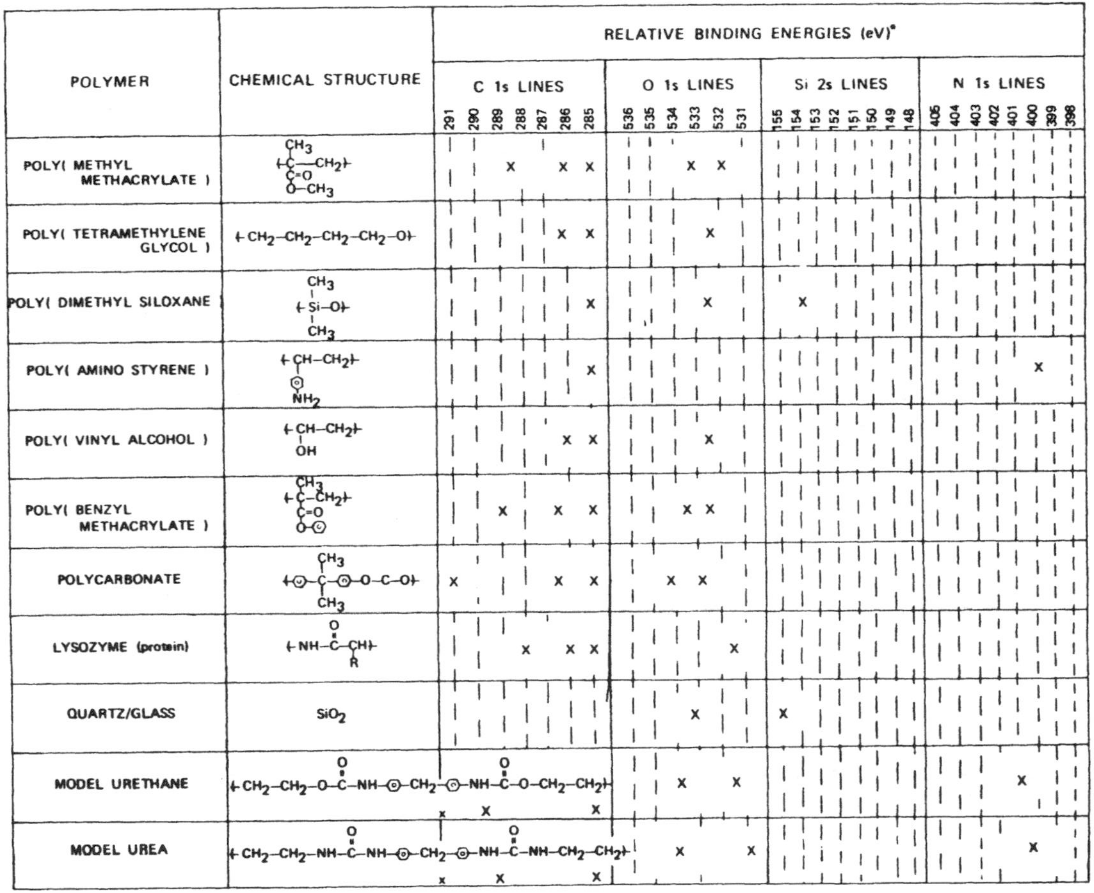

POLYMER	CHEMICAL STRUCTURE	RELATIVE BINDING ENERGIES (eV)*																												
		C 1s LINES							O 1s LINES						Si 2s LINES								N 1s LINES							
		291	290	289	288	287	286	285	536	535	534	533	532	531	155	154	153	152	151	150	149	148	406	404	403	402	401	400	399	398
POLY(METHYL METHACRYLATE)	$-(C(CH_3)(COOCH_3)-CH_2)-$			X			X	X				X	X																	
POLY(TETRAMETHYLENE GLYCOL)	$-(CH_2-CH_2-CH_2-CH_2-O)-$						X	X					X																	
POLY(DIMETHYL SILOXANE)	$-(Si(CH_3)_2-O)-$							X					X			X														
POLY(AMINO STYRENE)	$-(CH(C_6H_4NH_2)-CH_2)-$							X																				X		
POLY(VINYL ALCOHOL)	$-(CH(OH)-CH_2)-$						X	X					X																	
POLY(BENZYL METHACRYLATE)	$-(C(CH_3)(COOC_6H_5)-CH_2)-$			X			X	X				X	X																	
POLYCARBONATE	$-(C_6H_4-C(CH_3)_2-C_6H_4-O-CO-O)-$	X					X	X			X	X																		
LYSOZYME (protein)	$-(NH-CO-CHR)-$				X		X	X						X																
QUARTZ/GLASS	SiO_2											X			X															
MODEL URETHANE	$-(CH_2-CH_2-O-CO-NH-C_6H_4-CH_2-C_6H_4-NH-CO-O-CH_2-CH_2)-$	x	X					X			X			X													X			
MODEL UREA	$-(CH_2-CH_2-NH-CO-NH-C_6H_4-CH_2-C_6H_4-NH-CO-NH-CH_2-CH_2)-$	x		X				X			X			X														X		

*(small 'x' refers to the $\pi-\pi^*$ satellite peak)

FIGURE 14. A correlation chart for carbon, nitrogen, oxygen, and silicon in polymers and related materials taken from our own work. (See also Table 10.)

energy, $2p$ electrons. The $2p$ electrons of the pure germanium substrate are more affected by the thin oxide overlayer than are the $3d$ electrons. Thus, the oxide to element ratio for the $2p$ is greater than for the $3d$. As we shall see later, this effect can be useful in estimating the thickness of overlayer films.

Constant features present on the high binding side of each major peak may be plasmon loss structures, due to emitted photoelectrons which excite the crystal lattice (Figure 11). The crystal lattice excitations require discrete energies; thus the emitted photoelectrons which have excited these processes have a lower kinetic energy and thus appear on the spectrum at an increased binding energy. The lattice excitation is called a plasmon excitation or plasmon resonance, and result in plasmon loss structures or peaks in the XPS spectrum.[3]

4.2. ENERGY REFERENCING

Because the chemical shifts are not very large, it is important to carefully charge reference the spectra for chemical shift analysis. This is less important for simple elemental analysis because most of the photoelectron lines of the elements are clearly resolved and separated. The energy referencing problem is a particularly simple one in gas phase XPS, where the reference level is the vacuum level, defined as zero energy, resulting in the basic photoelectric effect equation described earlier [Eq. (1)]. Here the equation is given in slightly different notation, where the binding energy, E_B, is shown referenced to the vacuum level, E_B^V

$$E_B^V = h\nu - E_k \tag{6}$$

The situation is considerably different in the case of a solid sample which can have its own macroscopic electrical characteristics. Assuming the solid sample is a conductor, it can be shown[3] that

$$E_B^F = h\nu - E_k - \phi \tag{7}$$

where E_k is the kinetic energy seen by the spectrometer, ϕ is the work function of the spectrometer surface (the energy required to remove the most loosely bound electrons, which are located at the Fermi level), and E_B^F is the core electron binding energy referenced to the Fermi level of the sample. In solid state physics it is usual to reference the binding energies of electrons in solids to the work function or the Fermi level. The most loosely bound electrons in solids are said to reside at an energy level called the Fermi level. Therefore

$$E_B^V = E_B^F + \phi_s \tag{8}$$

The work function of the spectrometer, ϕ, is in general different from that of the sample, ϕ_s. Since the spectrometer and the sample are electrically connected, electrons will flow from one to the other until the Fermi positions are equal. In fact, a Fermi energy in a solid is essentially the chemical potential of the electrons in that solid. Electrons flow until the Fermi levels (chemical potentials) are equal. Equation (7) is a convenient result because it only requires the work function of the spectrometer, which is known. The work function of the sample is, in general, not known.

In the gas state there is no work function, because it is a solid state property and has no meaning for a single molecule. The same molecule in a condensed solid state experiences valence band broadening and the associated energy bands and energy levels, and has a Fermi level and a work function. Therefore, binding energy values referenced to the vacuum levels, such as in gas phase XPS, and to the Fermi level, as in solid XPS, are different by the work function of the solid, generally somewhere between 3–6 eV or more. There are also relaxation effects in the solid state which can significantly affect the deduced binding energies.

Generally, the work function of the spectrometer surface is well-known and most instruments have a calibration mechanism built into the electronics to permit one to dial in the appropriate spectrometer work function. Normally this is done by measuring a highly clean gold sample and recording the gold $4f^{7/2}$ line, which should appear at 83.8 eV (Figure 7b). Energy referencing is required in insulators due to charging effects, discussed earlier in Section 2.4.

4.3. INITIAL AND FINAL STATES

The sample is in some initial state prior to the XPS experiment. It is then bombarded by a beam of photons which perturb that state, causing photoelectron emission and leaving the atom in an unstable final state which must relax back to the ground state. For example, a $1s$ electron vacancy results in a $2s$ or $2p$ electron falling into the $1s$ shell, giving off energy in the process.

The photoemission event itself is extremely rapid and occurs in perhaps 10^{-16}–10^{-14} seconds. This is so rapid that a theorem of theoretical chemistry, Koopmans' theorem, is often invoked, which says that in that period of time we can assume that the core electrons are frozen in position, i.e., no relaxation occurs. The binding energies given in Table 3 all assume Koopmans' theorem is valid. The atom can relax during the process of photoemission or can relax after the photoemission event is completed. If it relaxes during the photoemission process then the relaxation energy will influence the kinetic energy of the emitted photoelectron and, therefore, its deduced binding energy. Such relaxation processes in polymers will be discussed later. Relaxation processes occurring during photoemission result in the emitted photoelectron having a decreased kinetic energy and, therefore, a higher deduced binding energy, resulting in peaks or features on the high binding energy side of the main unrelaxed photoelectron peak; such features are often called relaxation peaks or satellite peaks. These are very important in many classes of compounds, including polymers.

If the atom relaxes after the photoemission process, it usually does so by two mechanisms. Electrons from higher orbitals will fall into the lower now unoccupied orbital. In doing so, they must release energy, such as characteristic X-rays for that particular atom. This is the basis of X-ray fluorescence analysis, commonly done in electron microprobes or scanning electron microscopes equipped with X-ray fluorescence analytical equipment.

4.4. AUGER PROCESSES

Another way for the atom to relax is that, in the process of an outer orbital electron falling in to populate an inner unfilled orbital, the excess energy is transferred to another outer orbital electron, resulting in emission of that electron from the atom. This radiationless electron emission process is called the Auger (rhymes with Roget) process and such electrons are called Auger electrons.

In the range of atomic orbitals commonly examined by XPS, the dominant relaxation mode tends to be Auger emission, especially for polymers.[9] The Auger electron has a discrete kinetic energy and is detected by the analyzer in exactly the same manner as a photoelectron. Figure 15 presents a schematic view of the Auger emission process. Table 9 lists most of the key Auger peaks as they appear on an Al $K\alpha$ photoelectron spectrum [derived from Reference (9)].

Auger electrons are characterized by their kinetic energy, in contrast to photoelectrons, where binding energy is the quantity of interest. However, a typical XPS instrument is programmed to assume that all electrons are being produced by X-ray photon interaction (1487 eV in the case of an Al $K\alpha$ instrument). Thus, the computer automatically subtracts the binding energy and work function from the photon energy and deduces a "binding energy" for the Auger electron as well as for the photoelectrons. However, if the instrument were utilizing Mg $K\alpha$, the computer would be programmed to consider the photon energy at 1254 eV. The Auger electron would have the same kinetic energy, but the "binding energy" calculated would now be lower by 233 eV. Thus, the position of Auger peaks on different instruments may be different depending on the X-ray source used. This is strictly an artifact of the instrument. The Auger electron can only be characterized by its absolute kinetic energy and not by any "binding energy."

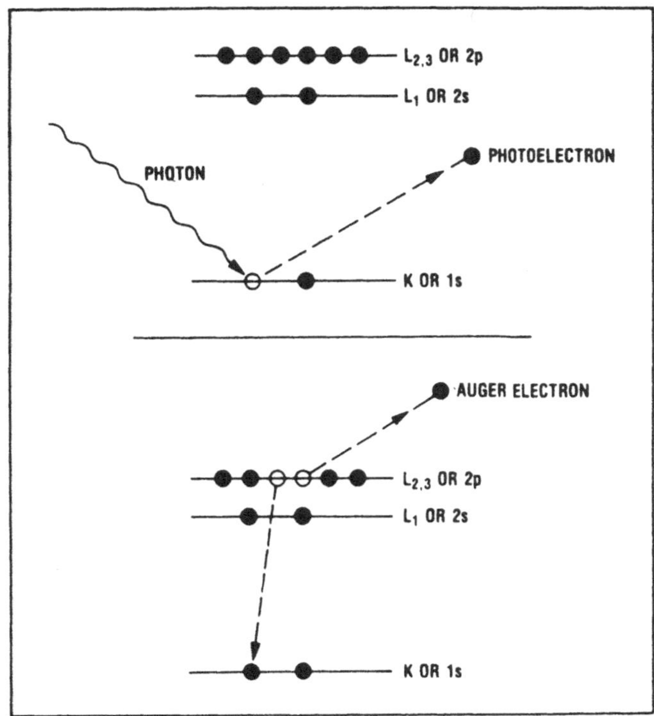

FIGURE 15. Comparison of the Auger and photoelectron emission processes and the atomic orbital basis for the nomenclature commonly used for Auger electrons. [Reprinted from Reference (9) p. 5, by permission.]

Auger electrons can be very useful with certain elements for quantitative analysis and, when utilized appropriately, for chemical shift and bonding analysis.

Figure 15 gives the nomenclature for Auger electrons (see also Tables 6 and 9). Auger emission is a more complex event than photoelectron emission; an Auger band usually consists of a series of lines of different kinetic energies reflecting the complexity of the Auger emission process. Certain *KLL* transitions are particularly well-resolved and intense; many of the other Auger lines are generally broad or so complex as to not be very useful.

The sodium *KLL* emission is particularly strong, sharp, and well-resolved in XPS and is often used for the detection of sodium, because the sodium $2p$ and $2s$ peaks are relatively insensitive and the sodium $1s$ peak is very broad and rides on a high background, making is somewhat difficult to detect. The magnesium *KLL* Auger band is also very useful and relatively easy to detect for similar reasons.

TABLE 9

Auger Peaks in AlKα Photoelectron Spectra (from Ref. 9)

Element	Atomic No.	Range (eV)	Auger Lines						
					KL_1L_1	KL_1L_{23}	$KL_{23}L_{23}$ c)		
Li	3								
Be	4								
B	5						1315		
C	6						1226		
N	7	12					1108		
O	8	10			1012	997	976		
F	9	8			878	859	832		
Ne	10	0			724	701	668		
Na	11	4			565	536	497		
Mg	12	8			384	350	305		
Al	13								
Si	14								
P	15								
			$L_3M_{23}M_{23}$ d)	$L_2M_{23}M_{23}$ d)	$L_3M_{23}M_{45}$ ^1P	$L_3M_{23}M_{45}$ ^3P	$L_2M_{23}M_{45}$ ^1P	$L_3M_{45}M_{45}$ e)	$L_2M_{45}M_{45}$
S	16	6	1336						
Cl	17	4	1304						
Ar	18	0	1270	1268					
K	19		1238	1236					
Ca	20		1197	1194					
Sc	21	8	1153		1125				
Ti	22	10	1106		1072				
V	23	6	1055		1017				
Cr	24		1000		962				
Mn	25	6	948		903			853	
Fe	26	6	892		841			786	
Co	27	7	837	830	779	774		716	701
Ni	28	7	781	775	715	709		643	626
Cu	29	5	719	712	649	641	629	570	550
Zn	30	7	662	655	585	576	562	498	475
Ga	31	7	601	594	517	508	490	422	395
Ge	32	10	538	530	448	438	417	346	315
As	33	11			376	365	341	266	132
Se	34	8			300	288	259	184	143
Br	35								
Kr	36								
Rb	37								
Sr	38								
Y	39								
			$M_{45}N_{23}V$	$M_5N_{45}N_{45}$	$M_4N_{45}N_{45}$	$M_{45}N_{45}V$	M_5VV	M_4VV	
Zr	40								
Nb	41		1321	1289					
Mo	42		1301	1266					
Tc	43		1280	1241					
Ru	44	4	1258	1214					
Rh	45	5	1235	1187					
Pd	46	4	1212	1161					
Ag	47	4		1136	1130				
Cd	48	5		1112	1105				
In	49	7		1086	1079				
Sn	50	7		1060	1052				
Sb	51	10		1036	1027				
Te	52	6		1008	998				
I	53	4		981	970				
Xe	54	4		957	944				
Cs	55			931	917				
Ba	56			904	890				
La	57			865					
Ce	58			827					

TABLE 9 (*continued*)

Element	Atomic No.	Range (eV)	Auger Lines $M_{45}M_{23}V$	$M_5N_{45}N_{45}$	$M_4N_{45}N_{45}$	$M_{45}N_{45}V$	M_5VV	M_4VV
Pr	59			788				
Nd	60			752				
Pm	61			714				
Sm	62			673				
Eu	63			635				
Gd	64			595				
Tb	65			568		426	265	235
Dy	66			527		375	195	155
Ho	67			490		325	142	100
Er	68			454		273	99	56
Tm	69							
Yb	70							
Lu	71							
Hf	72							
Ta	73							
W	74							
Re	75							
Os	76							
				$N_7O_{45}O_{45}$	$N_6O_{45}O_{45}$	$N_{67}O_{45}V$		
Ir	77							
Pt	78			1425				
Au	79			1417				
Hg	80			1409	1406			
Tl	81			1402				
Pb	82			1395	1392			
Bi	83			1388	1384			
Th	90			1333		1238		
U	92			1297		1203		
Np	93							
Pu	94							
Am	95							
Cm	96							
Bk	97							
Cf	98							

$1 = S_{1/2}$, $2 = P_{1/2}$, $3 = P_{3/2}$, $4 = D_{3/2}$, $5 = D_{5/2}$

[a] Lines enclosed in boxes are the most intense and are the most suitable for use of line energies in identifying chemical states.

[b] For brevity, $2p_3$ equals $2p^{3/2}$, $3d_5$ equals $3d_{5/2}$, etc.

[c] Includes KVV designation when L_{23} is not a core level.

[d] Designation is oversimplified.

[e] Includes LVV when M levels are not in core, and MVV when N levels are not in core.

The Auger kinetic energy is a reflection of the bonding environment just as are photoelectron binding energies. Wagner has noted that the Auger chemical shift in many cases can be larger than the photoelectron shift.[9,46,92-95] The final Auger state is doubly charged. The influence of screening and polarization on the measured energy is different for Auger and for photoelectron lines. Wagner's Auger Parameter is basically the difference between the photoelectron and Auger electron kinetic energies. One can correlate Auger parameter differences with different binding environments. A particular advantage of the Auger Parameter which Wagner has formulated is that the charging shift cancels out. Wagner *et al.*[46] have developed two-dimensional chemical state plots which greatly aid in differentiating the binding properties of many elements. Polymer applications of the Auger Parameter and two-dimensional state plots have been rather limited. These approaches should be helpful, particularly for oxygen and silicon,[93,95] where the photoelectron chemical shift is small or difficult to interpret.

One must also be concerned with mean free path effects in the use of Auger lines (see Section 7).

4.5. SHAKE-UP SATELLITES

One particularly important set of relaxation effects observed in polymers is the shake-up and shake-off phenomenon, illustrated in Figure 16. The right side of the figure illustrates normal photoionization, that is, the frozen core situation. The photoionization event results in a 1*s* electron being kicked out of the atom. The majority of photoionization events in organic molecules are of this type. However, under certain situations, some fraction of other final states may be observed. For example, the center of Figure 16 shows a 2*p* electron which has been promoted into a 3*p* level during the process of photoionization. The 2*p* electron is said to be "shaken-up" into the 3*p*. This takes a discrete amount of energy and that energy is

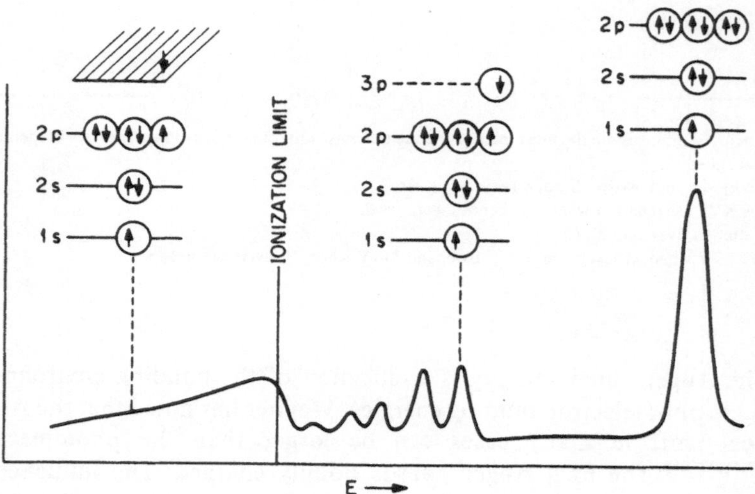

FIGURE 16. Shake-up and shake-off relaxation processes observed in photoelectron spectroscopy. These are commonly referred to as shake-up satellites and are always found on the high binding energy side of the main unrelaxed photoelectron line. [Reprinted from Reference (7), p. 242, by permission.]

basically taken from the kinetic energy of the 1*s* photoelectron which was emitted. The energy difference between the normal photoelectron peak, process at far right, and the shake-up peak at center is said to be the shake-up energy. The band resulting from this process is said to be a shake-up satellite of the main peak. One can envision shake-up processes into higher and higher normally unoccupied orbitals, each with a discrete energy; finally the electron could also be shaken free from the atom; then it is said to be shaken-off. It could, of course, be shaken-off with an increment of kinetic energy. Figure 16 diagrams the whole process in schematic terms. We see that there is a continuum of shake-off energies, a threshold for the shake-off process, and a series of shake-up satellites which can be present. Shake-up features are particularly easy to observe in gas phase XPS.[6]

Shake-up processes in synthetic polymers have been studied extensively by Clark, Dilks, and coworkers.[47-49] Their theoretical and experimental studies have shown that reasonably intense and useful shake-up structures are only observed from polymers which either have an unsaturated backbone or side chains with aromatic or other unsaturated character. The shake-up processes observed are largely due to $\pi-\pi^*$ molecular orbital transitions accompanying the core ionizations. This is a particularly important observation because the C-1*s* chemical shift between alkyl and unsaturated or aromatic bonds is relatively small and difficult to detect by XPS under normal conditions. The shake-up satellite, however, can be used as a diagnostic for the presence of unsaturated material at the polymer surface. There is a strong dependence of the shake-up intensity on substituent in *para*-substituted polystyrenes. A good summary of the available work to date on unsaturated shake-ups in polymers is available in the review by Dilks.[14] Table 10 gives the shake-up energy and intensity for a variety of polymers with unsaturated bonds (see also Figure 14). Figure 17 presents the C-1*s* shake-up spectrum for an unsaturated polymer, poly(benzyl methacrylate). The shake-up intensity (usually given as a percentage of the parent C-1*s* peak) and shake-up energy is highly dependent on polymer molecular structure (Table 10).

More data on other unsaturated systems is needed before shake-up features can be widely used to diagnose unsaturated character in polymer surfaces. Shake-up satellites are often difficult to observe immediately adjacent to the main peak because of other relaxation processes and final state effects; shake-ups more than 10 eV from the main peak are also difficult to see because they are largely masked by the tail from inelastic scattering events. Shake-up intensities rarely exceed 20% of the intensity of the parent peak, with 0 to 10% being more common.

Relaxation processes in XPS can be modeled from two sources, intramolecular or atomic relaxation and an interatomic term that includes all nearest neighbors [see Reference (3)]. Pireaux[50] has shown significant

TABLE 10

Core Level Binding Energy, Mean Shake-up Energy, and Total Shake-up Intensity as Percentage of the Mean Peak for a Range of Aromatic and other Unsaturated Polymers[a]

Polymer name	Chemical structure	$\pi-\pi^*$ as percent of aromatic component of total C-1s	Binding energy of $\pi-\pi^*$ (eV)	Binding energy (eV) C-1s 292	291	290	289	288	287	286	285	O-1s 534	533	532
Polystyrene	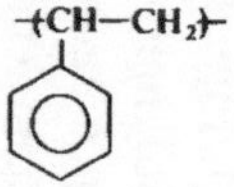 $-(CH-CH_2)-$ with phenyl ring	8.8%	291.7								x			
Poly(vinyl toluene)	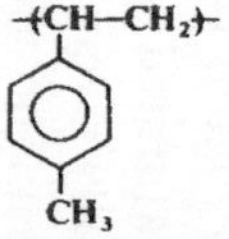 $-(CH-CH_2)-$ with phenyl ring, para-CH_3	9.4%	291.6								x			
Poly(isopropyl styrene)	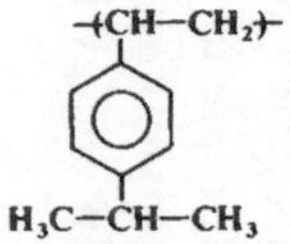 $-(CH-CH_2)-$ with phenyl ring, para-$H_3C-CH-CH_3$	7.7%	291.6								x			
Poly(alpha methyl styrene)	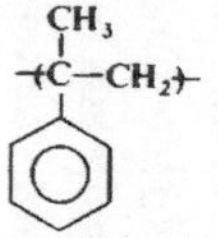 $-(C(CH_3)-CH_2)-$ with phenyl ring	7.5%	291.6								x			
Poly(4-Methoxystyrene)	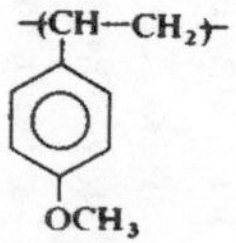 $-(CH-CH_2)-$ with phenyl ring, para-OCH_3	6.5%	291.7							x	x	x		
Poly(2,6-Dimethyl-1,4-Phenylene oxide)	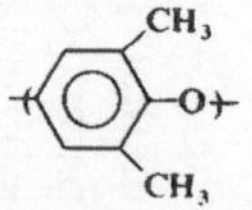 $-(C_6H_2(CH_3)_2-O)-$	9.5%	291.6							x	x		x	

[a] See also Figure 14.

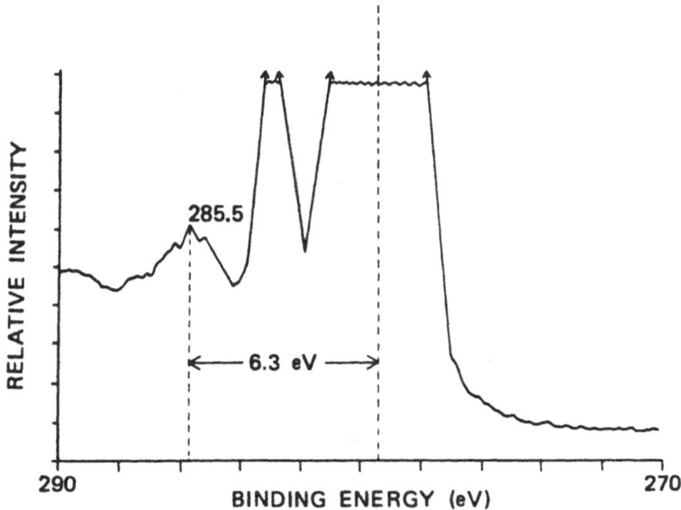

FIGURE 17. C-1*s* spectrum of poly(benzylmethacrylate) polymer containing aromatic and other unsaturated groups, showing the presence of a discrete shake-up satellite at 6.3 eV from the main peak. (See Table 10.)

binding energy shifts with carbon number in a series of homologous alkanes, attributing relaxation as the reason for the "chemical shifts" in these systems.

4.6. MULTIPLET SPLITTING

Multiplet splitting is an effect observed in atoms having unpaired electron spins in the ground state. Since there can be two different final states (Figure 18), there are two different energies observed. Thus, the band which is normally expected is now split into two bands, one for each of the final states. The peak separation observed is a function of the chemical environment of the unpaired *d* or *f* electrons. This effect is particularly pronounced for transition metals, such as chromium, manganese, iron, and cobalt because the 3*s* photoionization cross section is high. The effect is very important in catalysis, corrosion, and other studies where these metals are important. See References (3)-(11) for details.

4.7. MOLECULAR ORBITALS

The photoelectron spectrum in the region of 0–30 eV is commonly called the valence band region in the case of metals and semiconductors and the molecular orbital region in the case of organic materials and polymers. Here the XPS process excites weakly bonded electrons in the

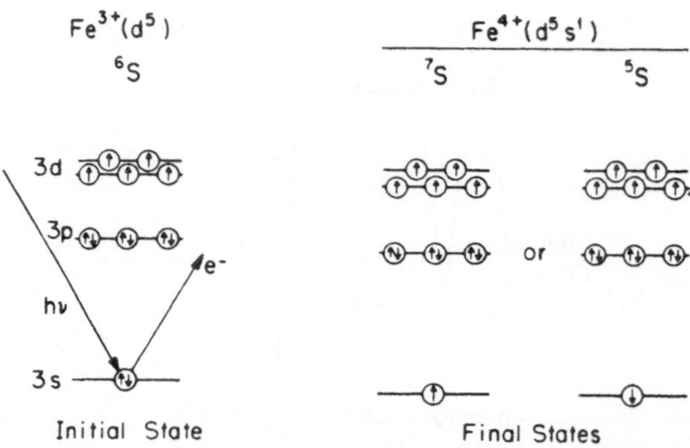

FIGURE 18. An illustration of multiplet or exchange splitting, observed primarily in *s* orbitals when outer *d* or *f* orbitals contain a large number of unpaired electrons. The two final states lead to two peaks, i.e., a peak splitting, whose separation is a function of the chemical environment of the unpaired *d* or *f* electrons. This effect is particularly useful for Cr, Mn, Fe, and Co as the 3*s* cross section is quite high. [Reprinted from Reference (7), p. 232, by permission.]

valence shells and molecular orbitals. The theoretical cross section for photoionization is maximized when the binding energy and the photon energy are roughly comparable. This is the region which is probed efficiently by ultraviolet photoelectron spectroscopy. Unfortunately, this region is difficult to analyze by XPS because of the low sensitivity. The advantage of XPS is the high photon energy available to probe inner core shells. One disadvantage in using a high-energy X-ray probe is that the cross section for molecular orbital and valence band interactions is very weak.

Good molecular orbital spectra from polymers by XPS requires extensive signal averaging and data accumulation times of the order of many hours. Nevertheless, the technique is widely used because it allows one to obtain molecular orbital and inner core level spectra on the same instrument and on the same sample, which is desirable for many studies. Clark and Thomas[51] have studied the molecular orbital spectra of polymers by the XPS technique as has Verbist, Pireaux, and coworkers[52,53] in Belgium.

Figure 19 presents molecular orbital spectra of selected polymers: polyethylene, poly(vinyl fluoride), and polystyrene. Because of the theoretical quantum chemistry expertise needed to properly interpret such spectra and primarily because of the long data acquisition times involved, molecular orbital spectra are not routinely used for surface analysis of polymers.

In a rather elegant study, Pireaux *et al.*[52] have shown that the second derivative of the loss structure of the main carbon-1*s* peak matches very nicely with the valence band molecular orbital spectrum from that same

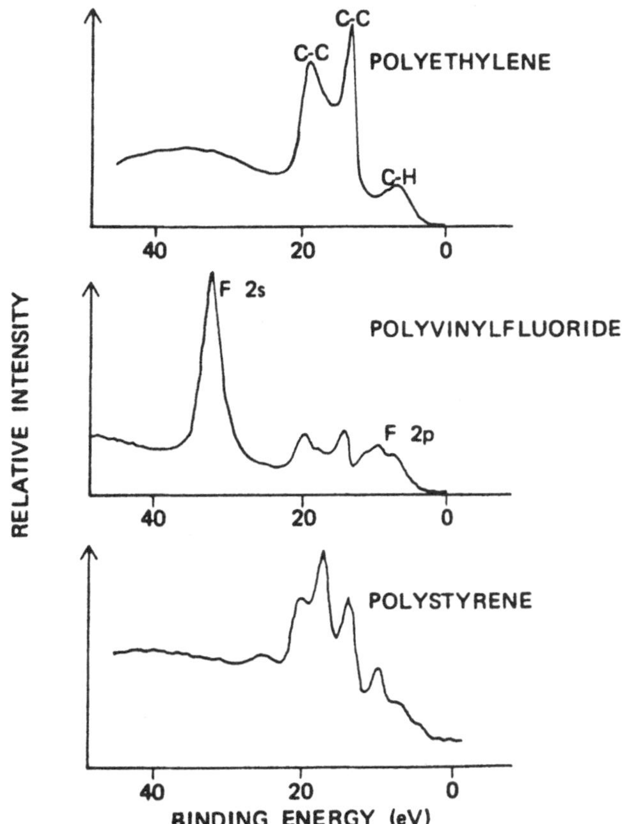

FIGURE 19. Valence band or molecular orbital spectra of polyethylene (PE), poly(vinyl fluoride) (PVF), and polystyrene (PS), showing the fluorine 2*s* feature at about 32 eV, the fluorine 2*p* at about 10 eV, the C—C bands at about 13 and 20 eV, and other features. Note the distinct appearance of the valence band spectrum which can be modeled and used to detect relatively subtle features which are almost impossible to detect from core level spectra alone. [Reprinted from Reference (98), pp. 332, 335, by permission.]

polymer, confirming that the shake-up processes in polymers mirror the molecular orbitals present.

Studies of the valence band region of metals and semiconductors by XPS methods are very important in research on the electronic structure and catalytic properties of such materials.

4.8. SUMMARY

1. Figure 20 summarizes the key features of an XPS spectrum; the photoelectron lines (Table 3) are intense, narrow, approximately

FEATURES OF XPS SPECTRA

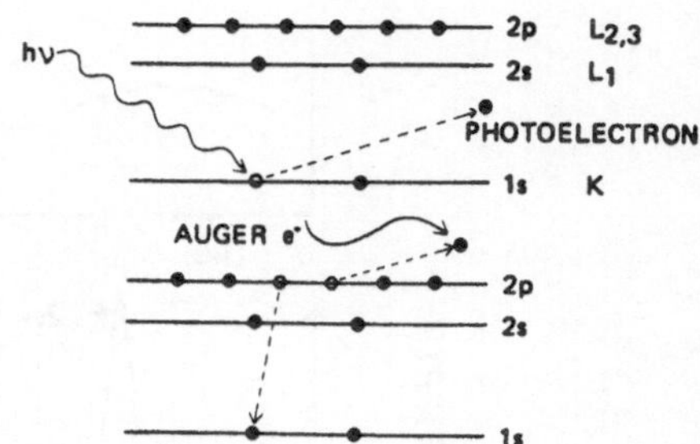

1) PHOTOELECTRON LINES:
   - INTENSE
   - NARROW
   - SYMMETRIC

2) AUGER LINES:
   - COMPLEX (GROUP)
   - BROAD
   - KLL, LMM, MNN, NOO
   - INDEPENDENT OF hν

3) X-RAY SATELLITES (to lower B.E.):
   - $K\alpha$ 3,4,5,6 ; $K\beta$

4) SHAKE-UP SATELLITES (to higher B.E.):
   - ENERGY "LOSS" PROCESS
   - EXCITED STATE

5) MULTIPLET SPLITTING:

6) ENERGY LOSS PROCESSES:
   - PLASMONS
   - BACKGROUND OR LOSS "TAIL"

7) VALENCE BANDS, MOLECULAR ORBITALS:

8) CHARGING EFFECTS:

FIGURE 20. Summary of main features of an XPS spectrum.

symmetric, generally Gaussian in shape, and easy to model and interpret, to a first approximation.

2. Auger lines (Table 9) are often complex, generally broad, and relatively difficult to interpret theoretically. In a limited number of cases the lines can be sharp, highly sensitive, and useful for elemental analysis (for example, sodium, magnesium, and zinc). Carbon, oxygen, and fluorine Auger lines generally appear in the XPS spectra of polymers. The Auger chemical shift is often useful for chemical state analysis.
3. If one is using a nonmonochromatic XPS instrument, the high photon energy X-ray satellites will lead to small peaks on the low binding energy side of each major photoelectron line. In many

instruments these X-ray satellites are calculated from the main photoelectron line and computer-subtracted from the spectra. This can sometimes be a problem if there is a small trace constituent whose normal line is in the position of an X-ray satellite feature of a major constituent. This problem is, of course, alleviated with the use of X-ray monochromator-equipped instruments as only the $K\alpha_{1,2}$ line is present and therefore the X-ray satellites do not appear.

4. Shake-up satellites are generally present in many organic polymer spectra if aromatic or other unsaturated groups are present. The shake-up satellites appear on the high energy side of the main peak, 6–12 eV from the parent peak for aromatic systems.
5. Other high-energy satellites may be present in certain samples, due to multiplet splitting and various ligand or coordination effects.
6. Discrete loss processes can be observed in many samples due to lattice or matrix excitation; such excitations in free-electron systems are called plasmon processes.
7. The low binding energy molecular orbital region is often of considerable usefulness, particularly the fluorine 2*s*, oxygen 2*s*, and carbon 2*s* and 2*p* features.
8. The electrical nature of the surface influences charging, charge equilibration, and energy referencing.

Therefore, there is clearly a very high information content in an XPS spectrum. To a first approximation it can be interpreted easily and straightforwardly. To maximize the information content and the analysis which can be derived from an XPS spectrum requires discussing each of the informational features in greater depth than is possible in this chapter.

5. FUNCTIONAL GROUP LABELING[54–60]

The chemical bonding information available from routine XPS analysis of polymer surfaces is often insufficient to distinguish between some of the functional groups which are expected to be present. For example, the chemical shifts for ethers and hydroxyl groups are nearly identical. Although such groups can be distinguished by attenuated total internal reflection infrared spectroscopy (ATR-IR) (Chapter 6), the surface sensitivity of ATR-IR is often insufficient for many surface analytical applications.

For these reasons there has been considerable interest in using functional group-specific reactions to produce products which can then be readily analyzed by XPS (Table 11). The major objective of these reactions is to label the functional group of interest with an element not normally present on the polymer surface and which has a high cross section for sensitive XPS detection. By careful reading of any one of the monographs available

TABLE 11
Surface Functional Group Reactions for XPS[a,b]

Surface functional group	Reagent and reaction condition (if available)	Product	Unique element	References and references cited therein
Unsaturated $>C{=}C<$	Br_2 in CCl_4	$>C(Br)-C(Br)<$	Br	55, 56
	$Hg(OAc)_2$	$>C(OR)-C(Hg(OAc))<$	Hg	55
	$Hg(CF_3CO_2)_2$ in CCl_3CH_2OH and benzene	$>C(Hg(CF_3CO_2))-C(OCH_2CCl_3)<$	Hg, F, Cl	56, 55
	OsO_4	$>C(-O)-C(-O)<$ cyclic osmate ester, OsO_2	Os	55
Carboxylic$-C(=O)OH$	NaOH in ROH or H_2O	$-C(=O)-O^-Na^+$	Na	55, 57, 56
	$BaCl_2$ in $Ba(OH)_2/H_2O$	$(-C(=O)-O^-)_2Ba^{++}$	Ba	55, 56

	$AgNO_3$ in H_2O	$-C(=O)-O^- Ag^+$	Ag	55, 60, 56, 88
	$TlOCH_2CH_3$	$-C(=O)-O^- Tl^+$	Tl	55, 61
	$CF_3CH_2OH/C_6H_{11}NCNC_6H_{11}$	$-C(=O)-O-CH_2CF_3$	F	55
	KOH in ROH then $C_6H_5CH_2Br$	$-C(=O)-O-CH_2-C_6F_5$	F	55, 56
	$(CF_3CO)_2O$ in pyridine/benzene	$-C(=O)-O-C(=O)-CF_3$	F	55, 56
	CF_3CH_2OH in CH_2Cl_2 pyridine/DDC	$-C(=O)-O-CH_2CF_2$	F	56
	HNO_3/KOH then $Ca(NO_3)_2$	$(-C(=O)-O^-)_2Ca^{++}$	Ca	85
Carbonyl $>C=O$	$NH_2NHC_6F_5$ in HCl/EtOH	$>C=NNHC_6F_5$	F, N	55, 57, 56
	$RNHNH_2$	$>C=NNHR$	N	55
	$H_2NNH-C_6H_4-Cl$ in acetone	$-C-NHN-C_6H_4-Cl$	Cl	88
Epoxy $>C(-O-)C<$	HCl (gas)	$-C(OH)-C-Cl$ (vicinal: $-\overset{OH}{C}-$ over $-C-Cl$)	Cl	55, 56

TABLE 11 (*continued*)

Surface functional group	Reagent and reaction condition (if available)	Product	Unique element	References and references cited therein
	$(CF_3CO)_2O$ (gas)	$-C\diagup O\overset{\overset{O}{\|}}{C}CF_3$ $\;\;\vert$ $-CO\underset{\underset{O}{\|}}{C}CF_3$	F	56
Hydroxyl—OH (Carbinol)	$OCN-C_6H_4-Cl$ in acetone	$-O-O-CHN-C_6H_4-Cl$	Cl	88
	$(CF_3CO)_2O$ (gas) (TFAA) $(CF_3CO)_2O$ in pyridine/ benzene	$-O\overset{\overset{O}{\|}}{C}CF_3$	F	55, 56, 60
	$CF_3CH_2CH_2Si(CH_3)_2NCOCH_3$ in pentane	$-OSi(CH_3)_2CH_2CH_2CF$	Si, F	56
	$ClCOC_6H_3(NO_2)_2$	$-O\overset{\overset{O}{\|}}{C}C_6H_3(NO_2)_2$	N	55
	$CBr_3CO_2H/C_6H_{11}NCNC_6H_{11}$	$-O\overset{\overset{O}{\|}}{C}CBr_3$	Br	55
	$(acac)_2Ti(Oi\text{-}Pr)_2$	$-CH_2-C-\underset{\underset{O-i-Pr}{\vert}}{Ti}-(acac)_2$	Ti	57
	C_3F_7COCl in *n*-hexane	$-O-\underset{\underset{O}{\|}}{C}-C_3F_7$	F	45

Functional group	Reagent	Product	Label	References
Sulfhydryl—SH	$AgNO_3$	$-S-Ag^+$	S	55
	$[C_6H_3(NO_2)_2(CO_2Na)-S]_2$	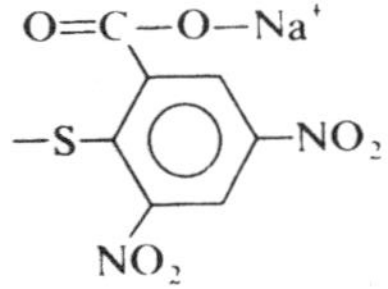 $-S-C_6H_2(NO_2)_2(C(=O)-O^-Na^+)$	N	55
Amine—NH_2	CS_2 (gas)	$-NHC(=S)SH$	S	55
	HCl (gas)	$-N^+H_3Cl^-$	Cl	56
	Acid Buffer	$-NH_3^+$	N^+ shift	86
	$HCOC_6F_5$ in pentane	$-N{=}CHC_6F_5$	F	55, 56
	$CH_3CH_2SCOCF_3$	$-NHCF_3$	F	55
—NR_2	$C_6F_5CH_2OC(O)Cl$	$-N(H)-C(=O)-O-CH_2-C_6F_5$	F	55
Acetate—$O-C(=O)-CH_3$	Convert to hydroxyl by base hydrolysis	Then use reactions for hydroxyl		45
Hydroperoxides—C—O—O—H	SO_2 (gas)	$-C-O-SO_3H$	S	87, 57
Silanol—Si—OH	$CF_3CH_2CH_2Si(CH_3)_2NCOCH_3$ (heptane reflux)	$-Si-O-Si(CH_3)_2-CH_2CH_2CF_3$	F	56
Reactions are available in discriminating between $>Si-OH$ and $>C-OH$				56

[a] References (55)–(57), 59, 61.

[b] Other functional group reactions may interfere, specificity must be verified. Note limitations and assumptions in text.

on organic functional group reactions,[54] one can devise a range of reactions which would appear to be suitable. A major problem, however, is that the surface reactivities of groups on solid surfaces are often considerably different from reactivities in homogeneous solution. One often finds that in order to carry out a reaction, the polymer should be above the glass transition temperature, thereby providing sufficient surface mobility to permit appropriate rearrangement and reactivity (Chapter 2). It is important in examining a surface functional group reaction to carefully evaluate time, temperature, and reactant concentration as variables.

In many cases one would like the reaction to be truly surface-specific and not penetrate at all into the bulk. One of the objectives of a study might be to separate surface groups from subsurface groups. Under certain conditions, the XPS technique is sensitive from 10 Å to 200 Å, depending on the photoelectron kinetic energy, its mean free path, and the photoelectron take-off angle (Section 7). If one is interested in the overall surface composition, including the subsurface region, one might select a derivatization reaction that would actually swell the polymer surface region and uniformly react with the surface. If one is interested only in the outermost 10 Å of the surface, then a set of reaction conditions must be selected which does not swell the polymer at all and does not permit significant reagent diffusion into the polymer surface.

Gas phase reactions are often desirable because they minimize the solvent and swelling problems which are common with liquid phase reactions. Depending on the size of the gas reactant and the permeability characteristics of the polymer surface, the surface/subsurface concern must be kept in mind.

It may also be possible to titrate the surface to determine charge or ionizable groups. This has been attempted[83] as a way to probe the acid–base character of a surface. The general procedure is to equilibrate the surface at some appropriate pH, and then expose it to a solution such as calcium nitrate, wherein the calcium would complex to the anionic sites. The solution is rinsed or blown off the surface by a jet of nitrogen and the sample analyzed by XPS. The calcium signal is plotted as a function of pH, resulting in a curve analogous to a titration curve. Sodium, calcium, silver, and other cations have been used.[85,86]

In some cases, a chemical shift may be directly observed, such as in the case of protonation of amine on the surface.[86]

It is important to note that the reactants listed in Table 11 have generally not been studied under a wide range of conditions or for a wide variety of surfaces. Most of the studies have been limited, qualitative studies. The establishment of these methods for routine surface analysis is far from proven or verified. Each individual investigator will have to conduct his/her own studies until the techniques are much more widely established.

The field of surface derivatization reactions for XPS analysis is a very new one. Table 11 summarizes most of the available literature on XPS derivatization reactions of polymer and related surfaces. Reaction conditions are given in sufficient detail in the cited references so that the investigator can design his own study. In view of the great success of derivatization reactions in other forms of spectroscopy and, in particular, in gas and liquid chromatography, it is expected that this approach will prove to be beneficial for routine XPS analysis. There is one caution, however. The fact that polymer surfaces respond to their environment means that the "labelled" surface may have reoriented itself during the reaction—thus the result of the labeling reaction may not truly represent the original surface [see Chapter 2 and Reference (58)].

6. DATA PROCESSING[62–66]

6.1. CURVE RESOLUTION

Practically all XPS instruments collect data digitally and are fully computer interfaced. In all forms of spectroscopy, chromatography, and related fields, one is concerned with the shape of a signal peak and methods of describing that peak. One is concerned with the overlap between two peaks and how to attempt to separate the peaks, if not by improving instrumental resolution, then by appropriate mathematical processing of the data. This is a particularly important exercise in XPS because the chemical shifts are not very large. A peak may consist of a number of components which we would like to resolve and separate.

The questions one needs to ask are: What is the shape of the peak? What is its area? What is its full width at half maximum (FWHM)? What is the background component? (Figure 7b).

The first consideration is how to subtract the background signal (see also Section 3.2.6). Although a sloping background function is generally adequate for polymers,[43–45] a more complex function is required[43,44] if the inelastic tail is particularly strong.

XPS peaks are generally Gaussian in shape. A typical Gaussian function is[63]:

$$y = A\,\mathrm{e}^{-ax^2} \tag{9}$$

where $a = 4 \ln 2/(\mathrm{FWHM})^2$. A typical Lorentzian is:

$$y = B/(1 + bx^2) \tag{10}$$

where $b = 4/(\mathrm{FWHM})^2$. The sum of two Gaussian peaks is given by[63]:

$$y = A_1\,\mathrm{e}^{-a_1(x-x_1)^2} + A_2\,\mathrm{e}^{-a_2(x-x_2)^2} \tag{11}$$

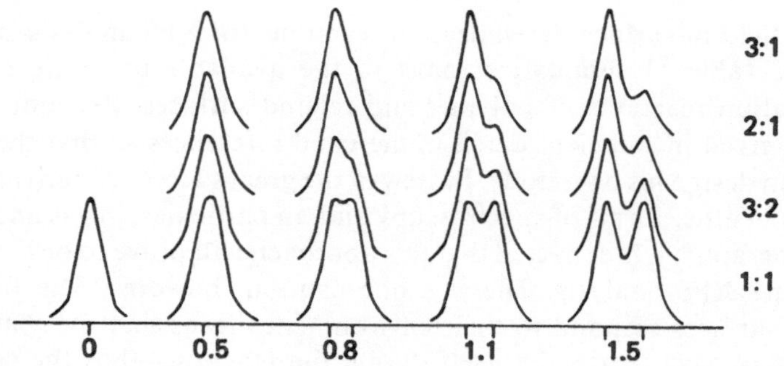

FIGURE 21. The superposition of two Gaussian peaks of different relative intensities and different energy separations. [Reprinted from Reference (6), App. B.]

where x_1 and x_2 are the component peak positions. Siegbahn in his second ESCA book[(6)] presented a very informative appendix which showed for Gaussian peaks what the overall peak shape would be for two components in various ratios and with various energy separations (Figure 21). It is assumed that the component peaks have a full width at half maximum of 1.0 eV, a reasonable value for a typical high-resolution spectrometer. It is clear that one gets a very adequate separation at about 1-eV difference between the two components if they are present at roughly comparable concentrations. If, however, one has a 3:1 concentration ratio with about a 1-eV separation, then the two components are much more difficult to separate; the small component simply becomes a distinct shoulder on a peak consisting largely of the major component.

Just as one can add two or more Gaussians to form a sum peak, given appropriate assumptions, one can also take a given spectral peak and attempt to deduce its subcomponents. This process is called curve resolution or curve fitting and is widely used in XPS and other low-resolution spectroscopies. Given a complex XPS peak, one can of course deduce many different combinations of subpeaks which will add up to give the recorded experimental peak. A variety of programs are available to perform these functions.[(62–64)]

To optimally apply curve resolution methods, one needs to know a good deal about the chemistry of the system and have some chemical intuition as to what is possible and reasonable. From the chemistry one can estimate the number of expected peaks in the spectrum. From reference compounds and reference spectra one obtains the binding energy, the peak shape, and the full width at half maximum for the component peaks. We then take each of the component peaks and add them together as in Eq. (11), adjusting the height of each component peak to get the best fit to the experimental data. This is, of course, a subjective procedure with many potential problems. The conventional peak fit computer routines go through multiple iterations until the least squares error between the calculated peak and the experimental peak is a minimum. Such programs are provided by most instrument manufacturers.[20]

One way to minimize the subjectivity in a curve resolution or peak fitting analysis is to take derivatives of the experimental peak to enhance shoulders, changes of slope, and other features which might be attributed to component peaks. As one can only successfully take derivatives from highly smooth continuous functions, we will first discuss smoothing briefly.

6.2. SMOOTHING

Because XPS equipment is generally digital in nature, one can easily take repetitive scans and signal average. The signal-to-noise level increases with the square root of the number of scans. It is always preferable to increase the signal-to-noise by increasing the data acquisition time if the sample and the budget can tolerate it. However, there are many situations in which the signal-to-noise is not sufficient and cannot be improved by experimental means. Under these conditions the data are generally smoothed. There are many types of smoothing functions and routines in the literature.[61,63] Briefly, smoothing amounts to taking a data point and comparing it with points on either side of it.[63] For example, a linear three-point smoothing equation could be

$$y_n = \frac{y_{n-1}}{4} + \frac{y_n}{2} + \frac{y_{n+1}}{4} \tag{12}$$

If one applies this three-point smoothing function to each point in the data, it is clear that noise fluctuations (high frequency) will tend to average out whereas the overall spectrum (low-frequency data) will not. This procedure inherently decreases resolution because one is looking at three data channels together rather than each one individually.

Rather than the linear smoothing function of Eq. (12), one can use a more complex smoothing function. One can determine the optimum smoothing point function by considering the resolution per data channel and the optimum resolution in an ideal spectral feature. For example, in the Hewlett-Packard instrument, one normally obtains high-resolution spectra by scanning a 20-eV window utilizing 125 data channels with an electronic "resolution," therefore, of about 0.16 eV per channel. Since the optimum resolution of the instrument is about 0.5 eV for a good spectral line, it is clear that one could use a five-point smooth without any significant loss in resolution. More rigorous calculation and experimental comparisons suggests that a seven- or nine-point smooth is optimum.[62] The smoothing function need not be linear as discussed above; it may be quadratic, polynomial, or even Gaussian. Conventional linear smoothing routines generally lead to a loss in the end data points. There are a number of current smoothing algorithms, however, which do not have this disadvantage.[65]

A smoothing operation is a convolution of the data. The data are processed or operated on by a smoothing function to reduce the high-frequency noise or statistical fluctuations without significantly affecting the information content of the signal (usually a low-frequency signal). This improves the signal-to-noise at a slight decrease in resolution. The rule of thumb, however, is never smooth unless that is the only way to improve the signal-to-noise ratio.

6.3. DERIVATIVES

Assuming one has a spectrum with adequate signal-to-noise, or can perform a smoothing routine to improve the signal-to-noise, one can derivatize the spectrum. Generally, the second derivative is taken, which is often helpful in determining the number of components in a peak and the binding energy for each component. For example, Figure 22 shows a two-component Gaussian with different component concentrations. The second derivative is also shown—note how the subcomponents can be readily separated.

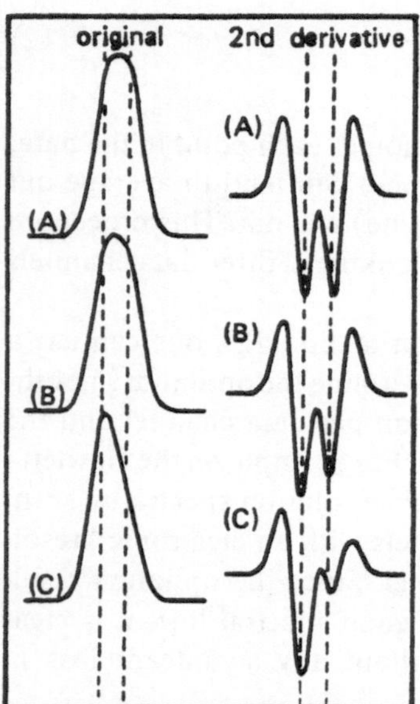

FIGURE 22. A two-component Gaussian, as in Figure 21, and its second derivative, showing the utility of the second derivative to identify subcomponents in a peak. Relative intensities: (A) 1 : 1, (B) 1 : 0.9, (C) 1 : 0.5. [Reprinted from Reference (43), p. 15, by permission.]

It is clear that derivatization of a multi-component spectrum leads one to deduce that the spectrum must be made up of appropriate subcomponents. The data must generally be smoothed between each derivatization. Given a second or even fourth derivative spectrum, one can often pick out the binding energy of appropriate subcomponents. This binding energy can then be specified in the curve resolution routine. The FWHM can also be specified, given the data for reference compounds run on that same instrument. Ideally, the derivatization process will result in binding energies of subcomponents which are in approximate agreement with what the investigator expects from his chemical intuition.

6.4. DECONVOLUTION

A major contributing factor to the poor resolution of XPS spectra is the line width of the X-ray photon source itself. Other components to the FWHM of an XPS peak include the intrinsic or inherent line width of the orbital itself and the analyzer resolution. The X-ray line width function is well-known for magnesium and aluminum Kα. It is usually Gaussian with a FWHM 0.70 to 0.85 eV, depending on the X-ray source. The analyzer resolution function is generally well-known or can be modeled.

The intrinsic shape of the orbital is the analytical information desired. The shape of an experimental XPS peak is a product of the intrinsic orbital shape, X-ray source function, and the analyzer resolution function. Recall from our smoothing discussion, we took experimental data and operated on it with a function to provide a result which had a better signal-to-noise. Deconvolution is the inverse of smoothing. To deconvolve an XPS peak we take an *assumed* true XPS peak, operate on it with a function which represents the X-ray line width and the analyzer transmission, and the product of that operation is the experimentally derived XPS peak. Therefore, one can devise a computer algorithm which basically assumes a peak, operates on it with the appropriate functions, and compares the result with the experimentally determined peak. One can then modify the assumed peak and iterate the process until the operation results in a peak as close as possible to the experimental peak. We then say that the assumed peak which leads to this result is, in essence, the true orbital shape and is the result of a deconvolution of the experimentally determined peak. Deconvolution results in an improvement in the resolution by mathematically removing those components of the experiment which tend to decrease resolution: the X-ray line width and the analyzer resolution functions. Deconvolution is closely related to smoothing. The computer searches to find a set of data which, after "smoothing" with the X-ray and analyzer functions, produces the experimental data. Several methods exist to perform this operation, including Fourier transform and numerical analysis procedures.

Deconvolution software is available in most modern XPS instruments, particularly those which do not employ an X-ray monochromator. Deconvolution has also been applied to X-ray monochromatized instruments.

A fairly recent development has been an application of the maximum entropy spectral estimation method, used in a number of other fields, to the deconvolution problem in X-ray photoelectron spectroscopy.[67] This method has the advantage that computation is numerically direct, does not involve trial and error optimization and is, therefore, much less subjective than other deconvolution procedures.

6.5. SUMMARY

1. Characterize the instrument. Are the experimental peak shapes Gaussian? What is the X-ray function for your instrument? What is the analyzer resolution function? What is the deconvolution function?
2. Obtain a library of spectra obtained on your instrument of appropriate model and reference compounds. Determine binding energies, peak shapes, and FWHM.
3. Take experimental data; background subtract.
4. Deconvolve that data (only if necessary).
5. Smooth if necessary (3- to 9-point smooth, depending on instrument).
6. Differentiate (first derivative).
7. Smooth (if necessary).
8. Differentiate (second derivative).
9. From the second derivative, select the number of component peaks and binding energies for each component.
10. From reference spectra estimate FWHM for each component peak.
11. Apply curve resolution routine, holding constant the energy position and FWHM of the components; adjust intensity of each component peak to obtain the best fit to the experimental peak.
12. Use the curve-resolved areas to quantitate the various components on the surface.

Appropriate strategies are discussed in several references.[12–15,21–23,43,44,62,63] It is important to note that deconvolution and curve resolution are two mathematically and analytically distinct processes. Note that the procedures which mathematically lead to an enhanced signal-to-noise result in decreased resolution, such as smoothing, and procedures which result in an increased resolution, such as derivatization, are paid for by a decrease in the signal-to-noise.

7. VARIABLE ANGLE METHODS

7.1. INTRODUCTION

Practical engineering surfaces are rarely uniform and homogeneous in composition or structure, as assumed in Section 3. Often the surface consists of a very thin film on a semi-infinite bulk substrate. The thin film may arise from mold release agents, anti-static agents, plasticizer, or low-molecular weight oligomers, in the case of polymers. In the case of a surface-modified polymer, such as after surface oxidation, the surface thickness involved may be as small as 5 to 10 Å and as large as thousands of Å.

It would be helpful to vary the sampling depth of the XPS method to enable nondestructive depth profiling from 10 Å to 200 Å. This is accomplished by the variable angle technique (Figure 23). By varying the angle between the collected photoelectrons and the surface, the electron will travel different distances before reaching the surface. It is clear that at low θ a smaller depth is sampled than at high θ (Figure 23). Therefore there is enhanced surface sensitivity at low θ and greatest bulk sensitivity at high θ.

In order to optimally apply this technique, some information on photoelectron mean free paths and XPS sampling depths must be available.

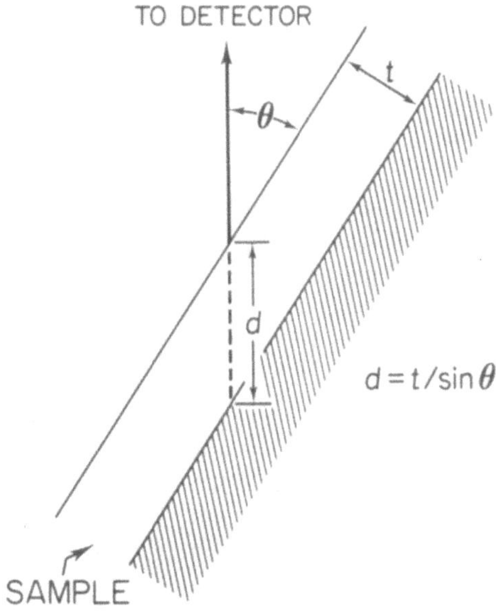

FIGURE 23. The variable angle XPS method, showing that as θ, the angle between the electron path to the detector and the surface of the sample, decreases, the effective electron travel distance in the sample increases, thus maximizing the surface sensitivity.

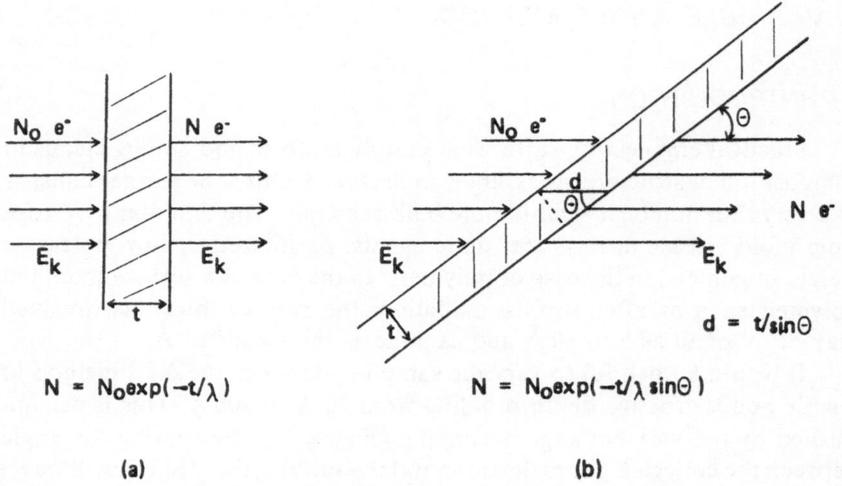

FIGURE 24. Definition and demonstration of electron mean free path, λ.

7.2. MEAN FREE PATHS*

The basis for the surface sensitivity of XPS is the fact that electrons do not travel very large distances in matter, due to inelastic scattering processes with the matrix or solid medium. Figure 24 formulates this effect. On the left side we assume a beam of N_0 monochromatic electrons with kinetic energy E_k traversing a sample of thickness t. We assume that N electrons of kinetic energy E_k have traversed the sample. $N_0 - N$ electrons have lost energy and are either retained within the sample or traverse with an energy less than E_k. This is analogous to mass attenuation of X-rays or to light attenuation in absorbing media. The basic expression is

$$N = N_0 \, e^{-t/\lambda} \tag{13}$$

where λ is defined as the mean free path, also calted the attenuation length or penetration length. Figure 24b shows that when θ is not equal to 90°, the electron path length, d, is greater than the thickness of the material, t, by the $1/(\sin \theta)$ term. The equation becomes

$$N = N_0 \, e^{-t/\lambda \sin\theta} \tag{14}$$

* See also Section 3.2.5.

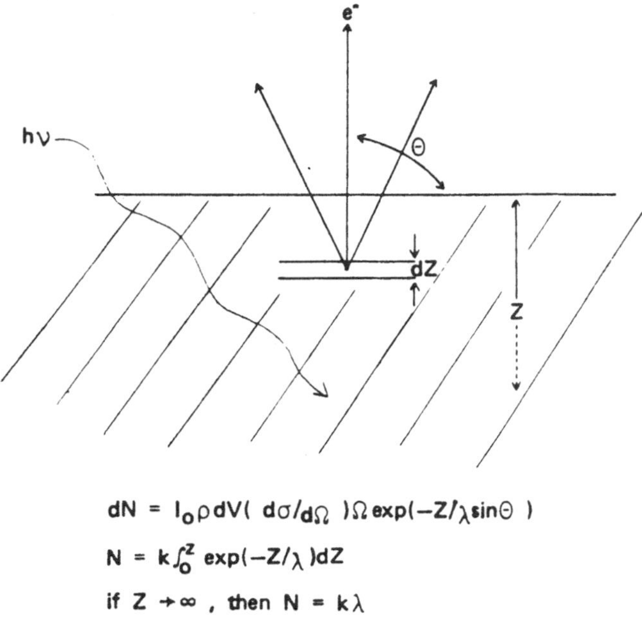

$$dN = I_0 \rho dV(d\sigma/d\Omega)\Omega \exp(-Z/\lambda \sin\Theta)$$

$$N = k\int_0^Z \exp(-Z/\lambda)dZ$$

$$\text{if } Z \to \infty, \text{ then } N = k\lambda$$

FIGURE 25. Illustration and derivation of photoelectron intensity on a clean semi-infinite homogeneous sample. [From Reference (3).]

The mean free path is that thickness ($t = \lambda$) when $N = N_0/e$; i.e., when 63% of the electrons have lost energy and no longer contribute to the no loss peak, E_k.

Now imagine that the thin film in Figure 24 is the *source* of electrons, such as in an XPS experiment. If $t \ll \lambda$, then all the photoelectrons generated reach the detector and $N = N_0$. As t increases, more photoelectrons are generated. One must consider the attenuation due to inelastic scattering of those photoelectrons generated below the surface. This can be set up as in Figure 25, derived in part from Fadley.[3] As $\int e^{-x}\,dx = -e^{-x}$, one can show that for an infinitely thick sample ($t \gg \lambda$) $N_0 \propto \lambda$, where N_0 denotes the intensity from an infinitely thick sample. For a sample of finite thickness, t,

$$N \propto \lambda(1 - e^{-t/\lambda}) \tag{15}$$

which, of course, reduces to $N \propto \lambda$ when $t \gg \lambda$. These relationships are summarized in Figure 26 for the general case where $\theta \neq 90°$ and for both uniform and patchy overlayers. N/N_0 is plotted as a function of t/λ in Figure 27. When $t = \lambda$, $N/N_0 = 0.63$ or $N = 63\%$ of N_0. When $t = 3\lambda$, 95% of N_0 is present. The sampling depth (three mean free paths) is defined as the depth from which 95% of the signal arises.

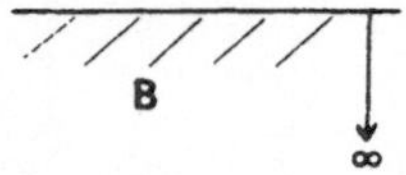

$$I_B^{\infty} \propto \lambda_B \tag{16}$$

$$\begin{aligned} I_B &\propto I_B^{\infty}(1 - \exp(-t_B/\lambda_B \sin\Theta)) \\ I_B &\propto \lambda_B(1 - \exp(-t_B/\lambda_B \sin\Theta)) \\ &\text{as } t \to 0,\ I_B \to 0 \\ &\text{as } t \to \infty,\ I_B \to I_B^{\infty} \end{aligned} \tag{17}$$

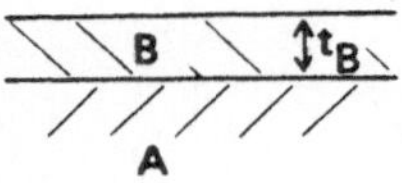

$$\begin{aligned} I_B &\propto \lambda_B(1 - \exp(-t_B/\lambda_B \sin\Theta)) \\ I_A &\propto \lambda_A \exp(-t_B/\lambda_B \sin\Theta) \end{aligned} \tag{18}$$

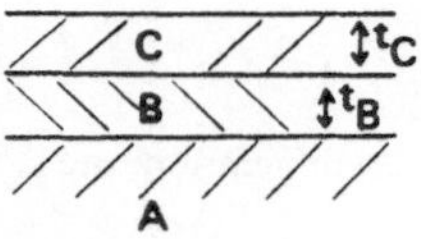

$$\begin{aligned} I_C &\propto \lambda_C(1 - \exp(-t_C/\lambda_C \sin\Theta)) \\ I_B &\propto \lambda_B(1 - \exp(-t_B/\lambda_B \sin\Theta))\exp(-t_C/\lambda_C \sin\Theta) \\ I_A &\propto \lambda_A \exp(-t_B/\lambda_B \sin\Theta)\exp(-t_C/\lambda_C \sin\Theta) \end{aligned} \tag{19}$$

FIGURE 26. (a) Uniform overlayer models for practical XPS surface analysis, including the appropriate equations for each component. Note that the interfaces between the layers are assumed to be sharp and that the individual layers are assumed to be completely homogeneous and uniform.

The mean free path, λ, is a function of kinetic energy. A good example of the importance of the mean free path kinetic energy dependence was given earlier with the case of the germaium $2p$ and $3d$ spectra (Figure 10).

Mean free path values are usually obtained by overlayer experiments, wherein one measures the decrease in signal from an infinite substrate as a function of the thickness of a deposited overlayer. The overlayers are usually deposited by Langmuir–Blodgett monolayer transfer[38,41] or by vapor phase methods.[40,83] Values of λ for organic polymers are somewhat controversial,[38–41] ranging from the relatively low values reported by Clark *et al.*[40,83] to the higher values reported by others.[38,49,41]

Wagner *et al.*[36] and Seah and Dench[35] recently reviewed mean free path data, reporting $\lambda \sim kE_k^{0.7}$ and $\lambda \sim 1.1E_k^{0.5}/\rho$ dependencies, respectively, where k is a material constant and ρ is the density.

Ashley and coworkers[69–72] have recently developed a theoretical approach to the problem, permitting the calculation of mean free paths for organic polymers. They used as input parameters the molecular weight of

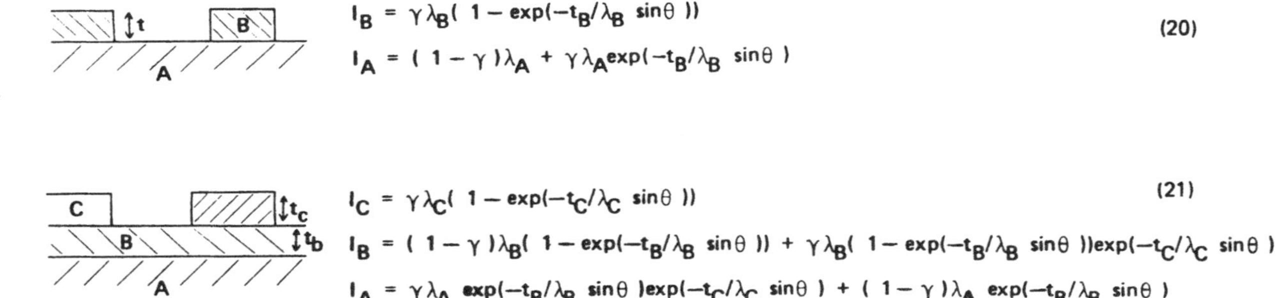

$$I_B = \gamma\lambda_B(1 - \exp(-t_B/\lambda_B \sin\theta)) \qquad (20)$$
$$I_A = (1 - \gamma)\lambda_A + \gamma\lambda_A \exp(-t_B/\lambda_B \sin\theta)$$

$$I_C = \gamma\lambda_C(1 - \exp(-t_C/\lambda_C \sin\theta)) \qquad (21)$$
$$I_B = (1 - \gamma)\lambda_B(1 - \exp(-t_B/\lambda_B \sin\theta)) + \gamma\lambda_B(1 - \exp(-t_B/\lambda_B \sin\theta))\exp(-t_C/\lambda_C \sin\theta)$$
$$I_A = \gamma\lambda_A \exp(-t_B/\lambda_B \sin\theta)\exp(-t_C/\lambda_C \sin\theta) + (1 - \gamma)\lambda_A \exp(-t_B/\lambda_B \sin\theta)$$

FIGURE 26. (b) Patchy overlayer models for practical XPS surface analysis. Patch assumed to cover an area fraction γ. The uncovered area fraction is $1-\gamma$. Note that no assumptions are made regarding the relative sizes of the patches except that the patch thickness must be small enough and the patches must be far enough apart so that shadowing and related effects can be neglected. [See References (3) and (75).]

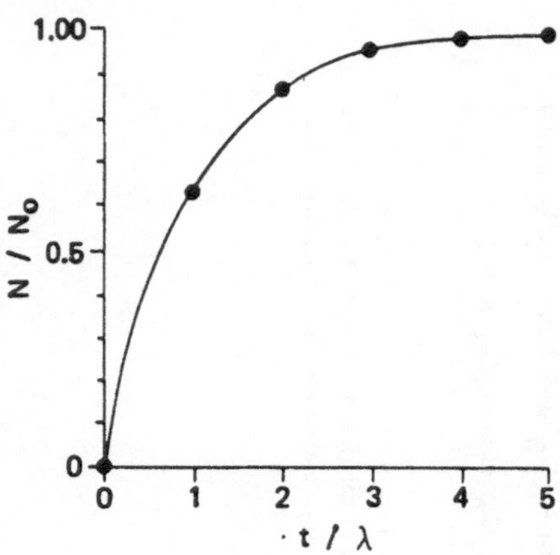

FIGURE 27. Demonstration of the sampling depth for a clean homogeneous semi-infinite sample. 95% of the signal comes from a sampling depth equivalent to three λ, 63% of the signal from a depth equal to λ.

the molecule or repeat unit (M), the number of valence electrons in the repeat unit (n), and the density (ρ). Their relationship is[69]:

$$\lambda = \frac{M}{\rho n} E_k \bigg/ (13.6 \ln (E_k) - 17.6 - 1400/E_k) \tag{22}$$

where E_k is the kinetic energy in eV. The expression is valid over the range from ~100 to 10,000 eV. It is derived from a theoretical model which utilizes the valence electron density in the solid as the key parameter influencing electron–solid interaction. This is a more reasonable scaling parameter than the density used by Seah and Dench.[35]

"The very large uncertainties in the experimental determination of electron mean free paths in solid organic insulators makes the calculation of these quantities from a simple formula both attractive and hard to verify." [Reference (70), page 372.] Given the uncertainties and controversy in the literature, and given the agreement of the Ashley values with the available data, the Ashley values are probably best until better and more extensive experimental data become available.

Table 12 presents M, ρ, and n values for typical polymers and the mean free path for several photoelectron energies of particular interest to biomedical polymer studies. Note that the λ values range from 10 to 45 Å, depending mainly on $M/\rho n$ and E_k. Table 13 presents the results of the Ashley equation for $M/\rho n$ values from 1.5 to 3.0. The role of sample density is clearly evident, as also pointed out by Seah and Dench.[35] Figure 28 plots the data of Table 13 for $M/\rho n$ = 1.50, 2.00, and 2.50.

Ashley has also considered non-organic solids, obtaining a relationship in agreement with Wagner *et al.*,[36] i.e.

$$\lambda(\text{Å}) \sim kE_k^P \tag{23}$$

TABLE 12

Ashley Parameters and Mean Free Path Values for Polymers and Biochemicals as a Function of Kinetic Energy [eV (Al $K\alpha$)][a]

					λ(Å)					
					Si-2*p*	C-1*s*	N-1*s*	O-1*s*	F-1*s*	
	Ashley parameters				100	285	400	530	686	BE
Material	ρ (g/ml)	M	n	$M/\rho n$	1383	1198	1083	953	797	E_k
Albumin	1.14	66,000	25,700	2.25	39	35	32	29	25	
Anthracene	1.25	178	66	2.16	37	33	31	28	24	
Bakelite	1.40	661	248	1.90	33	29	27	24	21	
Cellophane	1.50	162	64	1.69	29	26	24	22	19	
DNA	1.35	664	238	2.07	36	32	29	27	23	
Kapton	1.42	382	138	1.95	34	30	28	25	22	
Polyamides: Nylon 6/6	1.13	226	92	2.17	38	33	31	28	24	
Nylon 11	1.05	367	152	2.30	40	35	33	30	26	
Albumin	1.14	66,000	25,700	2.25	39	35	32	29	25	
Polyester-mylar	1.38	192	72	1.93	33	30	27	25	22	
Polyethylene	0.92	28	12	2.54	44	39	36	33	28	
Poly(methyl methacrylate)	1.19	100	40	2.10	36	32	30	27	23	
Polystyrene	1.05	104	40	2.48	43	38	35	32	28	

[a] From References (69) and (70).

TABLE 13
Ashley Mean Free Paths in Å for Various $M/\rho n$ Values[a]

Kinetic energy (Ev)	$M/\rho n$ 1.50	1.75	2.00	2.25	2.50	2.75	3.00
100	5	6	6	7	8	9	10
200	6	7	8	9	11	12	13
300	8	9	11	12	14	15	16
400	10	12	13	15	17	18	20
500	12	14	16	18	19	21	23
600	13	16	18	20	22	25	27
700	15	18	20	23	25	28	30
800	17	20	22	25	28	31	34
900	18	21	25	28	31	34	37
1000	20	23	27	30	33	37	40
1100	22	25	29	32	36	40	43
1200	23	27	31	35	39	42	46
1300	25	29	33	37	41	45	49
1400	26	31	35	39	44	48	53
1500	28	32	37	42	46	51	56

[a] Reference (69).

His k and p values are given in Table 14, together with λ values for the major photoelectron energies of interest. It is clear that the higher density metals all have about the same values, ~13 to 20 Å for the lines listed. The values for SiO_2 are relatively high (24-36 Å) and comparable to those observed for many polymers (albumin, Mylar, PMMA) (Table 12). Alumina

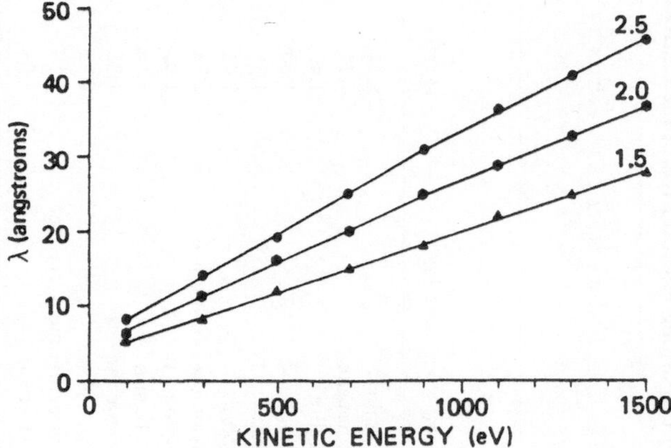

FIGURE 28. Mean free path as a function of kinetic energy for Ashley $M/\rho n$ values of 1.5, 2.0, and 2.5. This is the range expected for all common organic materials.

TABLE 14
Ashley Parameters and Mean Free Path Values for Common Non-organic Solids[a]

Material	ρ(g/ml)	200–400 eV		400–200 eV		$\lambda(\text{Å}) = KE_k^p$					
		K	p	K	p	Si-2p	C-1s	N-1s	O-1s	F-1s	
						100	285	400	530	686	BE
						1383	1198	1083	953	797	E_k
Al	2.71	0.204	0.657	0.102	0.773	27	24	23	20	18	
Al_2O_3	4.05	0.398	0.530	0.128	0.719	22	21	19	18	16	
Si	2.33	0.200	0.665	0.108	0.768	28	25	23	21	18	
SiO_2[b]	2.65	0.634	0.528	0.218	0.706	36	32	30	28	24	
Ni	8.91	0.318	0.536	0.123	0.694	19	17	16	14	13	
Cu	8.93	0.312	0.552	0.133	0.695	20	18	17	16	14	
Ag	10.50	0.330	0.545	0.125	0.707	21	19	17	16	14	
Au	19.30	0.338	0.530	0.131	0.688	19	17	16	15	13	
Ge	5.30	0.203	0.619	0.086	0.762	21	19	18	16	14	

[a] Reference (72).
[b] Crystalline SiO_2 has a density of 2.65; amorphous SiO_2 has density of about 2.3.

(Al_2O_3), because of its high density (4.05 g/ml), has low λ values, comparable to that of the higher density metals.

Mean free path data for higher kinetic energy photoelectrons, obtained using a titanium Kα source, have been reported by Clark *et al.*[73] One advantage of the use of higher-energy X-ray sources is the increased mean free paths due to the higher photoelectron kinetic energies, permitting depth profiling to greater depths than with conventional Mg or Al sources.

7.3. UNIFORM OVERLAYERS

Given the equations in Figure 26a and the Ashley mean free path values, one can now consider the modeling of a variable angle experiment. As it is difficult to obtain absolute intensity values, one generally uses the ratio of two peaks for the analysis.

For example, consider the adsorption of a protein on a glass or quartz surface. The substrate contains silicon, the overlayer contains nitrogen. There is no nitrogen on the substrate, nor any silicon in the protein overlayer. For both SiO_2 and protein (albumin) λ for Si-2p is ~36 Å and for N-1s in protein is ~28 Å. Now take the I overlayer/I substrate (I_B/I_A in Figure 26a) = $I_{N\text{-}1s}/I_{Si\text{-}1s}$, i.e., ratio of peak areas *after* correction for cross-section *and* throughput function (Section 3.2). For this example

$$\frac{I_B}{I_A} = \frac{\lambda_B(1 - \exp(-t_B/\lambda_B \sin\theta))}{\lambda_A \exp(-t_B/\lambda_B \sin\theta)} = \frac{\lambda_B}{\lambda_A}(e^{t_B/\lambda_B \sin\theta} - 1) \tag{24}$$

Figure 29 shows the result as a function of t_B and θ for the example given. By plotting the experimental data in Figure 29, one can deduce a best value for t_B, the thickness of the protein layer.

The case for two uniform overlayers is more difficult. Consider the special case of a protein on silicon. We must first characterize the thickness of the SiO_2 layer on silicon. This can be done using the elemental Si peak to represent the substrate and the chemically shifted Si peak to represent the oxide overlayer. A variable angle study of the Si/SiO_2 surface, given that the mean free paths for Si-2p in Si and SiO_2 are known (Table 14), allows one to determine the oxide thickness. Given that the oxide thickness is now known (t_B in Figure 26a), then one can plot I_C/I_B against θ, where I_C refers to the protein overlayer (Figure 26a) and I_B to the SiO_2 overlayer:

$$\frac{I_C}{I_B} \propto \frac{\lambda_C}{\lambda_B} \frac{\{1 - \exp[-t_C/(\lambda_C \sin\theta)]\}}{\{1 - \exp[-t_B/(\lambda_B \sin\theta)]\} \exp[-t_C/(\lambda_C \sin\theta)]} \tag{25}$$

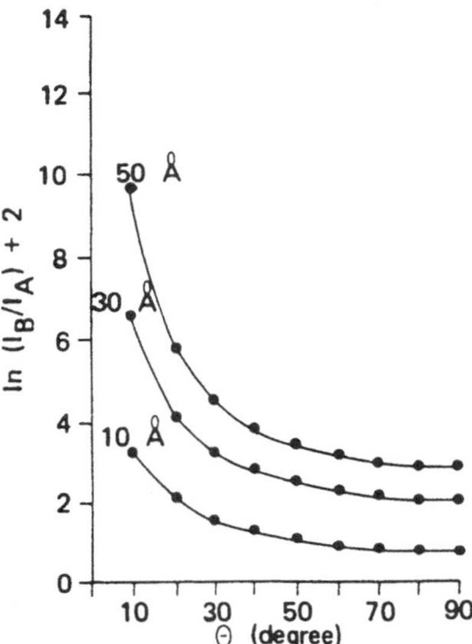

FIGURE 29. Log plot of the ratio overlayer to substrate intensity as a function of overlayer thickness, t, and θ. The substrate is assumed to be glass. The mean free path of the substrate signal, Si-2p, is 36 Å in both the substrate and the overlayer. The overlayer is assumed to be an adsorbed protein. The mean free path for the overlayer signal, N-1s, is 28 Å in the overlayer. This is an example of a case of a protein adsorbed on a glass or quartz substrate. Protein film thickness is assumed to be 10, 30, and 50 Å.

This function is plotted in Figure 30 for the case where $t_B = 20$ Å, λ_B is ~36 Å (Si-2p in SiO_2), λ_C is ~28 Å (N-1s in protein film) as a function of t_C, the protein overlayer thickness, from 10 to 50 Å. Note the expected enhancement in the protein overlayer signal at low θ. One can now plot experimental data as in Figure 30 and estimate a value for protein film thickness.

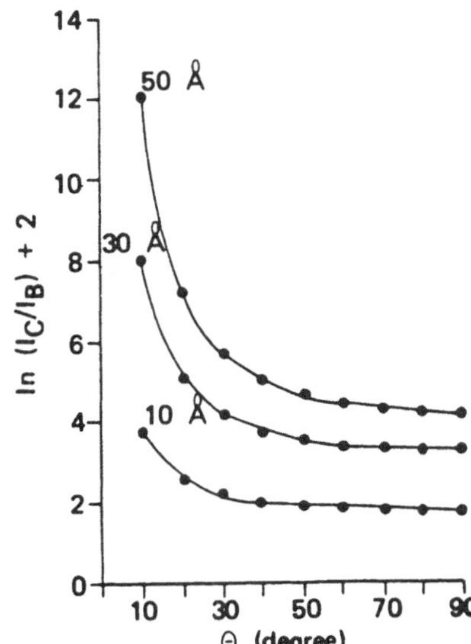

FIGURE 30. The two-film problem. Equation (25) is plotted as a function of thickness of the outermost film and θ. The example chosen is a protein overlayer on an SiO_2 film on a silicon substrate, for protein film thicknesses of 10, 30, and 50 Å. Note that we have plotted log N-1s/Si-2p where the Si 2p is from the SiO_2 overlayer.

7.4. PATCHY OVERLAYERS

Figure 26b shows the case for a single patchy overlayer on an infinite uniform substrate. We assume that the overlayer occupies an area fraction, γ. The area fraction uncovered is $(1 - \gamma)$. The intensities from the substrate and overlayer are given in Figure 26b. The overlayer/substrate intensity ratio is:

$$\frac{I_B}{I_A} = \frac{\gamma\lambda_B\{1 - \exp[-t_B/(\lambda_B \sin\theta)]\}}{(1-\gamma)\lambda_A + \gamma\lambda_A \exp[-t_B/(\lambda_B \sin\theta)]} \tag{26}$$

This function is plotted as a function of θ and γ in Figure 31 for the case of a patchy protein overlayer on a homogeneous substrate. We assume I_B is given by the N-1s of the protein ($\lambda_B \sim 28$ Å), I_A by Si-2p of glass ($\lambda_A \simeq 36$ Å), and $t = 10$, 30, and 50 Å.

The case for a patchy overlayer *on* a uniform overlayer *on* a uniform infinite substrate is also presented in Figure 26b, where γ is the area fraction of the outermost patchy layer, t_C is the patch layer thickness, and t_B is the uniform overlayer thickness.

7.5. SURFACE CONCENTRATION GRADIENTS

The two uniform overlayer problem (Figure 26a) can be generalized to n overlayers, i.e., the situation of a concentration gradient within the XPS analysis volume. Assume we have a polymer surface with the concentration profile shown in Figure 32. This problem has been addressed by Phillips *et al.*,[76] Payntor,[77] and Ratner *et al.*[78] Practical examples include surface oxidation, phase distribution in a block copolymer, such as block copolyetherurethanes,[78] and diffusion of an environmental component into the surface.

In applying the analysis to biomedical polymers, it is useful to consider several components. Two components could be oxygen (O-1s) and carbon (C-1s). Three components could be O-1s, N-1s, and C-1s for the case of a polyurethane surface. Five components could be alkyl carbon, ether carbon, carbonyl carbon, nitrogen, and ether oxygen, again for a polyetherurethane surface. Assume the concentration profiles in Figure 32. A practical example could be a polyetherurethane with roughly equal atomic and volume % hard and soft segments. Assume the soft segment dominates at the surface and the average bulk composition is observed within the XPS sampling depth. Given mass balance considerations, the hard segment must therefore be depleted in the subsurface region. One can take the curve-resolved ether carbon or ether oxygen signals to represent the soft segment and the N-1s to represent the hard segment. Given the assumed concentra-

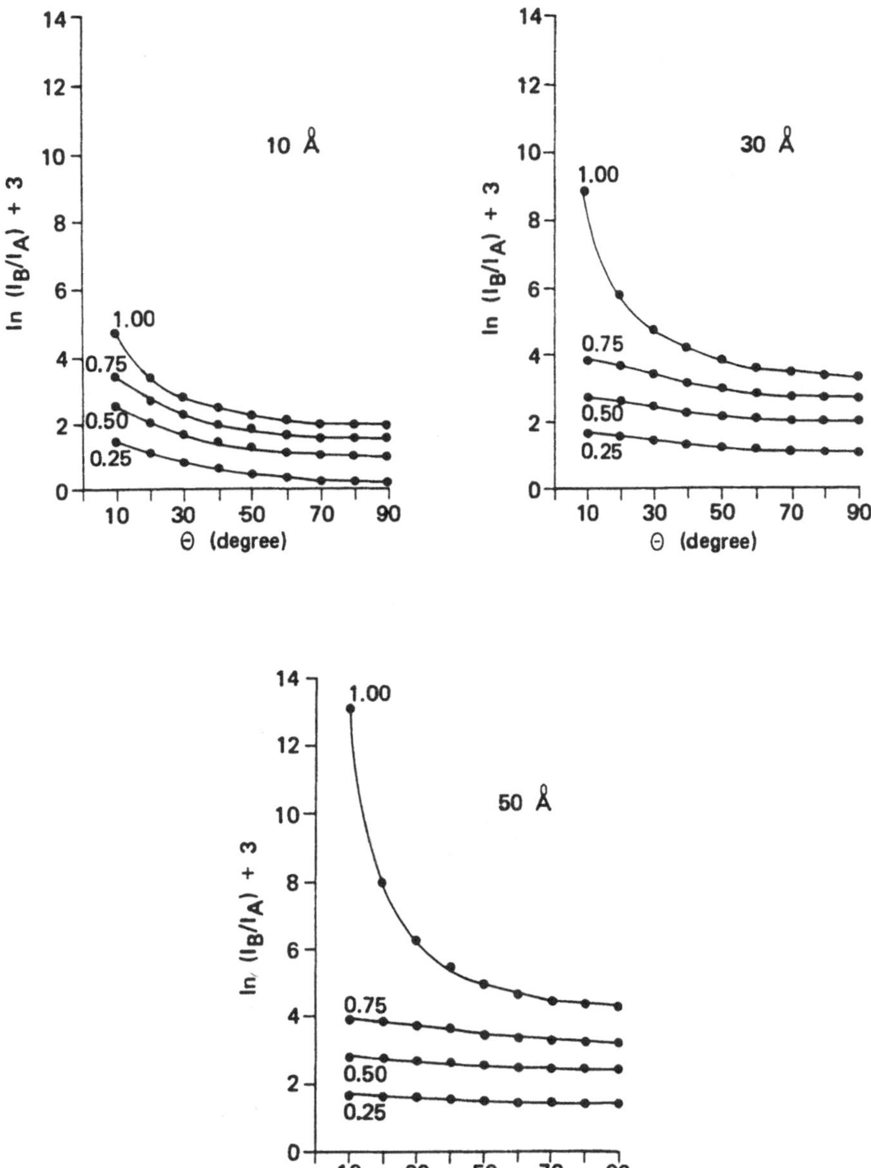

FIGURE 31. The patchy overlayer film function, Eq. (26), as a function of θ, γ (the area fraction covered), and patch thickness, t, for $t = 10$, 30, and 50 Å. Other conditions same as Figure 29. Area fractions covered are 0.25 to 1.0. This is an example of a protein adsorbing on a glass or quartz substrate, wherein the protein deposits as a patchy layer rather than as a uniform layer.

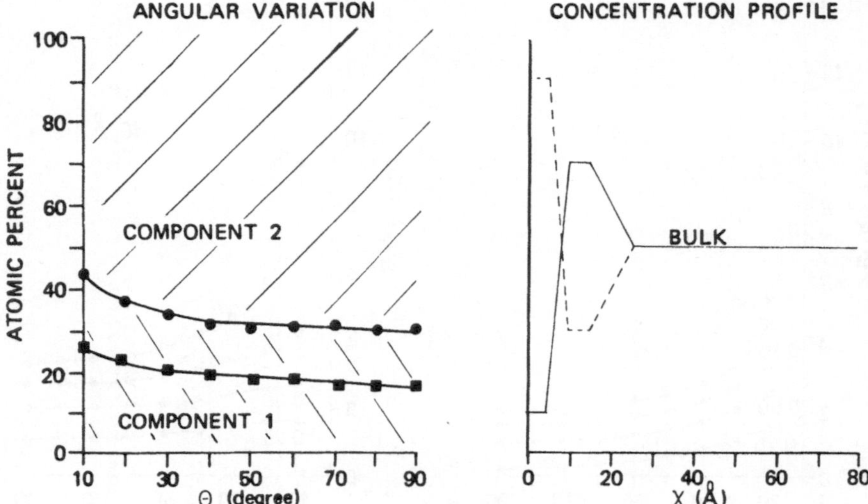

FIGURE 32. Expected or modeled concentration profiles for two components in an otherwise homogeneous sample. Left: the calculated angular distribution for each of the components using the Payntor algorithm [References (77), (78)]. The particular concentration profile assumed results in a particularly simple algorithm. Right: the cumulative concentration profile, again due to Payntor. The examples chosen are appropriate to a polyether urethane wherein one component represents the soft segment and the other the hard segment, showing, for example, soft segment excess at the near surface followed by soft segment depletion or hard segment excess in the subsurface region.

tion profile (Figure 32), one can calculate the angular distribution function [see References (77) and (78)]. We assume that the concentration profile is of the general form given in Figure 32, i.e., the surface concentration is constant up to some depth x_1, and the concentration is constant after some depth x_4. There are two linear gradient regions (x_1–x_2 and x_3–x_4) separated by a constant region (x_2–x_3). Such a profile results in a simple algorithm, as shown below. Recall from Figure 25,

$$I_i(\theta) \propto \int_0^x N_i(x)\, \mathrm{e}^{-x/\lambda_i \sin\theta}\, \mathrm{d}x \qquad (27)$$

We assume that although N varies as given in Figure 32, λ_i for each line is constant throughout the sample. N_i is defined as atomic % of component i. $I_i(\theta)$ is the corrected intensity (corrected for cross section, asymmetry function, and instrument throughput) for each component i at each angle

θ. Referring to Figure 32,

$$I(\theta) \propto \int_0^{x_1} (N_0) \exp(-x/\lambda \sin\theta)\, dx$$
$$+ \int_{x_1}^{x_2} \left[N_1 + \left(\frac{N_2 - N_1}{x_2 - x_1}\right)(x - x_1)\right] \exp(-x/\lambda \sin\theta)\, dx$$
$$+ \int_{x_2}^{x_3} N_2 \exp(-x/\lambda \sin\theta)\, dx$$
$$+ \int_{x_3}^{x_4} \left[N_3 + \left(\frac{N_4 - N_3}{x_4 - x_3}\right)(x - x_3)\right] \exp(-x/\lambda \sin\theta)\, dx$$
$$+ \int_{x_4}^{\infty} N_4 \exp(-x/\lambda \sin\theta)\, dx \tag{28}$$

where $N_0 = N_1$, $N_2 = N_3$ (Figure 32). By appropriate substitution, simplification, and integration, and by recalling that $\int e^{ax}\, dx = (1/a) \exp(ax)$ and $\int x\, e^{ax}\, dx = (1/a^2)(ax - 1) \exp(ax)$, it can be shown that[77]:

$$I(\theta) \propto N_1 \lambda \sin\theta + \lambda^2 \sin^2\theta \left\{ \left(\frac{N_4 - N_3}{x_4 - x_3}\right) \right.$$
$$\times [\exp(-x_3/\lambda \sin\theta) - \exp(-x_4/\lambda \sin\theta)]$$
$$\left. + \left(\frac{N_2 - N_1}{x_2 - x_1}\right)\Big([\exp(-x_1/\lambda \sin\theta) - \exp(-x_2/\lambda \sin\theta)]\right\} \tag{29}$$

Note that this simple form is only obtained for profiles of the type given in Figure 32.

The atomic percent of component i as a function of θ, $N_i(\theta)$ is given by

$$N_i(\theta) = \frac{I_i(\theta)}{\sum_i I_i(\theta)} \tag{30}$$

The $N_i(\theta)$ are then plotted as in Figure 32 (left), where $\sum N_i(\theta) = 100$, and compared to the experimentally determined $N_i(\theta)$. The assumed profile (Figure 32, right) is adjusted until the calculated and experimentally derived $N_i(\theta)$ match as closely as possible.

This method does not provide a unique solution—it provides one or more *assumed* profiles which agree with the data. Such assumed profiles must, of course, be firmly based on the physical and chemical realities of the sample. Ratner *et al.*[78] have applied this method to a series of polyurethane materials, concluding that the polyether soft segment tends to dominate the first 20 Å of the sample.

7.6. LIMITATIONS

Although variable angle methods are very informative, they are also very limited and depend on many assumptions. The most serious assumption is the mean free path. Other assumptions include surface homogeneity, the homogeneity and distribution of "patches," etc. A particular problem with variable angle methods is the expense. XPS is a slow data acquisition technique. Variable angle studies generally require five to ten times the instrument time of a conventional XPS study. One must also be concerned about X-ray radiation damage to the sample at such long exposures, particularly with nonmonochromatic instruments.

The fact that the instrument sensitivity function is a function of kinetic energy has already been noted. The data used in variable angle studies must be corrected for this effect. The instrument sensitivity function is also a function of sample angle, however, in part because the sample surface area (solid angle) seen by the detector is a function of angle. Therefore, for rigorous quantitative variable angle work such functions must be established and the data processed accordingly. The extensive variable angle work of Fadley and Baird[84] has resulted in such functions for the Hewlett-Packard monochromatic instrument.

The great advantage of the variable angle XPS method is that one can obtain a nondestructive depth profile for the first 50 Å of the sample. Using higher energy photon sources, such as Ti Kα, one can profile up to 200 Å in. Mean free paths for the 0- to 5-keV electrons produced by such sources are now available.[73]

Examples of variable angle methods applied to protein adsorption are given in the chapter by Payntor and Ratner in Vol. 2 of this series.

ACKNOWLEDGMENTS

Many coworkers, associates, and students over the years have contributed to my XPS and polymer surface education; particularly C. Fadley of the University of Hawaii, M. Kelly and L. Fay of Surface Science Labs, R. N. King of the Norton-Christensen Co., and G. Iwamoto now at Iowa State Univ. Hospital. R. N. King, G. Iwamoto, C. Doyle, P. Zajicek, I. Elliott, J. Warenski, R. Lowe, and P. Dryden have maintained and operated our surface analysis lab over the years. S. Hall, G. Iwamoto, L. Smith, R.

N. King, S. Hattori, P. Triolo, M. Davis, J. Hansen, and J. Chen have produced XPS-based theses. I also wish to thank my coworkers at the University of Utah, D. L. Coleman, D. E. Gregonis, R. Van Wagenen, L. M. Smith, and S. W. Kim, for stimulating discussions and assistance. Dr. R. Thomas supplied expanded versions of his polymer XPS correlation charts. I acknowledge discussions with R. Payntor and B. D. Ratner regarding their algorithm and models for surface concentration gradients. Much of this work has been supported by the NIH, NHLBI, and by University of Utah Faculty Research and Biomedical Science Support Grants.

REFERENCES

1. K. Keller, ed., *Guidelines for Physicochemical Characterization of Biomaterials*, NIH Publ. No. 80-2186 (1980).
2. K. Siegbahn, Electron spectroscopy for atoms, molecules, and condensed matter, *Science* **217**, 111 (1982).
3. C. S. Fadley, in: *Electron Spectroscopy* (C. R. Brundle and A. D. Baker, eds.), Vol. II, pp. 1-156, Pergamon Press, New York (1978).
4. H. Siegbahn and L. Karlsson, *Electron Spectroscopy for Atoms, Molecules, and Condensed Matter*, North-Holland, Amsterdam, in press (1985).
5. K. Siegbahn, C. Nordling, A. Fahlman, R. Nordberg, K. Hamrin, J. Hedman, G. Johansson, T. Bergmark, S.-E. Karlsson, I. Lindgren, and B. Lindberg, *ESCA—Atomic, Molecular and Solid State Structure Studied by Means of Electron Spectroscopy*, Nova Acta Regiae Societatis Sci. Upsallensis, Series 4, Vol. 20, pp. 1-282 (1967). (Also available as a U.S. Government Report, A.D. 844 315, October, 1968, available from National Technical Information Service—NTIS, Springfield, Va.)
6. K. Siegbahn, C. Nordling, VG. Johansson, J. Hedmark, P. F. Heden, K. Hamrin, U. Gelius, T. Bergmark, L. O. Werme, R. Manne, and T. Baer, *ESCA Applied to Free Molecules*, North-Holland, Amsterdam (1969).
7. T. A. Carlson, *Photoelectron and Auger Spectroscopy*, Plenum Press, New York (1975).
8. D. Briggs, ed., *Handbook of X-ray and Ultraviolet Photoelectron Spectroscopy*, Heyden and Son, Philadelphia (1977).
9. C. D. Wagner, W. M. Riggs, L. E. Davis, J. F. Moulder, and G. E. Muilenberg, *Handbook of X-ray Photoelectron Spectroscopy*, Perkin-Elmer Corp. (1979).
10. H. Windawi and F. F.-L. Ho, *Applied Electron Spectroscopy for Chemical Analysis*, Wiley, New York (1982).
11. C. R. Brundle and A. D. Baker, eds., *Electron Spectroscopy*, Vols. 1-5, Academic Press, New York (1977-1983).
12. D. T. Clark, ESCA applied to polymers, *Adv. Polym. Sci.* **24**, 125-188 (1977).
13. D. T. Clark and W. J. Feast, Application of ESCA to studies of structure and bonding in polymeric systems, *J. Macromol. Sci., Revs. Macromol. Chem.* **C12**, 191-286 (1975).
14. A. Dilks, XPS for investigation of polymeric materials, in: Reference 11, Vol. 4, pp. 227-359 (1981).
15. A. Dilks, Polymer surfaces, *Anal. Chem.* **53**, 802A-816A (1981).
16. B. D. Ratner, Application of XPS to biomedical polymers: A review, *Ann. Biomed. Eng.* **11**, 313-336 (1983).
17. J. A. Gardella, Jr, and D. M. Hercules, Comparison of static SIMS, ISS, and XPS for surface analysis of acrylic polymers, *Anal. Chem.* **53**, 1879-1884 (1981).

18. Dr. R. N. King (now at Norton-Christensen Co., Salt Lake City).
19. J. H. Weaver and G. Margaritonido, Solid state photoelectron spectroscopy with synchrotron radiation, *Science* **206**, 151-156 (1979).
20. Addresses of XPS manufacturers: (a) Surface Science Laboratories, 1206 Charleston Rd., Mountain View, CA 94043; (b) Kratos Scientific Instruments, Inc., 24 Booker Street, Westwood, NJ 07675; (c) V G Scientific Limited, The Birches Industrial Estate, Imberhorne Lane, East Grinstead, Sussex, England RH19 1QY; (d) Leybold-Heraeus, Inc., 5 Alder Drive, East Syracuse, NY 13057; (e) Perkin-Elmer, Physical Electronics Div., 6509 Flying Cloud Dr., Eden Prairie, MN 55344.
21. D. T. Clark and H. R. Thomas, ESCA: Core levels of simple homopolymers, *J. Polym. Sci., Chem.* **16**, 791-820 (1978).
22. H. R. Thomas, Ph.D. Dissertation, University of Durham (1977).
23. M. M. Millard, Fibers and Polymers, in: *Industrial Applications of Surface Analyses* (L. A. Casper and C. J. Powell, eds.), *Am. Chem. Soc. Symp. Ser.* **199**, 143-202 (1982).
24. B. D. Ratner, P. K. Weathersby, A. S. Hoffman, M. A. Kelly, and L. A. Scharpen, Biomaterial applications as studied by the ESCA technique, *J. Appl. Polym. Sci.* **22**, 643-664 (1978).
25. D. N. Hendrickson, Photoelectron Spectroscopy, Chapter in: R. S. Drago, *Physical Methods in Chemistry*, pp. 566-588, Saunders Publishing Co., Philadelphia (1977).
26. P. W. Atkins, *Physical Chemistry*, W. H. Freeman and Co., San Francisco (1978).
27. J. H. Scofield, Photoionization cross sections at 1254 and 1487 eV, *J. Electron Spectrosc.* **8**, 129-137 (1976).
28. C. J. Powell, Recent progress in quantifications of surface analysis techniques, *Applic. Surface Sci.* **4**, 492-509 (1980).
29. M. P. Seah, Quantitative analysis of surfaces by XPS, *Surface Interface Analysis* **2**, 222-239 (1980).
30. C. J. Powell and P. E. Larson, Quantitative analysis of surfaces by XPS, *Applic. Surface Sci.* **1**, 186-201 (1978).
31. C. J. Powell, in: *Quantitative Surface Analysis of Materials* (N. S. McIntyre, ed.), Am. Soc. Testing Materials Spec. Tech. Publ. **643**, 5-30 (1978).
32. G. K. Iwamoto, R. N. King, and J. D. Andrade in: *Photon, Electron, and Ion Probes of Polymer Structure and Properties*, (D. W. Dwight, T. J. Fabish, and H. R. Thomas, eds.), *Am. Chem. Soc. Symp. Ser.* **162**, 404-418 (1981).
33. W. J. Carter, G. K. Schweitzer, and T. A. Carlson, Model for quantitative analysis in XPS, *J. Electron Spectrosc.* **5**, 827-835 (1974).
34. R. F. Reilman, A. Msezane, and S. T. Manson, Relative intensities in photoelectron spectroscopy of atoms and molecules, *J. Electron Spectrosc.* **8**, 389-394 (1976).
35. M. P. Seah and W. A. Dench, Electron inelastic mean free paths, *Surface Interface Analysis* **1**, 2-11 (1979).
36. C. D. Wagner, L. E. Davis, and W. M. Riggs, Energy dependence of the electron mean free path, *Surface Interface Analysis* **2**, 53-55 (1980).
37. D. R. Penn, Quantitative chemical analysis by ESCA, *J. Electron Spectrosc.* **9**, 29-40 (1976).
38. S. M. Hall, J. D. Andrade, S. M. Ma, and R. N. King, Photoelectron mean free paths in Ba stearate layers, *J. Electron Spectrosc.* **17**, 181-189 (1979).
39. P. Cadman, S. Evans, G. Gossedge, and J. M. Thomas, Electron inelastic mean free paths in polymers, *J. Polym. Sci., Polym. Lett.* **16**, 461-464 (1978).
40. D. T. Clark, H. R. Thomas, and D. Shuttleworth, Electron Mean free paths in polymers, *J. Polymer Sci., Polym. Lett.* **16**, 465-471 (1978).
41. D. T. Clark, Y. C. T. Fok, and G. G. Roberts, Electron mean free paths in Langmuir-Blodgett multilayers, *J. Electron Spectrosc.* **22**, 173-185 (1981).
42. G. K. Wertheim and S. Hufner, ESCA and the quantitative analysis of metallic systems, *J. Inorg. Nucl. Chem.* **38**, 1701-1704 (1976).

43. A. Proctor and P. M. A. Sherwood, Data analysis techniques in XPS, *Anal. Chem.* **54**, 13-19 (1982).
44. W. L. Dunn and T. S. Dunn, An asymmetric model for XPS analysis, *Surface and Interface Analysis* **4**, 77-88 (1982).
45. J. F. M. Pennings and B. Bosman, Analysis of copolymer surfaces, *Colloid Polym. Sci.* **258**, 1099-1103 (1980).
46. C. D. Wagner, L. H. Gale, and R. H. Raymond, Two-dimensional chemical state plots, *Anal. Chem.* **51**, 466-482 (1979).
47. D. T. Clark, ESCA of Polymers, in: *Characterization of Metal and Polymer Surfaces* (L.-H. Lee, ed.), Vol. 2, pp. 5-51, Academic Press, New York (1977).
48. D. T. Clark, and A. Dilks, Shake-up Phenomena in Polymers, *J. Polym. Sci., Chem.* **14**, 533-542 (1976).
49. D. T. Clark, D. B. Adams, A. Dilks, J. Peeling and H. R. Thomas, Shake-up Phenomena in Polymer Systems, *J. Electron Spect.* **8**, 51-60 (1976).
50. J. J. Pireaux, Ph.D. Dissertation, University of Namur, Belgium (1976).
51. D. T. Clark and H. R. Thomas, Core and valence energy levels of a series of polyacrylates, *J. Polym. Sci., Chem.* **14**, 1671-1700 (1976).
52. J. J. Pireaux, J. Riga, R. Caudano, and J. Verbist, in: *Photon, Electron, and Ion Probes of Polymer Structure and Properties*, (D. W. Dwight, T. J. Fabish, and H. R. Thomas, eds.), Am. Chem. Soc. Symp. Ser. **162**, 169-202 (1981).
53. J. J. Pireaux, J. Riga, R. Caudano, J. J. Verbist, J. M. Andre, J. Delhalle, and S. Delhalle, Electronic structure of fluoropolymers, *J. Electron Spect.* **5**, 531-550 (1974).
54. S. Siggia and J. G. Hanna, *Quantitative Organic Analysis Via Functional Groups*, 4th ed., Wiley, New York (1978).
55. C. N. Reilley and D. S. Everhart, in: Reference 10, pp. 105-134 (1982), and *Anal. Chem.* **53**, 665-676 (1981).
56. D. E. Williams, Analysis of polymer surface functional groups, Abstracts, 182nd ACS National Meeting Am. Chem. Soc., p. INDE-23, unpublished preprint (1981).
57. D. Briggs and C. R. Kendall, Derivatization of discharge-treated LDPE, *Int. J. Adhesion and Adhesives*, 13-17 (January, 1982).
58. D. S. Everhart and C. N. Reilley, Functional group mobility, *Surface Interface Analysis* **3**, 126-133 (1981).
59. M. M. Millard and M. S. Masri, Protein functional group modification, *Anal. Chem.* **46**, 1820-1822 (1974).
60. R. A. Dickle, J. S. Hammond, J. E. de Vries, and J. W. Holubka, Surface derivatization of hydroxyl functional acrylic copolymers, *Anal. Chem.* **54**, 2045-2049 (1982).
61. C. D. Batich and R. C. Wendt, in: *Photon, Electron, and Ion Probes*, (D. W. Dwight, T. J. Fabish, and H. R. Thomas, eds.), Am. Chem. Soc. Symp. Ser. **162**, 221-236 (1981).
62. J. J. Pireaux, Smoothing and resolution enhancement, *Appl. Spectrosc.* **30**, 219-224 (1976).
63. A. E. Pavlath and M. M. Millard, XPS through even derivatives, *Appl. Spectrosc.* **33**, 502-509 (1979).
64. A. Proctor and P. M. A. Sherwood, Smoothing: An Extended Sliding Least-Squares Approach, *Anal. Chem.* **52**, 2315-2321 (1980).
65. J. E. Cahill, Derivative spectroscopy, *Am. Lab.* 79-85 (1979).
66. R. P. Vasquez, J. D. Klein, J. J. Barton, and F. J. Grunthaner, Maximum-entropy spectral estimation, *J. Electron Spect.* **23**, 63-81 (1981).
67. A. E. Hughes and C. C. Phillips, Study of transmission function of a commercial hemispherical electron energy analyzer, *Surface and Interface Analysis* **4**, 220-226 (1982).
68. T. A. Carlson, Basic assumptions and recent developments in quantitative XPS, *Surface and Interface Analysis* **4**, 125-134 (1982).
69. J. C. Ashley, Inelastic interactions of low-eenergy electrons with organic solids: Simple formulae, *IEEE Trans. Nucl. Sci.* **NS-27**, 1454-1458 (1980).

70. J. C. Ashley and M. W. Williams, Electron mean free paths in solid organic insulators, *Rad. Res.* **81**, 364-373 (1980).
71. J. C. Ashley, Simple model for electron inelastic mean free paths, *J. Electron Spect.* **28**, 177-194 (1982).
72. J. C. Ashley and C. J. Tung, Electron inelastic mean free paths in several solids, *Surface and Interface Analysis* **4**, 52-55 (1982).
73. D. T. Clark, M. M. Abu-Shbak, and W. J. Brennan, Electron mean free paths as a function of kinetic energy, *J. Electron spect.* **28**, 11-22 (1982).
74. W. A. Fraser, J. V. Florio, W. N. Delgass, and W. D. Robertson, Surface sensitivity and angular dependence of X-ray photoelectron spectra, *Surface Sci.* **36**, 661-674 (1973).
75. C. S. Fadley, Angular-dependent XPS, *Prog. Solid State Chem.* **11**, 265-343 (1976).
76. L. V. Phillips, L. Salvati, W. J. Carter, and D. M. Hercules, Quantitative Surface Analysis by ESCA, in: *Quantitative Surface Analysis of Materials* (N. S. McIntyre, ed.), pp. 47-63, ASTM STP 643, Am. Soc. Testing and Materials, Philadelphia (1978).
77. R. W. Payntor, Modification of the Beer-Lambert Equation for Application to Concentration Gradients, Ph.D. Thesis, University of Surrey, England, October, 1981: *Surface Interface Analysis* **3**, 186-187 (1981).
78. B. D. Ratner, R. W. Payntor, and H. R. Thomas, Polyurethane surfaces—an XPS study, Abstracts, 9th Ann. Meet. Soc. for Biomaterials, April 1983.
79. S. Suzer, Multiplets in atoms and ions displayed by photoelectron spectroscopy, *J. Chem. Educ.* **59**, 814-815 (1982).
80. H. R. Thomas and J. J. O'Malley, Surface studies on multi-component polymer systems by XPS, *Macromolecules* **12**, 323-329 (1979).
81. J. J. O'Malley, H. R. Thomas, and G. M. Lee, Surface studies on multi-component polymer systems by XPS, *Macromolecules* **12**, 996-1001 (1979).
82. I. Elliott, C. Doyle, and J. D. Andrade, Core level sensitivity factors for quantitative XPS, *J. Electron Spect.* **28**, 303-316 (1983).
83. D. T. Clark and H. R. Thomas, Electron mean free paths as a function of kinetic energy in polymeric films, *J. Polym. Sci., Chem.* **15**, 2843-2867 (1977).
84. R. J. Baird, Ph.D. Thesis, University of Hawaii (1977).
85. G. W. Simmons and B. C. Beard, Surface analysis applied to adhesion, Abstracts, Pittsburgh Conf. on Anal. Chem. and Applied Spectrosc., 212, 1981.
86. P. R. Moses, L. M. Wier, J. C. Lennox, H. O. Finklea, J. R. Lenhard, and R. W. Murray, XPS of alkylamine-silanes bound to metal oxide electrodes, *Anal. Chem.* **50**, 576-583 (1978).
87. D. T. Clark, Advances in ESCA applied to polymer characterization, *Pure Appl. Chem.* **54**, 415-438 (1982).
88. T. Ohmichi, H. Tamaki, H. Kawasaki, and S. Tatsuta, in: *Physicochemical Aspects of Polymer Surfaces*, (K. Mittal, ed.), pp. 793-800, Plenum Press, New York (1983).
89. A. Barrie, Instrumentation for electron spectroscopy, in: Reference 8, pp. 79-120 (1977).
90. H. Fellner-Feldegg, U. Gelius, B. Wannberg, A. G. Nilsson, E. Basilier, and K. Siegbahn, New developments in ESCA instrumentation, *J. Electron Spect.* **5**, 643-689 (1974).
91. G. E. McGuire, Instrumental in ESCA, in: Reference 10, pp. 1-18 (1982).
92. C. D. Wagner, Role of Auger lines in XPS, in: Reference 8, pp. 249-272 (1977).
93. C. D. Wagner, D. E. Passoja, H. F. Hillery, T. G. Kinisky, H. A. Six, W. T. Jansen, and J. A. Taylor, Auger and XPS of Al—O and Si—O compounds, *J. Vac. Sci. Technol.* **21**, 933-944 (1982).
94. C. D. Wagner, Combined use of Auger and photoelectron lines, *J. Electron Spect.* **10**, 305-315 (1977).
95. C. D. Wagner, D. A. Zatko, and R. H. Raymond, Use of the oxygen *KLL* Auger lines..., *Anal. Chem.* **52**, 1445-1451 (1980).

96. B. D. Cullity, *Elements of X-ray Diffraction*, Addison Wesley Publ. Co., Reading, Mass. (1956).
97. R. N. King, Ph.D. Dissertation, University of Utah (1980).
98. J. J. Pireaux, J. Riga, R. Caudano, J. J. Verbist, J. Delhalle, S. Delhalle, J. M. Andre, and Y. Gobillon, Polymer primary structures studied by ESCA, *Physica Scripta* **16**, 329-338 (1977).
99. J. D. Andrade, Surface analysis of materials, *Med. Dev. Diag. Ind.*, 22-23 (June, 1980).
100. P. K. Ghosh, *Introduction to Photoelectron Spectroscopy*, Wiley-Interscience, New York (1983).
101. D. A. Stephenson and N. J. Binkowski, XPS of silica, *J. Noncrystalline Solids* **22**, 399-421 (1976).

Note added in proof: An excellent new reference source is: D. Briggs and M. P. Seah, eds., *Practical Surface Analysis*, John Wiley, New York, 1983.

6

Surface Infrared Spectroscopy

Kristine Knutson and Donald J. Lyman

1. INTRODUCTION

Infrared spectroscopy traditionally involved determination of the electromagnetic radiation absorbed by an organic compound as a function of the wavelength of radiation transmitted through the entire sample thickness. The infrared region of electromagnetic radiation includes wavelengths from 7.8×10^{-5} to 1×10^{-1} cm, although the mid-infrared region generally studied includes wavelengths from 2.5×10^{-4} to 5×10^{-3} cm.

Surface infrared spectroscopy couples two powerful techniques, infrared spectroscopy and total internal reflection, to study the surface of a sample within a finite depth into the bulk. The electromagnetic radiation is totally internally reflected through an optically transparent material in contact with the sample. Total internal reflection can occur at the interface of the optically transparent material and the sample only if the index of refraction of the optically transparent material is greater (optically denser) than the index of refraction of the sample (optically rarer). Reflectivity of the interface is a measure of the interaction of an electromagnetic field established within the sample. The spectrum obtained by coupling infrared spectroscopy with internal reflection techniques is characteristic of the sample surface within a finite distance of the interface.

Molecular structure, inter- and intramolecular interactions, conformation, crystallinity, and orientation of organic compounds have been studied by infrared spectroscopy for decades. In addition, internal reflection as a physical phenomenon occurring with the reflection and refraction of electromagnetic radiation at an interface of two media of different indices of refraction has also been studied since the eighteenth century. The coupling

Kristine Knutson • Department of Pharmaceutics, *and Donald J. Lyman* • Department of Materials Science and Engineering, University of Utah, Salt Lake City, Utah 84112. All correspondence should be addressed to Kristine Knutson.

of spectroscopic techniques with internal reflection to obtain an optical spectrum of a sample was independently proposed by Harrick[1-4] and Fahrenfort[5,6] in the early 1960s. Hansen[7] also contributed to the understanding and development of the technique during the early 1960s. Applications of surface infrared spectroscopy have included surface spectra of bulk samples, spectra of tenths of micrograms of powdered samples without scattering, and spectra of organic compounds in solution.

While infrared spectra obtained by total internal reflection closely resemble the spectra obtained by transmission of electromagnetic radiation through the entire sample thickness, there are subtle differences between the spectra obtained by the two different techniques. The differences in the spectra become more apparent with an understanding of the interactions of the electromagnetic field with the sample. To understand the interactions and the resultant consequences, infrared spectroscopy and internal reflection will be briefly discussed. Through these discussions the magnitude of information derived from the coupling of infrared spectroscopy and internal reflection will be apparent. Specific applications of surface infrared spectroscopy in the area of polymers will then be discussed.

2. INFRARED SPECTROSCOPY

Covalently bonded atoms forming an organic molecule (low-molecular weight compounds or higher-molecular weight polymers) do not maintain fixed positions relative to one another but vibrate about an average interatomic distance. Absorption of electromagnetic radiation of the appropriate energy excites the atoms. The absorbed energy is then converted to vibrational and rotational motion. The resulting increase in molecular vibrations leads to changes of the interatomic distances and angles of the molecules. The absorbed energy associated with the vibrations of a covalently bonded molecule is described by quantum mechanics in terms of discrete vibrational energy levels as illustrated in Figure 1. The energy of the different vibrational levels can be expressed by

$$E = (n + \tfrac{1}{2})h\nu \qquad (1)$$

Quantum numbers n (integers 0, 1, 2, etc.) denote the particular energy level, h is Planck's constant (6.63×10^{-27} erg·sec), and ν is the vibrational frequency. The vibrational frequency is related to the speed of light ($c = 3 \times 10^{10}$ cm/sec) and to the wavelength of electromagnetic radiation (λ) initiating the particular excitation by $\lambda = c/\nu$. The wavelengths of radiation capable of exciting a molecule from one vibrational level to another include the infrared region. The energies absorbed (ΔE) by the molecules resulting

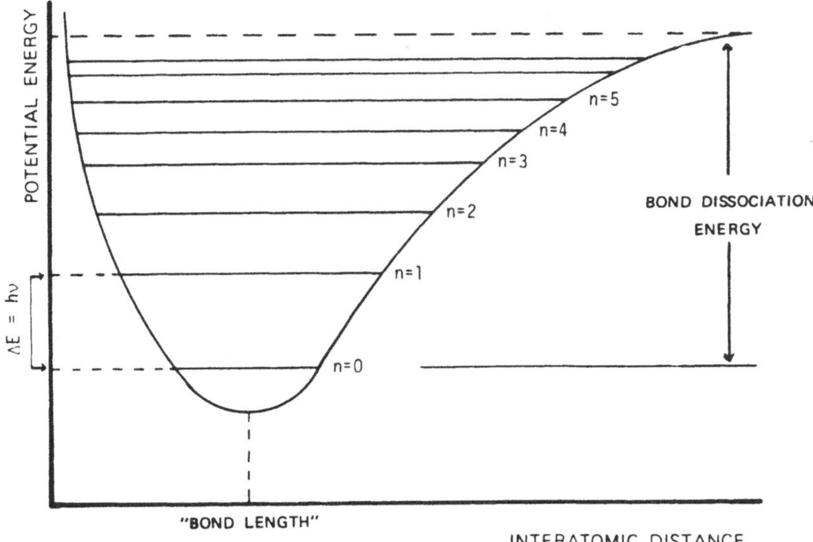

FIGURE 1. Energy level diagram describing energy of molecule versus interatomic distance for each quantum level.

in the vibrational motions within the molecules are several hundred times larger than those associated with rotational motions. Therefore, the discrete vibrational spectral lines are broadened into infrared bands by the associated rotational lines.

The frequency of electromagnetic radiation within the infrared region giving rise to the absorption process depends on the relative masses of the atoms, force constants of the bonds, and geometry of the atoms comprising the molecule. In order to describe such a complex situation, it is assumed the molecular vibrations move atoms toward and away from each other in simple harmonic motion. Such a system can vibrate in a vast number of complex patterns. Vibrational motions in a molecule having N atoms are described with the specification of three coordinates (such as the x, y, and z Cartesian coordinate system) for each nucleus. Therefore, a molecule having N atoms requires $3N$ coordinates and would have $3N$ degrees of freedom. However, not all of the coordinates or degrees of freedom are necessary to describe the vibrational motion of a molecule. Three of the coordinates describe the translational movement of the molecule as a rigid unit through space. Similarly, the rotational movement of the molecule as a rigid unit through space is also described by another set of coordinates. For example, a nonlinear molecule requires three additional coordinates: two angles describing the orientation of a line fixed through the molecule with respect to the coordinate system also fixed in space and another angle describing rotation about the fixed line. In contrast, a linear molecule requires only two additional coordinates to describe the rotational movement. Only one angle is needed to describe the orientation of the linear molecule with respect to the coordinate system fixed in space, while the second angle describes the rotation of the molecule about its axis. Therefore, there are $3N - 6$ coordinates or fundamental degrees of freedom for a nonlinear molecule or $3N - 5$ fundamental degrees of freedom for a linear molecule remaining from the original $3N$ fundamental degrees of freedom to describe the vibrational motions of the molecules.

The molecular vibrations are very complex with the majority of atoms within a molecular group generally contributing to a particular vibration. However, the influence that a molecular group within a polymer chain can have on the vibrational motions of a second group is a function of the interatomic distances separating the two groups. Therefore, some molecular vibrations of molecular groups within a polymer chain can be treated to a good approximation by considering the motions of only a few atoms and disregarding the remainder of the chain. The approximations are usually accomplished by considering the vibrations of selected bonds and the influence of the neighboring molecular groups.

Even though a nonlinear molecule has $3N - 6$ fundamental degrees of freedom available to describe the molecular vibrations, the infrared spectrum will not necessarily contain $3N - 6$ bands. A number of factors determine if a particular vibrational motion will result in a resolvable spectral band. Fundamental vibrational bands are located within particular wavenumber regions in the infrared spectrum of a molecule or molecular group characteristic of the wavelengths of infrared radiation associated with the energy that must be absorbed for the particular vibrational frequencies being excited. The absorption intensity of the band is a function of how effectively the infrared photon energy is transferred to the molecule. The transfer of energy depends on the change in the dipole moment of the molecule occurring as a result of the particular molecular vibration. Since the wavelengths of electromagnetic radiation in the infrared region are greater than the size of most molecules, the electric field of the photon in the vicinity of a molecule is considered uniform over the entire molecule. The electric field of the photon exerts forces on the molecular charges (electrons and protons) of the molecule or molecular group. In addition, the forces on opposite charges must be exerted in opposite directions. Therefore, the oscillating electric field of the photon exerts forces on the molecule inducing the molecular dipole moment to oscillate at the frequency of the photon. At certain frequencies the dipole moment and the atomic nuclei oscillate simultaneously, and these are the frequencies where vibration induces a change in the dipole moment of the molecule. Only spectral bands associated with these vibrations will be present in the infrared spectrum of the molecule since the molecular vibrations must cause a change in the dipole moment if absorption of the electromagnetic radiation within the infrared region is to occur. This requirement is known as the selection rule. The resultant infrared band is a fundamental vibrational band with an intensity proportional to the square of the change in dipole moment.

Additional spectral bands may be present in the spectrum that are not infrared-active fundamental bands. A band resulting from the summation of two or more different vibrational frequencies where the absorbed photon simultaneously excites the different vibrations is a "combination tone." A "difference tone" is an infrared band resulting from a difference between two frequencies where the molecule is already in one excited vibrational state and absorbs enough additional energy to raise to another excited vibrational state. "Overtones," the third type of additional bands, are multiples of a band at a given frequency.

Molecular vibrations can occur without resulting in infrared bands if the requirements of the selection rule are not upheld. In addition, factors other than the selection rule also serve to reduce the number of detectable bands in the infrared spectrum. For example, the wavelength of the electromagnetic radiation absorbed during a particular excitation may fall outside of the infrared region, or outside of the detection capabilities of the particular spectrometer in use. Since the intensity of the infrared band is related to the change in dipole moment, the change in dipole moment may not be large enough to result in a band detectable above the noise level of the spectrometer. Other vibrational motions may absorb close to the same wavelength resulting in overlapping bands. The ability to separate infrared bands arising from different vibrational motions depends on the wavelength difference between vibrations and instrumental capabilities in resolving such differences.

Two distinctive types of molecular vibrations for most molecules are stretching and bending. As an example, the stretching and bending vibrations of a methylene group are illustrated in Figures 2a and 2b. A few characteristic molecular group vibrations and the associated wavenumber regions in infrared spectra are tabulated in Table 1. There are many excellent references[8-15] describing the vibrational motions for various molecular groups and the wavenumber regions where the corresponding bands will appear in the infrared region.

Inter- or intramolecular interactions, conformation, crystallinity, and orientation of low-molecular weight organic compounds and polymer chains can influence the force constants and related vibrational motions of bonds. For example, hydrogen bonding can occur between molecular groups having a proton donor (X—H) and a proton acceptor (Y) where the s orbital of the proton donor overlaps effectively with the π orbital of the proton acceptor. The atoms X and Y are electronegative (i.e., oxygen or nitrogen)

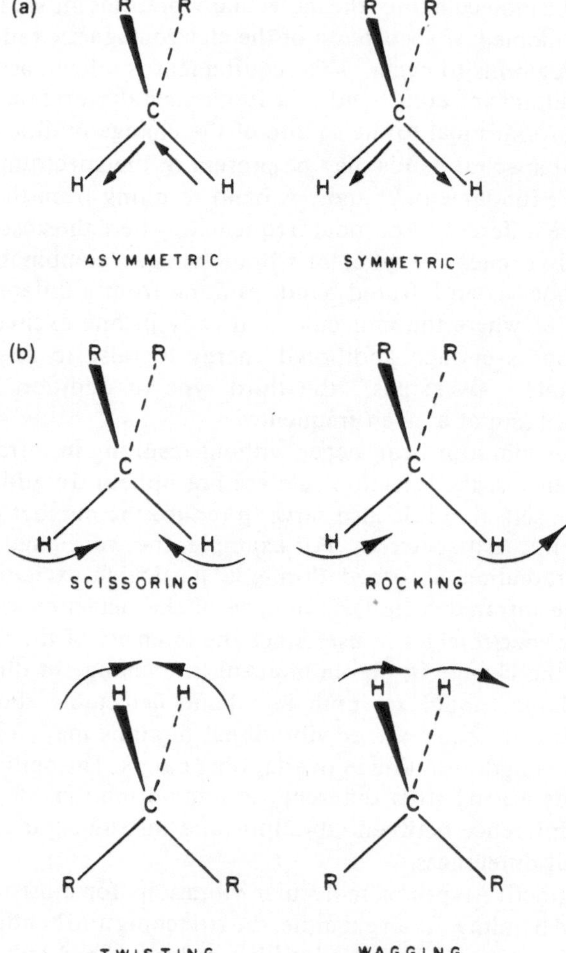

FIGURE 2. (a) Symmetric and asymmetric C—H stretching vibrations for CH_2 group. (b) In-plane (scissoring and rocking) and out-of-plane (twisting and wagging) bending vibrations for CH_2 group.

with Y possessing a lone pair of electrons. Hydrogen bonding alters the force constants of both groups, thus altering the frequencies and corresponding wavelengths of electromagnetic radiation absorbed due to the vibrational motions. The strength of the hydrogen bond is inversely proportional to the distance between X and Y, and is at a maximum when the proton donor group and the axis of the orbital of the lone pair of electrons of the proton acceptor group are collinear. The X—H stretching vibrations of the proton donor group participating in hydrogen bonding will absorb longer

TABLE 1
General Wavenumber Region of Selected Molecular Vibrations

Molecular vibrations	Wavenumber region (cm^{-1})
O—H, N—H stretching	2700-3100
C—H stretching	3200-2500
C=O stretching	1800-1600
C=N stretching	1700-1500
C=C stretching	1700-1500
N—H bending	1700-1400
C—H bending	1500-1300
O—H bending	1500-1200
C—O stretching	1300-900
C—N stretching	1300-900
C—C stretching	1200-800
N—H bending	900-700
C—H bending	900-600

wavelengths of radiation as reflected in a shift of the corresponding band in the infrared spectrum towards lower wavenumbers. In addition, the shifted band can experience an increased intensity and band broadening due to hydrogen bonding. However, the infrared bands arising from the X—H bending vibrations are characterized by a shift in the opposite direction toward shorter wavelengths (higher wavenumbers) due to hydrogen bonding of the proton donor group. The spectral shifts associated with the X—H bending vibrations are less pronounced than the corresponding shifts of bonds arising from the X—H stretching vibrations. The bands associated with stretching vibrations of the proton acceptor (Y) move to longer wavelengths (lower wavenumbers) with hydrogen bonding, although the shift is also less pronounced than the shifts noted in the associated X—H stretching bands. Similarly, alterations in force constants of molecular groups involved in other inter- and intramolecular interactions, conformational changes, orientation, and crystallinity can be reflected in the infrared spectrum by band shifting, broadening, or changes in band intensity.

3. PRINCIPLES OF INTERNAL REFLECTION

An understanding of internal reflections as a physical phenomenon begins with an appreciation of electromagnetic radiation described as electromagnetic waves and of the reflection and refraction properties of the radiation at an interface of two media having different indices of refraction. The vectorial properties of electromagnetic radiation are established with

Maxwell's equations in the presence of matter, describing the electromagnetic radiation in terms of waves. At an interface, the electromagnetic waves must be either entirely or partially reflected back into the original medium or refracted on into the second medium depending on the indices of refraction of the two media and the propagation parameters of the waves. The potential advantages and limitations of coupling infrared spectroscopy with internal reflection can be recognized once a foundation for the understanding of the internal reflection of electromagnetic waves as a physical phenomenon has been established.

3.1. ELECTROMAGNETIC RADIATION DESCRIBED AS WAVES

3.1.1. Vectorial Properties of Electromagnetic Waves

The state of excitation established by the presence of electromagnetic charges within a vacuum or non-vacuum medium is described by electromagnetic theory[16-22] as electromagnetic radiation. The electromagnetic radiation is represented by four vectors: the magnetic vector $\mathbf{H}$; magnetic induction vector $\mathbf{B}$; electric vector $\mathbf{E}$; and the electric displacement vector $\mathbf{D}$. In addition, the electric and electric displacement vectors are related through the dielectric or permittivity constant (ε) by $\mathbf{D} = \varepsilon\mathbf{E}$ if the medium is assumed to be isotropic. The magnetic and magnetic induction vectors are related through the permeability (μ) of the medium by $\mathbf{H} = \mathbf{B}/\mu$, although $\mathbf{B}$ has often been used interchangeably with $\mathbf{H}$ for nonmagnetic media ($\mu = 1$). The Maxwell equations for electromagnetic radiation in the presence of matter are given by Eqs. (2a)–(2d), where

$$\nabla \cdot \mathbf{D} = 4\pi\rho_f \tag{2a}$$

$$\nabla \cdot \mathbf{B} = 0 \tag{2b}$$

$$\nabla \times \mathbf{E} = -(1/c)(\partial\mathbf{B}/\partial t) \tag{2c}$$

$$\nabla \times \mathbf{H} = (1/c)(\partial\mathbf{D}/\partial t) + (4\pi/c)\mathbf{j}_f \tag{2d}$$

ρ_f represents the free charge density and $\mathbf{j}_f$ the free current density. However, the assumed absence of free charges ($\rho_f = 0$) and free currents ($\mathbf{j}_f = 0$) within the medium simplifies Maxwell's equations to a form similar to the presence of electromagnetic radiation within a vacuum. Maxwell's equations must hold true for every point in the continuous regions of the medium. Wave equations in terms of $\mathbf{E}$ or $\mathbf{H}$ are derived from Maxwell's equations by taking the curl of either the curl of $\mathbf{E}$ or $\mathbf{H}$. The partial differential wave equations for **E** and **H** with respect to time and position are given in Eqs. 3a and 3b, respectively.

$$\nabla^2 \mathbf{E} = \left[\frac{\varepsilon\mu}{c^2}\right]\left[\frac{\partial^2 \mathbf{E}}{\partial t^2}\right] \tag{3a}$$

$$\nabla^2 \mathbf{H} = \left[\frac{\varepsilon\mu}{c^2}\right]\left[\frac{\partial^2 \mathbf{H}}{\partial t^2}\right] \tag{3b}$$

The speed of the radiation within the medium (v) is related to the speed of light (c) through the permeability and permittivity of the medium [Eq. (4)].

$$v = c/(\varepsilon\mu)^{1/2} \tag{4}$$

In addition, the permeability and permittivity of the medium can also be equivalently expressed in terms of the index of refraction of the particular medium [Eq. (5)].

$$(\varepsilon\mu)^{1/2} = \eta \tag{5}$$

Therefore, **E** and **H** satisfy the same partial differential wave equation. In addition, each directional component (i.e., E_x and H_x in the x-direction) must also satisfy the same partial differential wave equation. The wave equations and their solutions are not independent because the conditions established by Maxwell's equations in a vacuum must hold true.

The electromagnetic waves are represented by three transverse vectors: the propagation vector $\mathbf{k} = \hat{\mathbf{i}}k_x + \hat{\mathbf{j}}k_y + \hat{\mathbf{k}}k_z$; the electric vector **E**; and the magnetic vector **H**. The propagation vector travels in the direction the electromagnetic waves are propagating with **E** and **H** being normal to **k** and to each other. The three vectors **E**, **H**, and **k** form a right-handed Cartesian coordinate system.

Solutions to the wave equations for **E** and **H** must be functions whose second derivative has the same form as the original function. One solution to the wave equation for **E** [Eq. (3a)] or **H** [Eq. (3b)] is in the form of a cosine function [Eq. (6)].

$$\cos(\omega t - \mathbf{k}\cdot\mathbf{r} + \Theta) \tag{6}$$

The angular frequency (ω) as defined in Eq. (7) is related to the speed of the electromagnetic waves traveling in a non-vacuum medium and the magnitude of the propagation vector.

$$\omega = v|\mathbf{k}| \tag{7}$$

If the electromagnetic waves are traveling in a vacuum, then the speed of the waves would be the speed of light. The direction of travel of the electromagnetic waves is described by the propagation vector $\mathbf{k}$, while the position of the waves is described by the position vector $\mathbf{r} = \hat{\mathbf{i}}x + \hat{\mathbf{j}}y + \hat{\mathbf{k}}z$. The phase constant is Θ. Another solution to the partial differential wave equations is in the form of a complex exponential function having the argument given in Eq. (8).

$$\exp\left[i(\omega t - \mathbf{k} \cdot \mathbf{r} + \Theta)\right] \tag{8}$$

However, the complex exponential solutions to the wave equations have real and imaginary portions, and either portion would also be a solution to the wave equation by the superposition principle. Therefore, equivalent real portions of the trigonometric and exponential solutions to the wave equations can be expressed for $\mathbf{E}$ and $\mathbf{H}$ as given in Eqs. (9a) and (9b), where $\mathbf{E}_0$ and $\mathbf{H}_0$ are constant vectors.

$$\mathbf{E} = \mathbf{E}_0 \cos(\omega t - \mathbf{k} \cdot \mathbf{r} + \Theta) = \mathbf{E}_0 \exp\left[i(\omega t - \mathbf{k} \cdot \mathbf{r} + \Theta)\right] \tag{9a}$$

$$\mathbf{H} = \mathbf{H}_0 \cos(\omega t - \mathbf{k} \cdot \mathbf{r} + \Theta) = \mathbf{H}_0 \exp\left[i(\omega t - \mathbf{k} \cdot \mathbf{r} + \Theta)\right] \tag{9b}$$

3.1.2. Energy Flux of Electromagnetic Radiation

The electromagnetic waves are assumed to be traveling in an isotropic, nonconducting medium in order to simplify the mathematical representation of the waves. The medium is implied to be nonconducting from the previous assumption of the absence of free charges or currents.

The Poynting theorem in electromagnetic theory[16-22] defines the flow of electromagnetic energy per unit time passing a unit surface area by the Poynting vector ($\mathbf{S}$). The Poynting vector is defined in terms of $\mathbf{E}$ and $\mathbf{H}$ for electromagnetic radiation in the presence of matter as given in Eq. (10).

$$\mathbf{S} = (c/4\pi)(\mathbf{E} \times \mathbf{H}) \tag{10}$$

The magnitude of the Poynting vector is a measure of the amount of electromagnetic radiation crossing a unit area per unit time. In addition, the time average of the Poynting vector is the average flow of electromagnetic radiation past a unit surface area normal to the propagation direction of the electromagnetic waves. Therefore, the magnitude of the time average Poynting vector ($S = |\langle \mathbf{S} \rangle|$) is the average radiant energy flux per unit surface area. However, the expression for $\mathbf{S}$ can also be written directly in terms of $\mathbf{E}$. If $\mathbf{E}$ is expressed in trigonometric form, then the time average over many cycles of $\cos^2(\omega t - \mathbf{k} \cdot \mathbf{r} + \Theta)$ is a half ($\frac{1}{2}$). The resultant form of S is given in Eq. (11).

$$S = (c\eta/8\pi\mu)|\mathbf{E}_0|^2 \tag{11}$$

3.2. REFLECTION AND REFRACTION OF ELECTROMAGNETIC WAVES AT AN INTERFACE

Assume electromagnetic waves traveling initially in an isotropic, homogeneous medium having index of refraction η_c encounter an interface with a second isotropic, homogeneous medium having index of refraction η_s. At the interface, a portion of the incident electromagnetic waves (subscript i) is reflected back into the initial medium (subscript r) and a portion is refracted or transmitted on into the second medium (subscript t). The interface is assumed to be in the xy-plane and the plane of incidence is the zy-plane. Two independent cases arise as **E** and **H** are resolved into their respective components with the incident electromagnetic waves propagating in the z-direction and intersecting the interface at $z = 0$. Parallel polarization occurs when **H** is perpendicular to the zy-plane of incidence, while perpendicular polarization occurs when **E** is perpendicular to the zy-plane of incidence. The boundary conditions that must be met at the interface imply only the tangential components of **E** and **H** are continuous across the interface. These conditions are expressed in Eqs. (12a) and (12b) and illustrated in Figure 3 for parallel polarization.

$$E_{yi} + E_{yr} = E_{yt} \tag{12a}$$

$$H_{xi} + H_{xr} = H_{xt} \tag{12b}$$

Similar expressions for the tangential components of **E** and **H** across an interface are written in Eqs. (13a) and (13b) for perpendicular polarization.

$$E_{xi} + E_{xr} = E_{xt} \tag{13a}$$

$$H_{yi} + H_{yr} = H_{yt} \tag{13b}$$

Perpendicular polarization of the electromagnetic waves is illustrated in Figure 4.

Portions of the incident electromagnetic waves traveling along **k** at an incident angle (θ) with the z-axis can be reflected at the interface at an angle equal to the incident angle or can be refracted at an angle different from the incident angle. Snell's Law of Refraction [Eq. (14)] describes the

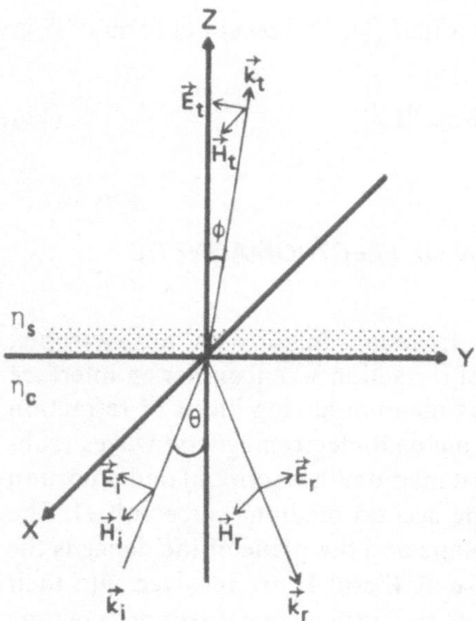

FIGURE 3. Electric and magnetic vector components for reflection and refraction of parallel-polarized incident electromagnetic radiation.

relationship between the incident angle (θ) and the refracted angle (ϕ) as a function of the indices of refraction of the two media.

$$\eta_c \sin \theta = \eta_s \sin \phi \tag{14}$$

The portions of the incident electromagnetic waves reflected back into the initial medium and refracted forward into the second medium are

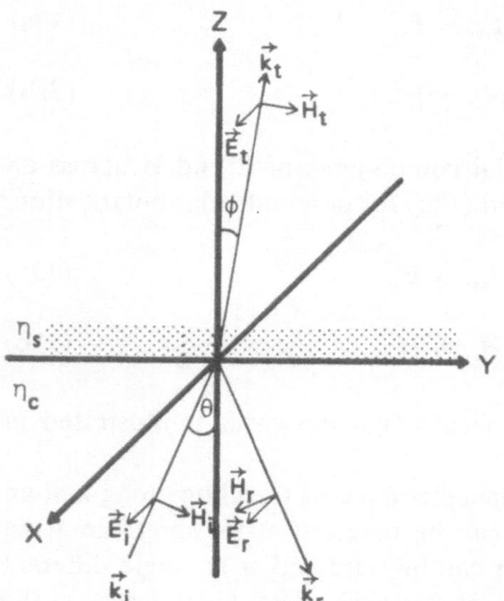

FIGURE 4. Electric and magnetic vector components for reflection and refraction of perpendicular-polarized incident electromagnetic radiation.

determined by considering the boundary conditions. The media are assumed not only to be homogeneous and isotropic, but also to have zero conductivity. Since the media are nonmagnetic, the error in assuming the permeabilities of the two media are unity ($\mu_c = \mu_s = 1$) is negligible. The expressions for the reflected and refracted portions of the incident electromagnetic waves are known as Fresnel equations. The Fresnel equations for parallel polarization can be derived from the boundary conditions and Snell's Law of Refraction in terms of the tangential components of **E**. The reflected and refracted portions of the plane-polarized electromagnetic waves are represented as ratios of the reflected and refracted components of **E** to the incident component of **E** as given in Eqs. (15a) and (15b), respectively.

$$\frac{E_{r\parallel}}{E_{i\parallel}} = \frac{\tan(\theta - \phi)}{\tan(\theta + \phi)} \tag{15a}$$

$$\frac{E_{t\parallel}}{E_{i\parallel}} = \frac{2\sin\phi\cos\theta}{\sin(\theta + \phi)\cos(\theta - \phi)} \tag{15b}$$

The corresponding Fresnel equations for perpendicular polarization can be derived to give Eqs. (16a) and (16b), representing the reflected and refracted portions of plane-polarized incident electromagnetic waves, respectively.

$$\frac{E_{r\perp}}{E_{i\perp}} = \frac{-\sin(\theta - \phi)}{\sin(\theta + \phi)} \tag{16a}$$

$$\frac{E_{t\perp}}{E_{i\perp}} = \frac{2\sin\phi\cos\theta}{\sin(\theta + \phi)} \tag{16b}$$

3.3. TOTAL INTERNAL REFLECTION

Total internal reflection is a special case of the general phenomenon of reflection and refraction of incident electromagnetic radiation at an interface.[1-7,23-59] If the index of refraction of the initial medium is greater than that of the second medium ($\eta_c > \eta_s$) and the incident angle is greater than a critical angle [$\theta > \sin^{-1}(\eta_s/\eta_c)$], then the incident electromagnetic waves are refracted tangential to the *xy* interface plane. The refracted angle becomes imaginary as verified by substitution of the known indices of refraction into Snell's Law of Refraction and of the sine of the incident angle into the trigonometric identity $\sin^2\phi + \cos^2\phi = 1$. The equality is rearranged to give the cosine of the refracted angle in terms of the indices of refraction and the incident angle [Eq. (17)].

$$\cos\phi = \pm i(\eta_c/\eta_s)[\sin^2\theta - (\eta_s/\eta_c)^2]^{1/2} \tag{17}$$

Only the negative sign is used in further derivations in order to have the refracted electric vector amplitude decreasing with increasing distances in the z-direction into the second medium.

The boundary conditions established for reflection and refraction at an interface must still hold for the special case of total internal reflection. In addition to these general boundary conditions, additional boundary conditions include the requirements that the incident waves are totally reflected at the interface and the refracted angle becomes imaginary. The Fresnel equations for total internal reflection consist of only the expressions for the reflected portion of the incident electromagnetic waves after taking into consideration the boundary conditions for total internal reflection. The relationships are derived from the Fresnel equations for the general boundary conditions written in terms of the trigonometric functions of single angles. The sine of the refracted angle is then expressed in terms of the indices of refraction of the two media and the sine of the incident angle from Snell's Law. The imaginary value of the cosine of the refracted angle is also substituted into the relationships. The resultant Fresnel equations for parallel and perpendicular polarization, taking into consideration the additional boundary conditions for total internal reflection, are written as Eqs. (18a) and (18b), respectively.

$$\frac{E_{r\parallel}}{E_{i\parallel}} = \frac{\cos\theta + i(\eta_c/\eta_s)^2[\sin^2\theta - (\eta_s/\eta_c)^2]^{1/2}}{\cos\theta - i(\eta_c/\eta_s)^2[\sin^2\theta - (\eta_s/\eta_c)^2]^{1/2}} \tag{18a}$$

$$\frac{E_{r\perp}}{E_{i\perp}} = \frac{\cos\theta + i[\sin^2\theta - (\eta_s/\eta_c)^2]^{1/2}}{\cos\theta - i[\sin^2\theta - (\eta_s/\eta_c)^2]^{1/2}} \tag{18b}$$

The square of the reflection coefficients for either state of polarization is the *reflectivity*, just as the square of the transmission coefficients is the *transmittivity*. The sum of the reflectivity and transmittivity for either state of polarization must equal unity with reflection and refraction at an interface. However, total internal reflection implies that the incident electromagnetic waves are totally reflected. Therefore, the reflectivities for parallel and perpendicular polarization should equal one. The squares of the Fresnel equations for total internal reflection are completed by multiplying Eqs. (18a) and (18b) by the corresponding complex conjugates. The resultant expressions for the reflectivities are indeed equal to one for both states of polarization, thus verifying total internal reflection of the incident electromagnetic waves.

3.3.1. Presence of an Electromagnetic Field in the Second Medium

In spite of total internal reflection, an electromagnetic field is present in the second, less dense medium.[1-7,16-59] The electromagnetic radiation does not in theory penetrate the second medium during total internal reflection, but flows along the boundary in the plane of incidence. However, at the point of incidence a small amount of radiation does penetrate into the second medium due to diffraction at the edges of the incident beam.[19,59] While the radiation emerges from the plane of incidence back into the first medium a short distance from the point of incidence,[3,4,7,19,22,23,26-34,58,59] there is an absence of radiation flow across the interface between the points of incidence and emergence of the electromagnetic radiation. A phase difference between the incident and reflected electromagnetic waves occurs in order to uphold the boundary conditions establishing total internal reflection. In addition, the phase difference is not identical for both states of polarization.[16-22,58,59] The electromagnetic field in the second medium arises from the presence of the diffracted radiation.

Evanescent Wave. The electromagnetic field established within the second medium is represented by an evanescent wave whose amplitude decreases to zero within a few wavelengths. The refracted portion of the electromagnetic waves can represent the electromagnetic field in the second medium during total internal reflection if the appropriate boundary conditions for total internal reflection are maintained.[16-22] The refracted portion of **E** is expressed in complex, exponential form [Eq. (9a)]. The boundary conditions for total internal reflection are taken into consideration by completing the dot product of **k** and **r**, with the refracted portion of **k** then expressed in terms of the corresponding x, y, and z components to give Eq. (19a).

$$\mathbf{E}_t = \mathbf{E}_{0t} \exp\left[i(\omega t - y|\mathbf{k}| \sin\phi - z|\mathbf{k}| \cos\phi + \Theta)\right] \tag{19a}$$

The trigonometric functions of the imaginary refracted angle ϕ are expressed in terms of real constants and variables. First, $\sin\phi$ is substituted with a corresponding real expression as determined from Snell's Law [Eq. (14)] in terms of the incident angle and the indices of refraction of the two media. The imaginary function of $\cos\phi$ [Eq. (17)] is then substituted into the expression. The relationship for $\mathbf{E}_t$ can then be separated into real and imaginary portions. In addition, the magnitude of **k** is equivalently expressed in terms of the angular frequency and the speed of the electromagnetic waves in the second medium [Eq. (7)]. Rearrangement and further substitution of the speed of the waves in the second medium in terms of the speed of light in a vacuum [Eq. (4)] simplifies the resultant expression. In the real portion of the relationship, the unknown angular frequency of the electromagnetic waves is expressed in terms of known wavelengths of electromagnetic radiation to describe $\mathbf{E}_t$ in the form of Eq. (19b).

$$\begin{aligned}\mathbf{E}_t = \mathbf{E}_{0t} &\exp\{i[\omega t - y(\omega\eta_c/c) + \Theta]\}\\ &\times \exp\{-z(2\pi\eta_c/\lambda)[\sin^2\theta - (\eta_s/\eta_c)^2]^{1/2}\}\end{aligned} \tag{19b}$$

The electromagnetic field in the second medium, as expressed in terms of the refracted portion of the electromagnetic waves with the appropriate boundary conditions for total internal reflection taken into consideration, is indeed an inhomogeneous wave propagating along the interface in the zy plane of incidence. The electromagnetic field in the second medium is real in the z-direction and imaginary in the y-direction.

The amplitude of the electromagnetic wave does not change discontinuously at the interface. Therefore, the continuity of **E** is maintained as required by the general boundary conditions for reflection and refraction at an interface. However, the amplitude of the real portion of the refracted electromagnetic waves is periodic and decays exponentially with distance z into the second medium.[1,2,16-22,27,29,36,39,55,57-59] Therefore, the real portion of the electromagnetic field in the second medium as described by Eq. (19b) is an evanescent wave. The evanescent wave can be described in simpler form by Eq. (20a), where δ is the decay factor given in Eq. (20b).

$$\mathbf{E}_t = \mathbf{E}_{0t} \exp(-z/\delta) \tag{20a}$$

$$\delta = \lambda/2\pi\eta_c[\sin^2\theta - (\eta_s/\eta_c)^2]^{1/2} \tag{20b}$$

$\mathbf{E}_{0t}$ is the initial constant vector of the electromagnetic field originating at the interface within the second medium. The initial constant vector can be derived from the refracted portion of the incident electromagnetic waves being reflected at $z = 0$, as expressed in terms of the transmission coefficients after taking into consideration the additional boundary conditions associated with total internal reflection. However, the transmission coefficients are not identical for parallel and perpendicular plane-polarized incident electromagnetic radiation. Therefore, the initial constant vector for the evanescent wave is also a function of polarization. $\mathbf{E}_{0t}$ for parallel polarization of the incident electromagnetic waves is given by Eq. (21a).

$$\mathbf{E}_{0t\parallel} = \frac{2\cos\theta[2\sin^2\theta - (\eta_s/\eta_c)^2]^{1/2}}{[(\eta_s/\eta_c)^4\cos^2\theta + \sin^2\theta - (\eta_s/\eta_c)^2]^{1/2}} \tag{21a}$$

The corresponding relationship for $\mathbf{E}_{0t}$ with perpendicular polarization is given by Eq. (21b).

$$\mathbf{E}_{0t\perp} = \frac{2\cos\theta}{[1 - (\eta_s/\eta_c)^2]^{1/2}} \tag{21b}$$

While the initial constant vector of the evanescent wave is a function of polarization, the decay of the evanescent wave is a function of the wavelength of radiation, the incident angle, and the indices of refraction of the two media. The influence of these variables will be discussed in Section 3.3.4.

3.3.2. Sampling the Electromagntic Field in the Second Medium

The evanescent wave characterizes the electromagnetic field established within the second medium; therefore, the wave can potentially be very informative about the structure of the second medium. Less than total internal reflection occurs by coupling to the evanescent wave, as indicated by the reflectivity of the interface becoming less than unity. Coupling to the evanescent wave can involve either a redirection or absorption of all or a portion of the radiant energy associated with the electromagnetic field within the second medium.

Frustrated total reflection is a "lossless" coupling to the evanescent wave involving a redirection of the wave. The coupling is termed "lossless" because no radiation associated with the electromagnetic field is lost or absorbed due to the coupling mechanism. There are two frustrated total reflection coupling mechanisms.[20,58,60] One mechanism of redirecting the evanescent wave involves the placement of a third optically transparent medium within a few wavelengths of the interface separating the first and second media to form a second interface between the second and third media. The third medium often has the same index of refraction as the first medium. The second coupling mechanism to redirect the evanescent wave involves changing the index of refraction of the second medium with the appropriate choice of incident angle.

Attenuated total reflection is a "lossy" coupling to the evanescent wave involving a loss of radiant energy associated with the electromagnetic field.[20,58,60] The incident electromagnetic radiation is not totally reflected back into the initial medium at the interface due to an absorption of radiation by the second medium. Polymers are "lossy" materials because some radiation is irreversibly lost due to absorption by the polymer during total internal reflection when the second medium is a polymer. The evanescent wave amplitude is decreased without distortion or becomes attenuated as a result of reflection at the interface. Substitution of a complex index of refraction for the second medium into the reflectivity expressions for parallel and perpendicular polarization derived under the boundary conditions for total internal reflection can give the radiant energy loss per reflection due to the second medium being "lossy." Therefore, attenuated total reflection is the coupling mechanism most often used with total internal reflection when a polymer is to be used as the second medium. However, polymers are only weakly absorbing materials. As a result, multiple internal reflections are often used in conjunction with attenuated total reflection to amplify the effects of only small decreases in reflectivity per reflection. The geometry of the first medium, the internal reflection element, determines the number of reflections.[29,32,34,36,58,59]

3.3.3. Sampling Depth of Electromagnetic Field in the Second Medium

Sampling the electromagnetic field that is represented by the evanescent wave as a function of depth has been discussed in the literature from two viewpoints.[3-7,24,25,27,29,33,35-40,42,55-59] One viewpoint discusses the decrease in reflectivity of the interface associated with total internal reflection in terms of an effective thickness (d_e).[3,4,7,33,35,36,47,56,58,59] The second viewpoint regards the decrease in reflectivity in terms of the decrease in radiant energy as a function of distance into the second medium.[5,6,24,25,27,36,37,39,40,42,57] A depth of penetration (d_p) term is utilized in the discussions associated with both viewpoints of the sampling depth. However, the depth of penetration term has also been defined differently in discussions by various investigators using the same viewpoint of sampling depth. Therefore, it is necessary to understand not only the differences between the two viewpoints of sampling depth, but also the depth of penetration term when comparing results or discussions by different investigators.

3.3.3.1. Effective Thickness. Effective thickness would be the equivalent sample thickness in transmission infrared spectroscopy (transmission of the infrared radiation through the entire thickness of the bulk sample) resulting in the same absorbance as obtained by the decrease in reflectivity per reflection during total internal reflection.[3,4,7,33,35,36,40,42,45,56,58,59] The effective thickness is related to the initial constant vector of the electromagnetic field originating in the second medium and a depth of penetration term as described by Eq. (22).

$$d_e = (\eta_s/2\eta_c \cos\theta)|\mathbf{E}_{0t}|^2 d_p \tag{22}$$

The magnitude of $\mathbf{E}_{0t}$ is a function of the state of polarization as given by Eqs. (21a) and (21b) for parallel and perpendicular polarization, respectively.

If the actual thickness of the second medium is less than the distance the electromagnetic field extends in the z-direction into the second medium, then the depth of penetration term can be replaced by the actual thickness of the medium. The electromagnetic field is considered constant through the entire thickness of the film. However, the effective thickness becomes a function of the index of refraction of the third medium into which the electromagnetic field must extend.[29,36,58,59]

If the thickness of the second medium is greater than the distance the electromagnetic field extends in the z-direction into the second medium, then the electromagnetic field can no longer be considered uniform through the thickness of the medium. The evanescent wave representing the electromagnetic field within the second medium decreases in amplitude as a function of distance into the medium. The depth of penetration term in the effective thickness relationship [Eq. (22)] describes the decay of the evanescent wave.(3,4,29,33,35,36,56,58,59)

Harrick(3,4,33,35) originally defined the depth of penetration term as the distance z into the second medium required for the amplitude of the evancescent wave to decay to half of the value of the initial constant vector of the wave originating at the interface within the second medium. The depth of penetration can be derived from the evanescent wave and the associated decay factor [Eqs. (20a) and (20b)] as outlined in Eqs. (23a) to (23c). First, the evanescent wave relationship is rearranged to describe the decay of the wave in terms of the change in the amplitude of the evanescent wave at some distance z into the second medium ($\mathbf{E}_t$) in terms of the amplitude of the initial constant vector of the wave originating at the interface ($\mathbf{E}_{0t}$) as expressed by the ratio $\mathbf{E}_t/\mathbf{E}_{0t}$.

$$\mathbf{E}_t/\mathbf{E}_{0t} = \exp(-z/\delta) = 1/2 \tag{23a}$$

The relationship is then solved to determine the distance z into the second medium where the evanescent wave has decayed to the defined value as a function of the decay factor (δ).

$$z = \ln(2)\delta \tag{23b}$$

Finally, the distance z defined as the depth of penetration is described in terms of real experimental variables according to the decay factor as given by Eq. (20b).

$$d_p = \ln(2)\lambda/2\pi\eta_c[\sin^2\theta - (\eta_s/\eta_c)^2]^{1/2} \tag{23c}$$

However, the depth of penetration has been redefined since the late 1960s.(29,36,56,58,59) The second version of the depth of penetration definition is the distance z into the second medium where the amplitude of the evanescent wave has decayed to 1/e (rather than half) of the amplitude of the initial constant vector originating at the interface within the second medium. The second definition of the depth of penetration is derived in terms of the distance z and decay factor in a similar manner as the first definition to give Eq. (24).

$$d_p = \lambda/2\pi\eta_c[\sin^2\theta - (\eta_s/\eta_c)^2]^{1/2} \tag{24}$$

The depth of penetration as derived in either version of the definition [Eqs. (23c) and (24)] does not describe the actual distance z the evanescent wave extends into the second medium. However, the term does define the distance z into the second medium at which the amplitude of the evanescent wave has decayed to a predesignated value of the amplitude of the initial constant vector originating at the interface. In addition, the rate of decay is a function of the decay factor (δ).

The effective thickness for a second medium having a thickness greater than the distance z by which the electromagnetic field extends into the second medium can now be determined by substituting the appropriate depth of penetration term into Eq. (22). In addition, the amplitude of the evanescent wave as a function of the state of polarization can be taken into consideration by substitution of the appropriate expression for the initial constant vector for either state of polarization [Eqs. (21a) and (21b)] into Eq. (22). Therefore, the effective thicknesses for parallel and perpendicular polarization where the amplitude of the evanescent wave is defined to decay to half of the amplitude of the initial constant vector originating at the interface are given by Eqs. (25a) and (25b), respectively.

$$d_{e\parallel} = \frac{\cos\theta(\eta_s/\eta_c)[2\sin^2\theta - (\eta_s/\eta_c)^2]\ln(2)\lambda}{\pi\eta_c[(\eta_s/\eta_c)^4\cos^2\theta + \sin^2\theta - (\eta_s/\eta_c)^2][\sin^2\theta - (\eta_s/\eta_c)^2]^{1/2}} \tag{25a}$$

$$d_{e\perp} = \frac{\cos\theta(\eta_s/\eta_c)\ln(2)\lambda}{\pi\eta_c[1 - (\eta_s/\eta_c)^2][\sin^2\theta - (\eta_s/\eta_c)^2]^{1/2}} \tag{25b}$$

If the second definition of the depth of penetration term is substituted into the effective thickness relationship, the effective thicknesses for parallel and perpendicular polarization are given by Eqs. (26a) and (26b), respectively.

$$d_{e\parallel} = \frac{\cos\theta(\eta_s/\eta_c)[2\sin^2\theta - (\eta_s/\eta_c)^2]\lambda}{\pi\eta_c[(\eta_s/\eta_c)^4\cos^2\theta + \sin^2\theta - (\eta_s/\eta_c)^2][\sin^2\theta - (\eta_s/\eta_c)^2]^{1/2}} \tag{26a}$$

$$d_{e\perp} = \frac{\cos\theta(\eta_s/\eta_c)\lambda}{\pi\eta_c[1 - (\eta_s/\eta_c)^2][\sin^2\theta - (\eta_s/\eta_c)^2]^{1/2}} \tag{26b}$$

The sampling depth as discussed in terms of an effective thickness is a function of the state of polarization, wavelength of infrared radiation, incident angle, and the indices of refraction of the two media. The influence of these variables on the effective thickness, as well as the two definitions of the depth of penetration, will be discussed in greater detail in Section 3.3.4.

3.3.3.2. Relative Penetration. Besides effective thickness, the sampling depth of the electromagnetic field in the second medium as represented by the evanescent wave has been discussed in the literature from another viewpoint. The decrease in reflectivity of the interface per reflection is considered by the second viewpoint in terms of a relative penetration distance required for the intensity of the radiant energy associated with the electromagnetic field within the second medium to decay to a defined value.[5,24,25,27,36-39,42,57] The magnitude of the Poynting vector [Eq. (10)] prior to time averaging is a measure of the intensity of the radiant energy crossing a unit area per unit time. In addition, the magnitude of **S** [Eq. (11)] is also proportional to $\eta|\mathbf{E}_t|^2$ where η is the index of refraction of the particular medium through which the radiant energy is traveling. Because the electromagnetic field is present only through the second medium, the decay of radiant energy associated with the electromagnetic field can be represented by the magnitude of the Poynting vector as approximated by the square of the magnitude of the evanescent wave equation [Eq. (20a)]. Therefore, the decay of the electromagnetic field within the second medium is expressed in terms of the magnitude of the evanescent wave constant vector per unit magnitude of the initial constant vector originating at the interface as given by Eq. (27).

$$\left|\frac{\mathbf{E}_t}{\mathbf{E}_{0t}}\right|^2 = \exp(-2z/\delta) \tag{27}$$

In the previous viewpoint of sampling depth, the effective thickness was a function of a depth of penetration term defining the decay of the electromagnetic field in terms of the amplitude of the evanescent wave. However, the second viewpoint of sampling depth defines the decay of the intensity of the radiant energy associated with the electromagnetic field directly in terms of a relative penetration also known as the depth of penetration. In addition, the depth of penetration term in the second viewpoint is a function of $|\mathbf{E}_t|^2$ rather than the amplitude of $\mathbf{E}_t$.[5,6,24,25,27,36,37,39,40,42,57]

Similar to the depth of penetration term in the first viewpoint of the sampling depth, two definitions of the depth of penetration term have also been used in the second viewpoint of the sampling depth. In addition, the difference between the two definitions involves the predesignated decay of the intensity of the radiant energy associated with the electromagnetic field. One definition of the depth of penetration term in the second viewpoint of sampling depth is the distance z into the second medium required for the intensity of the radiant energy associated with the electromagnetic field to decay to half the initial value of the evanescent wave originating at the interface.[5,6,38] The depth of penetration term is derived in a similar manner as in the previous viewpoint of sampling depth, as outlined in Eqs. (28a) to (28c) by designating the decay point of the intensity of the radiant energy, solving Eq. (28a) for z, and then substituting the expression for the decay factor [Eq. (20b)] in terms of real, known variables.

$$\left|\frac{\mathbf{E}_t}{\mathbf{E}_{0t}}\right|^2 = \exp(-2z/\delta) = 1/2 \tag{28a}$$

$$d_p = z = [\ln(2)\delta]/2 \tag{28b}$$

$$d_p = \ln(2)\lambda/4\pi\eta_c[\sin^2\theta - (\eta_s/\eta_c)^2]^{1/2} \tag{28c}$$

The second definition of the depth of penetration term in the particular viewpoint of sampling depth under consideration redefines the term as the distance z into the second medium required for the intensity of the radiant energy to decay to $1/e$ (rather than half) of the initial value originating at the interface.[24,25,27,37,39,42] The depth of penetration is derived from Eq. (27) according to the second definition to give Eq. (29).

$$d_p = \lambda/4\pi\eta_c[\sin^2\theta - (\eta_s/\eta_c)^2]^{1/2} \tag{29}$$

The sampling depth as discussed in terms of a relative penetration is defined directly in terms of the depth of penetration term. The depth of penetration is essentially a decay function defining the distance into the second medium required for the intensity of the radiant energy associated with the electromagnetic field within the second medium to decay to a predesignated value. The term is directly proportional to the wavelength of infrared radiation, incident angle, and indices of refraction of the two media as discussed in Section 3.3.4. While not directly apparent, the term is indirectly a function of the state of polarization of the incident electromagnetic waves. The change in magnitude of the evanescent wave is considered per unit magnitude of the initial constant vector originating at the interface within the second medium, although the initial constant vector is a function of the state of plane polarization of the incident electromagnetic waves as given in Eqs. (21a) and (21b).

3.3.4. Sampling Depth as a Function of Experimental Variables

Sampling depth of the electromagnetic field within the second medium as considered from either of the two viewpoints is a function of the experimental variables including the particular wavelength of radiation, incident angle, indices of refraction, and the state of polarization of the incident electromagnetic radiation. The effective coupling of the combined influence of the individual experimental variables allows the technique to range from surface to subsurface sensitivity. However, the relative influence of the experimental variables can be illustrated from one of the two viewpoints of sampling depth. Therefore, the sampling depth as a function of the experimental variables will be discussed for the individual variables using the second viewpoint of the sampling depth in terms of a relative thickness as expressed by the depth of penetration term in Eq. (29) in Section 3.3.4.1. The different viewpoints of sampling depth and the associated definitions of the depth of penetration term will then be compared in Section 3.3.4.2 using one particular set of values for the experimental variables.

3.3.4.1. Influence of Experimental Variables. Synthetic polymers may not have the necessary materials and optical properties to rigidly uphold some of the assumptions involved with several of the derivations for reflection and refraction at an interface. Both media were assumed to be homogeneous, isotropic, and nonconducting with an absence of free, unbound currents and charges. In addition, the second medium was assumed to only very weakly absorb infrared radiation.

During attenuated total reflection studies of polymers, the polymer is the second, less dense medium. Most synthetic polymers are weakly absorbing, nonconducting materials. However, the anisotropy and homogeneity of polymers can differ between polymers of varying chemical structure, as well as between different samples of the same polymer. The indices of refraction (η_s) for several polymers of different chemical structure are tabulated in Table 2. The indices of refraction range from 1.38 to 1.63, although an average value of the index of refraction of the second medium (η_s) is assumed to be 1.5 for the following evaluations of the sampling depth.

The first medium in attenuated total reflection studies is the internal reflection element. Crystalline materials used as internal reflection elements have included germanium (Ge), silicon (Si), zinc selenide (ZnSe), and thallium bromide-thallium iodide (KRS-5). The indices of refraction and transmission regions of these materials are tabulated in Table 3. Crystalline materials cut as internal reflection elements have indices of refraction (η_c) greater than the indices of refraction (η_s) of most synthetic polymers. Decreasing the index of refraction of the first medium by using a different crystalline material increases the sampling depth into the second medium if other experimental variables are held constant.

It is also important to recognize the sampling depth is not constant over the entire mid-infrared region. Polymers have molecular vibrational modes absorbing between 4000 and 400 cm^{-1}. In addition, the crystalline materials used as internal reflection elements may either totally absorb infrared radiation on the edges of the mid-infrared region or have characteristic vibrational modes absorbing infrared radiation in portions of the region. Infrared bands associated with some internal reflection elements may be intense and can potentially mask weaker bands associated with the polymer absorbing within the same portion of the mid-infrared region. In addition, the sampling depth is directly proportional to the particular

TABLE 2
Indices of Refraction of Synthetic Polymers

Polymer	Index of refraction (η_s)[a]
Polyacrylonitrile–butadiene–styrene	1.54
Polyacetal	1.48
Polyacrylonitrile	1.52
Polyamide-6	1.53
Polyamide-6,6	1.48–1.58
Poly-*trans*-1,4-butadiene	1.52
Polybutadiene–styrene	1.56
Polycarbonate	1.59
Polychloroprene	1.56
Polyethylene (amorphous)	1.49
Polyethylene (crystalline)	1.52–1.58
Poly(ethylene -terephthalate)	1.58–1.64
Polyisoprene	1.52
Polymethylmethacrylate	1.49
Polyoxymethylene	1.49–1.55
Polypropylene	1.49
Polystyrene	1.59–1.60
Polysulfone	1.63
Polytetrafluoroethylene	1.38
Polyurethane	1.50–1.60
Poly(vinyl acetate)	1.47
Poly(vinyl chloride)	1.54–1.55
Silicone rubber	1.43

[a] Reference (61).

wavelength of infrared radiation being reflected with attenuated total reflection [Eq. (29)]. Therefore, the sampling depth increases with longer wavelengths (lower wavenumbers) over the mid-infrared region.

The dependence of sampling depth on the wavelength of infrared radiation has two subsequent effects on the infrared spectra obtained by

TABLE 3
Materials and Optical Properties of
Internal Reflection Elements

Crystalline material	Index of refraction (η_c)	Transmission region (cm^{-1})
Germanium (Ge)	4.00	4000–900
Silicon (Si)	3.42	4000–1500
Zinc selenide (ZnSe)	2.42	4000–700
Thallium bromide–thallium iodide (KRS-5)	2.35	4000–400

attenuated total reflection as compared with transmission spectra of the same sample if all other variables are held constant. The first effect is that two infrared bands of equal absorbance (intensity) absorbing at different wavelengths in transmission spectra will have unequal absorbance in the attenuated total reflection spectra. In particular, the infrared bands at higher wavelengths (lower wavenumbers) will have higher absorbance than bands absorbing at lower wavelengths (higher wavenumbers) because of the difference in sampling depth. The second effect of the dependence of sampling depth on wavelength of infrared radiation is that single, symmetric bands in transmission spectra appear asymmetric in attenuated total reflection spectra. The higher wavelength (low wavenumber) side of the single infrared band will have a higher absorbance than the lower wavelength (higher wavenumber) side of the band.

The sampling depths are plotted in Figure 5 for the four different crystalline materials commonly used as internal reflection elements. The index of refraction of the second medium (η_s) is assumed to be 1.50 and the incident angle (θ) is assumed to be 45°. The sampling depth for each material is plotted over the approximate wavenumber region the particular material transmits infrared radiation (Table 3). The sampling depths increase for internal reflection elements having lower indices of refraction. Also,

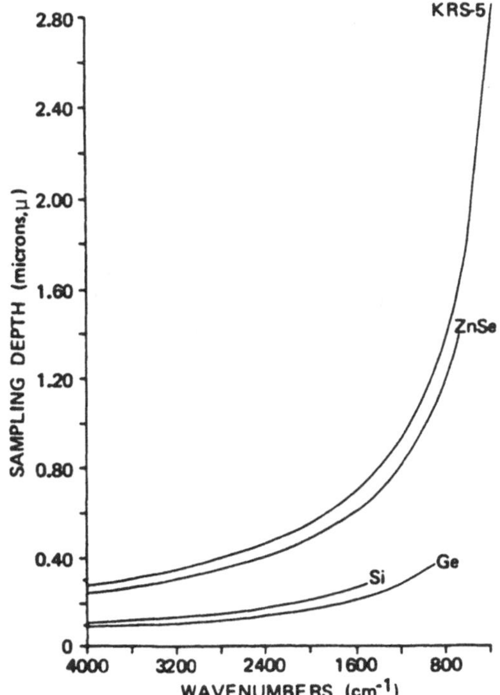

FIGURE 5. Sampling depth versus wavenumber curves for KRS-5, ZnSe, Si, and Ge internal reflection elements having 45° incident angles.

sampling depths increase at higher wavelengths (lower wavenumbers) for a particular crystalline material. For example, the germanium element having the highest index of refraction ($\eta_c = 4.00$) has sampling depths ranging from 0.083 microns (μ) [830 angstroms (A)] at 4000 cm^{-1} to 0.369 μ (3690 A) at 900 cm^{-1}. In contrast, the KRS-5 element having the lowest index of refraction ($\eta_c = 2.35$) has sampling depths ranging from 0.278 μ (2780 A) at 4000 cm^{-1} to 2.782 μ (27,820 A) at 400 cm^{-1}. The variations in sampling depths for internal reflection elements having similar indices of refraction can be relatively minor considering the theoretical assumptions made in the derivation of the sampling depth equations, as well as possible experimental errors. Possible experimental errors include variations in internal reflection element geometry (size and endface angle), precise alignment of the element at the particular optics angle, uniformity of structure within the surface regions along the length of the polymer sample, and uniformity of contact between the sample and internal reflection element. The small difference of 0.07 between the index of refraction of KRS-5 ($\eta_c = 2.35$) and the index of refraction of zinc selenide ($\eta_c = 2.42$) may lead to a difference in sampling depths of less than 0.040 μ (400 A) at 4000 cm^{-1}. As in sampling depth, the variations in sampling depths also decrease between crystalline materials having higher indices of refraction. Only a 0.022 μ (220 A) variation in sampling depths at 4000 cm^{-1} occurs with a 0.58 difference between indices of refraction for silicon ($\eta_c = 3.42$) and germanium ($\eta_c = 4.00$). If only minor fluctuations in sampling depths occur between two different crystalline materials to be used as the internal reflection element, then the transmission region, mechanical properties, and chemical resistance of the materials become important factors in choosing the particular material to be used. KRS-5 and zinc selenide have longer transmission regions than germanium and silicon (Table 3). In addition, germanium and silicon are hard, brittle, and chemically inert materials. Zinc selenide is considered relatively soft and brittle, although the material is resistant to water and many solvents. However, H_2Se can be generated if zinc selenide is exposed to some acids. Compared to zinc selenide, KRS-5 is relatively soft and pliable. KRS-5 is resistant to some solvents (e.g., methanol, toluene and acetone), but is attacked by warm water, bases, or acids. Therefore, the decrease in sampling depth with zinc selenide as compared to KRS-5 may be irrelevant considering the increased hardness and chemical resistance of zinc selenide as compared to KRS-5. However, if the vibrational modes of interest associated with the polymer sample absorb between 900 and 400 cm^{-1}, the longer transmission region of KRS-5 will become a major consideration in choosing the appropriate crystalline material to be used as the internal reflection element.

Sampling depths can also be varied for a particular internal reflection element to decrease or increase surface sensitivity by changing the incident

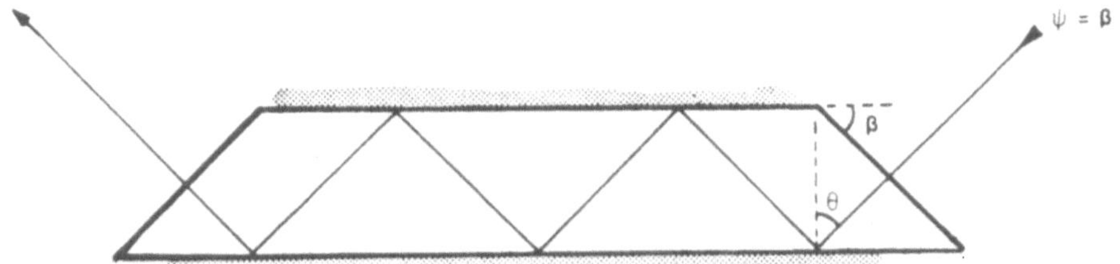

FIGURE 6. Multiple internal reflection of electromagnetic radiation within internal reflection element with equal variable angles ($\beta = \psi = \theta$).

angle (θ). The optics of internal reflection cells are designed to have either fixed or continuously varied incident angles. Fixed incident angles are usually 30°, 45°, and 60°. Continuously varied incident angles range from under 20° to nearly 90° depending on the particular cell design. However, the incident angle is actually a function of two variable angles. The endfaces of internal reflection elements are cut at a particular endface angle (β), usually 30°, 45°, or 60° depending on the crystalline material and the desired surface sensitivity. The second variable angle is the optics angle (ψ), which is determined by the optical alignment of the internal reflection cell. Optics angles can be either fixed or continuously varied depending on the particular design of the internal reflection cell.

If the two variable angles are identical, then the particular angle is the incident angle ($\theta = \beta = \psi$). The electromagnetic waves entering the internal reflection element are normal to the endface of the element if the endface and optics angles are equal as illustrated in Figure 6. The incident angle is then the angle between the z-axis and the ir..ident electromagnetic waves striking the interface between the internal reflection element and the polymer sample. Figure 7 illustrates the sampling depths for the four crystalline materials over the mid-infrared region with 30°, 45°, and 60° incident angles. For a particular crystalline material, the sampling depths are increased with lower incident angles. For example, the sampling depth of germanium at 4000 cm^{-1} varies from 0.064 μ (640 A) for a 60° incident angle to 0.083 μ (830 A) for a 45° incident angle and to 0.150 μ (1500 A) for a 30° incident angle. However, similar sampling depths can also be obtained with different combinations of crystalline materials and incident angles. For example, a silicon internal reflection element with an incident angle of 30° has essentially the same sampling depth over the 4000 to 1500 cm^{-1} region as a zinc selenide element with an incident angle of 45°. The sampling depths are also similar over the 4000 to 900 cm^{-1} region for a germanium element with a 30° incident angle as compared to either a zinc selenide or KRS-5 element with 60° incident angles.

Not all crystalline materials generally used as internal reflection elements can be functional elements at all of the common incident angles. As

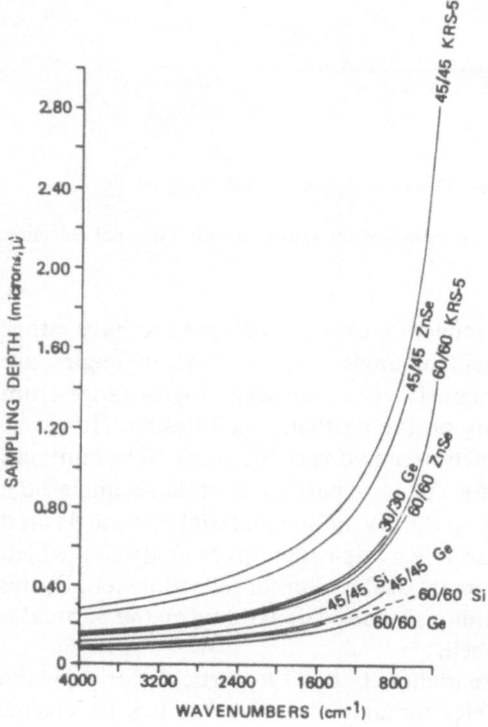

FIGURE 7. Sampling depth versus wavenumber curves for KRS-5, ZnSe, Si, and Ge internal reflection elements having 30°, 45°, and 60° incident angles.

discussed in Section 3.3., total internal reflection of the incident infrared radiation occurs only if the incident angle is greater than the critical angle. If the index of refraction of the second medium is assumed to be 1.50, the critical angles for the four crystalline materials are those tabulated in Table 4. The critical angles for zinc selenide and KRS-5 are greater than 30°; therefore, total internal reflection does not occur if these materials are to be used as internal reflection elements at incident angles below the critical angle. The sampling depth curves for the crystalline materials at incident angles below the critical angle are not included in Figure 7.

TABLE 4
Critical Angles of Internal Reflection Elements

Crystalline material	Indices of refraction	Critical angle
Ge	4.00	22.0°
Si	3.42	26.0°
ZnSe	2.42	38.3°
KRS-5	2.35	39.7°

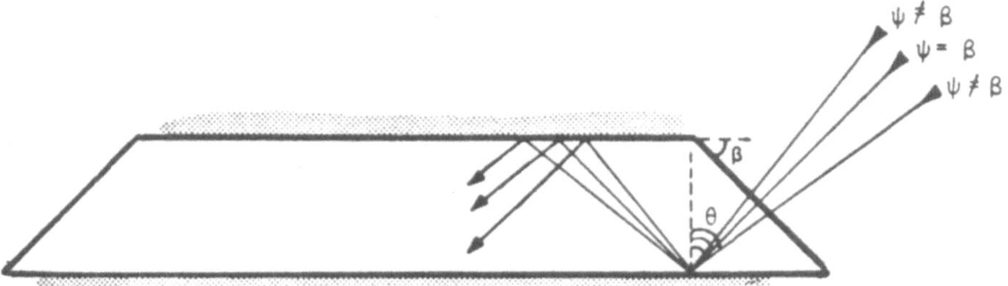

FIGURE 8. Multiple internal reflection of electromagnetic radiation within internal reflection element with unequal variable angles ($\beta \neq \psi$).

The sampling depth can also be varied by altering either of the two variable angles. In this case, the electromagnetic waves are no longer normal to the endface of the internal reflection element as the waves enter the element. Figure 8 illustrates the effect of unequal variable angles on the incident angle at the interface between the internal reflection element and the polymer. The incident angle is a function of the two variable angles (endface and optics) and the index of refraction of the crystalline material being used as the internal reflection element as described by Eq. (30).

$$\theta = \beta - \sin^{-1}\left[\frac{\sin(\beta - \psi)}{\eta_c}\right] \tag{30}$$

The incident angles for the four crystalline materials with different variable angles are tabulated in Table 5.

TABLE 5
Incident Angles for Internal Reflection Elements Having Different Variable Angles and Indices of Refraction

Endface angle (β) at optics angle (ψ) (β/ψ)	Incident angles for different materials (°)			
	Ge	Si	ZnSe	KRS-5
60/60	60.0	60.0	60.0	60.0
60/45	56.3	55.7	53.9	53.7
60/30	52.8	51.6	48.1	47.7
45/60	48.7	49.3	51.1	51.3
45/45	45.0	45.0	45.0	45.0
45/30	41.3	40.7	38.9	(38.7)[a]
30/60	37.2	38.4	41.9	42.3
30/45	33.7	34.3	(36.1)[a]	(36.3)[a]
30/30	30.0	30.0	(30.0)[a]	(30.0)[a]

[a] Incident angle below critical incident angle.

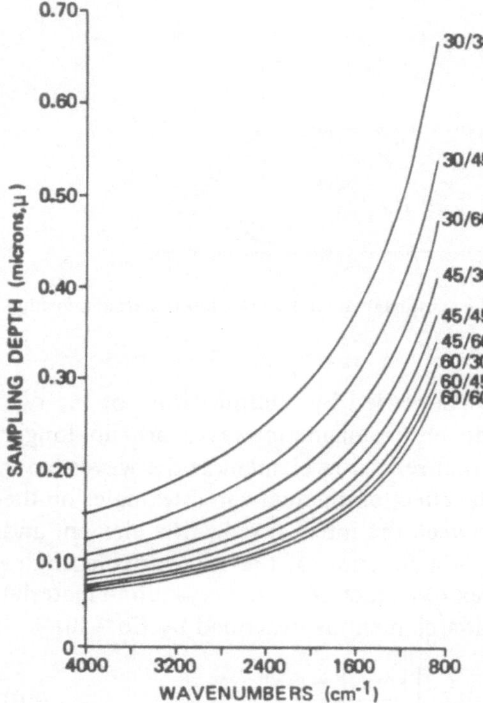

FIGURE 9. Sampling depth versus wavenumber curves for Ge internal reflection elements having endface angles β aligned at optics angles ψ (β/ψ).

The sampling depth is more sensitive to variations in the endface angle (β) for the two materials (germanium and silicon) having higher indices of refraction. Figure 9 illustrates the sampling depths for a germanium element having an endface angle β aligned at an optics angle ψ as denoted by β/ψ. The sampling depth can be significantly changed for a particular optics angle by varying the endface angle. As an example, an optics angle set at 45° gives sampling depths at 4000 cm^{-1} of 0.067 μ (670 A), 0.083 μ (830 A), and 0.122 μ (1220 A) for endface angles of 60°, 45°, and 30°, respectively. In contrast, the influence of varying the optics angle while maintaining the same endface angle is less significant for the germanium element. The sampling depths at 4000 cm^{-1} with a 45° endface angle are 0.076 μ (760 A), 0.083 μ (830 A), and 0.092 μ (920 A) for 60°, 45°, and 30° optics angles, respectively.

For the two crystalline materials (zinc selenide and KRS-5) having lower indices of refraction, sampling depth becomes as sensitive to the optics angle as to the endface angle. In addition, the sensitivity of the sampling depth to the optics angle becomes more significant with lower variable angles. For example, Figure 10 illustrates the sampling depths for zinc selenide elements having different endface angles aligned at various

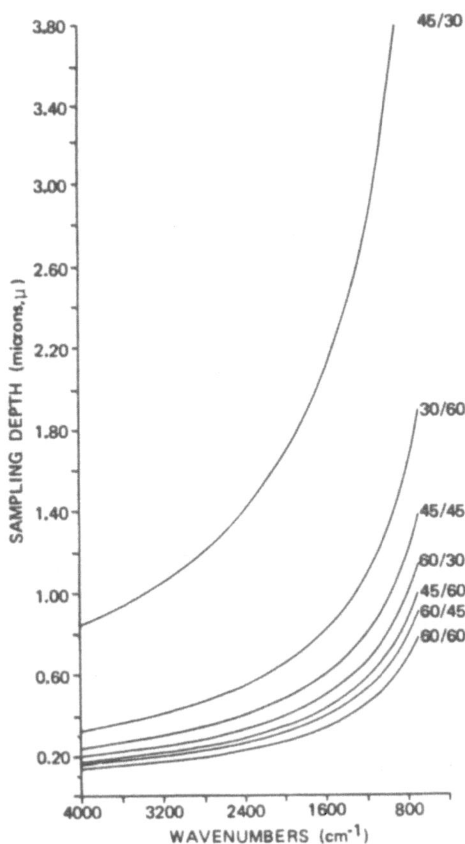

FIGURE 10. Sampling depth versus wavenumber curves for ZnSe internal reflection elements having endface angles β aligned at optics angle ψ (β/ψ).

optics angles. The sampling depths at 4000 cm^{-1} for a 60° optics angle are 0.136 μ (1360 A), 0.174 μ (1740 A), and 0.330 μ (3300 A) for 60°, 45°, and 30° endface angles. While holding the endface angle constant at 60° and varying the optics angle, the sampling depths at 4000 cm^{-1} are only 0.136 μ (1360 A), 0.159 μ (1590 A), and 0.200 μ (2000 A) at 30°, 45°, and 60° optics angles, respectively. The increasing influence of the optics angle on the sampling depth becomes more apparent when the particular variable angle being held constant is a lower angle. For example, zinc selenide elements with 60° and 45° endface angles have sampling depths at 4000 cm^{-1} ranging from 0.159 μ (1590 A) to 0.242 μ (2420 A) at a 45° optics angle. Infrared radiation is not transmitted by the zinc selenide internal reflection element with a 30° endface angle at a 45° optics angle because the incident angle is below the critical angle. However, the incident angle for zinc selenide is not below the critical angle when the endface angle is held constant at 45° and the optics angle is 30°. Therefore, the sampling depths at 4000 cm^{-1}

are 0.174 μ (1740 A), 0.242 μ (2420 A), and 0.845 μ (8450 A) for a zinc selenide element with a 45° endface angle at 60°, 45°, and 30° optics angles, respectively. The limitations of using crystalline materials with lower indices of refractions become more restrictive when the variable angle being held constant is 30°. While the crystalline materials with lower indices of refraction have a wide range of sampling depths with subsurface sensitivities, the number of effective variable angle combinations giving an incident angle above the critical angle is limited. At a constant 30° optics angle, the sampling depths for zinc selenide at 4000 cm^{-1} are 0.200 μ (2000 A) with a 60° endface angle and 0.845 μ (8450 A) with a 45° endface angle. A variable angle combination of a 30° optics angle with a 30° endface angle results in an incident angle below the critical angle of incidence. In contrast, a constant 30° endface angle results in a sampling depth at 4000 cm^{-1} of 0.330 μ (3300 A) with a 60° optics angle. However, the incident angle for zinc selenide with a constant endface angle of 30° at either a 45° or 30° optics angle is less than the critical angle of incidence.

The influence of experimental variables on the sampling depth of the evanescent wave in the second medium can be qualitatively summarized for the individual experimental variables. First, sampling depth increases with higher wavelengths (lower wavenumbers) and internal reflection elements having lower indices of refraction. Second, the sampling depth increases for a particular wavelength and crystalline material with lower incident angles.

3.3.4.2. Influence of Sampling Depth Viewpoint. The sampling depth of the electromagnetic field within the second medium is discussed in the literature[3-7,24,25,27,29,33,35-40,42,55-59] and in Section 3.3.3. in terms of either an effective thickness or a relative penetration depth into the medium. When the second medium is thicker than the sampling depth, either definition involves a depth of penetration term. Both viewpoints of the sampling depth have been defined in terms of real experimental variables including the wavelength of infrared radiation, indices of refraction of the two media, and the incident angle. The influence of these four experimental variables on the sampling depth was discussed in Section 3.3.4.1. as related to the second viewpoint of sampling depth, with the relative depth of penetration defined by Eq. (29). However, the sampling depth defined by either viewpoint will be influenced by the four experimental variables to the same relative degree as discussed in the previous section for the specific viewpoint of the sampling depth. Therefore, the discussions in the previous section provided the basis for the different viewpoints of sampling depth and corresponding depth of penetration terms to be discussed using only one particular set of experimental variables. The sampling depths and associated depth of penetration terms will be compared assuming germanium (η_c = 4.00) is used as the internal reflection element with an incident angle of 45° ($\theta = \beta = \psi = 45°$).

Sampling depth as defined from the first viewpoint discusses the decrease in reflectivity associated with total internal reflection in terms of an effective thickness (Section 3.3.3.1.). The effective thickness is the equivalent sample thickness in transmission infrared spectroscopy that results in the same absorbance as obtained per single reflection in attenuated total reflection. Sampling depths defined in terms of effective thickness for sample thicknesses greater than the depth the electromagnetic field extends into the second medium also involve a depth of penetration term as given in Section 3.3.3.1. by Eq. (22). Therefore, the depth of penetration terms must be compared prior to comparing the two effective thicknesses.

The depth of penetration term (d_p) associated with the effective thickness is defined as the distance into the second medium at which the amplitude of the evanescent wave decays to either half or 1/e of the amplitude of the initial constant vector originating at the interface [Eqs. (23c) and (24)]. The two depth of penetration terms are plotted in Figure 11 for a germanium internal reflection crystal at a 45° incident angle. The amplitude of the evanescent wave decays to half of the amplitude of the initial constant vector originating at the interface as a function of wavelength within distances of 0.115 μ (1150 A) at 4000 cm^{-1} to 0.511 μ (5110 A) at 900 cm^{-1}. However, the evanescent wave decays to 1/e of the amplitude of the initial

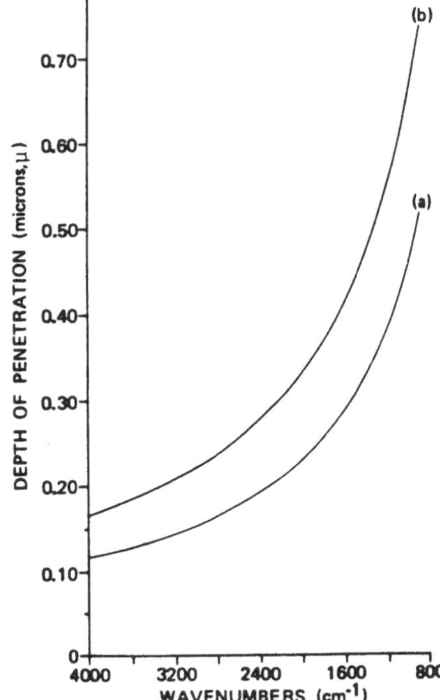

FIGURE 11. Depth of penetration versus wavenumber curves for Ge internal reflection elements having 45° incident angles. (a) Depth of penetration defined as distance into second medium required for amplitude of evanescent wave to decay to half of initial constant vector [Eq. (23c)]. (b) Depth of penetration defined as distance into second medium required for amplitude of evanescent wave to decay to 1/e of initial constant vector [Eq. (24)].

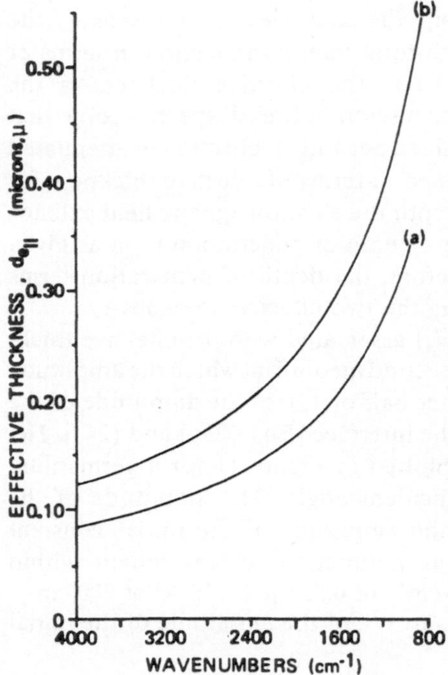

FIGURE 12. Effective thickness versus wavenumber curves for Ge internal reflection elements having 45° incident angles with parallel polarization. (a) Effective thickness [Eq. (25a)] with associated depth of penetration defined as distance into second medium required for amplitude of evanescent wave to decay to half of initial constant vector. (b) Effective thickness [Eq. (26a)] with associated depth of penetration defined as distance into second medium required for amplitude of evanescent wave to decay to 1/e of initial constant vector.

constant vector originating at the interface within distances of 0.1660 μ (1660 A) at 4000 cm^{-1} to 0.7370 μ (7370 A) at 900 cm^{-1} for the same crystalline material and incident angle. Therefore, the two definitions of the depth of penetration differ by the constant ln (2).

In addition to the depth of penetration term, the effective thickness is also a function of the state of polarization of the initial electromagnetic waves originating at the interface because the initial constant vector of the electromagnetic field at the interface is not assumed to be unity in these particular definitions of sampling depth. Sampling depths defined in terms of effective thickness for parallel polarization and the two depths of penetration [Eqs. (23c) and (24)] in terms of the experimental variables are given in Eqs. (25a) and (26a). The effective thicknesses for parallel polarization are illustrated in Figure 12 for a germanium internal reflection element at a 45° incident angle over the 4000 to 900 cm^{-1} region of the infrared spectrum. Similar effective thicknesses for perpendicular polarization are given in Eqs. (25b) and (26b) for the two depths of penetration. The effective thicknesses for perpendicular polarization are plotted in Figure 13, also for a germanium internal reflection element at a 45° incident angle. The two effective thickness functions for each state of polarization differ only by the constant ln (2). However, parallel polarization requires a larger effective

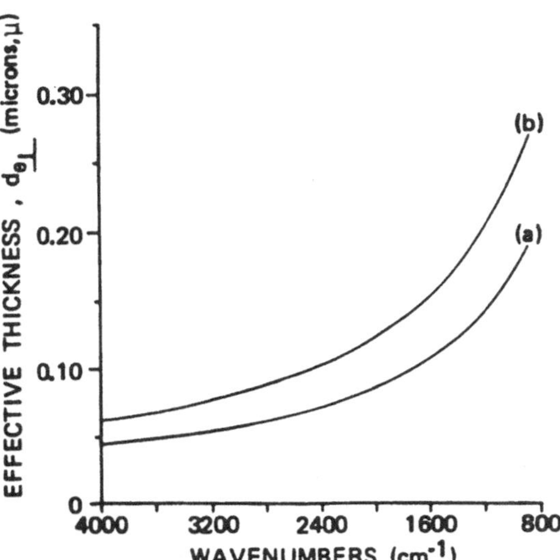

FIGURE 13. Effective thickness versus wavenumber curves for Ge internal reflection elements having 45° incident angles with perpendicular polarization. (a) Effective thickness [Eq. (25b)] with associated depth of penetration defined as distance into second medium required for amplitude of evanescent wave to decay to half of initial constant vector. (b) Effective thickness [Eq. (26b)] with associated depth of penetration defined as distance into second medium required for amplitude of evanescent wave to decay to 1/e of initial constant vector.

thickness to attain an equivalent decay of the amplitude of the evanescent wave than does perpendicular polarization. For example, the effective thickness defined in terms of the second definition of the depth of penetration (amplitude of the evanescent wave decays to 1/e of initial constant vector) is 0.123 μ (1230 A) for parallel polarization and only 0.061 μ (610 A) for perpendicular polarization at 4000 cm^{-1}.

The second viewpoint of sampling depth discusses the decrease in reflectivity at the interface as related to the electromagnetic field intensity within the second medium (Section 3.3.3.2). Sampling depth is defined in terms of a relative depth of penetration that describes the distance within the second medium necessary for the intensity of the radiant energy associated with the electromagnetic field to decay to half or 1/e of initial intensity at the interface [Eqs. (28c) and (29)]. It is important to recognize the intensity of the radiant energy is determined per unit magnitude of $\mathbf{E}_{0t}$ at the interface, rather than by the actual amplitude of $\mathbf{E}_{0t}$ as associated with the effective thickness viewpoint of sampling depth. Therefore, the second viewpoint of the sampling depth as a relative depth of penetration is only a function of the experimental variables of wavelength, indices of refraction of the two media, and the incident angle. Unlike the sampling depth from the first viewpoint as an effective thickness, the second viewpoint of the term is not directly a function of the state of polarization. The two sampling depths in terms of the relative depth of penetration are defined directly as the depth of penetration in Eqs. (28c) and (29). The relative depths of penetration are illustrated in Figure 14 for a germanium internal reflection element at

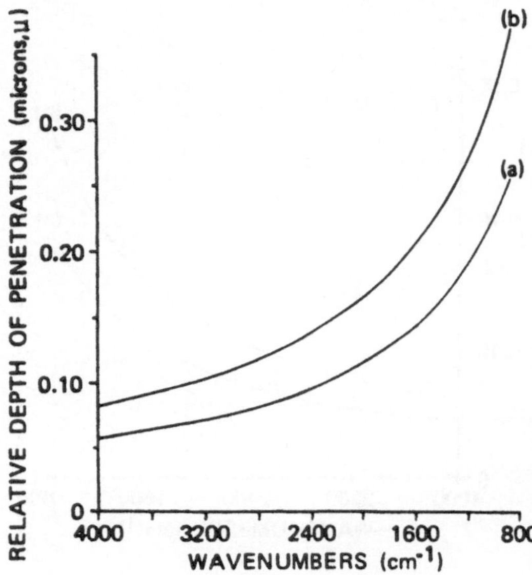

FIGURE 14. Relative depth of penetration versus wavenumber curves for Ge internal reflection elements having 45° incident angles. (a) Relative depth of penetration [Eq. (28c)] defined as distance into second medium required for intensity of electromagnetic field to decay to half of initial intensity. (b) relative depth of penetration [Eq. (29)] defined as distance into second medium required for intensity of electromagnetic field to decay to $1/e$ of initial intensity.

a 45° incident angle. The distances into the second medium required for the intensity of the radiant energy associated with the electromagnetic field within the second medium to decay to a half and $1/e$ of the initial value originating at the interface are $0.058\ \mu$ (580 A) and $0.083\ \mu$ (830 A) at 4000 cm^{-1}, respectively. The two relative depths of penetration differ only by the constant term ln (2).

The two viewpoints of sampling depth involve a difference in definition, although neither definition determines the actual depth the electromagnetic field extends into the second medium for a given set of experimental variables. The sampling depths are proportional to the distance into the second medium required for either the amplitude of the evanescent wave or the intensity of the radiant energy associated with the electromagnetic field to decay to a predesignated value from the initial values at the interface. The depth of penetration terms associated with the two viewpoints of sampling depth are the decay functions defining the decay of the electromagnetic field to the predesignated values of either half or $1/e$. Figure 15 illustrates the sampling depths from the first viewpoint in terms of the effective thicknesses as a function of the state of polarization and from the second viewpoint in terms of the relative depths of penetration for a germanium internal reflection element at a 45° incident angle. The amplitude of the evanescent wave and the intensity of the radiant energy associated with the electromagnetic field are also defined to decay to $1/e$ of the original values at the interface according to Eqs. (25b), (26b), and (29). The effective thicknesses at 4000 cm^{-1} are $0.123\ \mu$ (1230 A) for parallel polarization and

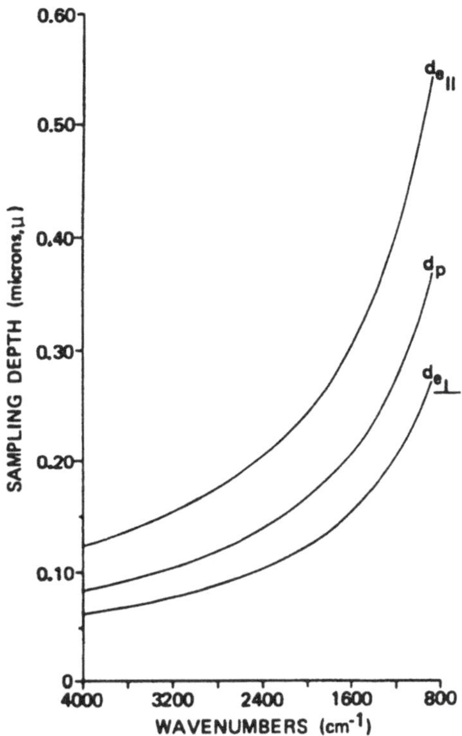

FIGURE 15. Sampling depth versus wavenumber curves for Ge internal reflection elements having 45° incident angles; ($d_{e\perp}$) effective thickness [Eq. (26b)] for perpendicular polarization; (d_p) relative depth of penetration [Eq. (29)]; ($d_{e\parallel}$) effective thickness (Eq. (26a)] for parallel polarization.

0.061 μ (610 A) for perpendicular polarization of the incident electromagnetic waves. The relative depth of penetration at 4000 cm^{-1} is 0.083 μ (830 A). The effective thickness for parallel polarization is greater than the relative depth of penetration, while the effective thickness for perpendicular polarization is less than the relative depth of penetration.

4. POLYMER APPLICATIONS OF TOTAL INTERNAL REFLECTION INFRARED SPECTROSCOPY

Polymers have been studied by infrared spectroscopy using total internal reflection techniques since the mid-1960s. Infrared spectroscopy probes the molecular structure, crystallinity, conformation, and secondary bonding of organic compounds, including polymers. Total internal reflection expands the capabilities of infrared spectroscopy to include samples that are difficult or impossible to study by conventional transmission techniques. For example, the technique is particularly useful in studying surfaces of samples that are too thin, thick, or opaque for transmission studies. Also, microgram quantities of samples can be studied by total internal reflection if insufficient material is present for forming either potassium bromide (KBr) pellets or films. Infrared spectroscopy coupled with total internal reflection can probe surface molecular structure, orientation, and morphology of polymers.

4.1. SURFACE MOLECULAR STRUCTURE

Surface structures of polymers can be studied by total internal reflection techniques if the sample is too thick, thin, or opaque for transmission studies. The technique can also be used to elucidate possible differences in surface structure as compared to either surface structures of other samples, other surfaces of the same sample, surfaces after surface treatments, or the bulk structures. A major requirement for total internal reflection is the ability to establish sufficient contact between the polymer sample and the internal reflection element. Insufficient sample contact results in scattering of the incident infrared radiation leading to insufficient radiation reaching the spectrometer detector to obtain a usable spectrum. Hirschfeld[(62)] discussed methods for preparing samples for which it is difficult to obtain sufficient contact with an internal reflection element. Other investigators[(63,64)] have also discussed relationships between sample contact and the pressure holding the sample against the internal reflection element.

Thick or opaque samples can absorb essentially all of the incident infrared radiation in transmission studies. Multiple internal reflection techniques for studying the sample surface may provide some spectral information not available from transmission studies of the bulk; however, the surface structures studied by total internal reflection can differ from the bulk structures. In addition, microgram quantities of samples can be studied on an internal reflection element, rather than as either KBr pellets or films. In these cases, multiple attenuated total reflection techniques are useful because the resultant spectra are the coaddition of the spectral information from multiple reflections, as compared to the spectral information obtained from a single transmission through the sample. As an example, Garton *et al.*[(65)] studied polymer surface coatings on glass fibers using total internal reflection. Sufficient sample contact was difficult to achieve with the internal reflection element because of the high modulus of the glass fibers coupled with low concentrations of the surface coatings. Therefore, the surface coatings were dissolved off the fibers. Films were then solvent cast from the dissolved coating solution onto the internal reflection element and subsequently studied. The technique proved satisfactory in providing sufficient spectral information to enable the chemical structures of the coatings to be identified and compared. In another investigation using internal reflection, Nel'son and coworkers[(66)] were able to study the structural changes associated with vulcanization of polystyrene–butadiene rubbers filled with carbon black. Such samples can be difficult or impossible to study by normal transmission infrared spectroscopy due to the opacity of the samples.

Transmission studies of thin polymer films can result in fringing along the baseline. Fringes occur with constructive and destructive interference of components of the propagating electromagnetic waves multiply reflected within the sample film with either the first transmitted component or the first reflected component of the waves originating at the front surface of the film. Fringing does not occur with total internal reflection because the interactions with the absorbing polymer medium occur with the nonpropagating evanescent wave of the electromagnetic field established within the second medium, rather than with the propagating incident electromagnetic waves.[(58)]

Besides samples difficult to study by transmission infrared spectroscopy, total internal reflection techniques have also been used to study differences in chemical structures between two surfaces of the same film, the surface structure as compared to bulk structure, as well as the effects of surface treatments. For example, Mirabella[(63)] used total internal reflection to determine the vinyl acetate content of poly(ethylene-vinylacetate) copolymers and the methyl content of polyethylene. Chemical composition of the surfaces of rubber/resin films was shown by Whitehouse *et al.*[(67)] to vary with the bulk resin concentration, thickness of the film, and the time allowed for establishing equilibrium after the film was dried. Garton *et al.*[(68,69)] studied the effect of chlorinated hydrocarbon coupling agents when forming mica-polypropylene composites. The depth of absorption of silicone oil into the bulk structure of polystyrene exposed to silicone oil was measured by Hirschfeld.[(39)]

The chemical structure of polymer surfaces can be altered under controlled conditions such as glow discharge (also known as plasma or corona discharge). The polymer is exposed to a known electrical charge under a gaseous atmosphere at known pressures for varying lengths of time. Nguyen *et al.*[(70)] used total internal reflection techniques to determine the optimum experimental parameters for achieving particular structural changes on polyethylene surfaces. The influence of various gases on the surface modification of polyethylene has also been investigated by Blythe and coworkers.[(71)] An interesting study on the surface structure of low-density polyethylene after fluorination during glow discharge was reported by Anand *et al.*[(72)] The investigators found that the degree of crystallinity of the polyethylene was altered depending on whether the sample was placed within an aluminum cage or not during exposure. Doughty and Pantelis[(73)] studied the surface effects of poly(vinylidene fluoride) with glow discharge to support a mechanism of dipole orientation as the origin of the piezoelectric properties of the polymer.

Chemical structures of polymer surfaces have also been altered by grafting different monomers onto the backbone of the polymer chain. Ultraviolet and ionizing radiation are often used to initiate the grafting. Yamakawa *et al.*[74,75] studied the surface modification of polyethylene by UV radiation-induced grafting of methyl acrylate. Besides being able to determine the surface structure by total internal reflection, the investigators studied the relationship between the differences in surface structures and the bond strengths of the resultant adhesive materials.

Chemical and mechanical degradation of polymers can result in surface alterations, as well as alterations in the bulk structure. Oxidation is one of the modes of degradation often leading to surface alterations. As an example, Hawkins[76] studied the oxidation of polyolefins in contact with metal surfaces. The existence of a metal complex leading to the oxidation of the polymer surface in contact with the metal was investigated by exposing the polyolefins to metal surfaces for varying lengths of time. Surfaces exposed to the metal were compared to opposite surfaces of the same sample. The metal complex was identified by its infrared spectrum. In addition, the influence of carbon black on the oxidation process was also investigated. Photooxidation mechanisms were studied by Ito and Porter[77] and by Carlsson and coworkers.[78-81] The initiation processes of oxidation on the surface were investigated, as well as the extent of structural alterations into the bulk structures of various polymers. Polymers can experience structural alterations when exposed to such environmental factors as UV radiation or water. The degradation of poly(vinyl chloride) as a function of exposure time to UV radiation was explored by Decker and Balandier.[82] Levy *et al.*[83] determined the effects of moisture in epoxy resins when exposed to various environments, while the hydrolysis of resin-coated poly(ethylene terephthalate) fibers was analyzed by Carlsson and Milnera[84] and compared to that of noncoated fibers. Possible modes of degradation under soil burial conditions were determined for polyolefins, poly(ethylene terephthalate), and polyamides by Colin *et al.*[85] Besides the chemical environment, the mechanical environment of a polymer can also bring about structural alterations of the surface or bulk. The influence of internal sliding friction as a form of mechanical degradation was reported to alter surface structures of high-density polyethylene, polypropylene, and nylon 6 by Krzeminski and Wiechowicz-Kowalska.[86] Molding processes, as well as environmental conditions, were determined to alter the surface structures of poly(acrylonitrile-butadiene-styrene) by Wyzgoski.[87] The degree of polybutadiene unsaturation as a function of sampling depth was determined after exposing poly(acrylonitrile-butadiene-styrene) to various atmospheres and molding processes by using different internal reflection elements and incident angles.

An interesting investigation of surface degradation using total internal reflection was conducted by Stupp and Carr.[88,89] The authors used the technique to explore the chemical origin of thermally stimulated discharge currents on the surfaces of polyacrylonitrile films. The charges were believed to be associated with degradation along the backbone chain and the possible generation of cyanide ions.

4.2. SURFACE ORIENTATION

Total internal reflection studies of anisotropic polymer samples with plane-polarized electromagnetic radiation can provide information complementary to that obtained by dichroic transmission studies. The theoretical treatment for total internal reflection using anisotropic films was developed by Flournoy and coworkers[90–93] from Maxwell's equations in a similar manner to the development of total internal reflection involving isotropic materials in Section 3. The absorption coefficients of the second medium can be determined for x, y, and z coordinates from the theoretical considerations. The reflectivity of the interface is a function of the x-component of the absorption coefficient with perpendicular polarization (using the coordinate axes in Section 3), while the reflectivity is a function of the y- and z-components of the absorption coefficients with parallel polarization. In contrast, dichroic studies by transmission infrared spectroscopy involve only x- and y-components of the absorption coefficients.

Comparisons between orientation and crystallinity within surface and bulk structures of polymers have been made using polarized internal reflection and transmission infrared spectroscopy. Flournoy and Schaffers[92] applied their theoretical considerations to studies of oriented polypropylene films. Their studies indicated the degree of surface orientation was qualitatively similar to the degree of bulk orientation. Sung[94] has since shown the influence of fabrication methods on the surface orientation of polypropylene. Surface and bulk orientations were similar for uniaxially drawn polypropylene; however, the degree of orientation on the surface was greater than that through the bulk of injection-molded polypropylene. Hobbs and coworkers[95] determined the degree of crystallinity in polypropylene samples by studying the orientation of undrawn, uniaxially, and biaxially drawn polypropylene. The degree of crystallinity within the surface structures obtained by polarized internal reflection was similar to the bulk crystallinity of the different samples as determined by other methods. In addition, Trott[96] used the technique to investigate why two isotactic polypropylene samples having the same initial degree of crystallinity would have vastly different elongations at rupture. The studies indicated a different orientation was present within the surfaces of the two samples prior to elongation, despite apparent similarities in the crystallinities of the two samples.

Molecular orientation and crystallinity have been studied with polarized internal reflection techniques in other polymers besides polypropylene. Tshmel *et al.*[97] have studied poly(vinyl alcohol), polytetrafluorethylene, poly(ethylene terephthalate), polyethylene, polypropylene, polyoxymethylene, poly(vinyl chloride), copolystyrene-methylmethacrylate, polyamide, and polyimides. The internal reflection studies indicated a higher concentration of polyolefin end groups present on the surface as compared to the bulk polymers. Polyethylene, polypropylene, poly(ethylene terephthalate), nylon 6, and polyimides exhibited higher orientation in the surfaces than through the bulk polymers. Surface and bulk studies of oriented polyethylene and nylon 6 films showed a *trans* isomer was predominate on the surface as compared to the bulk. In contrast, the surfaces of nonoriented poly(ethylene terephthalate), polyethylene, and poly(vinyl alcohol) showed higher concentrations of *gauche* conformations than the bulk. Spontaneous orientation on the surface of polyamide acid films occurred simultaneously with the imidization of the films to form the polyimides at higher temperatures. Stas'kov *et al.*[98] were able to determine a preferential orientation of backbone axes along the direction normal to the film surface of extruded poly(ethylene terephthalate). In addition, there appeared to be higher crystallinity within the bulk polymer as compared to the surface regions of the extruded samples. Druschke *et al.*[99] characterized the molecular order and orientation existing within the surface of aromatic polyamide fibers by coupling polarized internal reflection techniques with deuterium exchange. Molecular orientation within the surface of nylon 6,6 fibers was compared to the bulk as a function of draw ratios by Barr.[100] Surface dichroic ratios were related to the bulk dichroic ratios for draw ratios greater than 1.4; however, the surface of the fibers appeared to have greater orientation than the bulk for draw ratios of 1.4 and lower.

4.3. SURFACE MORPHOLOGY

Copolymers have two or more chemically different repeat units covalently bonded in an ordered or random repeat pattern along single polymer chains. For example, AB block copolymers have a segment or block comprised of repeat units A covalently bonded to another segment comprised of repeat units B. However, multiple block $(AB)_n$ copolymers have lower molecular weight segments of A and B repeating n times along the polymer chains. One segment of the copolymer is usually viscous or rubbery at the use temperature and the other segment is glassy or semicrystalline. Copolymers can form two phase morphologies due to the chemical and steric incompatibilities of the chemically different blocks or segments. The glassy block known as the hard segment forms the domains, while the rubbery block known as the soft segment forms the matrix. The unusual range of physical and chemical properties often associated with many copolymers results from the morphological separation of the different blocks. In addition, there can be a gradation of morphologies within the surface regions as compared to the inner bulk regions of a particular sample. Differences in surface and bulk morphologies can be particularly important in biomedical applications of copolymers. Bulk morphologies contribute to the general physical and mechanical properties of the copolymer, while the surface morphologies are important in interactions with blood and tissues. Surface infrared spectroscopy has become a vital tool in understanding the surface morphologies of copolymers have possible biomedical applications.

Avcothane® is a commercially available polyether-urethane and polysiloxane copolymer used in many cardiovascular products, such as intraaortic balloon pumps. There have been a number of different formulations of Avcothane® available over the years making direct comparisons of structure and morphology between different studies often difficult. However, most investigations have compared the relative concentrations of the polyether-urethane component to the polysiloxane component present within the surface regions formed in contact with either a substrate or the air atmosphere. ·Nyilas and Ward[(101)] studied the anisotropic distribution of the polysiloxane components within the surface regions as a function of fabrication variables. The air-facing and substrate-facing surfaces of Avcothane® 51 were shown to have quite different infrared spectra as obtained by total internal reflection. The polysiloxane components were present in higher concentrations within the substrate-facing surfaces. Sung and coworkers[(102,103)] have also studied air-facing and substrate-facing surfaces of Avcothane®. These studies were consistent with those by Nyilas and Ward,[(101)] showing higher concentrations of polysiloxane components on the substrate-facing surfaces than on the air-facing surfaces with similar internal reflection elements and incident angles. However, the soft or polyether segments of the polyether-urethane were also determined to be present in slightly higher concentrations on the air-facing surfaces as compared to the substrate-facing surfaces. To investigate possible anisotropy as a function of sampling depth into the polymer surfaces, a barrier film was interposed between the polymer surface being studied and the internal reflection element. The thickness of the barrier film influences the sampling depth of the evanescent wave into the polymer surfaces. The previous trend was reversed with the air-facing surfaces having significantly higher concentrations of polysiloxane components as compared to the substrate-facing surfaces. The air-facing surfaces also exhibited higher concentrations of the polyether segments of the polyether-urethane within the narrower surface regions obtained with the barrier film (approximately 0.8 μ) as compared to the deeper surface regions obtained without the barrier film (approximately 1.5 μ). To determine the anisotropy as a function of lateral position along the polymer films, different samples were taken from various locations and studied. The lateral morphologies were apparently homogeneous with the compositions of the different components of the copolymer qualitatively similar as a function of lateral position. Kellner and Unger[64] studied the optimization of the experimental variables including internal reflection element material, sample size, sample location on the internal reflection element, and the applied pressure contacting the sample to the element in order to quantitatively determine relative ratios of the major components of a polyurethane–polysiloxane copolymer within the surface regions.

Biomer® is another commercially available polyether–urethane–urea copolymer used in cardiovascular applications, such as the artificial heart. The copolymer is also available in a variety of forms. Sung and coworkers[102,103] have studied the solution form of Biomer® in conjunction with their studies of Avcothane®. The Biomer® films cast from dimethyl acetylamide onto glass plates exhibited less anisotropy within the surface regions for substrate-facing and air-facing surfaces than Avcothane®. The air-facing surfaces had approximately the same concentrations of soft segments as the substrate-facing surfaces. However, the concentrations of the soft segments increased slightly on the air-facing surfaces within the narrower surface region obtained using the barrier films. Lelah and coworkers[104] investigated the surface morphologies of Biomer® in solution and extruded forms. Infrared spectra indicated the two different forms of Biomer® were not only different in morphological structures, but also had very different chemical structures.

Knutson and Lyman[105,106] have studied block copolyether–urethane–ureas having different molecular weight polyether segments. By varying internal reflection elements and incident angles, the surface morphologies were shown to vary as a function of sampling depth from the surfaces into the inner bulk structures. The studies also showed the urea domains were well-separated from the polyether matrix through urethane interfacial regions. However, the urethane groups within the interfacial regions were mixing with the polyether matrix and possibly with the urea domains. The degree of mixing between the urethane groups and the polyether segments within the interfacial regions varied as a function of sampling depth for a copolymer having a particular polyether molecular weight, and also as a function of the molecular weight of the polyether segments. There was less mixing between the urethane groups and polyether segments within the urethane interfacial regions, characteristic of a narrower interface, in the surface regions as compared to the inner bulk regions of copolymer samples having the same molecular weight polyether segments. In addition, the urethane interfaces within the surface regions narrowed due to decreased mixing of the urethane groups with the polyether segments, as compared to the inner bulk of the copolymers, with decreasing molecular weight polyether segments.

Stupp *et al.*[107] studied the surface morphologies of copolyether-urethane-ureas as a function of the substrate material used in the solvent casting of the copolymers. The copolymers were cast onto either glass or poly(ethylene terephthalate) sheets. Polyether-urethane-urea surfaces cast against the glass plates had higher concentrations of polyether segment. However, the copolymer surfaces cast against the poly(ethylene terephthalate) sheets had significantly higher concentrations of urethane-urea segments. An oxygen bridge structure on the surfaces of the glass plates was believed to result in the increased concentrations of polyether segments on the copolymer surfaces cast against the plates. The ability of the urethane and urea secondary amino groups to hydrogen bond with the carbonyl groups on the surfaces of the poly(ethylene terephthalate) sheets, as well as the aromatic structure of poly(ethylene terephthalate), was believed to be responsible for the higher concentrations of urethane-urea segments on the copolymer surfaces cast against the poly(ethylene terephthalate) sheets.

5. SUMMARY

Surface infrared spectroscopy couples the analytical method of infrared spectroscopy with the physical phenomenon of total internal reflection to enable the molecular vibrations within the surface regions of polymers to be studied. The study of molecular vibrations absorbing infrared radiation can provide molecular structure, conformation, crystalline, and secondary bonding information about the polymer surface regions.

Total internal reflection is a special case of the general physical phenomenon of reflection and refraction of electromagnetic radiation at an interface of two media having different indices of refraction. During total internal reflection of the electromagnetic radiation within the infrared region, the incident electromagnetic waves are entirely reflected back into the initial medium. However, an electromagnetic field is established within the second medium, as represented by an evancescent wave, due to diffraction at the edges of the incident radiation at the interface. The evanescent wave decays with distance into the second medium as dictated by a decay function. The decay function is a function of the experimental variables including the indices of refraction of the two media, the particular wavelength of radiation, and the incident angle.

The distance the electromagnetic field extends into the second medium is discussed in terms of a sampling depth from two different viewpoints. The sampling depth is not the actual distance the electromagnetic field extends into the second medium, but defines the distance in the z direction from the interface within the second medium necessary for the evanescent wave to decay to a predesignated value. The first viewpoint of the sampling depth discusses the decay in terms of an effective thickness. Effective thickness is the equivalent thickness the same sample studied by transmission infrared spectroscopy must be in order to have the equivalent absorbance as obtained per single reflection with total internal reflection techniques. An associated depth of penetration term defined from the decay function describes the distance into the sample required for the amplitude of the evanescent wave to decay to either half or 1/e of the amplitude of the initial constant vector originating at the interface. The effective thickness is a function of the state of polarization of the incident electromagnetic radiation, indices of refraction of the two media, the incident angle, and the particular wavelength of radiation being studied. The second viewpont of sampling depth describes the decay of the intensity of the radiant energy associated with the electromagnetic field in terms of a relative depth. The relative depth is defined directly as a depth of penetration term describing the distance required for the intensity of the radiant energy associated with the electromagnetic field to decay to either half or 1/e of the initial value at the interface. However, the decay of the intensity of the radiant energy is defined per unit magnitude of the evanescent wave initial constant vector originating at the interface. Therefore, the relative depth of penetration is only directly a function of the indices of refraction of the two media, the incident angle, and the particular wavelength of radiation.

Surface infrared spectroscopy can vary from surface to subsurface sensitivity by altering the sampling depth of the electromagnetic field within the second medium. The sampling depth can be altered primarily by varying the index of refraction of the initial medium (internal reflection element, η_c) and the incident angle. Since most polymers have indices of refraction around 1.5, the index of refraction of the second medium (η_s) does not significantly influence the sampling depth. In general, lower indices of refraction of the internal reflection element increase the sampling depth, while lower incident angles also increase the sampling depth according to either viewpoint.

Surface chemical and morphological structures of polymers have been investigated using surface infrared spectroscopy. In addition, the coupling of surface infrared spectroscopy with the effective use of polarized incident radiation has enabled orientation and crystallinity of polymer surfaces to be compared to the orientation and crystallinity in the bulk.

ACKNOWLEDGMENT

This work was supported in part by the National Science Foundation, Grant DMR 80-05499, Polymer Program.

REFERENCES

1. N. J. Harrick, Use of infrared absorption in germanium to determine carrier distribution for injection and extraction, *Phys. Rev.* **103**, 1173-1181 (1956).
2. N. J. Harrick, Effect of the metal-to-semiconductor potential on the semiconductor surface barrier height, *J. Phys. Chem. Solids* **8**, 106-108 (1959).
3. N. J. Harrick, Surface chemistry from spectral analysis of totally internally reflected radiation, *J. Phys. Chem.* **64**, 1110-1114 (1960).
4. N. J. Harrick, Study of physics and chemistry of surfaces from frustrated total internal reflections, *Phys. Rev. Lett.* **4**, 224-226 (1960).
5. J. Fahrenfort, Attenuated total reflection, A new principle for the production of useful infra-red reflection spectra of organic compounds, *Spectrochim. Acta* **17**, 698-709 (1961).
6. J. Fahrenfort and W. M. Visser, On the determination of optical constants in the infrared by attenuated total reflection, *Spectrochim. Acta* **18**, 1103-1116 (1962).
7. W. N. Hansen, A new spectrophotometric technique using multiple attenuated total reflection, *Anal. Chem.* **35**, 765-766 (1963).
8. P. C. Painter, M. M. Coleman, and J. L. Koenig, *The Theory of Vibrational Spectroscopy and Its Applications to Polymeric Materials*, John Wiley and Sons, New York (1982).
9. N. L. Alpert, W. E. Keizer, and H. A. Szymanski, *IR Theory and Practice of Infrared Spectroscopy*, Second Edition, Plenum Press, New York (1970).
10. R. T. Conley, *Infrared Spectroscopy*, Second Edition, Allyn and Bacon, Boston (1972).
11. N. B. Colthup, L. H. Daly, and S. E. Wiberly, *Introduction to Infrared and Raman Spectroscopy*, Second Edition, Academic Press, New York (1975).
12. R. M. Silverstein, G. C. Bassler, and T. C. Morrill, *Spectroscopic Identification of Organic Compounds*, Third Edition, John Wiley and Sons, New York (1974).
13. R. Zbinden, *Infrared Spectroscopy of High Polymers*, Academic Press, New York (1964).
14. L. J. Bellamy, *The Infrared Spectra of Complex Molecules*, Volume One, Third Edition, Chapman and Hall, London (1975).
15. L. J. Bellamy, *Advances in Infrared Group Frequencies*, Volume Two of *The Infrared Spectra of Complex Molecules*, Chapman and Hall, London (1978).
16. M. Born and E. Wolf, *Principles of Optics: Electromagnetic Theory of Propagation, Interference and Diffraction of Light*, Fifth Edition, Pergamon Press, New York (1975).
17. D. W. Dearholt and W. R. McSpadden, *Electromagnetic Wave Propagation*, McGraw-Hill, New York (1973).
18. R. H. Good, Jr. and T. J. Nelson, *Classical Theory of Electric and Magnetic Field*, Academic Press, New York (1971).
19. G. R. Fowles, *Introduction to Modern Optics*, Holt, Rinehart and Winston, New York (1968).
20. M. V. Klein, *Optics*, John Wiley and Sons, New York (1970).
21. R. W. Ditchburn, *Light*, Third Edition, Academic Press, New York (1976).
22. M. A. Plonus, *Applied Electromagnetics*, McGraw-Hill, New York (1978).
23. J. Fahrenfort and W. M. Visser, Remarks on the determination of optical constants from ATR measurements, *Spectrochim. Acta* **21**, 1433-1435 (1965).
24. W. N. Hansen, Expanded formulas for attenuated total reflection and the derivation of absorption rules for single and multiple ATR spectrometer cells, *Spectrochim. Acta* **21**, 815-833 (1965).
25. W. N. Hansen, T. Kuwana, and R. A. Osteryoung, Observation of electrode-solution interface by means of internal reflection spectrometry, *Anal. Chem.* **38**, 1810-1821 (1966).
26. W. N. Hansen, On the determination of optical constants by a two-angle internal reflection method, *Spectrochim. Acta* **21**, 209-210 (1965).
27. W. N. Hansen, Electric fields produced by the propagation of plane coherent electromagnetic radiation in a stratified medium, *J. Opt. Soc. Am.* **58**, 380-390 (1968).

28. N. J. Harrick, Optical spectrum of the semiconductor surface states from frustrated total internal reflections, *Phys. Rev.* **125**, 1165-1170 (1962).
29. N. J. Harrick, Electric field strengths at totally reflecting interfaces, *J. Opt. Soc. Am.* **55**, 851-857 (1965).
30. N. J. Harrick, Vertical double-pass multiple reflection element for internal reflection spectroscopy, *Appl. Opt.* **5**, 1-3 (1966).
31. N. J. Harrick, Introductions to polymer symposium, Reflectance spectroscopy, *J. Colloid Interface Sci.* **47**, 591-594 (1974).
32. N. J. Harrick, Multiple reflection cells for internal reflection spectrometry, *Anal. Chem.* **36**, 188-191 (1964).
33. N. J. Harrick, Use of frustrated total internal reflection to measure film thickness and surface reliefs, *J. Appl. Phys.* **33**, 2774-2775 (1962).
34. N. J. Harrick, Double-beam internal reflection spectrometer, *Appl. Opt.* **4**, 1664-1665 (1965).
35. N. J. Harrick, Total internal reflection and its application to surface studies, *Ann. N.Y. Acad. Sci.* **101**, 928-959 (1963).
36. N. J. Harrick and F. K. du Pré, Effective thickness of bulk materials and of thin films for internal reflection spectroscopy, *Appl. Opt.* **5**, 1739-1743 (1966).
37. T. Hirschfeld, High sensitivity attenuated total-reflection spectroscopy, *Appl. Spectrosc.* **20,** 336-338 (1966).
38. T. Hirschfeld, Accuracy and optimization of the two prism technique for calculating the optical constants from ATR data, *Appl. Spectrosc.* **24,** 277-282 (1970).
39. T. Hirschfeld, Subsurface layer studies by attenuated total reflection Fourier transform spectroscopy, *Appl. Spectrosc.* **31**, 289-292 (1977).
40. T. Hirschfeld, Procedures for attenuated total reflection study of extremely small samples, *Appl. Opt.* **6**, 715-718 (1967).
41. T. Hirschfeld, Determination of optical constants by ATR measurements, *Spectrochim. Acta* **22**, 1823-1824 (1966).
42. T. Hirschfeld, Relationships between the Goos-Hänchen shift and the effective thickness in attenuated total reflection spectroscopy, *Appl. Spectrosc.* **31**, 243-244 (1977).
43. N. Bloembergen and P. S. Pershan, Light waves at the boundary of non-linear media, *Phys. Rev.* **128**, 606-622 (1962).
44. N. Bloembergen and C. H. Lee, Total reflection in second-harmonic generation, *Phys. Rev. Lett.* **19**, 835-837 (1967).
45. F. Goos and H. Hänchen, Ein neuer und fundamentaler versuch zur total-reflexion, *Ann. Physik* **1**, 333-346 (1947).
46. B. R. Horowitz and T. Tamir, Lateral displacement of a light beam at a dielectric interface, *J. Opt. Soc. Am.* **61**, 586-594 (1971).
47. H. K. V. Lotsch, Beam displacement at total reflection: The Goos-Hänchen effect, I., *Optik* **32**, 116-137 (1970).
48. H. K. V. Lotsch, Beam displacement at total reflection: The Goos-Hänchen effect, II., *Optik* **32**, 189-204 (1970).
49. H. K. V. Lotsch, Beam displacement at total reflection: The Goos-Hänchen effect, III., *Optik* **32**, 299-319 (1971).
50. H. K. V. Lotsch, Beam displacement at total reflection: The Goos-Hänchen effect, IV., *Optik* **32**, 553-569 (1971).
51. H. K. V. Lotsch, Reflection and refraction of a beam of light at a plane interface, *J. Opt. Soc. Am.* **58**, 551-561 (1968).
52. H. Kogelnik and H. P. Weber, Rays, stored energy, and power flow in dielectric waveguides, *J. Opt. Soc. Am.* **64**, 174-185 (1974).
53. R. H. Renard, Total reflection: A new evaluation of the Goos-Hänchen shift, *J. Opt. Soc. Am.* **54**, 1190-1197 (1964).

54. A. W. Snyder and J. D. Love, Goos-Hänchen shift, *Appl. Opt.* **15**, 236-238 (1976).
55. H. G. Tompkins, The physical basis for analysis of the depth of absorbing species using internal reflection spectroscopy, *Appl. Spectrosc.* **28**, 335-341 (1974).
56. P. A. Wilks, Jr., Internal reflection spectroscopy I: Effect of angle of incidence change, *Appl. Spectrosc.* **22**, 782-784 (1968).
57. J. Fahrenfort, in: *Infra-red Spectroscopy and Molecular Structure, An Outline of the Principles* (M. Davies, ed.), pp. 377-404, Elsevier, New York (1963).
58. N. J. Harrick, *Internal Reflection Spectroscopy*, John Wiley and Sons, New York (1967).
59. G. Kortüm, *Reflectance Spectroscopy: Principles, Methods, Applications*, Springer-Verlag, New York (1969).
60. Standard definitions of terms and symbols relating to molecular spectroscopy, *Annual Book of ASTM Standards* **E 131-81**, 1-7 (1981).
61. J. Brandrup and E. H. Immergut, ed., *Polymer Handbook*, 2nd edition, John Wiley and Sons, New York (1975).
62. T. Hirschfeld, Solution for the sample contact problem in ATR, *Appl. Spectrosc.* **21**, 335-336 (1967).
63. F. M. Mirabella, Jr., Quantitative analysis of polymers by attenuated total reflectance Fourier-transform infrared spectroscopy: Vinyl acetate and methyl content of polyethylenes, *J. Polym. Sci., Phys. Ed.* **20**, 2309-2315 (1982).
64. R. Kellner and F. Unger, IR-spektroskopische charakterisierung der blutverträglichkeit von polyurethan-silicon-copolymeren, *Z. Anal. Chem.* **283**, 349-355 (1977).
65. A. Garton, A. Stolow, and D. M. Wiles, Infrared spectroscopic characterization of surface coatings on glass fibres, *J. Mater. Sci.* **16**, 3211-3214 (1981).
66. K. V. Nel'son, T. G. Arkatova, and V. M. Zolotarev, Study of attenuated multiple total internal reflectance infrared spectra of elastomers filled with carbon black, *Polym. Sci. U.S.S.R.* **15**, 560-566 (1973).
67. R. S. Whitehouse, P. J. C. Counsell, and G. Lewis, Composition of rubber/resin adhesive films: 1. Surface composition as determined by ATR spectroscopy, *Polymer* **17**, 699-704 (1976).
68. A. Garton, S. W. Kim, and D. M. Wiles, Chlorinated hydrocarbon coupling agents for mica-polypropylene composites, *J. Appl. Polym. Sci.* **27**, 4179-4189 (1982).
69. A. Garton, S. W. Kim, and D. M. Wiles, Modification of the interface morphology in mica-reinforced polypropylene, *J. Polym. Sci., Lett. Ed.* **20**, 273-278 (1982).
70. L. T. Nguyen, N. H. Sung, and N. P. Suh, Determination of optimum glow discharge parameters based on ATR-FTIR spectra, *J. Polym. Sci., Lett. Ed.* **18**, 541-548 (1980).
71. A. R. Blythe, D. Briggs, C. R. Kendall, D. G. Rance, and V. J. I. Zichy, Surface modification of polyethylene by electrical discharge treatment and the mechanism of autoadhesion, *Polymer* **19**, 1273-1278 (1978).
72. M. Anand, R. E. Cohen, and R. F. Baddour, Surface modification of low density polyethylene in fluorine gas plasma, *Polymer* **22**, 361-371 (1981).
73. K. Doughty and P. Pantelis, Surface effects in corona-charged polyvinylidene fluoride, *J. Mater. Sci.* **15**, 974-978 (1980).
74. S. Yamakawa, Surface modification of polyethylene by radiation-induced grafting for adhesive bonding, I. Relationship between adhesive bond strength and surface composition, *J. Appl. Polym. Sci.* **20**, 3057-3072 (1976).
75. S. Yamakawa, F. Yamamoto, and Y. Kato, Surface modification of polyethylene by radiation-induced grafting for adhesive bonding. 2. Relationship between adhesive bond strength and surface structure, *Macromolecules* **9**, 754-758 (1976).
76. W. L. Hawkins, Recent advances in mechanisms for the stabilization of polyolefins, *J. Polym. Sci., Symp.* **57**, 319-328 (1976).
77. M. Ito and R. S. Porter, The effects of oxygen diffusion on the surface photooxidation of polystyrene, *J. Appl. Polym. Sci.* **27**, 4471-4476 (1982).

78. D. J. Carlsson, L. H. Gan, and D. M. Wiles, Photodegradation of aramids. II. Irradiation in air, *J. Polym. Sci., Chem. Ed.* **16**, 2365–2376 (1978).
79. D. J. Carlsson and D. M. Wiles, Photooxidation of polypropylene films. IV. Surface changes studied by attenuated total reflection spectroscopy, *Macromolecules* **4**, 174–179 (1971).
80. D. J. Carlsson and D. M. Wiles, Photooxidation of polypropylene films. V. Origin of preferential surface oxidation, *Macromolecules* **4**, 179–184 (1971).
81. D. J. Carlsson and D. M. Wiles, The photodegradation of polypropylene films. II. Photolysis of ketonic oxidation products, *Macromolecules* **2**, 587–597 (1969).
82. C. Decker and M. Balandier, Degradation of poly(vinyl chloride) by UV radiation. I. Kinetics and quantum yields, *Eur. Polym. J.* **18**, 1085–1091 (1982).
83. R. L. Levy, D. L. Fanter, and C. J. Summers, Spectroscopic evidence for mechanochemical effects of moisture in epoxy resins, *J. Appl. Polym. Sci.* **24**, 1643–1664 (1979).
84. D. J. Carlsson and S. M. Milnera, Hydrolysis of resin-coated poly(ethylene terephthalate) yarns, *J. Appl. Polym. Sci.* **27**, 1589–1600 (1982).
85. G. Colin, J. D. Cooney, D. J. Carlsson, and D. M. Wiles, Deterioration of plastic films under soil burial conditions, *J. Appl. Polym. Sci.* **26**, 509–519 (1981).
86. J. L. Krzeminski and E. Wiechowicz-Kowalska, Structural changes in the surface layer of plastics as a result of sliding friction in a metal-thermoplastic system, *Polym. Eng. Sci.* **21**, 594–602 (1981).
87. M. G. Wyzgoski, Effects of oven aging on ABS, poly(acrylonitrile-butadiene-styrene), *Polym. Eng. Sci.* **16**, 265–269 (1976).
88. S. I. Stupp and S. H. Carr, Chemical origin of thermally stimulated discharge currents in polyacrylonitrile, *J. Polym. Sci., Phys. Ed.* **15**, 485–499 (1977).
89. S. I. Stupp and S. H. Carr, Spectroscopic analysis of electrically polarized polyacrylonitrile, *J. Polym. Sci., Phys. Ed.* **16**, 13–28 (1978).
90. P. A. Flournoy, Applications of attenuated-total-reflection spectroscopy to absolute intensity measurements, *J. Chem. Phys.* **39**, 3156–3157 (1963).
91. P. A. Flournoy, Attenuated total reflection from oriented polypropylene films, *Spectrochim. Acta* **22**, 15–20 (1966).
92. P. A. Flournoy and W. J. Schaffers, Attenuated total reflection spectra from surfaces of anisotropic, absorbing films, *Spectrochim. Acta* **22**, 5–13 (1966).
93. J. K. Barr and P. A. Flournoy, in: *Physical Methods in Macromolecular Chemistry*, Volume 1 (B. Carroll, ed.) pp. 109–164, Marcel Dekker, New York (1969).
94. C. S. P. Sung, A modified technique for measurement of orientation from polymer surfaces by attenuated total reflection infrared dichroism, *Macromolecules* **14**, 591–594 (1981).
95. J. P. Hobbs, C. S. P. Sung, K. Krishnan, and S. Hill, Characterization of surface structure and orientation in polypropylene and poly(ethylene terephthalate) films by modified attenuated total reflection IR dichroism studies, *Macromolecules* **16**, 193–199 (1983).
96. G. F. Trott, Orientation on elongation at rupture of two injection-molded isotactic polypropylenes, *J. Appl. Polym. Sci.* **14**, 2421–2425 (1970).
97. A. E. Tshmel, V. I. Vettegren, and V. M. Zolotarev, Investigation of the molecular structure of polymer surfaces by ATR spectroscopy, *J. Macromol. Sci., Phys.* **B21**, 243–264 (1982).
98. N. I. Stas'kov, V. I. Golovachev, and S. S. Gusev, Anisotropy of IR absorption of an extruded polyethylene terephthalate film, *Polym. Sci. U.S.S.R.* **19**, 2628–2634 (1977).
99. F. Druschke, H. W. Siesler, G. Spilgies, and H. Tengler, Molecular order and orientation in aromatic polyamide fibers by internal reflection spectroscopy and wide angle X-ray diffraction, *Polym. Eng. Sci.* **17**, 93–95 (1977).
100. J. K. Barr, Molecular orientation of fibres by polarized internal reflexion spectroscopy, *Nature* **215**, 844 (1967).

101. E. Nyilas and R. S. Ward, Jr., Development of blood-compatible elastomers. V. Surface structure and blood compatibility of Avcothane® elastomers, *J. Biomed. Materials Res., Symp.* **8**, 69-84 (1977).
102. C. S. Paik Sung, C. B. Hu, E. W. Merrill, and E. W. Salzman, Surface chemical analysis of Avcothane® and Biomer® by Fourier transform IR internal reflection spectroscopy, *J. Biomed. Materials Res.* **12**, 791-804 (1978).
103. C. S. Paik Sung and C. B. Hu, Surface chemical analysis of segmented polyurethanes. Fourier transform IR internal reflection studies, in: *Multiphase Polymers* (S. L. Cooper and G. M. Estes, eds.), *Adv. Chem. Ser.* **176**, 69-82 (1979).
104. M. D. Lelah, L. K. Lambrecht, B. R. Young, and S. L. Cooper, Physicochemical characterization and *in vivo* blood tolerability of cast and extruded Biomer®, *J. Biomed. Materials Res.* **17**, 1-22 (1983).
105. K. Knutson and D. J. Lyman, Morphology of block copolyurethanes. II. FTIR and ESCA techniques for studying surface morphology, in: *Biomedical and Dental Applications of Polymers* (C. G. Gebelein and F. F. Koblitz, eds.), pp. 173-188, Plenum Press, New York (1980).
106. K. Knutson and D. J. Lyman, The effect of polyether segment molecular weight on the bulk and surface morphologies of copolyether-urethane-ureas, in: *Biomaterials: Interfacial Phenomena and Applications* (S. L. Cooper and N. A. Peppas, eds.) *Adv. Chem. Ser.* **199**, 109-132 (1982).
107. S. I. Stupp, J. W. Kauffman, and S. H. Carr, Interactions between segmented polyurethane surfaces and the plasma protein fibrinogen, *J. Biomed. Materials Res.* **11**, 237-250 (1977).

The Contact Angle and Interface Energetics

Joseph D. Andrade, Lee M. Smith, and Donald E. Gregonis

1. INTRODUCTION

Information on the outermost few angstroms of solid surfaces is very difficult to obtain. One of the most sensitive methods known for obtaining true surface information is solid/liquid/vapor (S/L/V) or solid/liquid/liquid (S/L/L) contact angles. These methods are unique in that the equipment required is relatively simple and inexpensive. Although interpretation of the results obtained is dependent on a number of assumptions, each of which is somewhat controversial, a first-order interpretation is possible and has proven to be very useful in practically all areas of surface science and engineering. Most of the surface science texts briefly referred to in Chapter 1 contain one or more chapters on surface tension, capillarity, or contact angle methods. In addition, a number of the monographs and review serials cited in Chapter 1 also contain chapters on the contact angle technique.

In this chapter we concentrate on measurements using water as a probe liquid and on the role of water hydration time, surface dynamics, and hysteresis on the surface properties. Water is chosen as the probe because it is the simplest biologically relevant environment. The reader is referred to the sources in Table 1, Chapter 1, for treatments of other areas of contact angle science, capillarity, surface and interfacial tension, and interface thermodynamics.

Other sources include the textbooks by Aveyard and Haydon[1] and Davies and Rideal.[2] The text by Bikerman[3] is particularly good with

Joseph D. Andrade, Lee M. Smith, and Donald E. Gregonis • Departments of Bioengineering, Materials Science and Engineering, and Pharmaceutics, University of Utah, Salt Lake City, Utah 84112.

regard to a critique of the major assumptions in surface science. A number of key review articles have been consulted in preparing this chapter.[4-10] In addition to some basic background in surface and polymer chemistry, the reader is expected to have read Chapters 1 and 2 of this volume.

2. FUNDAMENTALS

2.1. SURFACE AND INTERFACE CONVENTIONS, DEFINITIONS, AND SEMANTICS

The basis of the contact angle technique is the three-phase equilibrium which occurs at the contact point at the solid/liquid/vapor or solid/liquid/liquid interface. This equilibrium is normally considered in terms of the surface and interfacial tensions or surface and interfacial free energies present. The various thermodynamic terms and quantities involved must therefore be properly defined and understood.

The terms surface free energy, surface tension, and surface stress are routinely used in the description of surfaces. The word surface used here should be interpreted as interface.

The major thermodynamic quantity which characterizes a surface or an interface is the reversible work, γ, to create unit area of surface at constant temperature (T), volume (V), and chemical potential of component $i(\mu_i)$.

This quantity is not equal to the surface free energy except under certain conditions. It is not surface stress. Perhaps one hundred years before the development of the energy concept, the term surface tension was used to describe the contractile nature of surface films; i.e., their tendency to minimize surface area. This term became so entrenched in the literature that it is widely used today. Thus γ is widely called the surface tension, though the meaning of those words may have little physical significance in many situations.

We will call the surface thermodynamic quantity "gamma", γ, with the understanding that it is some quantity characterizing the thermodynamic property of an interface.

The γ of a newly created surface is defined as:

$$\gamma \equiv \mathrm{d}w/\mathrm{d}A \tag{1}$$

i.e., the specific surface work to form dA new surface area.

The thermodynamics of interfaces (Table 1) is exactly the same as the thermodynamics of homogeneous systems except that the work term of

TABLE 1

Summary of Basic Reversible Thermodynamic Equations for Single-Component, Multicomponent, and Surface-Containing Systems[a]

Quantity	Single-component "surface-free" system	Open or multiple component system—no surface	Open system with surface[b]
Internal Energy, E	$\mathrm{d}E = T\,\mathrm{d}S - P\,\mathrm{d}V$	$\mathrm{d}E = T\,\mathrm{d}S - P\,\mathrm{d}V + \sum_i \mu_i\,\mathrm{d}N_i$	$T\,\mathrm{d}S - P\,\mathrm{d}V + \sum_i \mu_i\,\mathrm{d}N_i + \gamma\,\mathrm{d}A$
Enthalpy, H	$\mathrm{d}H = T\,\mathrm{d}S + V\,\mathrm{d}P$	$\mathrm{d}H = T\,\mathrm{d}S + V\,\mathrm{d}P + \sum_i \mu_i\,\mathrm{d}N_i$	$T\,\mathrm{d}S + V\,\mathrm{d}P + \sum_i \mu_i\,\mathrm{d}N_i + \gamma\,\mathrm{d}A$
Helmholtz free energy, F	$\mathrm{d}F = -S\,\mathrm{d}T - P\,\mathrm{d}V$	$\mathrm{d}F = -S\,\mathrm{d}T - P\,\mathrm{d}V + \sum_i \mu_i\,\mathrm{d}N_i$	$-S\,\mathrm{d}T - P\,\mathrm{d}V + \sum_i \mu_i\,\mathrm{d}N_i + \gamma\,\mathrm{d}A$
Gibbs free energy, G	$\mathrm{d}G = -S\,\mathrm{d}T + V\,\mathrm{d}P$	$\mathrm{d}G = -S\,\mathrm{d}T + V\,\mathrm{d}P + \sum_i \mu_i\,\mathrm{d}N_i$	$-S\,\mathrm{d}T + V\,\mathrm{d}P + \sum_i \mu_i\,\mathrm{d}N_i + \gamma\,\mathrm{d}A$
		$\left(\dfrac{\partial G}{\partial N_i}\right)_{T,P,N_j} \equiv (\mu_i)_{T,P,N_j}$ = chemical potential	$\mathrm{d}N_i/\mathrm{d}A \equiv \Gamma_i$ = surface excess $\sum \mu_i\,\mathrm{d}N_i = \sum \mu_i \Gamma_i\,\mathrm{d}A$

[a] In the absence of other forms of work, e.g., electrical, magnetic, gravitational, etc.

[b] For more than one type of surface, replace $\gamma\,\mathrm{d}A$ by $\sum_i \gamma_i\,\mathrm{d}A_i$.

conventional thermodynamics must include all of the γdA components for the heterogeneous (interface-containing) systems. In systems where charges or electrical potentials are present, the electrical work must also be included. This is usually done via the electrochemical potential.

The creation of a new area of surface, dA, may cause a flow of dN molecules to or from the surface region, which leads to a surface excess (or deficiency) of component i:

$$\Gamma_i \equiv dN_i/dA \equiv \text{surface excess of component } i \tag{2}$$

The surface excess concept is widely used in the study of adsorption.

Recall from basic thermodynamics, the chemical potential of component i, μ_i, is given as:

$$\mu_i = \left(\frac{\partial F}{\partial N_i}\right)_{T,V,N_j,A}$$

$$\mu_i = \left(\frac{\partial G}{\partial N_i}\right)_{T,P,N_j,A}$$

$$\mu_i = \left(\frac{\partial E}{\partial N_i}\right)_{S,V,N_j,A} \tag{3}$$

where P = pressure, F = Helmholtz free energy, G = Gibbs free energy, E = internal energy, S = entropy, N_j = the number of molecules other than type i, and A = the surface area.

Noting that $dN_i = \Gamma_i dA$ and applying basic thermodynamics, one can develop expressions for surface energy, surface Gibbs free energy, and surface Helmholtz free energy (there is often confusion between these terms):

$$(dE)_V = T\,dS + \gamma\,dA + \sum \mu_i\Gamma_i\,dA$$

$$(dF)_{T,V} = \gamma\,dA + \sum \mu_i\Gamma_i\,dA$$

$$(dG)_{T,P} = \gamma\,dA + \sum \mu_i\Gamma_i\,dA \tag{4}$$

The corresponding specific energy and specific free energies are:

$$\left(\frac{dE}{dA}\right)_{S,V} \equiv e_s = \gamma + \sum \mu_i\Gamma_i \equiv \text{specific surface energy}$$

$$\left(\frac{dF}{dA}\right)_{T,V} \equiv f_s = \gamma + \sum \mu_i \Gamma_i \equiv \text{specific surface Helmholtz free energy}$$

$$\left(\frac{dG}{dA}\right)_{T,P} \equiv g_s = \gamma + \sum \mu_i \Gamma_i \equiv \text{specific surface Gibbs free energy} \tag{5}$$

At constant T, P, S, V, and if $\Gamma_i = 0$, $\gamma = e_s = f_s = g_s$, but only under these conditions.

In general, the surface Helmholtz and Gibbs free energies and the surface "internal" energy are different quantities. Because pressure is generally a more constant parameter than volume in heterogeneous systems, Helmholtz free energy is usually preferred over Gibbs free energy.

The process of forming a new surface can be split into two parts:

1. The phase must be cleaved to expose the new surface.
2. Atoms in the surface plane rearrange to assume their equilibrium positions.

In a multicomponent system, part 2 may also be combined with the migration of bulk atoms to or from the interface, i.e., the development of surface excesses or deficiencies, Γ_i. In a liquid, parts 1 and 2 occur nearly instantaneously. In a solid, part 2 may occur very slowly or not at all.

In a one-component-system, $\Gamma_i = 0$ unless there is such a restructuring around the interface so as to significantly change the density of the phase near the surface.

Therefore, at constant density in a single-component system,

$$\gamma = f_s = g_s$$

or γ is both the specific Gibbs free energy and the specific Helmholtz free energy. Thus, γ can be called the specific surface free energy. At constant S (no restructuring), γ will also be the specific surface energy.

We will use a term, γ, a thermodynamic property of an interface—it is not in general g_s, f_s, or e_s, though it may be equal to one, both, or all three of these quantities under certain conditions. Gamma, the specific surface work, is commonly called the "surface tension." We will refer to surface tension and surface free energy as synonymous with the understanding that we really mean specific surface Helmholtz free energy. Units are ergs/cm^2 or millijoules (mJ)/m^2, although dynes/cm or milli-Newtons (mN)/m are very common and equivalent.

Gamma is the work necessary to form or create unit area of new surface. Surface stress is the work necessary to stretch or compress an existing surface. In a liquid this cannot be done without causing more atoms to join the surface; i.e., creating new surface. Hence in liquids the surface stress is gamma (the reversible work to form new surface); i.e., surface tension. In a solid this is not so.

The lack of mobility of atoms in a solid means we can stretch it without causing bulk atoms to join the surface planes. Thus, we may stretch or compress the surface of a solid without changing the number of atoms in the surface, only their distances of separation, thus producing a surface stress.

Adsorption is a concentration of some species in the interfacial zone, resulting in a decrease in the interfacial free energy. Adsorption can often be mechanistically treated in terms of the residual or unsatisified bonding capacity at the interface.[76] The adsorbed material takes up some of the residual bonding capacity, resulting in a decreased surface energy.

From basic thermodynamics, using either the Gibbs or Guggenheim treatment of interfaces, one can show that

$$(\mathrm{d}\gamma)_T = -\sum_i \Gamma_i \, \mathrm{d}\mu_i \tag{6}$$

where $\Gamma_i \equiv N_i^s/A$, the surface excess of component i. Equation (6) is the Gibbs adsorption equation, a fundamental equation of surface chemistry which relates a surface excess (or deficiency) of components with the change in surface free energy. The Gibbs adsorption equation is, in essence, the Gibbs–Duhem equation of classical thermodynamics applied to a system containing an interface.

If $\Gamma_i > 0$, i.e., $N_i^s > 0$, i is concentrated in the surface zone and we call this adsorption.

If $\Gamma_i < 0$, $N_i^s < 0$, or there is a deficiency of i in the surface zone and we call this negative adsorption.

Since the work to form new surface is γ, the work of cohesion, W_c, of a pure phase is $2\gamma_A$ (Figure 1):

$$W_c = 2\gamma_A \tag{7}$$

When we put two different phases, A and B, in contact, there will be some adherence or interaction energy between them, which is called the work of adhesion, W_{AB} (Figure 1). We can define the interfacial free energy, γ_{AB} as

$$\gamma_{AB} = \gamma_A + \gamma_B - W_{AB} \tag{8}$$

i.e., the work to form each of the surfaces, γ_A and γ_B, less the mutual adhesion or interaction between them, W_{AB}. For two identical phases which adhere in perfect registry, $\gamma_A = \gamma_B$ and $W_{AB} = W_c$, and therefore $\gamma_{AB} = 0$. Often the interfacial (Helmholtz) free energy is called the interfacial tension.

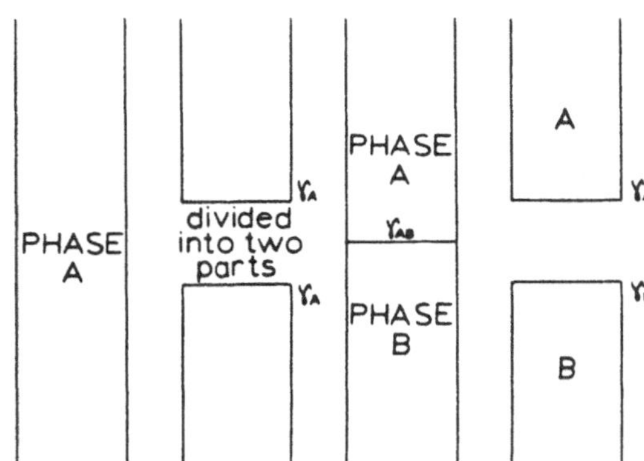

FIGURE 1. Work of cohesion on the left, $W_C = 2\gamma_A$. Work of adhesion on the right, $W_{AB} = \gamma_A + \gamma_B - \gamma_{AB}$.

A newly formed surface may restructure with time and will reach some equilibrium with its surroundings. Assuming a solid surface in a pure vapor environment, vapor molecules may adsorb on the surface. The free energy of a solid surface newly formed in ultra high vacuum is often called γ_{S^0}. The same surface after equilibration with a vapor has a free energy γ_{SV}. $\gamma_{SV} < \gamma_{S^0}$ and the difference is defined as the spreading pressure, π, of vapor, V, on solid surface, S:

$$\pi \equiv \gamma_{S^0} - \gamma_{SV} \equiv \text{spreading pressure} \tag{9}$$

In a complex multicomponent environment, such as typical laboratories, a variety of interactions will occur, leading to an interfacial hierarchy (Figure 2). Such hierarchies are more complex with high-energy

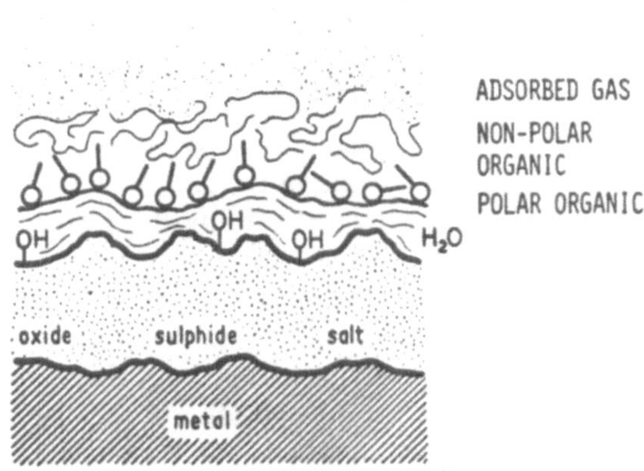

FIGURE 2. Hierarchy of spontaneously adsorbed layers on a metal or other high-energy surface. [Redrawn after References (75) and (76).]

surfaces, such as pure metals or ceramics, than with the much lower-energy surfaces of organic polymers.

2.2. TWO-PHASE EQUILIBRIA—THE LIQUID/VAPOR AND LIQUID/LIQUID INTERFACES

Most basic texts on surface chemistry cover this topic.[1-3] Many reviews are available[11-14] as well as an excellent film.[15] We are basically concerned with the interface between a liquid and its vapor or ambient atmosphere. The surface free energy or surface tension of the liquid is of interest, often as a function of temperature and vapor pressure of the liquid. There are really no direct means of making the measurement, although progress has been made in direct theoretical calculations[16] (see also Chapter 11).

Surface atoms or molecules are in an asymmetric force field (Figure 3a). To produce new surface, work must be done against the unbalanced attractive forces which oppose the movement of molecules from the interior to the surface.[11] Because of the work (γ), the surface has a higher free energy than the interior. Consider Figure 3b, where two surfaces are created by rupturing a column of liquid by applying work to break and separate the attractions across the plane.

Both models of producing new liquid/vapor interface (Figure 3) do not provide a direct explanation of surface tension. The work done to produce new surface is stored as retrievable energy. The system "... is opposing the change by a force against which the external work has to be done."[11] But why "tension"? This "... requires some unfamiliar thinking about hydrostatics, and the high state of pressure (thousands of atmospheres) which molecular attractions produce in a liquid but which cannot be directly measured..."[11] If a molecule in or near the surface is acted upon only by the forces shown in Figure 3a, it would leave the surface and return to the bulk reducing the density in the surface region. This would continue until the reduced concentration in the surface sets up a pressure gradient which stops further net migration. At this equilibrium state there are fewer molecules per unit volume in the surface than in the bulk; thus lower density means a lower pressure, putting the surface layer in a state of tension compared with the bulk.

Another model which has commonly been used to qualitatively explain the state of tension is shown in Figure 3c, the so-called pulley tension model. The net inward attractive force tends to stretch the bonds in the surface region, putting it in a state of tension.

Although surface tension is an important mechanical concept, for our purposes the major quantity of interest is the surface free energy. Under many practical conditions, as indicated earlier, the surface tension and the

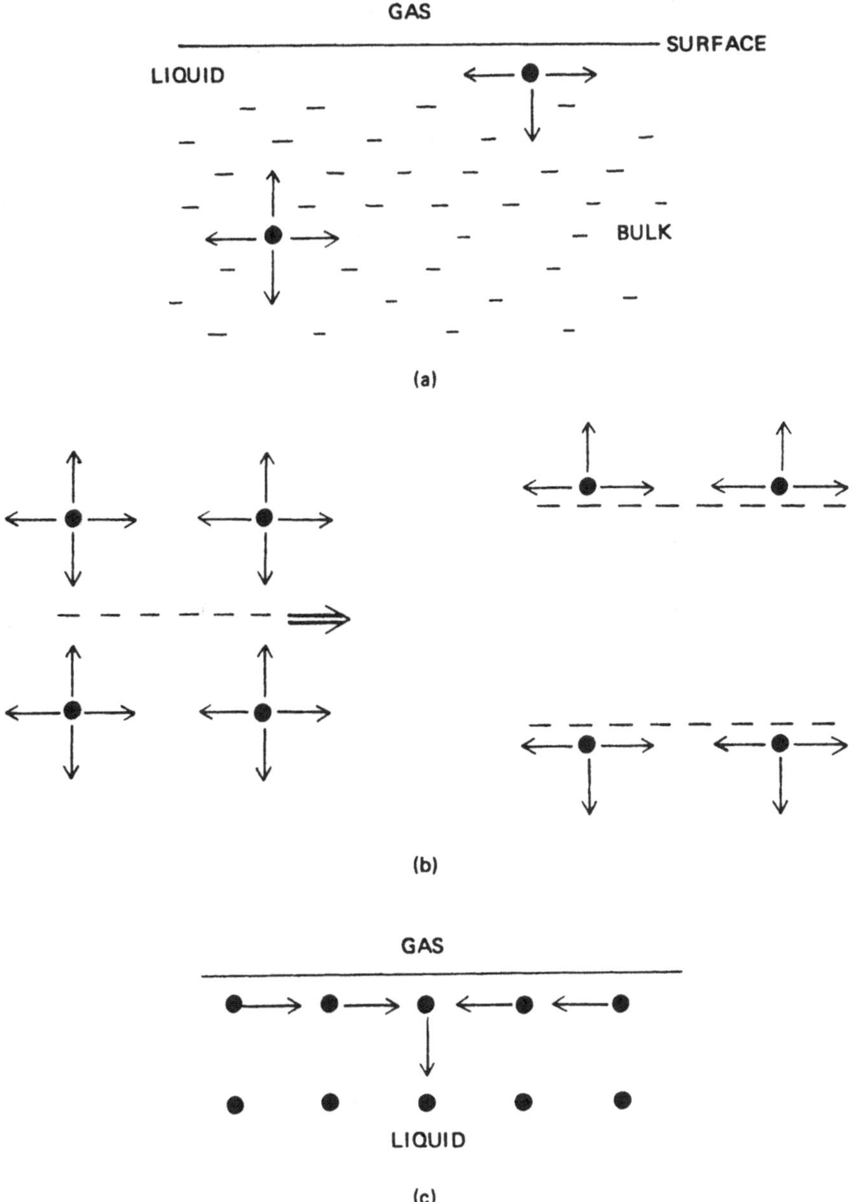

FIGURE 3. (a) Representation of forces due to intermolecular interactions on single molecules in the bulk and in the surface of a liquid. Attractive forces only are shown. [Redrawn from Reference (11).] (b) The creation of surfaces by rupture of a liquid; compare with Figure 1, work of cohesion. [From Reference (11).] (c) The so-called pulley model for demonstrating tension effects in a surface.

surface Helmholtz free energy are equivalent and will be treated as such for the rest of this chapter. It is important to note that the equality of surface tension and surface free energy is not generally true for systems other than a pure liquid in contact with its vapor.

The techniques for measurement of liquid–vapor interfacial tension are well-known and are covered in practically all basic texts on surface science. The Wilhelmy plate technique for the measurement of surface tension will be discussed in Section 3.2. after presentation of the basics of the solid–liquid–vapor interface. Such surface tension measurements are easy to perform, given careful attention to cleanliness and purity.

The study of aqueous systems is particularly prone to contamination because most organic impurities in water adsorb at the water–vapor interface due to the high interfacial tension driving force. Water used for surface measurements must be doubly distilled with a final distillation out of glass or quartz. Water purified by reverse osmosis, by ion exchange columns, or by various charcoal treatments is often unsatisfactory. Very low-conductivity water of the type commonly employed for microcircuit and microelectronic use is often not suitable for surface tension studies, because of the presence of trace organic contaminants. A good check of the purity of water for surface tension studies is a direct measurement of its surface tension and a comparison of the measured values with those generally accepted in the literature.[(24,25)]

The surface tension is a thermodynamic quantity. As it is a function of temperature, measurements must be made under carefully controlled temperature conditions. Values as a function of temperature are readily available.[(24,25)]

The liquid–vapor interface, particularly the water–vapor interface, is useful as a model system for the study of protein adsorption and will be briefly discussed in Vol. 2.

The liquid–liquid interface, primarily the water–apolar liquid interface, is of considerable interest as a model system. A thorough review of liquid–liquid interfacial tensions, thermodynamics, experimental methods, and as a model system for the study of adsorption is available[(17)] and will not be further discussed here. It is important to note, however, that the liquid–liquid interface is an excellent model system for the study of protein adsorption and other biological interactions at relatively mobile interfaces. The interfacial tension can be measured by a variety of methods and a change in the interfacial tension upon adsorption can be monitored. The water-oil interface can also be studied in a Langmuir–Blodgett monolayer trough facility although clearly such studies are more complicated than comparable studies using the water–air interface.

2.3. THREE-PHASE EQUILIBRIA

2.3.1. Ideally Deformable Phases: the L/L/V Interface

Consider the case of three ideally deformable phases in equilibrium, e.g., three immiscible liquids or two liquids and a vapor (Figure 4). Figure 4 top can be represented as a triangle, commonly called Neumann's triangle. The general expression for such a triangle is[18]

$$\frac{\gamma_{23}}{\sin \theta_1} = \frac{\gamma_{13}}{\sin \theta_2} = \frac{\gamma_{12}}{\sin \theta_3} \tag{10}$$

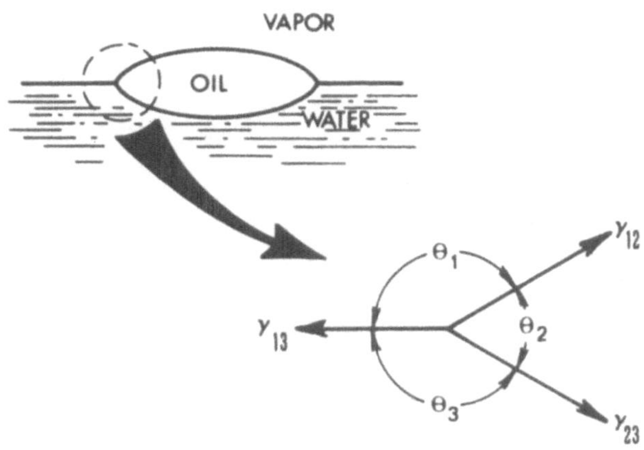

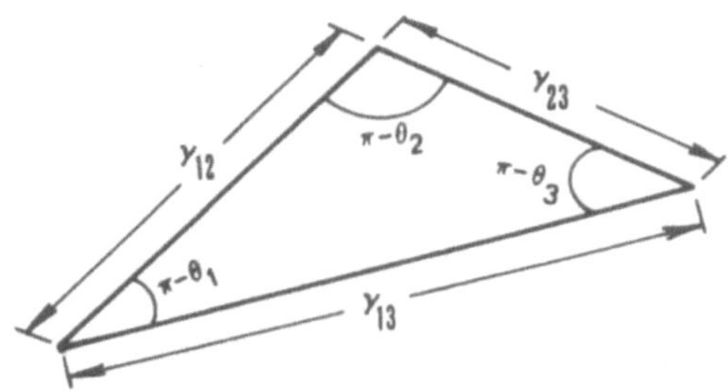

FIGURE 4. Top: An oil droplet at a vapor/water interface. All three phases are ideally deformable and achieve the equilibrium configuration shown. *Bottom*: Neumann's triangle, presenting in simple form the relationships among the phases. [From References (18), (22).]

or equivalently

$$\gamma_{12} + \gamma_{23} \cos \theta_2 + \gamma_{13} \cos \theta_1 = 0 \tag{10a}$$

$$\gamma_{23} + \gamma_{13} \cos \theta_3 + \gamma_{12} \cos \theta_2 = 0 \tag{10b}$$

$$\gamma_{13} + \gamma_{12} \cos \theta_1 + \gamma_{23} \cos \theta_3 = 0 \tag{10c}$$

The symbols are defined in Figure 4.

2.3.2. Ideally Nondeformable Phases: The S/L/V Interface

The interface between an ideally rigid nondeformable solid and an ideally deformable liquid and liquid vapor is illustrated in Figure 5. Figure 5 illustrates an energy balance approach to the three-phase equilibrium which results in Young's equation. Referring to Figure 5 let there be a small displacement, dx. Then

$$(\mathrm{d}F)_{T,V,n} = \gamma_{SV}\,\mathrm{d}x - \gamma_{SL}\,\mathrm{d}x - \gamma_{LV}\,\mathrm{d}x \cos \theta \tag{11}$$

where F is the Helmholtz free energy, γ is the interfacial free energy for the solid–vapor (γ_{SV}), liquid–vapor (γ_{LV}), and solid–liquid (γ_{SL}) interfaces, and θ is the equilibrium contact angle at the three-phase junction. At equilibrium, d$F = 0$, or

$$\gamma_{SV} - \gamma_{SL} = \gamma_{LV} \cos \theta \tag{12}$$

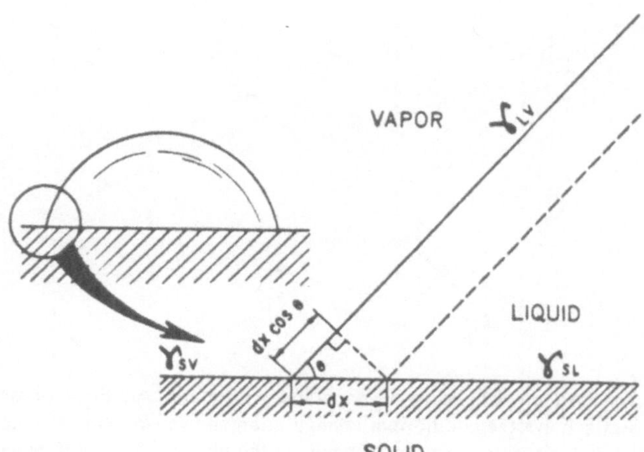

FIGURE 5. The contact angle geometry used in deriving Young's equation. Symbols are defined in the text.

This is Young's equation, one of the fundamental equations of surface chemistry.

There is an implicit assumption in the derivation of Eq. (12). It is assumed that the solid is sufficiently rigid that vertical displacement of the solid is negligible. This assumption is not generally valid for gels or other highly deformable solids. Clearly, Young's equation is just a special case of Eq. (10) and (10c) with $\theta_1 + \theta_2 = 180° = \theta_3$.

The problem is: are the angles measured in a conventional contact angle experiment the true equilibrium contact angles?[4,18] Bikerman has questioned the use of Eq. (12) with low-modulus solids.[3] He demonstrated that the solid does not necessarily remain planar when in contact with a high surface tension liquid drop.[19] Rigorous treatments of the extent of deformation at the solid-liquid-vapor boundary have been given by Lester[18] and Rusanov[20,21] and discussed by Andrade *et al.*[22] The basic problem is illustrated in Figure 6. There is considerable direct evidence for surface tension-induced deformation of solids.

Bikerman has shown[19] that a mercury-aqueous gel interface can produce a deformation ridge on the gel 40 μm high. Though this seems large, the gel modulus was 6×10^4 dyn/cm^2 and γ_{SL} for mercury is roughly 450 dyn/cm. Using Rusanov's treatment,[20,22] we get $z_2 = 0.05$ cm or roughly 500 μm for this system. Thus, Bikerman's observation is within an order of

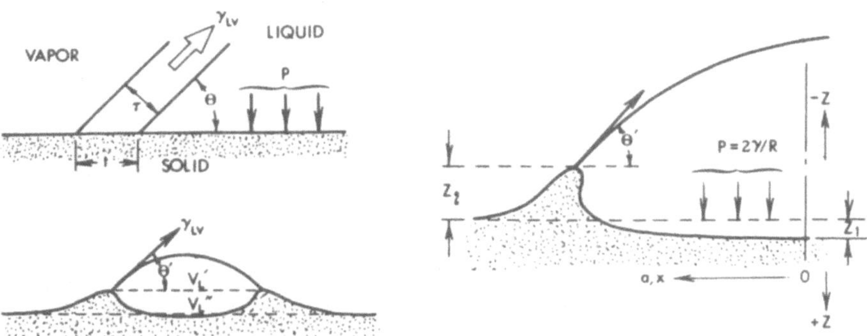

FIGURE 6. Rusanov's model for contact angle-induced deformation. *Left top*: The geometry at the zone of the three-phase contact; τ is the thickness of the liquid-vapor interface, t is its intersection with the solid surface, where $t = \tau/\sin\theta$, θ is the contact angle. *Left bottom*: Exaggerated drop profile after the solid has deformed and equilibrium has been achieved. V_L' is the volume of the portion of the drop above the horizontal line shown; V_L'' is the volume below that line. The contact angle with respect to the horizontal has decreased from θ to θ'. *Right*: Schematic detail of the deformation in the three-phase region. Z_1 is the vertical deformation of the solid inside the drop; Z_2 is the maximum height or rise at the drop periphery. [Redrawn from References (20) and (22).]

magnitude of Rusanov's treatment. Actually, Bikerman suggests that E in the surface region is greater than 6×10^4 (owing to dehydration of the gel surface). If this is so, E could easily be larger than 6×10^4, which would give z_2 closer to Bikerman's data. Although Bikerman's numbers are reasonable, his treatment has been criticized by Rusanov.[20]

It is clear that contact angle measurements on a highly deformable solid must be treated with caution. Unfortunately, a good useful model for contact angle-induced deformation is not available. Our operating rule of thumb, based on Rusanov's calculations[20,21] and our observations,[22] suggests that solids with a modulus of elasticity greater than about 3.5×10^5 dyn/cm^2 should not lead to very large errors in the measured contact angle due to deformation. Such a modulus corresponds, for example, to about a 40% water content poly(hydroxyethyl methacrylate) material. Materials with greater water contents and/or lower moduli need to be treated with some suspicion regarding contact angle measurements, unless account is taken of the deformation effects. We need to point out, however, that most materials containing 20 to 30% water and greater already show a receding contact angle near zero with water due to the hydrophilic component dominating the interface under these conditions.[22] This is due to some of the surface dynamics arguments presented in Chapter 2.

2.4. EQUILIBRIUM

The key assumption in any thermodynamic treatment is equilibrium. It is assumed that Young's equation is thermodynamically valid because the three phases are in true equilibrium at the contact points. As is evident from Chapter 2, polymer materials are rarely, if ever, in bulk equilibrium and the same is true of their surfaces. Thus, one must expect time, temperature, and environment dependent changes in the contact angle. Such changes in the contact angle allow important deductions to be made with regard to surface dynamics, water equilibration, and the general surface properties of materials under water.

2.5. SURFACE AND INTERFACIAL TENSION MEASUREMENT

Each of the classic textbooks on polymer surface chemistry (Table 1, Chapter 1) have at least one chapter on capillarity and surface tension. Although there are many methods available for measuring the surface tension or surface free energy of liquid surfaces, including capillary rise, pendant drop, DuNuoy ring, and others, the only one discussed here is the Wilhelmy plate technique,[26] because it is a highly accurate and precise method, and because it is the basis of the dynamic contact angle hysteresis technique which will be discussed later.

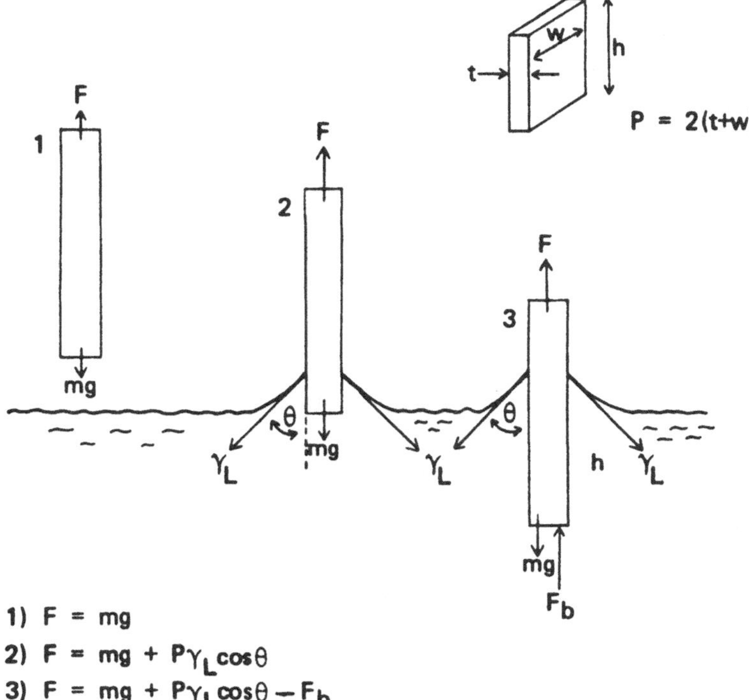

FIGURE 7. The Wilhelmy plate method for the measurement of surface tension and contact angle. Notation is described in the text. Condition 1 is the plate in air prior to contact with liquid. Condition 2 is the plate just at the liquid-air interface. The total force measured by the balance is now mg plus the downward pull of the surface tension around the perimeter. Condition 3 is the same as 2, but now, because the plate is immersed in the liquid some distance, there is a buoyancy force which must be considered.

Figure 7 presents the basic principles of the Wilhelmy plate technique for the measurement of surface tension. One takes a totally wetting thin plate such as a freshly acid-cleaned glass microscope coverslip, attaches it to a recording balance (generally an electrobalance), and records the mass of the plate. A container of clean liquid whose surface tension is to be measured is raised until the liquid touches the bottom of the plate, at which point the mass recorded by the balance increases due to the vertical pull of the surface tension. As one continues to advance the liquid over the plate, the measured mass decreases due to buoyancy forces. The surface tension is calculated from the force on the balance and the dimensions of the plate assuming total wetting, i.e., $\theta = 0°$. The basic equation is:

$$F = mg + p\gamma_{LV} \cos \theta - F_b \tag{13}$$

where F = total force recorded by the balance, m = mass of the plate, g = acceleration due to gravity, p = perimeter of the plate, γ_{LV} = surface tension of the liquid, θ = contact angle, and F_b = buoyancy force, given by

$$F_b = \rho_L V_{\text{immersed}} g = \rho_L ghtw \tag{14}$$

where ρ_L = density of the liquid, g = acceleration due to gravity, V = volume of plate immersed in the liquid and $V = htw$; w = width of the plate, t = plate thickness, and h = immersion depth.

A typical x-y recorder output of such a measurement, with the balance output (force) feeding the y-axis and immersion depth feeding the x-axis, is given in Figure 8, where the results of an immersion and emersion (withdrawal) of the plate are shown, showing no hysteresis.

Standard values of surface tensions of liquids are tabulated in many chemical handbooks,[24,25] texts, monographs, and reviews. Values for water are readily available, including measurements as a function of temperature.[27,28]

Interpretation of a contact angle measurement on a solid assumes that the surface tension of the probe liquid is well-known. The Wilhelmy plate method is a means to accurately measure the surface tension of the liquid *and* the $L/S/V$ contact angle.

Liquid/liquid interfaces are important as model systems. For example, the oil/water interface may be a reasonably good model for a flexible apolar

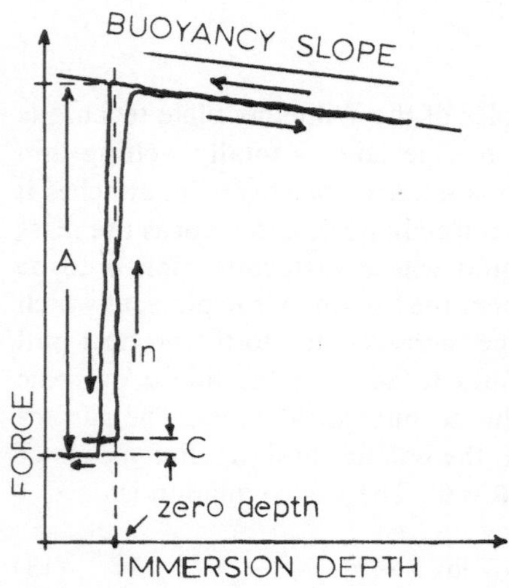

FIGURE 8. Wetting characteristics for a highly clean surface, radio frequency glow-discharged mica, prehydrated in water. Note that there is no hysteresis present; the advancing and receding slopes and angles are identical. The vertical axis is the force measured by the balance, the horizontal axis is the immersion depth. The force, A, is used in the equation given in the text to calculate the surface tension of the liquid, assuming total wetting of the solid.

hydrocarbon polymer interfacing with water, such as the poly(1-hexene)/water interface. The liquid silicone/water interface may be important in understanding the properties of poly(dimethylsiloxane)/water interfaces. For example, changes in water/oil interfacial tensions have been used as a sensitive probe of protein adsorption from the water phase to the water/oil interface by Bagnall *et al.*[(23)]

3. CONTACT ANGLE MEASUREMENT

3.1. ASSUMPTIONS

A straightforward interpretation of the contact angle can usually be made if the following assumptions are valid:

1. The solid surface is rigid, immobile, and nondeformable. In practical terms this means that the surface modulus of elasticity should be greater than 3.5×10^5 dyn/cm^2.
2. The solid surface is highly smooth. The result of surface roughness is hysteresis in the contact angle which will be discussed later. Most theoretical models and experimental studies suggest that if the surface is smooth at the 0.1 μ level this effect is negligible. Thus, a surface which appears glassy smoth to the eye or optically smooth is generally satisfactory.
3. The solid surface is uniform and homogeneous. This basically means there is no patchwise heterogeneity in the solid surface, i.e., the surface does not consist of regions which vary greatly in surface free energy, such as hydrophobic and hydrophilic regions or patches on the surface. Dirty surfaces are often of a patchy character. Polymer blends and block copolymer surfaces are known to have two or more phases present, each of which may have different properties.
4. The liquid surface tension is well-known and constant and does not change during the time course of the experiment.
5. The solid surface does not interact in any way with the liquid other than the three-phase equilibrium noted earlier, i.e., the liquid does not swell the solid surface, not even in the surface region. The liquid must not cause extraction or partitioning of material from the solid phase into the liquid phase, which could then change either the liquid surface tension or the solid/liquid interfacial tension. This is a difficult constraint for many materials.
6. The liquid spreading pressure [Eq. (9)] on the solid is zero. This means that liquid vapor does not adsorb on the solid surface to change the solid surface free energy. High-energy liquids such as

water do not generally adsorb on very low-energy solids, such as many synthetic polymers; thus, the spreading pressure in many of these systems can be approximated to zero.

7. It is also generally assumed that the solid surface is so rigid and immobile that the surface groups cannot reorient or re-equilibrate in response to changes in environment, i.e., polymer surface dynamics, as outlined in Chapter 2, does not exist.

Each of these assumptions will be challenged later in this chapter and will generally be shown to result in a contact angle hysteresis which is either time invariant (true hysteresis) or hysteresis whose nature changes with time and therefore represents a kinetic re-equilibration or related phenomenon rather than a true hysteresis metastable phenomenon.

A very simple and sensitive test has been devised by R. E. Baier to test the assumption[20,30] of non-extraction of the solid by the liquid. Figure 9 is a standard plot of the contact angles obtained for a series of pure wetting liquids on Teflon. To perform the test for surface contaminants, a drop of pure liquid—for example, water—is placed on smooth Teflon FEP film; its contact angle is recorded—for water this would be about 110°. The material having unknown surface properties is touched to the drop; if there are no transferable surface-active constituents, the contact angle of the drop remains unchanged—a good test of the surface cleanliness is thereby com-

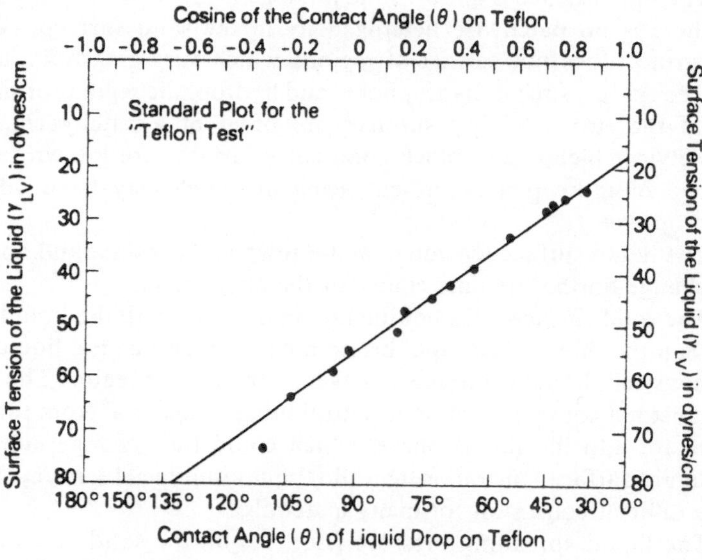

FIGURE 9. A standard plot for the "Teflon test," The contact angles of Teflon with each of a series of pure liquids are shown. [From References (29) and (30).]

pleted. If surface-active agents are present, the drop profile changes instantaneously to exhibit a new equilibrium contact angle reflecting its new surface tension. The surface tension which corresponds to the new contact angle can be deduced from the standard Teflon plot. The difference between the surface tension of the pure liquid and the surface tension of the contaminated liquid represents the surface activity (i.e., the spreading pressure) of the agent.

One can also expose the liquid to the polymer for an appropriate time and then simply make a direct surface tension measurement of the liquid. If there has been any substantial change in the liquid surface tension as a result of exposure to the polymer, then something is being extracted from the polymer into the liquid, resulting in a decreased surface tension of the liquid.

It is important to note that practically all commercial polymer materials must be exhaustively purified, extracted, or otherwise cleaned; if not, surface-active extractables will be present which make contact angle measurements on such surfaces difficult to interpret.

3.2. *METHODS*

The most common methods of measuring the contact angle include:

1. Direct microscopic measurement of the three-phase interface with a goniometer or protractor which directly measures the angle.
2. Measurements of the dimensions of a drop profile on a surface from which the contact angle can be calculated from spherical trigonometric relationships.
3. Measurement of the diameter of a drop of known volume on a surface.
4. Rise in a capillary or on a vertical plate of a liquid of known surface tension.
5. The DuNouy ring technique.
6. Wilhelmy plate procedure described earlier with a liquid of known surface tension.

We shall concentrate here primarily on the Wilhelmy plate method of measurement.

The Wilhelmy plate method has been well described primarily for the measurement of liquid surface tension and L/L interfacial tension.[6,8,12,16] Johnson and Dettre[8] have developed the method for the direct measurement of $L/S/V$ contact angles.

The principles of the method have already been described [Figure 7 and Eqs. (13) and (14)]. After careful measurement of the liquid surface tension with a full wetting plate, one then mounts the solid surface of

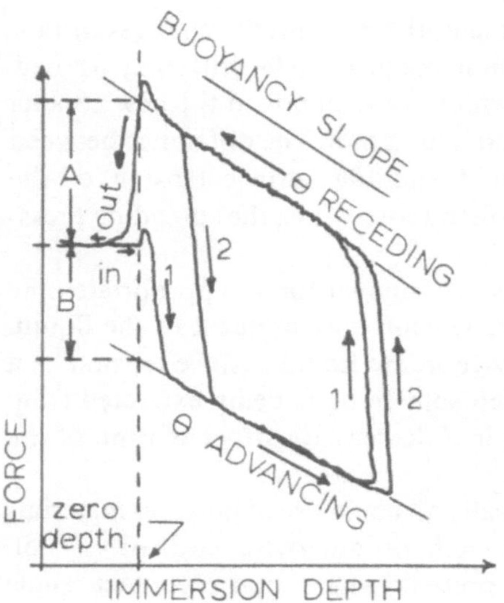

FIGURE 10. A typical stable Wilhelmy plate hysteresis loop for a polybutadiene sample. The buoyancy slope is parallel in both the advancing and the receding cases. The sample was taken through two successive cycles to demonstrate reproducibility. Displacement A is used to compute the receding angle; displacement B is used to compute the advancing angle. The data were obtained at the point of zero depth of immersion so the buoyancy effect can be neglected. [From Reference (32).]

interest on the recording balance and obtains a plot (Figure 10). Cos θ can be directly calculated from Eq. (13). Note that Figure 10 shows hysteresis, i.e., the force measured on immersion is less than that measured on emersion. Most polymer surfaces show such hysteretic behavior. The contact angle deduced on immersion is the advancing contact angle, θ_A, that on emersion is the receding angle, θ_R. The difference, $\Delta\theta$, or

$$\Delta\theta = \theta_A - \theta_R \tag{15}$$

is called the contact angle hysteresis.

Additional examples and a discussion of contact angle hysteresis will be given later.

The Wilhelmy plate technique has the advantage that many different shapes of materials can be measured, including rods or fibers,[33] tubes, circular plates such as contact lens blanks,[34] etc. One disadvantage is that both surfaces of the plate or sample should be identical; otherwise the plot is difficult to interpret.

It is important to note the difference between a true static and a dynamic contact angle measurement. Measurement at a solid/liquid interface which is *not* in motion is inherently a static contact angle measurement. If the interface is in motion, it is a dynamic contact angle measurement. The Wilhelmy plate technique is normally a dynamic measurement although the motion of the liquid with respect to the plate can be stopped and a static measurement made. Although there has been considerable discussion regarding static and dynamic measurements, for low immersion velocities with normal low viscosity liquids on relatively rigid surfaces, there is no difference between static and dynamic values of the contact angle.

3.1. CRITICAL SURFACE TENSION, γ_c

The most popular and widely used method of characterizing the surface properties of polymers has been to measure the critical surface tension, a term and technique pioneered by Zisman and his coworkers.[7]

The critical surface tension, γ_c, is a useful parameter obtained by measurement of the contact angle of each of a series of homologous liquids on the test material. A graphical treatment of the contact angle data (Figure 11) is used to deduce γ_c. For most liquids the cos θ values fall along the same straight line or cluster closely around it in a narrow rectilinear band.

The critical surface tension values are empirically related to the surface constitution. Even small changes in the outermost atomic layer are reflected in a change of critical surface tension. For example, a simple hydrocarbon surface (like that of polyethylene) exhibits contact angles leading to a critical surface tension of about 31 dyn/cm. Replacement of hydrogen atoms in the surface by fluroine atoms gradually decreases the values to 19 dyn/cm (as

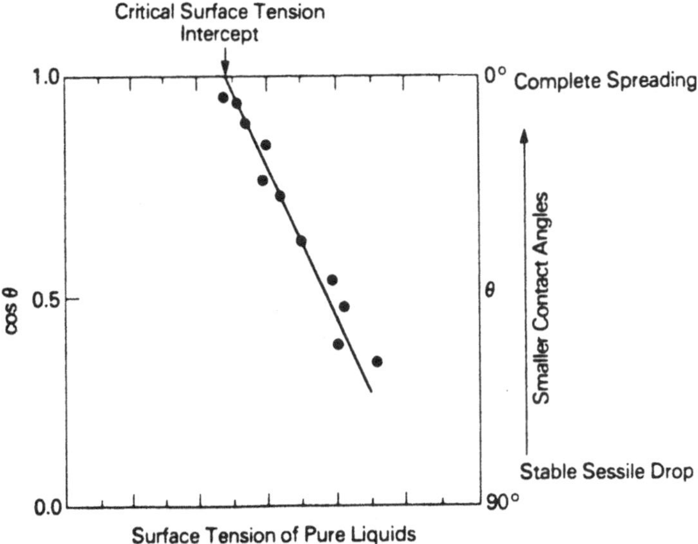

FIGURE 11. A typical Zisman γ_c plot. The cosines of the contact angle for a range of pure liquids on a given solid are plotted against the liquid surface tensions. The critical surface tension is given by the intercept at cos $\theta = 1$ and is defined as the surface tension of that liquid which would just totally spread on the solid surface. This is an empirical measure related to the surface free energy of the solid and is called the critical surface tension for wetting of that particular solid. [From Reference (30); see also Reference (7).]

observed with polytetrafluoroethylene). Conversely, replacement of surface hydrogen with chlorine atoms leads to an increased critical surface tension approaching 41 dyn/cm [as with poly(vinyl chlorides)]. A complete discussion and tabulated values are given in Reference (7).

It is important to note that γ_c values are determined from advancing contact angle measurements using liquids selected to minimize swelling or penetration into the polymer surface. Water, alcohols, and related hydrogen-bonding liquids are often difficult to use in this test because of the extensive interaction and penetration effects observed. Many investigators have argued that the critical surface tension for wetting is more a measure of the apolar or dispersion force contribution to the surface free energy rather than of the total surface free energy of the polymer. Others have cautioned that this measurement is a measurement of the surface properties of the polymer in equilibrium with air or vapor rather than any measure of the surface properties in more biological environments, such as water. One problem with the test is that it requires a large series of highly purified specialty liquids to properly make the measurements; these liquids are often not readily available to investigators. Nevertheless, the test has proven to be extremely useful among the biomedical materials community as well as in other areas where polymer surface properties and quality control are important and correlate with in-use performance. Values are tabulated in the *Polymer Handbook*[31] for a wide range of polymers.

We will not further discuss nor utilize this technique here, because one of the major objectives and goals in this chapter is to characterize the polymer/water interface, since that is the relevant interface for biomedical polymer purposes. This is not possible by the critical surface tension technique.

3.4. CAPTIVE BUBBLE (UNDERWATER) METHOD

The captive bubble method consists of immersing the solid in a liquid of interest, water in our case, and then introducing an air bubble or an immiscible liquid droplet (such as *n*-octane) at the interface (Figure 12). The angle can be measured directly via a goniometer or calculated from measurement of the dimensions of the bubble.

The bubbles of air or liquid are introduced onto the immersed sample surface using a microliter syringe equipped with a U-shaped needle. The highest purity octane and dodecane available (99% or better) should be used. Surface and interfacial tensions must always be checked. The height and diameter are measured by use of micrometers which manipulate a stage holding the sample immersed in water. The air bubbles are observed through a microscope using a 20× objective with a long working distance and a 15× eye-piece equipped with a fine crosshair reticle. The bubble is manipulated

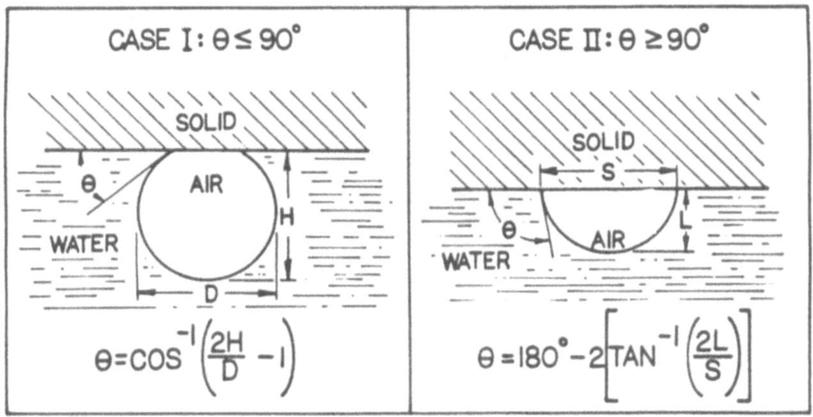

FIGURE 12. The captive bubble technique for underwater contact angle measurements. After the solid is fully immersed in water, an air bubble or octane droplet is introduced and allowed to contact the solid-water interface. As shortly after contact as possible, the dimensions of the droplet are measured and the angle θ calculated as shown above. [From References (30), (22), (65), (66), and (73).]

across the crosshairs and the dimensions are read directly from the micrometers. The sample box is back-illuminated by a variable light source. The bubble volume is approximately 0.1 to 0.2 μl. The bubble volume is minimized in order to avoid buoyancy and gravity effects. The bubbles are applied to the surface by forming a bubble at the tip of a microsyringe needle and then tapping the syringe body with the index finger to allow the bubble to dislodge and float from the needle tip up to the water sample surface. Figure 12 schematically illustrates the geometry of the contact angle measurement. The contact angles are calculated using the equations

$$\theta = 180° - 2\left[\tan^{-1}\left(\frac{2L}{S}\right)\right]$$

for angles greater than 90°

$$\theta = \cos^{-1}\left(\frac{2H}{D} - 1\right)$$

for angles less than 90°.

The sample is fully equilibrated with water before applying the air bubble, whose dimensions are then measured as soon as possible after surface attachment. The angle measured is fairly close but not necessarily equal to the receding water contact angle, as measured by Wilhelmy plate.[67]

Clearly the γ_c method, which relies mainly on advancing angles, and the captive bubble in water method, which measures only receding angles, are both very limited in information content. A complete hysteresis loop containing both θ advancing and θ receding is much more informative.

4. CONTACT ANGLE HYSTERESIS

4.1. BACKGROUND

It is commonly observed that if one measures the contact angle of a liquid drop being advanced slowly over a polymer surface and then makes the measurement with the drop receding over the previously liquid-contacted surface, the two contact angle measurements are different. The difference in the advancing and the receding contact angle, Δ, is commonly called contact angle hysteresis. A surface–liquid pair which meets all the assumptions discussed earlier (Section 3.1.) should not exhibit contact angle hysteresis, i.e., it is in true thermodynamic equilibrium and there is only one thermodynamically definable angle of contact. In experimental situations where all of the assumptions discussed above have been met, only one contact angle is observed and hysteresis is not present. However, most systems do not meet one or more of the assumptions listed and, therefore, contact angle hysteresis is the rule rather than the exception. Rather than being a problem, as it is commonly portrayed in classical surface chemistry textbooks, contact angle hysteresis is a rich source of information on the behavior of polymer–liquid interfaces.

The contact angle hysteresis measurement is commonly made by the methods of contact angle measurement discussed above. One can advance a drop over a solid surface by inclining the surface, that is, making an incline plane and adjusting the angle of incline until the drop moves slowly over the polymer surface or just begins to move. Then one can measure the advancing angle on the low end of the plane and the receding angle on the upper portion of the plane. This method is discussed in all surface chemistry textbooks and will not be further discussed here.

Another method is to increase the volume of the drop such as by use of a motorized syringe, forcing the drop to advance on the solid. The receding measurement is made by decreasing the volume of the drop and making the appropriate measurement.

Sometimes the receding angle is measured by simply letting the drop evaporate with time. This, of course, has the disadvantage that any small impurities present in the liquid or at the solid surface will now begin to concentrate in the drop and can contribute artifacts to the measurement.

Probably the simplest, most straightforward, and most informative way to make the contact angle hysteresis measurement is again via the Wilhelmy plate technique previously described. In this method the contact angle is not measured directly, but is deduced from a force measurement as a function of immersion depth in the liquid. As the solid can either be immersed or emersed from the liquid one inherently has both the advancing and receding angles as a function of depth of the plate. This angle is averaged, of course, around the periphery of the plate at each particular immersion depth. The result of such a measurement is generally a hysteresis loop as discussed earlier. Each of the classical assumptions involved in contact angle interpretation discussed earlier will now be approached from the point of view of contact angle hysteresis measurements.

4.2. HYSTERESIS (Table 2)

There are two classes of hysteresis[8,35,36]

1. True or thermodynamic hysteresis, where the hysteresis curve is reproducible over many cycles and is independent of time or frequency.
2. Kinetic hysteresis, where the curve changes with time or frequency.

There are many examples of both types of hysteresis.[35,36] The interpretation of true thermodynamic hysteresis is generally based on the concepts of metastable states and microscopic domains undergoing cooperative non-equilibrium transitions. Kinetic hysteresis is generally due to slow equilibrium times.

If a system exhibits a set of scanning curves (Figure 10), independent of immersion or emersion depth or conditions, then the loops represent true hysteresis.

If multiple cycles of measurement are not repeatable, as in Figure 13, then kinetic hysteresis is expected. Figure 13 shows the case of an initially dry polymer which swells in water with time, changing both its weight and its wettability. The Wilhelmy plate hysteresis loops are now time- and position-dependent.

Johnson[8] and Smith[32] have seen many systems such as this. Often in this situation the buoyancy slopes are not parallel, indicating that the solid phase is interacting with the wetting medium. One can usually distinguish kinetic effects from true hysteresis by varying the time of the measurement. If the interaction between the solution and the solid phase is either very fast or extremely slow in comparison with the time of measurement, then there is usually no difference in contact angle and no hysteresis will be observed. In Figure 13 the immersion and withdrawal rate is approximately 40 mm per minute, which allows up to one and one-half to two minutes for the solution to interact with the hydrogel. If one were to speed up the

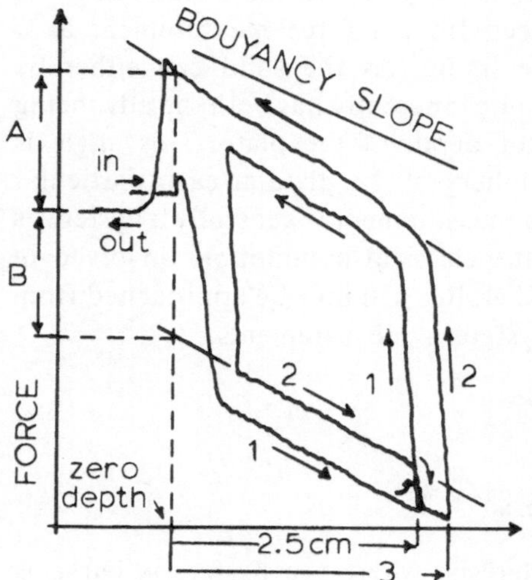

FIGURE 13. An example of contact angle hysteresis which is due to kinetic or time-dependent effects. In this particular case an unhydrated poly(hydroxyethyl methacrylate) gel was immersed in water and the Wilhelmy plate experiment carried out. The gel is swelling during the course of the experiment. As it swells, its mass, dimensions, and wettability change. The result is that the recorded force increases with time in contact with water. Note that on the second cycle the gel was immersed a short distance beyond the immersion distance for the first cycle, i.e., dry gel is seeing water for the first time. The plot on cycle 2 is parallel to that of cycle 1 for that portion of the gel which was still initially dry. [From Reference (32) by permission.]

process, the slopes of the curves would change, allowing less time for the water to swell the gel solid phase.

4.3. ROUGHNESS

Refer to Figure 14 and Reference (8). A drop on a tilted rough surface shows apparent angles θ_a and θ_r, but the true angle of contact is θ_0. Clearly, the macroscopic angles measured can be very different from the actual microscopic or true angles.

Wenzel (see Reference 8) attempted to relate contact angles to true or total area, A, and to apparent or geometric area, A'. He defined roughness, r, as

$$r = A/A' \tag{16}$$

As $A \geq A'$, $r \geq 1$.

Wenzel further said

$$\cos\theta = r\cos\theta_0 \tag{17}$$

where θ is the observed macroscopic angle and θ_0 is the true microscopic angle. It is clear that as r increases for a given θ_0, θ increases or decreases (Figure 15). Unfortunately, Wenzel's treatment doesn't consider metastable states on the surface induced by roughness.

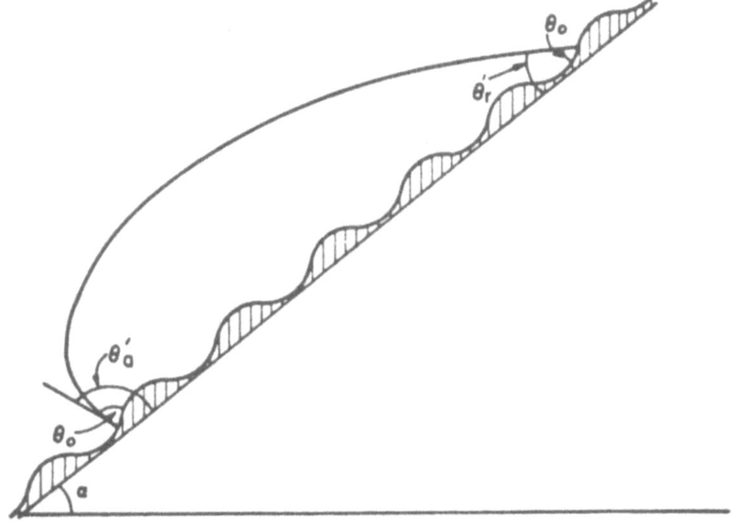

FIGURE 14. A drop on an inclined plane which is an ideally rough surface. The front and rear of the drop make identical thermodynamic contact angles with the surface, i.e., θ_0. However, because of the surface texture, the angles appear to the observer as θ_a for the front surface and θ_r for the rear surface. θ_a is the measured advancing angle, θ_r is the measured receding angle. [From Reference (8) by permission.]

Although the effects of surface roughness on contact angle hysteresis have been modeled by a number of groups,[37-41] it is often difficult to relate those results to practical situations. Perhaps the best general study is that of Johnson and Dettre in which a theroetical study was coupled with various experimental measurements.[37,38] The basic conclusion is that as roughness increases, the hysteresis increases (Figure 15); the advancing angle increases with increasing roughness and the receding angle decreases with increasing roughness. The Johnson and Dettre theoretical results[37] have been nicely verified by direct experiment.

There is a general consensus that, in addition to the heights of the asperities, their shape and distribution is also of major importance, i.e., the surface texture is just as important as the absolute surface roughness. This has been demonstrated by a number of groups.[39-41] The old rule of thumb that asperities in the range of 0.1 to 0.5 μ or below do not lead to significant hysteresis is probably still generally valid, particularly for surfaces whose textures are more or less uniform.

If the surface becomes too rough, then capillary and wicking phenomena become important, resulting in total wetting and spreading. Spontaneous wicking becomes possible when

$$r = 1/\cos\theta_0(\theta_0 < 90°) \tag{18}$$

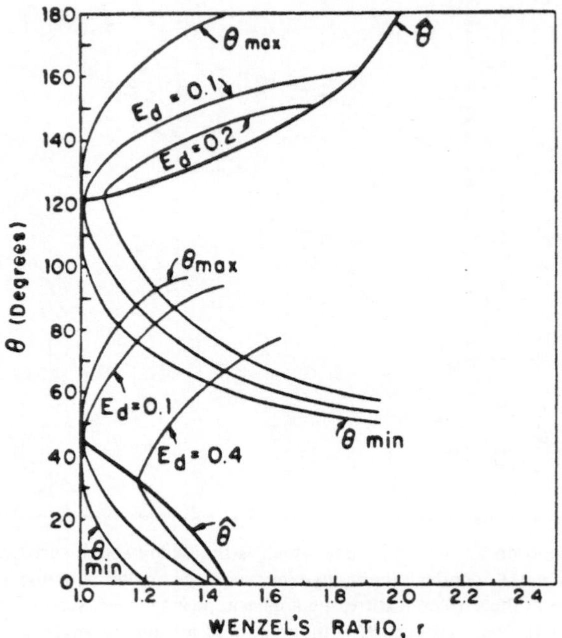

FIGURE 15. Result of the Johnson and Dettre model of the effect of roughness on contact angle. The contact angle is plotted vs. the roughness factor as given by Wenzel's ratio, Eq. (15). Note that θ_{adv} increases with roughness until one reaches a roughness ratio which leads to a composite interface (defined in the text). θ_{rec} decreases until the composite condition is reached. [From Reference (8), p. 103, with permission.]

If θ_0 is high then the liquid may not be able to penetrate into cracks and crevices. Such an interface which contains entrapped air is called a composite interface. The transition to a composite surface is shown in Figure 15.

It is difficult to measure contact angles without introducing some type of random vibration into the system. The degree of random vibration affects the measured contact angles. It is important to reduce inherent mechanical vibrations to a minimum.

Surface roughness produces contact angle hysteresis of the true or thermodynamic type. True surface roughness should not be responsible for kinetic or time-dependent hysteresis.

4.4. SURFACE HETEROGENEITY

Figure 16 shows a two-phase, heterogeneous surface. The small area or discontinuous phase is shown to be relatively non-wetting whereas the

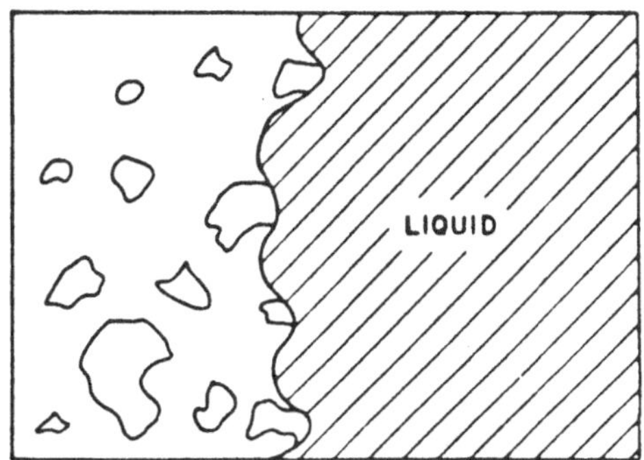

FIGURE 16. Schematic version of a liquid spreading on a heterogeneous surface consisting of small low-energy or hydrophobic domains in a matrix of high-energy or more hydrophilic material. The hydrophobic domains act as pinning points which retard the liquid from advancing over the heterogeneous surface, thus leading to an advancing angle higher than otherwise expected. [From Reference (8) by permission.]

continuous phase is partially wetting. As the liquid advances, the liquid edge tends to stop or be held up by the low-wetting islands. By viscosity coupling this tends to prevent the liquid from spreading over the high-energy phase. The result is an advancing contact angle which is high due to the pinning effect of the low-wetting, high-angle phase.

Once the entire surface is covered with liquid and one makes a receding measurement, there is a tendency for the high-energy, low-angle phase to keep the liquid from being pulled away. Now the high-energy regions function as pinning points to keep the liquid from receding. The result is that the advancing angle tends to represent the low-energy (non-wetting) phase; the receding angle represents the high-energy (wetting) phase, producing a contact angle hysteresis.

Cassie (see Reference 8) has argued that the equilibrium contact angle of a heterogeneous surface, θ, is the weighted average of the angles on each of the two phases:

$$\cos\theta = Q_1 \cos\theta_1 + Q_2 \cos\theta_2 \qquad (19)$$

where Q_1 and Q_2 are the area fractions with contact angles θ_1 and θ_2, respectively. This model does not consider hysteresis effects.

These effects have been modeled[(42,44)] and experimentally analyzed.[(43)] Figure 17 is the result of Johnson and Dettre's model[(42)] for different drop energies. Clearly the receding angle approaches θ_2 and the advancing angle approaches θ_1. Johnson and Dettre conclude[(8)]:

1. Advancing angles are more reproducible on predominantly low-energy surfaces whereas receding angles are more reproducible on predominantly high-energy surfaces.
2. Advancing contact angles alone are not a reliable measure of surface coverage. Thus, 10% and 90% coverage (for example, by a low-energy monolayer) give about the same advancing angle. Similar considerations apply for receding angles.

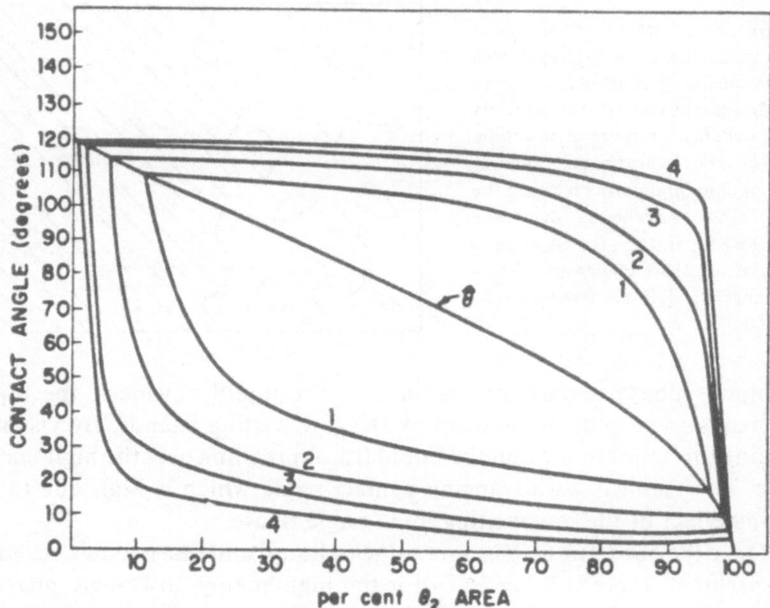

FIGURE 17. Result of the Johnson and Dettre model for contact angles on heterogeneous surfaces. Note that the advancing angle stays high until the high-energy phase is over 90% of the surface area. The advancing angle represents the low-energy or hydrophobic component of the surface. Note that the receding angle drops and already approaches the value of the high-energy phase even when the high-energy phase is only approximately 5% of the surface. Curves 1, 2, 3, and 4 indicate decreasing drop energies where drop energy is related to the intrinsic energy of the drop and its ability to overcome the metastable states present in the surface region. [From Reference (8) by permission.]

3. An advancing angle is a good measure of the wettability of the low-energy part of the surface and receding angle is more characteristic of the high-energy part.

Neumann and Good[(44)] conclude that for patches $<0.1\ \mu$ in size, heterogeneity should make a negligible contribution to hysteresis.

Contact angle hysteresis due to surface heterogeneity is true thermodynamic hysteresis. Kinetic effects may also be present if the proportion of high- or low-energy phase at the surface can change in different environments (see Chapter 2).

4.5. DEFORMATION

If the solid surface is not truly rigid and is deformable under the action of the surface forces at the three-phase boundary, then the measured static contact angle is not the true thermodynamic or equilibrium angle. This we

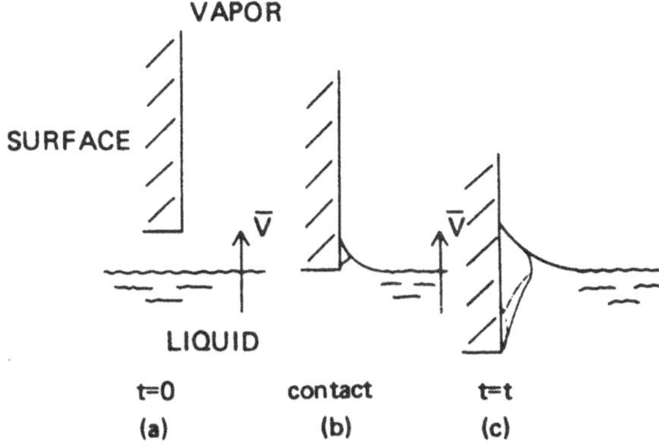

FIGURE 18. Contact angle-induced surface deformation superimposed on a moving liquid line, leading to a deformation ripple which relaxes after the liquid front passes. $\bar{V}$ is the velocity of the liquid over the solid surface (see text).

already recognize. If on top of this effect we now impose a motion of the liquid with respect to the solid or vice versa and we make the very reasonable assumption that the deformation of the solid is time-force-temperature dependent, then we expect the behavior shown in Figure 18, where a deformation ripple may move along with the liquid front. As the liquid front passes, the deformation will relax. A different deformation ripple will result from the receding liquid front. Thus, one has a very complex problem in which the measured contact angle hysteresis is a function of the mechanical properties of the surface and its time-temperature relaxation characteristics, as well as of the dynamics of solid or liquid motion.

The problem of modeling such a deformation is very difficult, because most of the continuum mechanics assumptions are simply not valid on the molecular level, where the deformation is occurring.

The question of liquid motion over a solid surface has been considered in some detail,[(46-48)] including liquid viscosity effects[(47)] and the question of slip at the wall.[(48)] It is generally assumed that classical hydrodynamic arguments and assumptions are not applicable within 1000 Å of the interface.[(47)]

Attempting to couple a moving contact line with the deformation of the solid and the relaxations which may occur is a very complex problem which, to our knowledge, has not yet been treated.

These effects should result in a kinetic, time-dependent component to the hysteresis. One possible example is given in Figure 19. Here we see a series of Wilhelmy plate loops using water as the liquid on a poly(vinyl pyrrolidone) (PVP)-coated surface. Note that the receding angle is ***higher*** than the advancing angle and that this effect appears to be related to the

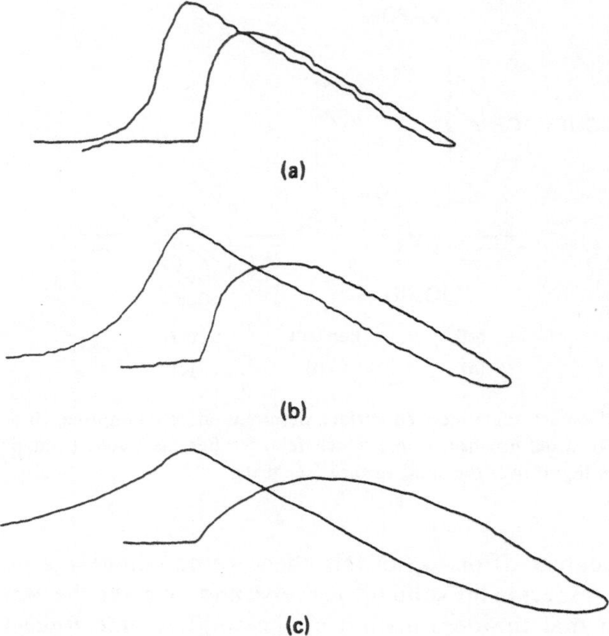

FIGURE 19. Wilhelmy plate plot of a fully hydrated poly(vinyl pyrrolidone) coating on glass. Note that the advancing angle is *lower* than the receding angle, producing a negative hysteresis according to Eq. (13). After examining all possible explanations for this effect, the only thing remaining is contact angle-induced deformation coupled with the velocity of the liquid front, as outlined in Figure 18. This situation has not been modeled. Note that the hysteresis appears to increase (in absolute value) as one goes to faster speeds: (a) 1.34 mm/min., $\Delta\theta = 5°$; (b) 2.00 mm/min., $\Delta\theta = 13°$; (c) 4.00 mm/min., $\Delta\theta = 24°$.

immersion/emersion velocity. PVP is a high water content, deformable gel. We have considered practically all causes of hysteresis in analyzing these data and could exclude all but the deformation/motion considerations in this section. The problem has not been modeled, however. What is seen in Figure 19 may, or may not, be related to deformation.

4.2. SWELLING AND PENETRATION

Another major cause of contact angle hysteresis is penetration of liquid into the surface region of the polymer. Timmons and Zisman observed that apparent penetration of water molecules into a surface can cause significant contact angle hysteresis "... whereas the advancing drop usually moved over a hydrophobic surface free of water, the receding drop moved over a composite hydrophobic–hydrophilic surface in which the intermolecular pores were saturated with water."[49] Timmons and Zisman[49] studied the role of molecular volume and size on contact angle hysteresis using a fluorinated monolayer. The hysteresis was related to the molecular volume of the liquid. They recommended the use of liquids such as methylene iodide which do not penetrate readily to avoid the penetration hysteresis effect. The use of liquids with molecular volumes greater than 125 cc/g-mole (*n*-alkanes, for example) results in little hysteresis. Water is a particular problem because at 18 cc/g-mole it has the lowest molecular volume of any common wetting liquid and readily penetrates into even well-packed surfaces. Although ethylene glycol (56 cc/g-mole) showed considerable hysteresis, glycerol (73 cc/g-mole) behaved as a low-hysteresis liquid. They argued that water behaves as if it were associated in clusters containing about six water molecules.[49] Others have also shown that water readily penetrates into poly(dimethylsiloxane)[50] and lipid multilayers.[51] As a polymer swells, its contact angle with the swelling liquid drops. Such swelling effects have been modeled by Good and coworkers.[52]

The water penetration effect can be very rapid for very thin films. Figure 20 shows a series of water hysteresis loops on thin polystyrene films coated

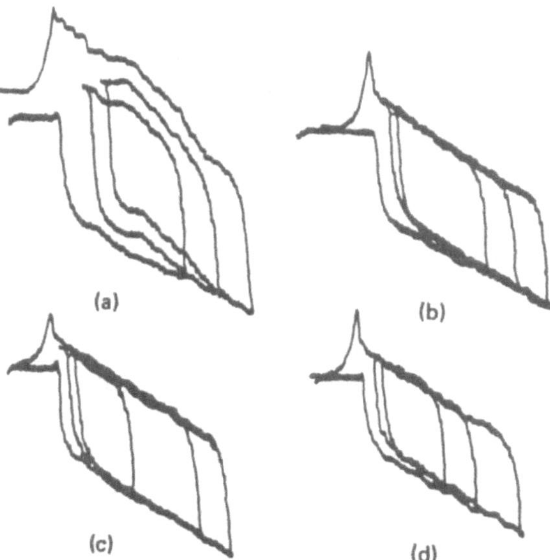

FIGURE 20. Wilhelmy plate plots of polystyrene films on glass. (a) A 150-Å film showing extensive water penetration into the film, resulting in decreasing advancing and receding angles with time. It is likely that portions of the film are actually being removed in this case. (b) a 280-Å film; the loops are now reversible and reproducible. (c), (d) As one goes to thicker films, 600 and 1000 Å, respectively, the curves settle down to a stable limiting hysteresis. Simple multiple cycle Wilhelmy plate plots such as these are useful in determining the integrity and stability of a thin polymer film.

on glass. The very thin film (150 Å) shows a series of loops characteristic of significant changes in the film due to interactions with the water phase. A film ~280 Å thick is well-behaved; the hysteresis loops are reproducible and stable. Thicker films behave similarly. The hysteresis observed with the thick films may be (and probably is) due to water penetration of the outermost regions of the film, as well as to some reorientation effects (next section).

It is clear from these results that water as well as other very small molecules are particularly effective in penetrating organic surfaces. The more water which penetrates, of course, the greater the concentration of aqueous component. One can think of the surface as heterogeneous on a microscopic level with a hydrophilic (water) component. Therefore, one must expect contact angle hysteresis on the basis of the heterogeneous surface argument alone. This effect may not necessarily be due to any specific interaction between water and the solid phase, but rather is due to a concentration or entropically driven diffusion in response to the infinite concentration gradient initially present. In addition, however, water or other probe liquids may interact directly with the polymer phase, and one can now begin to consider an enthalpic contribution to the overall interaction. In the case of water this would generally require polar groups present on the polymer surface. Such interactions will serve to draw more water into the surface region, thereby augmenting the penetration effect discussed above. In addition, if the polymer surface region is relatively mobile or dynamic, then the hydrophilic groups present can orient (Chapter 2) to optimally interact with the water phase, thereby reducing the interfacial tension.

4.7. SURFACE REORIENTATION AND MOBILITY

Langmuir was perhaps the first to suggest[61] that contact angle hysteresis could be due to functional group "flipping" on the surface. Polymer surface molecules exhibit motions and relaxations related to the well-known bulk relaxation phenomena in polymers. It has been well demonstrated (see Chapter 2) that such motions result in polymer surfaces orienting in various environments to minimize their surface or interfacial free energies.[53-56]

Adamson and coworkers have shown that polyethylene surfaces are modified structurally due to adsorbate interactions.[57] Another rigid polymer, polystyrene, appears to reorient in contact with water[58] (see also Figure 20). Interfacial tension studies at liquid siloxane-water interfaces argue for a siloxane orientation which leads to optimum interaction with water, thereby minimizing the interfacial tension.[59] Contact angle studies on poly(dimethylsiloxane) surfaces show substantial hysteresis.

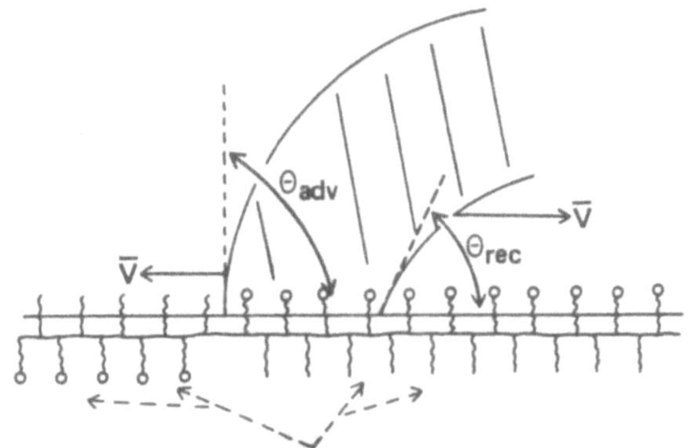

FIGURE 21. The Holly and Refojo[60] model of polymer surface reorientation or relaxation. The ○ indicate hydrophilic pendant groups, the ⁓ indicate hydrophobic pendant groups. The hydrophobic groups orient towards the outer surface in air, minimizing the surface free energy; the hydrophilic groups orient towards the water phase in water, minimizing the interfacial free energy.

Holly and Refojo, in studying the wetting properties of soft contact lenses, concluded that the surfaces orient in air with the hydrophobic methacrylate backbone exposed, minimizing the surface free energy, while underwater the hydroxyl-containing pendant groups orient towards the water phase, minimizing the interfacial free energy[60] (Figure 21). The work of Holly and Refojo with aqueous gels stimulated considerable interest and activity in the study of surface dynamics of biomedical polymers.

Figure 22 shows water contact angle hysteresis of poly(isobutyl methacrylate) films as a function of water equilibration time up to 24 hours.[62] Note the significant decrease in the receding angle, which is roughly linear for the first two hours. Whether this is due to a slow surface orientation process or due to slow water penetration into the surface region is not known. Most likely both effects are operative. Careful analysis of such data in terms of changes in θ_{adv} and θ_{rec} with water contact time may aid in sorting out the effects[62] (Figure 23).

The bottom line is that polymer–water interfaces restructure to minimize the interfacial free energy, both by polymer reorientation and water penetration effects, producing a kinetic-dependent hysteresis. These effects can and do occur in rigid, hydrophobic polymers (see Chapter 2).

4.8. SURFACE ENTROPY

Consider a series of surfaces of essentially identical surface energy, such as a surface of methyl groups, another of ethyl groups, and others of

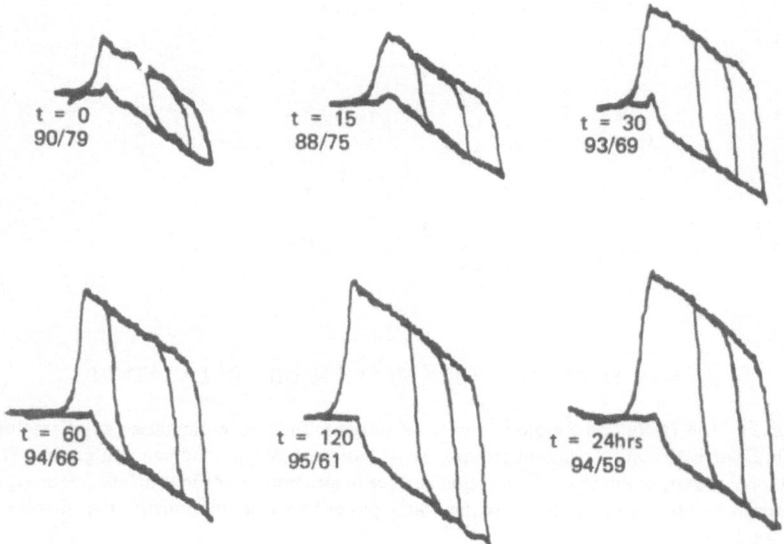

FIGURE 22. Wilhelmy plate hysteresis loops for poly(isobutyl methacrylate) films on glass as a function of water hydration time (min). The advancing/receding angles at zero hydration time are 90/79; hysteresis is at a minimum. As time increases hysteresis increases, primarily due to a decreasing receding contact angle. This effect is probably due to water penetration into the surface region, as well as to hydrophilic group reorientation at the interface.

propyl, butyl, hexyl groups, etc. If these surfaces have the same approximate packing, one would expect no differences among them in contact angles nor any significant contact angle hysteresis, except for a solvent penetration effect. Such a series of surfaces will exhibit different surface mobilities, however. Bulk mobilities increase as the chain length increases until one

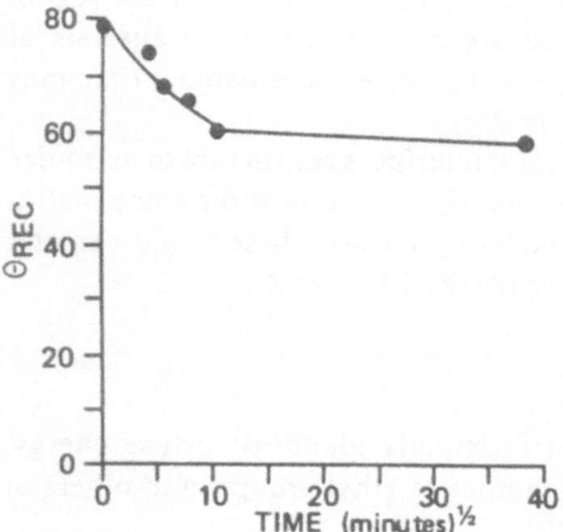

FIGURE 23. The receding angle data of Figure 22 plotted as a function of $(\text{time})^{1/2}$. The linear plot for the first 100 min. suggests a diffusion-driven water penetration process.

reaches octyl or 2-ethylhexyl (the optimum side chain length for plasticization in bulk polymers). If the mobilities increase with chain length, then there should be a surface configurational entropy effect which increases with chain length. This must, of course, influence the overall surface and interfacial free energies and thus may affect the measured contact angles.

A similar problem has been studied by Kessaissia *et al.*, using alkyl chains grafted onto silica,[63,64] with the number of alkyl chains per unit area constant at ~two chains/nM^2. The influence of the polarity of the silica surface is masked when the alkyl chain length exceeds ten carbon atoms. The C-20 surface behaves as a polyethylene surface. Although the contact angle for water increases with number of carbons and levels off as the chain length approaches 20 carbons, no receding angles were measured; thus, nothing can be said about hysteresis. It is interesting to note that surface mobility transition temperatures, observed by gas chromatography,[64] decrease as the number of carbon atoms increase. (See also Chapter 2.)

To our knowledge, there is no study relating contact angle hysteresis to side chain length, motions, or entropy.

In addition to the alkyl-grafted silica system,[63,64] suitable model polymers are available to test the surface motion/entropy hypothesis: polyethylene, polypropylene, polybutene, poly(methyl pentene), polyhexene, and higher poly(1-enes).

Water Wilhelmy plate studies of clean polyhexene surfaces show a very high hysteresis. X-ray photoelectron spectra show no evidence of surface oxidation or hydrophilicity. These are apparently smooth, clean, hydrophobic surfaces. Perhaps the hysteresis is due to surface configurational entropy effects.

Clearly, as surface motions increase, the tendency for liquid penetration also increases. Thus, the surface entropy effect may be difficult to separate from liquid penetration effects.

Liquid structuring and orientation at the interface also introduce an entropy term which should be considered.[69]

5. INTERFACIAL ENERGETICS

A large literature is available which attempts to determine the polar and apolar components of surface and interfacial tensions of liquids in polymers by appropriate assumptions and approximations [see discussion and references in Reference (68), also (65) and (66)]. This area was stimulated by the work of Fowkes in 1964 who suggested that the dispersion force contribution to the work of adhesion at interfaces could be approximated by the geometric mean of the dispersion force contribution of the surface tensions of the two interacting phases. (See Chapter 9.) Others extended this technique to the polar or hydrogen bonding contribution to the interfacial tension by assuming the geometric mean approximation applies for those interactions as well,[65,66,68] which Fowkes and others have clearly pointed out is invalid. Fowkes and coworkers have been developing means to estimate the polar contribution to interfacial interactions, using a partial acid-base or electron donor-acceptor approach. This approach is discussed in detail in Chapter 9. In light of the success and development of the partial electron donor-acceptor approach, all of the older work in the literature, including our own,[65,66] which depends on geometric mean or related approximations in the treatment of polar interactions of polymers in aqueous media must be reevaluated and reconsidered. Fortunately, this is becoming possible through the methods developed by Fowkes and coworkers (see Chapter 9).

6. CHARGE EFFECTS

The contact angle is sensitive to the surface charge nature of a surface. Classical electrocapillarity theory[1-3,68] gives

$$\frac{\partial \gamma}{\partial E} = -\sigma \tag{20}$$

where γ is the surface tension, E the interface electrical potential, and σ the surface charge density.

Holly[71] combined Eq. (20) with Young's equation [Eq. (12)] and applied contact angles as a probe of the surface charge density of polymer electrets. He saw 3-6° decreases in water contact angle with increasing surface charge density in various polymer electrets. Weidert and Stupp[74] recently showed contact angle changes of up to 10° for tetraethylene glycol dimethacrylate on glass electrets. These and other studies have been reviewed.[72] King [73] reviewed the effects of charge on contact angles, correlating his results with streaming potential data.

The overall conclusion is that for the surface charge densities commonly found in surface oxidized polymers or charged copolymers, the expected water contact angle change is a few degrees or less. Even for highly charged electrets the change is probably no more than 5–10°. Thus it is not likely that contact angle will be useful as a measure of surface charge or polarization of polymers.

However, an electric field can affect the surface tension of a liquid. As the surface tension changes, the wetting characteristics of the liquid change.

TABLE 2
Sources of Contact Angle Hysteresis

General assumption	Specific assumption	Effect on hysteresis	Time dependent
Surface is smooth	Surface must be smooth at the 0.1 to 0.5 μ level	$\Delta\theta$ increases with increasing roughness (θ_{adv} increases and θ_{rec} decreases with increasing roughness)	No
Surface is heterogeneous	Surface must be homogeneous at the 0.1 μ level and above	θ_{adv} dependent on low-energy phase; θ_{rec} dependent on high-energy phase	No
Surface is nondeformable	Modulus of elasticity in surface $>\sim 3 \times 10^5$ dyn/cm^2	Not known	Yes—due to surface deformation/relaxation effects
Wetting liquid does not penetrate surface	Liquid molecular volume > 60-70 cc/g-mole	Increased liquid penetration leads to increased hysteresis	Yes—due mainly to diffusion
Surface does not reorient	Reorientation time $\gg$time of measurement	Increased tendency to orient leads to increased hysteresis	Yes
Surface immobile, therefore, surface entropy is constant	Configurational entropy independent of local environment	Unknown—but probably increase in hysteresis as surface mobility increases	Yes

This effect has now been applied to the development of an "electrowetting switch."[77]

7. CONCLUSIONS

Advancing and receding water contact angle measurements provide considerable insight into the character and properties of polymer surfaces in air and in water environments.

The important assumptions involved in contact angle measurement and analysis are fairly well understood. Given proper selection of materials and measurement conditions, the angles generated can be readily interpreted.

Contact angle hysteresis is also well understood (Table 2). There are two classes of hysteresis: thermodynamic (due to metastable states) and kinetic. Classical thermodynamic hysteresis produces stable and reproducible hysteresis loops (scanning curves). Kinetic hysteresis shows changes in the hysteresis loops as a function of rate measurement. Classical thermodynamic hysteresis is due to surface roughness, heterogeneity, and possibly surface entropy and surface deformation. Kinetic hysteresis is due to swelling and penetration effects, surface mobility and reorientation, and possibly surface deformation.

The Wilhelmy plate contact angle hysteresis method produces sequential scanning curves or hysteresis loops which can be interpreted in terms of surface mobility and reorientation, solvent penetration, and intrinsic wettability, both in air and underwater. As few workers have used this method, detailed analyses and modeling of the loops are not yet available. As the technique becomes more widespread and detailed analyses are performed, we can expect more and more information to be extracted from this method.

Water is a difficult liquid to use for contact angle studies because of its small molecular volume, resulting in penetration and local swelling of most surfaces. However, water is the key constituent of all biological environments. Thus it must be the liquid of choice for contact angle and interface studies of polymers and other materials destined for biomedical applications.

Due to specific ion interactions at surfaces, and possibly surface charge and pH effects, it can be expected that an appropriate buffered salt solution may even be preferable to water for contact angle studies of biomedical materials.

Certainly water and aqueous solutions are the most relevant liquids to use for polymer surface characterization if we expect the results to correlate with biological interactions, which occur in aqueous solutions.

ACKNOWLEDGMENTS

Our contact angle and interface characterization studies have been supported by NIH Grants HL18519 and HL26469. Dr. R. N. King and S. M. Ma were key figures in the development and modeling of the underwater captive bubble method. Discussions over the years with F. J. Holly, R. E. Baier, A. W. Neumann, and F. M. Fowkes, and correspondence with J. J. Bikerman and A. I. Rusanov are gratefully acknowledged. Discussions with Dr. S. M. Ma and Dr. P. Blatz on surface deformation are acknowledged.

REFERENCES

1. R. Aveyard and D. A. Haydon, *Intro to Principles of Surface Chemistry*, Cambridge Univ. Press, London (1973).
2. J. J. Davies and E. K. Rideal, *Interfacial Phenomena*, 2nd Edition, Academic Press, New York (1963).
3. J. J. Bikerman, *Physical Surfaces*, Academic Press, New York (1970).
4. A. W. Neumann, Contact angles and their temperature dependence, *Adv. Colloid Interface Sci.* **4**, 105-192 (1974).
5. R. J. Good, Contact angles and the surface free energy of solids, *Surface Colloid Sci* **11**, 1-30 (1979).
6. A. W. Neumann and R. J. Good, Techniques of measuring contact angles, *Surface Colloid Sci.* **11**, 31-92 (1979).
7. W. A. Zisman, Relation of the equilibrium contact angle to liquid and solid constitution, in: *Contact Angle, Wettability, and Adhesion* (F. M. Fowkes, ed.), *Adv. Chem. Ser.* **43**, 1-51 (1964).
8. R. E. Johnson and R. Dettre, Wettability and contact angles, *Surface Colloid Sci.* **2**, 85-153 (1969).
9. M. C. Phillips and A. C. Riddiford, Contact angles and the free surface energies of solids, *Wetting*, S.C.I. Monograph No. 25, pp. 31-56.
10. G. E. P. Elliott and A. C. Riddiford, Contact angles. *Rec. Prog. Surface Sci.* **2**, 111-128 (1964).
11. R. C. Brown, Surface tension of liquids, *Contemp. Physics* **15**, 301-327 (1974).
12. J. F. Padday, Surface tension. *Surface Colloid Sci.* **1**, 39-149 (1969).
13. R. Defay and G. Petre, Dynamic surface tension. *Surface Colloid Sci.* **3**, 27-82 (1969).
14. M. V. Berry, Molecular mechanism of surface tension. *Phys. Educ.* **6**, 79-84 (1971).
15. L. Trefethen, Film: Surface tension in fluid mechanics. *Encyclopedia Britannica and National Committee for Fluid Mechanics*, 1969.
16. S. Ono and S. Kondo, *Molecular Theory of Surface Tension*, Springer-Verlag, New York (1960).
17. R. Aveyard and B. Vincent, Liquid-liquid interfaces. *Prog. Surface Sci.* **8**, 59-102 (1977).
18. G. K. Lester, Contact angles of liquids at deformable solid surfaces. *J. Colloid Interface Sci.* **16**, 315-326 (1961).
19. J. J. Bikerman, *Contribution to the Thermodynamics of Surfaces*, privately published, 1961, Chaps. 3 and 4.
20. A. I. Rusanov, Theory of wetting of elastically deformed bodies, *Colloid J. USSR* **37**, 614-641 (1975).

21. A. I. Rusanov, Thermodynamics of deformable solid surfaces, *J. Colloid Interface Sci.* **63**, 330–345 (1978).
22. J. D. Andrade, R. N. King, D. E. Gregonis, and D. L. Coleman, Surface characterization of PHEMA. Contact angle methods in water, *J. Polym. Sci. Symp.* **66**, 313–336 (1979).
23. R. D. Bagnall, J. A. D. Annis, and P. A. Arundel, ... Adsorption of plasma proteins on hydrophobic surfaces, *J. Biomed. Materials Res.* **12**, 653–663 (1978).
24. R. C. Weast, ed., *CRC Handbook of Chemistry and Physics*, 60th Ed., CRC Press, Boca Raton, Florida (1980).
25. J. A. Dean, ed., *Lange's Handbook of Chemistry*, 11th Ed., McGraw-Hill, New York (1973).
26. J. E. Lane and D. O. Jordan, ... Surface tension by means of a vertical plate balance, *Aust. J. Chem.* **23**, 2153–2170 (1970).
27. G. Loglio, A. Ficalbi, and R. Cini, Surface tension-temperature coefficients for water, *J. Colloid Interface Sci.* **64**, 198 (1978).
28. K. Johansson, ... Accurate surface temperature ... $d\gamma/dT$ for water ... *J. Colloid Interface Sci.* **48**, 176–177 (1974).
29. R. E. Baier, V. L. Gott, and A. Feruse, Surface Chemical evaluation ..., *Trans. Am. Soc. Art. Int. Org.* **16**, 50–57 (1979).
30. K. E. Keller, ed., *Guidelines for Physicochemical Characterization of Biomaterials*, NIH Publ. No. 80-2186 (Sept., 1980).
31. J. Brandrup and E. H. Immergut, eds., *Polymer Handbook*, 2nd Ed., Wiley, New York 1975.
32. L. Smith, C. Doyle, D. E. Gregonis, and J. D. Andrade, Surface oxidation of cis-trans polybutadiene, *J. Appl. Polym. Sci.* **26**, 1269–1276 (1982).
33. R. L. Bendure, Dynamic adhesion tension measurement, *J. Colloid Interface Sci.*, **42**, 137–144 (1973).
34. L. M. Smith, L. Bowman, and J. D. Andrade, Contact angle analysis of hydrated contact lenses, *Proc. Int. Symp. on Contact Lenses and Artificial Eyes*, Durham, England, July 12–15, 1982, pp. 279–286.
35. E. Neumann, Molecular hysteresis and its cybernetic significance. *Angew. Chem. Int. Ed. Engl.* **12**, 356–369 (1973).
36. D. H. Everett, Adsorption hysteresis, in: *Solid–Gas Interface* (A. Flood, ed.), Vol. 2, pp. 1055–1113, Dekker, New York (1967).
37. R. E. Johnson and R. H. Dettre, Contact angle hysteresis. I., in: *Contact Angle, Wettability, and Adhesion* (F. M. Fowkes, ed.), *Adv. Chem. Ser.* **43**, 112–135 (1964).
38. R. H. Dettre and R. E. Johnson, Contact angle hysteresis. II., in: *Contact Angle, Wettability, and Adhesion* (F. M. Fowkes, ed.), *Adv. Chem. Ser.* **43**, 136–144 (1964).
39. E. Bayramli, T. G. M. van de Wen, and S. G. Mason, Tensiometric studies on wetting, *Can. J. Chem.* **59**, 1954–1961 (1981).
40. J. D. Eick, R. J. Good, and A. W. Neumann, Thermodynamics of contact angles. II., *J. Colloid Interface Sci.* **53**, 235 (1975).
41. J. F. Oliver, C. Huh, and S. G. Mason, Effects of solid surface roughness on wetting, *Colloids Surfaces* **1**, 79–104 (1980).
42. R. E. Johnson, Jr. and R. H. Dettre, Contact angle hysteresis. III, *J. Phys. Chem.* **68**, 1744–1750 (1964).
43. R. H. Dettre and R. E. Johnson, Jr., Contact angle hysteresis. IV., *J. Phys. Chem.* **69**, 1507–1515 (1965).
44. A. W. Neumann and R. J. Good, Thermodynamics of contact angles. I. *J. Colloid Interface Sci.* **38**, 341–358 (1972).
45. A. B. D. Cassie, Contact angles, *Disc. Faraday Soc.* **3**, 11–15 (1948).
46. E. B. Dussan, Spreading of liquids on solid surfaces, *Ann. Rev. Fluid Mech.* **11**, 371–400 (1979).
47. R. J. Hansen and T. Y. Toong, Dynamic contact angle and its relationship to forces of hydrodynamic origin, *J. Colloid Interface Sci.* **37**, 196–207 (1971).

48. E. Rukenstein and C. S. Dunn, Slip velocity during wetting of solids. *J. Colloid Interface Sci.* **59**, 135–138 (1977).
49. C. O. Timmons and W. A. Zisman, Effect of liquid structure on contact angle hysteresis, *J. Colloid Interface Sci.* 165–171 (1966).
50. M. C. Phillips and A. C. Riddiford, Dynamic contact angles. II., *J. Colloid Interface Sci.* **41**, 77–85 (1972).
51. S. Windreich and A. Silberberg, Interaction of lipid multilayers with water, *J. Colloid Interface Sci.* **77**, 427–434 (1980).
52. R. J. Good and E. D. Kotsidas, Contact angles on swollen polymers, *J. Adhesion* **10**, 17–24 (1979).
53. J. F. M. Pennings and B. Bosman, Relaxation of the surface energy of solid polymers, *Colloid Polym. Sci.* **257**, 720–724 (1979).
54. A. Baszkin, M. Nishino, and L. Terminassian-Saraga, S-L adhesion of oxidized polyethylene films. *J. Colloid Interface Sci.* **54**, 317–322 (1976).
55. T. J. McCarthy, Polymer surface modification by diffusion of functional groups, *Organic Coatings and Applied Polymer Science Preprints*, **48**, 520–522 (1983).
56. H. Yasuda, A. K. Sharma, and T. Yasuda, Effect of orientation and mobility of polymer molecules at surfaces on contact angle hysteresis, *J. Polymer Sci, Phys.* **19**, 1285–1291 (1981).
57. J. Tse and A. W. Adamson, Adsorption and contact angle, *J. Colloid Interface Sci.* **72**, 515–523 (1979).
58. R. J. Good and E. D. Kotsidas, Contact angle of water on polystyrene. *J. Colloid Interface Sci.* **66**, 360–362 (1978).
59. M. J. Owen, Surface activity of silicones, *Ind. Eng. Chem., Prod. Res. Develop* **19**, 97–103 (1980).
60. F. J. Holly and M. F. Refojo, Wettability of hydrogels, *J. Biomed. Materials Res.* **9**, 315–326 (1975).
61. I. Langmuir, Overturning and anchoring of monolayers, *Science* **87**, 493–500 (1938).
62. A. Okawa, B. S. Thesis, Department of Materials Science and Engineering, University of Utah, June, 1983.
63. Z. Kessaissia, E. Papirer, and J.-B. Donnet, Surface energy of silicas grafted with alkyl chains of increasing lengths, *J. Colloid Interface Sci.* **82**, 526–533 (1981).
64. Z. Kessaissia, E. Papirer, and J.-B. Donnet, Molecular transitions of alkyl chains grafted onto silicas, *J. Colloid Interface Sci.* **79**, 257–263 (1981).
65. J. D. Andrade, S. M. Ma, R. N. King, D. E. Gregonis, Contact angles at the S–W interface, *J. Colloid Interface Sci.* **72**, 488–494 (1979).
66. R. N. King, J. D. Andrade, S. M. Ma, D. E. Gregonis, and L. R. Brostrom, Interfacial tension of acrylic gel-water interfaces, *J. Colloid Interface Sci.*, in press (1985).
67. D. E. Gregonis, R. Hsu, D. E. Buerger, L. M. Smith, and J. D. Andrade, Wettability of polymers and hydrogels in *Macromolecular Solutions* (R. B. Seymour and G. A. Stahl, eds.), pp. 120–133, Pergamon, London (1982).
68. A. W. Adamson, *Physical Chemistry of Surfaces*, 4th ed., Wiley, New York (1982).
69. A. I. Bailey, A. G. Price and S. McKay, Interfacial energies of . . . films of fatty acids, *Spec. Disc. Faraday Soc.* **1**, 118–127 (1970).
70. A. M. Schwartz, Contact angle hysteresis, *J. Colloid interface Sci.* **75**, 404–408 (1980).
71. F. J. Holly, Contact angle . . . as indicator of surface polarization, *J. Colloid Interface Sci.*, **61**, 435–437 (1977).
72. J. A. Finch and G. W. Smith, Contact angles and wetting, in: *Anionic Surfactants* (E. H. Lucassen-Reynders, ed.), pp. 317–382, Dekker, New York (1981).
73. R. N. King, Surface characterization of synthetic polymers, Ph.D. Thesis, University of Utah, Spring, 1980.
74. S. C. Weidert and S. I. Stupp, Electrostatic interfacial phenomena, *Polymer Prepr. Am. Chem. Soc., Div. Polym. Chem.* **24**, 217–218 (1983).

75. F. R. Eirich, Factors in interface conversion, in: *Interface Conversion for Polymer Coatings* (P. Weiss and G. D. Cheever, eds.), pp. 350-378, Amer. Elsevier, New York (1968).
76. J. D. Andrade, Interfacial phenomena and biomaterials, *Med. Inst.* **7**, 110-120 (1973).
77. J. L. Hackel, S. Hackwood, J. J. Veselka, and G. Beni, Electrowetting switch for multimode optical fibers, *Appl. Opt.* **22**, 1765-1770 (1983).

Note added in proof: A dynamic Wilhelmy plate contact angle balance is now commercially available (Wet-tek; Biomaterials International, Inc., Salt Lake City, Utah).

8

Interfacial Electrochemistry of Surfaces with Biomedical Relevance

J. Lyklema

1. INTRODUCTION

The present chapter is concerned with the electrochemical characterization of biomedical polymers. Together with such properties as the relative hydrophobicity/hydrophilicity and the chemical composition of the surface, the presence of electrical charges is an important additional feature, determining the suitability of materials for artificial implants, dialysis or perfusion membrane supports, and other medical applications.

More specifically at issue are now the following questions:

1. What is the origin of the charges on the surface and how much is there?
2. How is this charge affected by pH, ionic strength, admixtures, and other relevant parameters?
3. How is the charge distributed?
4. What can be said about the ensuing electrical potentials?
5. What experimental techniques are available to obtain information and what kind of information do they produce?

Answering questions like the ones posed above takes us to the domain of electrical double layers or, more generally, to that of interfacial electrochemistry. This is a well-established discipline, providing the roots for the understanding of colloid stability, electrode processes, corrosion, etc. Since the earlier works by Helmholtz, about a century ago, very detailed double layer pictures have been developed, many of them inspired by

J. Lyklema • Laboratory for Physical and Colloid Chemistry of the Agricultural University, De Dreijen 6, 6703 BC Wageningen, The Netherlands.

high-precision measurements on model systems, notably the mercury-solution interface.

Double layers on biomedical materials, polymeric substances in particular, are another matter. As a rule, such surfaces are much less well characterized, the surface properties are not always sufficiently reproducible, and the availability of experimental techniques is more limited. For instance, most biomedical polymers are unsuitable to act as electrodes in electrochemical cells, so that there is no way to measure directly such important characteristics as double layer capacitance and surface potential.

Despite these limitations, the importance of biomaterials in practice is so great that serious attempts to describe its interfacial electrochemistry deserve commendation. In trying to elaborate on this, use can be made of the collective double layer experience obtained with many other systems. Various facts are so well-established that they may be taken as also applicable to biopolymers, even if this cannot be supported by direct experimental evidence. Because of this, we shall from time to time give examples obtained with surfaces beyond the domain of biomedical polymers.

A particular position is taken by the so-called latices (or latexes). They are polymer colloids: finely dispersed polymer particles in (aqueous) solutions, and often homodisperse (all particles of the same size). Latices resemble biopolymeric materials, both with respect to their chemical composition and surface properties, but they have the great advantage of offering a large specific surface area (of the order of several tens or hundreds of m^2 per sample). Because of that, analytical determination of adsorbed amounts, ionic or nonionic, is often feasible in cases where, with sheets of the corresponding biopolymer, the total adsorbed amount would be too low to be analytically detectable, because of the low surface area available.

As it is the purpose of this chapter to present the general basic picture, we shall refrain from treating specific samples in detail. Rather, the pictures to be developed are likely to be of wider validity, within the restraints set by the model parameters.

2. BASIC ELECTRICAL PARAMETERS

The most fundamental electrostatic parameters characterizing charged interfaces are the *charge* and the *potential.* Derived quantities are the capacitance and electric field strength.

2.1. CHARGE

There are various mechanisms by which surfaces in contact with an aqueous solution can acquire a charge, but only two of them are really significant in the case of biomedical polymers, viz. dissociation of surface groups and specific adsorption of ions. The former mechanism can apply if there are dissociable groups on the polymers, for instance carboxyl, sulfate, sulfonate, phosphate, or amino groups. Such groups may have been introduced intentionally, for instance by conducting the polymerization process in such a fashion as to achieve a product of specific properties or, alternatively, they may be inadvertently present on the surface, as a result of, for instance, oxidation steps during the preparation stage. Adsorption of ions can also lead to charging of the surface provided this adsorption is *specific*, meaning that the adsorption forces are in part of a nonelectrical nature, so that, say anions, can overcome opposing negative potentials and by their adsorption render the potential more negative. Non-specific or *generic*, adsorption of ions would never lead to a net charge on the surface, because equal amounts of cations and anions would always accumulate. A clear example of specific adsorption is that of ionic surfactants, where the surfactant ion has a very great affinity for surfaces. In fact, many biological materials are charged and surface active, and so is heparin.

It follows from the foregoing that more likely than not, a biomedical polymer in contact with an aqueous solution is charged: there will always be some dissociable groups and it is likely that some of the various ionic species in the solution adsorb more than others. It follows also that the charge on the surface in some cases will depend on conditions such as pH or ionic strength, whereas in other cases it will not. For instance, if this charge is due to sulfate groups, which are strongly adsorbed, it will be insensitive to pH, whereas the charge on carboxylic surfaces depends on pH. Charge due to ion adsorption tends to increase with the activity of that ion, up to saturation of the surface.

Biopolymers bearing only one type of dissociable group (e.g., $-NH_2$ or $-COOH$ groups) can be charged either positively or negatively. However, if there are both basic and acid groups (say, $-NH_2$ and $-COOH$ groups), the surface is amphoteric. At high pH it is negatively charged, at low pH positive. There is a certain pH where the surface charge is zero. This pH is called the *point of zero charge* (p.z.c.). Such a zero point can also be established in the case where the charge is due to competitive specific adsorption of cations and anions; then the p.z.c. occurs at a certain combination of concentrations of these ions in solution rather than at a particular value of pH. However, as the dissociation/association of surface groups is itself influenced by ions (they screen the resulting charges and compete with H^+ and OH^- ions), the p.z.c. usually is a function of pH and the nature and concentration of the electrolyte.

There is another zero point, called the *isoelectric point* (i.e.p.). The isoelectric point and point of zero charge are different notions. We will consider this distinction in Section 5.

A biopolymer surface together with the adjacent part of the solution must always be electroneutral, except under nonequilibrium conditions with nonconducting systems, which are virtually not encountered in medical practice. If the surface is charged due to the splitting off of, say, a proton, this proton remains in the neighborhood of the surface because it is electrostatically attracted by it. How far it goes into the solution depends on the conditions and will be discussed in Section 4, but in the thermodynamic sense it remains a bound ion. Similar reasoning applies to charging by specific adsorption of ions.

It follows immediately that the biopolymer-solution interface is the seat of an *electrical double layer* consisting of a surface charge and an equal but opposite charge in the solution, the *countercharge.*

Customarily, surface charges are not expressed as such, but as *surface charge densities*, symbol σ_0, that is, the charge per unit area. The usual unit is $\mu C\ cm^{-2}$ ($1\ \mu C\ cm^{-2} = 10^{-2}\ C\ m^{-2}$). The charge itself is an extensive parameter. It may be added that often the term "surface charge" is used where "surface charge density" is meant.

The defining of the surface charge density automatically takes us to the question of the surface heterogeneity. Only if the charge is smeared out over the surface or, for that matter, if for all practical purposes it may be treated as if it is smeared out, does it make sense to define a surface charge density. In practice the mercury-solution interface is an acceptable representative of this category. Here, σ_0 is determined by an excess or deficit of electrons, and since mercury is an excellent conductor, the charge can move extremely fast along the surface. For metallic implant materials the same reasoning applies, but for the more common polymeric ones the surface charges are usually covalently bound and not laterally movable. In other words, they are *localized.* In terms of σ_0 this means that on a molecular scale the surface is electrostatically heterogeneous; there are patches with very high σ_0 interspersed with patches of low, or even zero, σ_0. It is difficult to predict the consequences of such heterogeneity for medical practice. One can imagine, for instance, that the adsorption of a negatively charged protein on a surface with a few widely spread and strongly negative sites will proceed in such a fashion that the first molecules avoid those sites and accumulate at the uncharged patches of the surface between them. Later adsorbing molecules will then attach onto the charged sites, provided they can overcome the Coulombic repulsion. If, however, the surface charges are evenly distributed with mutual distances short in comparison to the diameter of an adsorbing protein molecule, each molecule "sees" some charges, and in that case the surface is homogeneous on the scale of the protein molecule.

The outcome of this discussion is that the necessity to account for electrical heterogeneity is virtually a matter of scale. For some purposes it has to be taken into account; for others it may be safely ignored. Theoretical treatments of heterogeneity are by their very nature cumbersome and often of little general validity, since they usually require some assumption on the distribution of the charged sites. Because of that we shall discuss homogeneous surfaces only.

To obtain some feeling for the orders of magnitude: a saturated monolayer of fully dissociated ionic surfactants would bear a surface charge density of about 70-80 $\mu C\ cm^{-2}$. For biomedical polymers values of a few up to a few tens of $\mu C\ cm^{-2}$ are usually measured. In some instances values exceeding the maximally attainable surface charges are observed; this then points to the fact that the charge is not a surface charge in the strict sense of the word, but that also charged sites are found within the solid at some distance from the surface. A substance having such properties is so to speak pervious to charges, and the ensuing double layer type is referred to as a "porous double layer." In the theoretical part we shall return to this concept (Section 4.3.).

Anticipating the discussion of various methods to measure surface charges, it is good to be aware of the fact that charges as high as several tens or $\mu C\ cm^{-2}$ give rise to such high potentials that counterions are automatically attracted and strongly bound. Borrowing a term from polyelectrolyte theory, this may be called *counterion condensation.* A number of experimental methods, notably the electrokinetic ones (Section 5), do not measure σ_0 proper but the charge due to the surface and a fraction of the counterions, this fraction increasing with increasing σ_0. As a consequence, the electrokinetically obtained surface charge σ_{ek} is often much lower than σ_0. Typically, σ_{ek} never exceeds a few $\mu C\ cm^{-2}$.

2.2. POTENTIAL

The potential $\phi(\mathbf{r})$ at a place $\mathbf{r}$ is defined as the electrical work involved in bringing a unit charge e from infinity to $\mathbf{r}$. (The notation $\mathbf{r}$, i.e., the vector $\mathbf{r}$, is very generally used to indicate a point in space that has the coordinates r_1, r_2, and r_3 with respect to an agreed origin.) This definition of potential is theoretically adequate, but it can give rise to great confusion if measurements are to be interpreted. For one thing, potentials can only be defined with respect to the potential at some pre-agreed reference point, so that all reported potentials are virtually relative potentials. A second and greater problem is that in practice one cannot transport *solely* a unit charge; it will always be bound to a molecule to give an ion. Transporting a unit charge can only be achieved by simultaneously transporting matter and, hence, the total work contains, in addition to the sought electrical contribution, also "chemical" work needed to displace and/or rearrange other molecules. We use the term "chemical" here not in the strict sense of chemical bonds that are broken and reformed, but very widely as due to changes in any bond between molecules, such as H-bridges, van der Waals interactions, etc., and it also includes entropic terms if the ordering of the system is changed by the transport. Briefly, "chemical" encompasses everything in addition to the purely Coulombic work $z_iF\phi(x)$, where z_i is the valency of ion i (sign included) and F the Faraday constant (96.485 C mol^{-1}). Basically, the issue can be summarized by stating that one can only measure electrochemical work and that there is no unambiguous way to split it up into a purely electrical and a purely chemical contribution.

This problem is not new. In fact, we are used to breaking up electrochemical potentials $\tilde{\mu}_i$ ions i into a chemical potential μ_i and a Coulombic term

$$\tilde{\mu}_i(x) = \mu_i + z_iF\phi(x) \tag{1}$$

$$\tilde{\mu}_i(x) = \mu_i^0 + RT \ln a_i + z_iF\phi(x) \tag{2}$$

where a_i is the activity of i. Equation (2) shows clearly the thermodynamic situation. Single ionic activities are experimentally inaccessible quantities and, therefore, thermodynamically undefined, as are electric potentials inside condensed phases.

Although absolute potentials cannot be established, potentials as such play an important role in the description of double layers and, therefore, the question of under what conditions *relative* estimates are obtainable must now be answered. One case is that in which we are confident that the transport of matter does not involve chemical work. This is particularly the case when the transported ion remains in the same phase, not crossing phase boundaries. In particular, one may assume that the structure of liquid water remains unaltered at distances further than a few molecular diameters from the polymer surface (in a few instances, very thick adsorbed water layers have been postulated and often contested, a matter that we shall exclude from the present discussion). Consequently, moving ions in the outer, solution-side part of the electrical double layer involves no chemical work and, provided the double layer is sufficiently extended to be felt so far from the surface, the only work involved is the electrical contribution. As will be discussed in more detail in Sections 3.2 and 4.1, this is by definition the *diffuse part* of the double layer, so that we can state that in the diffuse double layer part potentials are well-defined and measurable with respect to the bulk of the liquid.

A gratifying circumstance is that also so-called *electrokinetic* or *ζ-potentials* belong to this category (see Section 5).

What has been said for liquid water applies *mutatis mutandis* to the bulk of the polymer. However, all of this does not apply to the inner region of the double layer, both on the aqueous and on the polymer side, because in this region ion transport involves chemical contributions. In summary, potentials inside both phases are accessible with respect to the bulks of these phases, provided measurements are made far enough from the interface to neglect the influence of the second phase on the structure of the first. It follows immediately that the absolute potential difference ${}^{\alpha}\Delta^{\beta}\phi$ between adjacent phases α and β is also inaccessible.

For a few substances there is a way out of the last mentioned difficulty, namely if an electrode can be made out of the solid phase. In this case an electrochemical cell can be set up consisting of this electrode and a reference (e.g., a calomel electrode) in a solution, and then the potential difference between the two electrodes can be measured. Although such measurements do not yield absolute potential differences between electrode material and solution (one measures this quantity in addition to the potential drop due to the reference), they are nevertheless very useful because the variation of ${}^{\alpha}\Delta^{\beta}\phi$ with conditions can sometimes be measured in some detail. Representative examples are the glass electrode and the Ag/AgI electrode, responding according to Nernst's law to pH and pAg, respectively. Since biomedical polymers are unsuitable electrode materials (because of their low conductivity), this alternative does not appear viable for the systems under discussion. In summary, there seems to be no way of measuring the potential difference ϕ_0 between the polymer surface and the bulk of the solution. The only path left open is to obtain ϕ_0 by developing some theoretical picture. However, it is perhaps after all not so serious that ϕ_0 is unmeasurable since, if a quantity is thermodynamically inaccessible, it cannot play a role as such in the phenomenological description of experiments.

For further discussions on potentials near interfaces see References (1) and (2). In most of the sections below, we will return to the above problem. Potentials in double layers characteristically are of the order of a few tens to a few hundreds of mV.

2.3. OTHER IMPORTANT ELECTRICAL CHARACTERISTICS

The *capacitance* is a derived quantity and generally defined as the quotient of charge and potential. For double layers a capacitance per unit area is used or, more precisely, there are two of them, namely the *integral capacitance*

$$K = \sigma_0/\phi_0 \tag{3}$$

and the *differential capacitance*

$$C = \mathrm{d}\sigma_0/d\phi_0 \tag{4}$$

There are two reasons for defining C in addition to K: (a) in a number of experiments one does not measure σ_0 as a function of ϕ_0 but rather it is established by how much σ_0 is varied if ϕ_0 is altered, and (b), closely connected with (a): more often than not ϕ_0 is inaccessible, whereas changes of ϕ_0 can be measured (see Section 2.2).

The two capacitances are interrelated:

$$K = \frac{1}{\phi_0}\int C \, \mathrm{d}\phi_0 \tag{5}$$

$$C = K + \phi_0 \frac{\mathrm{d}K}{\mathrm{d}\phi_0} \tag{6}$$

where the integration in Eq. (5) has to start from some pre-established zero point, for which the p.z.c. (Section 2.1) is a suitable choice. If the capacitance is a constant (independent of σ_0 and ϕ_0), $C = K$.

Capacitances for double layers are usually expressed in $\mu\mathrm{F\,cm^{-2}}$. Typical values for K and C are several tens of $\mu\mathrm{F\,cm^{-2}}$.

Working with capacitances is expedient for two reasons. First, for a number of materials it is much easier to measure the (differential) capacitance than the charge, particularly by using capacitance bridges. Integration of Eq. (4) then gives directly $\sigma_0(\phi_0)$, except for a constant. For biomedical polymers this has, as far as the author is aware, not yet been done systematically.

Second, capacitances can offer interpretational advantages in that they are a measure of the *screening* (or shielding) of charges. Briefly, if charges are not screened at all, their introduction into a given material leads to the creation of a high local potential which inhibits further charging. In that case the capacitance of the system is low. Screening of charges can be achieved by polarizing the medium around them, e.g., by orientation of dipoles (as in water) or by co-uptake of counterions. If a charge is screened this means that electrostatically its effect is not, or not so strongly, felt. Many charges can then be accumulated without increasing the potential too much and the capacitance is high in that case.

Screening of charges by free ions is one of the main themes of double layer theory. Obviously the extent of screening, and hence the double layer capacitance, is strongly related to the distribution of ions. Typically, capacitances increase with increasing electrolyte concentration.

Screening of charges by the polarization of the medium is governed by the *dielectric permittivity* $\varepsilon\varepsilon_0$ of the medium. Here, ε_0 is the permittivity of free space, a universal quantity equal to $8.854 \times 10^{-12}\,\mathrm{C\,V^{-1}\,m^{-1}}$, and ε is the relative dielectric constant. Polar liquids like water have high values for ε (about 78 at room temperature); therefore, charges in it are well screened. In fact, the high ε for water is the main reason that electrolytes do dissolve and dissociate in it so well. Apolar media without dipoles can only polarize by displacing the centers of positive and negative charges inside the molecules with respect to each other. For those media, ε is much lower and equal to n^2, where n is the index of refraction. The bigger the molecules, the better the polarizability. For biomedical polymers, depending on their composition, ε will have intermediate values, ranging from 4–6 for entirely apolar, water-free materials to perhaps 20 or more for water-rich gel-like materials.

The relationship between capacitance and dielectric permittivity is very simple for a charge-free flat plate condenser of plate distance d:

$$C = K = \varepsilon\varepsilon_0/d \tag{7}$$

showing C and ε to be proportional. If there are free ions between the plates, these participate in the screening and C is higher than the value that would be obtained from Eq. (7).

Another important electrical parameter is the *electric field strength* $E = \mathrm{d}\phi/\mathrm{d}x$ (in order not to complicate the mathematics unnecessarily we shall in this chapter only consider flat surfaces, so that the potential varies only in the direction x normal to the surface; deviations due to bending or curving the polymer surface play a role only if the radius of curvature becomes comparable with the double layer thickness, which is not usually encountered in practice). Unlike σ_0, ϕ_0, and C, E is a vector. By definition, at any point in space (indicated by $\mathbf{r}$ in general or by x for flat geometry) the field strength is the electrical force exerted on a unit charge. Part of the relevance of E as a double layer characteristic is that it determines the extent of orientation of water dipoles. The stronger E, the stronger this orientation and hence the stronger the deviation of the water "structure" from that in the absence of a field. Such deviations can become substantial in the inner part of a double layer and in water vicinal to ions. For our purposes, there are two consequences.

(a) If water dipoles are oriented in a strong field, they can no longer adjust to other charges, so that they behave to those other charges as if they have a poor polarizing capacity, that is, a low ε. This phenomenon is called *dielectric saturation.* For it to occur, field strengths of 5×10^5 V cm^{-1} and higher are needed. In double layers such high values are usually attained in the inner double layer part but not in the diffuse region. The consequence for our purposes is that in the inner part ε is typically between 8 and 20, whereas in the outer or diffuse part it is 78, with a relatively sharp transition range between them. For double layer studies, dielectric saturation around dissolved ions plays a significant role only at very high electrolyte concentrations, beyond the range usually encountered. We shall not discuss it here.

(b) Orientation of water near surfaces has hydrodynamical implications in that tangential flow is inhibited. This may be instrumental in explaining the generally observed feature that in electrokinetic experiments (i.e., experiments involving the tangential motion of liquid and surface with respect to each other) the boundary between movable and stationary phase does not coincide with the solid–liquid interface, but is displaced over a few molecular diameters into the aqueous solution. This boundary is known as the "*slipping plane*" (see Section 5.).

The description of electrical fields in space belongs to the domain of vector field analysis. Of the various theorems and equations of general validity that have been derived in this theory, the following special case of *Gauss's law* applies to the field strength immediately adjacent to a charged surface

$$E = \left(\frac{\mathrm{d}\phi}{\mathrm{d}x}\right)_{x\to 0} = -\frac{\sigma_0}{\varepsilon\varepsilon_0} \tag{8}$$

This is a general law. It does not matter whether or not σ_0 is a real surface charge or due to charges which penetrate to some depth into the polymer. Also, Eq. (8) can be applied at any position x' in the double layer, in the sense that all the charge per unit area to the left of x' is identified as σ_0 and the field strength is then $E(x')$.

Note that, if the double layer would simply consist of a surface charge σ_0 and an equal but opposite charge at distance d, $\mathrm{d}\phi/\mathrm{d}x$ can be replaced by $(\phi_0 - 0)d$ (because the second plane we have bulk conditions and the potential is there at its reference value, i.e., zero), so that Eq. (8) becomes identical to Eq. (7).

The final electrical parameter that we wish to mention is the *space charge density* $\rho(x)$. It is the bulk equivalent of the surface charge density σ_0. The quantity ρ can vary widely with x. For $x\to\infty$, that is, in the heart of the electrolyte, $\rho \to 0$. Also in the bulk of the polymer it is zero, but in between ρ can assume high values.

Vector field expressions have also been derived for ρ. Important for the description of double layers is *Poisson's law*, which for flat geometry reads as follows:

$$\frac{\mathrm{d}^2\phi}{\mathrm{d}x^2} = \frac{\mathrm{d}E}{\mathrm{d}x} = -\frac{\rho}{\varepsilon\varepsilon_0} \tag{9}$$

The relevance of Eq. (9) for double layer theory is that if something can be said about the distribution of charges, that is, if ρ is known as a function of x or as a function of ϕ, Eq. (9) can be integrated to give the potential distribution $\phi(x)$. Note that, although in regions of zero potential

ρ must be zero, the converse is not necessarily true: if $\rho = 0$ (as it is between the plates of a condenser), E is constant, meaning that the potential drop is linear with distance.

3. DOUBLE LAYER MODELS

In this section we shall discuss some double layer models phenomenologically. Theories and elaborations follow in Section 4. The purpose is twofold: (a) to become acquainted with the various models and their limitations, and (b) to investigate their applicability to biomedical polymers.

3.1. THE MOLECULAR CONDENSER

This is the simplest imaginable double layer. It is depicted in Figure 1. The surface charge is thought to be negative, and it is really a surface charge, and the distance in the solution (x) is measured with respect to a plane through these charges ($x = 0$). Inside the solid (here used as a general term for the biomedical material) the potential is constant and equal to ϕ_0. [Only if the solid is polarized by an outer source would ϕ inside the solid vary with distance; this variation would then be linear because inside the solid $\rho = 0$ (Eq. (9).] All counterions are at $x = d$. In reality such a situation

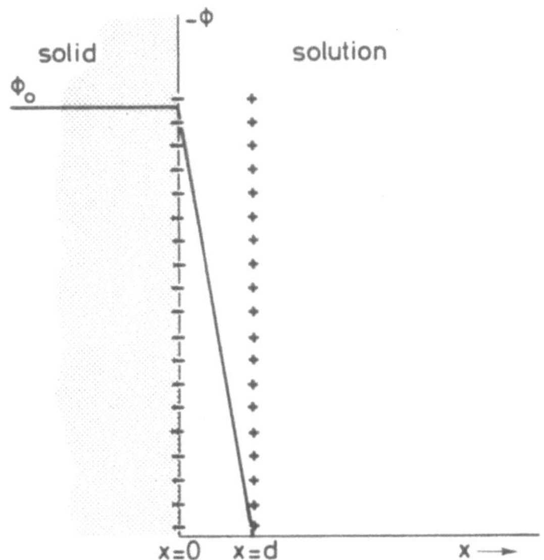

FIGURE 1. A molecular condenser or Helmholtz double layer.

could arise if the electrostatic attraction of these ions to the surface were so strong that they would approach it to a distance of closest approach. This distance could be determined by the radius of the ions assuming them to be either hydrated or not, i.e., d in Figure 1 would be equal to the sum of the radii of surface- and counterion, with or without hydration water in between.

Double layers like those in Figure 1 are seldom met in practice, but as a limiting case they deserve attention. About a century ago, Helmholtz first used such double layer models to describe electrokinetic phenomena, and therefore it is appropriate to refer to them alternatively as "*Helmholtz layers.*"

As already pointed out in Section 2.3, such double layers can be described by Eqs. (7) and (8).

3.2. THE DIFFUSE DOUBLE LAYER

Figure 2 gives a representation of the diffuse double layer. This model was developed independently by Gouy[3,4] and Chapman[5] and is therefore also known as the "*Gouy-Chapman (GC) layer*" or briefly as the "*Gouy layer.*"

The underlying picture is that counterions are not only electrostatically attracted by the surface but also subject to thermal motion. The former force tends to accumulate all counterions at the distance of closest approach, as in Figure 1, whereas the latter tries to spread all countercharge evenly through the solution. The result of these counteracting tendencies is a diffuse type of double layer in which the countercharge density $\rho(x)$ decreases

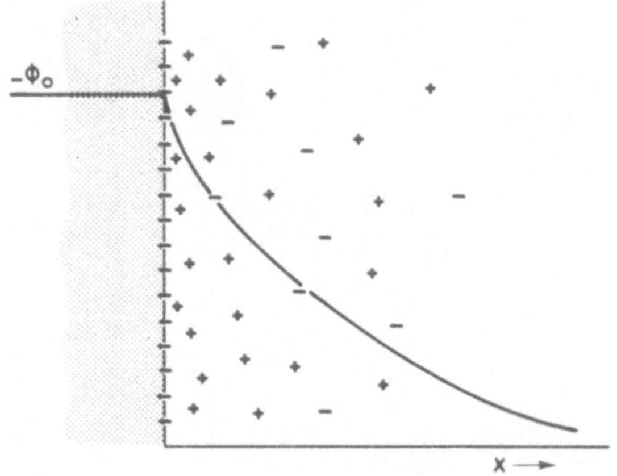

FIGURE 2. A purely diffuse double layer.

gradually to zero at large x. The final situation is analogous to the barometric pressure distribution in the earth's atmosphere.

The GC picture introduces some essentially new features. Most important is that the *distribution* of ions now plays an essential role. In turn, this implies that entropic contributions are accounted for, whereas the purely electrostatic Helmholtz layer (Figure 1) could be treated in terms of energy only. A second point is that co-ions now also play a role (co-ions are ions with the same charge sign as the surface, i.e., anions in our example): at any x, $\rho(x)$ is made up of an excess (as compared with the bulk) of counterions and a deficit of co-ions. Figure 3 sketches these distributions. Area B is the excess of counterions (cations in the example). Likewise, A is the deficit of co-ions. It is customary to call the amount of cations accumulated in excess of the bulk the adsorbed amount or briefly "the adsorption." Similarly, the shortage of anions is called their *negative adsorption*. Common or "positive" adsorption is experimentally verified as depletion of the adsorbing substance in the bulk. Likewise, negative adsorption leads to an expulsion of ions which, because of the electroneutrality of double layers as a whole, is felt as an increase in the concentration of electrolyte in the bulk due to the charged particles. This increase is known as the *Donnan effect*, and its very existence (for instance as verified in membrane equilibria) is proof of at least the qualitative correctness of this aspect of the GC picture.

There are other physical phenomena supporting the slow decay of ϕ with x under certain conditions. One of them is colloid stability. Because the double layer can have a considerable thickness, two colloid particles of the same surface charge, dispersed in aqueous solution, repel each other

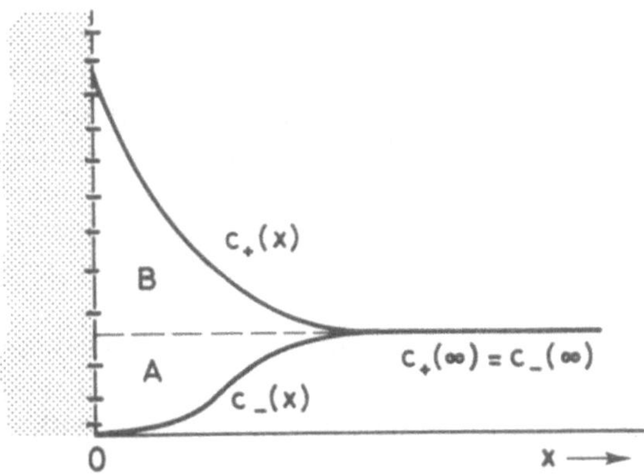

FIGURE 3. Charge compensation in a diffuse double layer on a negative surface. B = excess of cations (positive adsorption) and A = deficit of anions (negative adsorption) together constitute the countercharge.

already at relatively large distance of separation, thus inhibiting aggregation and flocculation. Similar features are operative when blood proteins, thrombocytes, and other cells from the blood approach charged biomedical polymer surfaces. Also, the reduction of electrokinetic potentials with increase in the electrolyte concentration can be explained with a diffuse double layer picture, although alternative interpretations are also viable.

According to present-day insight, diffuse double layers in the GC sense are also limiting cases, the theory working only at low potentials and low electrolyte concentrations, far below physiological ionic strengths. In practice the outer part of the double layer obeys the GC theory very well, whereas the inner part rather resembles a Helmholtz type condenser. Combining and integrating these two pictures has been achieved by Stern.

3.3. THE GOUY–STERN DOUBLE LAYER

In the mathematical treatment of diffuse double layers, Gouy and Chapman introduced quite a few simplifications. Two of them are rather restrictive, viz. the neglect of the finite sizes of the ions, and that of specific adsorption. The first is perhaps the more serious one: as the ions are considered to be point charges, they can theoretically accumulate without bounds in the interior double layer part, and this gives rise to excessively high σ_0 and C values, by many times exceeding what is usually measured. Specific adsorption, if occurring, would favor the accumulation of counterions close to the surface, and consequently reduce the potentials further away from the surface. Modifying the entire GC theory by incorporating these two corrections is not impossible, but it is mathematically rather unwieldy. Stern's idea was an effective compromise in that these two corrections were realized only in the first ionic layers, that is, in that part of the double layer where their effect is maximal.[6] As these corrections automatically reduce the remaining potentials in the outer part to relatively low values, no further corrections for the outer part are needed. The result is a double layer consisting of an inner part, resembling the Helmholtz molecular condenser and an outer part, obeying the Gouy theory.

Figure 4 gives a sketch for two degrees of approximation: (top) only ion size effect considered, and (bottom) ion size plus specific adsorption considered.

Both models have in common that in the immediate vicinity of the surface there is a charge-free layer, inaccessible to ions because of their finite size. In this layer the potential decays linearly with distance until, at some distance d, a value ϕ_d is reached, the *Stern potential.* Beyond $x = d$ the distribution obeys GC theory, i.e., in both pictures the diffuse charge, σ_d, is, according to GC theory, related to ϕ_d. The difference between the two models is that in the top diagram *all* countercharge is assumed to obey

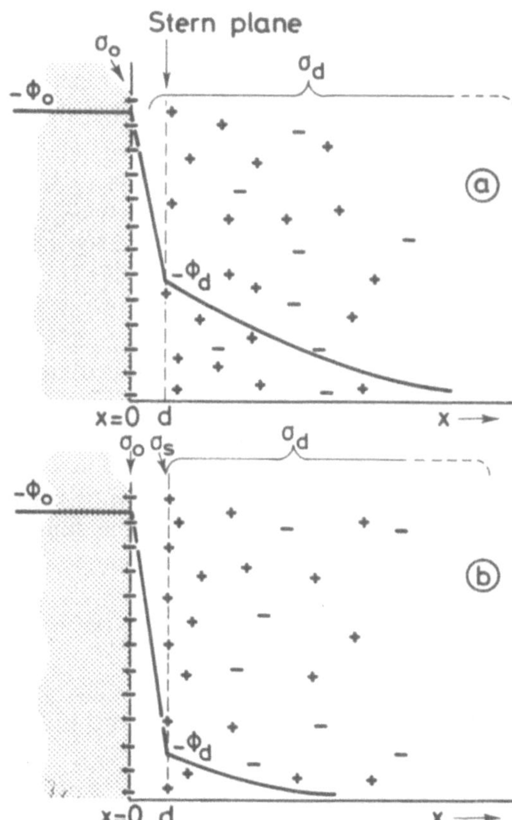

FIGURE 4. Gouy-Stern double layer without specific adsorption (top) and with specific adsorption in the Stern plane (bottom).

GC theory (in that case $\sigma_0 = -\sigma_d$), whereas in the bottom figure a part σ_s of the charge is thought to adsorb specifically [the charge balance then reads $\sigma_0 = -(\sigma_d + \sigma_s)$].

For practical purposes, the Gouy-Stern picture is an excellent compromise between theoretical rigor and experimental accessibility. Many features of relevance to biomedical polymers can be satisfactorily interpreted in terms of this model. Accepting that such surfaces are, by relative standards, not good double layer models, the availability of electrochemical data as a rule is too limited to require much more sophisticated double layer models, and therefore we shall not pursue attempts to refine the models. However, one inconsistency must be mentioned. In the picture of Figure 4 (bottom) the plane at $x = d$ has a dual physical meaning: it marks the position at which ions are specifically adsorbed and the distance where the diffuse part of the double layer starts—that is, the distance at, and beyond which, there is *no* specific adsorption. This is conflicting. It would be more logical to distinguish two planes, one close to the surface where specific adsorption occurs, and another one a little further out where the diffuse double layer starts. In fact, such a picture has been proposed and developed by Grahame.[7] He called the two planes the *inner Helmholtz plane* (iHp) and the *outer Helmholtz plane* (oHp), respectively. The Stern theory ignores the difference between iHp and oHp. In Figure 4 the inconsistency is seen in the steeper decrease of ϕ in the charge-free region in the second case: this is demanded by the fact that ϕ_d is lower (because, due to specific adsorption of part of the countercharge, the diffuse fraction is less), but according to Eq. (8) the slope is determined by σ_0 only and, in this quantity is determined by dissociation of surface groups, it is invariant. In Grahame's picture this difficulty is resolved by distinguishing two consecutive charge-free layers, each with its own linear slope, one between $x = 0$ and the iHp and the other betwen the iHp and the oHp.

For a Gouy–Stern type double layer three (differential) capacitances can be distinguished: (1) the overall capacitance $C = \mathrm{d}\sigma_0/\mathrm{d}\phi_0$; (2) the capacitance of the diffuse part of the double layer, $C_d = \mathrm{d}\sigma_d/\mathrm{d}\phi_d$; and (3) the capacitance of the molecular condenser, which we define as

$$\partial\sigma_0/\partial(\phi_0 - \phi_d) \tag{10}$$

If C_s is constant, $C_s = K_s = \varepsilon_s\varepsilon_0/d$, as in Eq. (7), where ε_s is the relative dielectric constant of the Stern layer. It can be proven that

$$\frac{1}{C} = \frac{1}{C_s} + \frac{1}{C_d}\left(1 + \frac{\partial\sigma_s}{\partial\sigma_0}\right) \tag{11}$$

This equation shows that specific adsorption increases C (note that $\mathrm{d}\sigma_s/\mathrm{d}\sigma_0 < 0$). In the absence of specific adsorption (Figure 4, top) Eq. (11) reduces to the familiar equation for two capacitances in series:

$$\frac{1}{C} = \frac{1}{C_s} + \frac{1}{C_d} \tag{12}$$

showing that the overall capacitance is determined by the smaller of the two contributing capacitances. This is in fact one of the main virtues of the Stern theory: excessively high C_d values in GC theory are nullified because they are in series with a lower capacitance.

Another feature of the Stern theory that may be relevant for biomedical polymers is the possibility of *superequivalent adsorption*, meaning that more counterions are specifically adsorbed than are needed to compensate the surface charge, i.e., $|\sigma_s| > |\sigma_0|$. As, by definition, specific adsorption is adsorption involving other forces besides the Coulombic ones, this is possible in

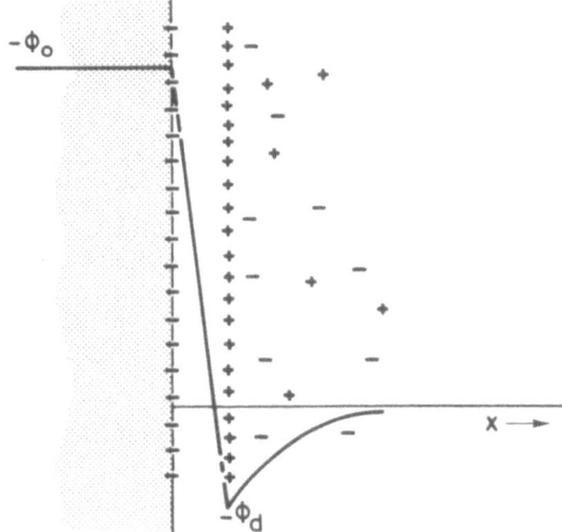

FIGURE 5. Gouy-Stern double layer with superequivalent counterion adsorption.

practice. In fact, it is very common for hydrolyzable multivalent cations and for several anions. In this case the potential distribution assumes the characteristic shape sketched in Figure 5. The potential at the Stern plane (or, rather, at the iHp and consequently also at the pHp) now has a sign opposite to that of the surface. Experimental procedures that measure the surface charge *plus* the specifically bound countercharge would in the case of Figure 5 yield a positive charge even if σ_0 is negative. All electrokinetic methods belong to this category. The phenomenon that the "surface" charge measured by such techniques changes sign with increasing counterion adsorption is known as *charge reversal.* However, this term is sloppy in that only the charge of the diffuse part reverses sign, σ_0 retaining the sign it had before. If superequivalent adsorption occurs, the sign of σ_d is the same as that of σ_0.

The importance of this phenomenon for medical practice is that the electrostatic component of the adhesional energy between implant materials and blood colloids can be drastically affected by adsorption of charged molecules.

3.4. POROUS DOUBLE LAYERS

The double layer pictures of Figures 1, 2, 4, and 5 involved differences with respect to the charge distribution in the aqueous part of the double layer. It was always assumed that the charge on the solid was concentrated on its surface. However, this applied only to a limited number of cases, notably to free charges in highly conducting materials such as metals. More likely the polymer contains covalently bound charged groups that cannot

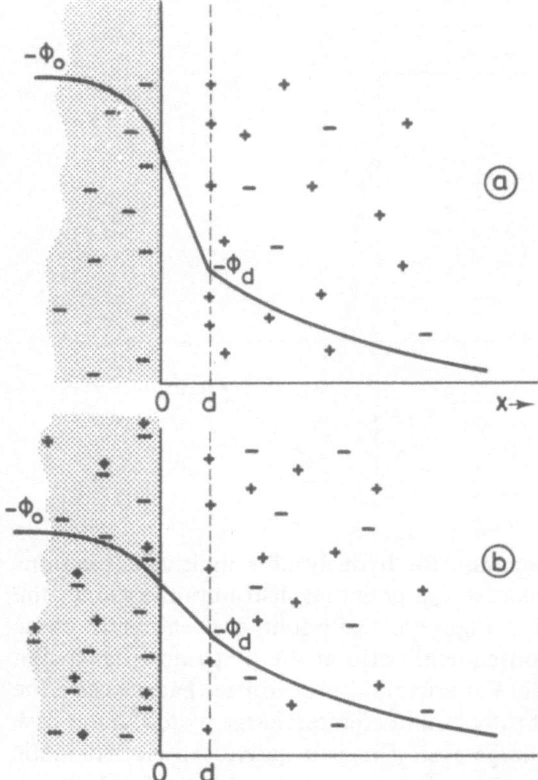

FIGURE 6. Porous double layer on the solid side with Gouy-Stern double layer on the solution side in the absence (a) and presence (b) of counterion penetration into the solid.

move at all or it is gel-like with restricted mobility. At any rate, it is likely that the charge is not concentrated at the surface proper but extends to a certain depth into the solid.

Two examples are visualized in Figure 6. Figure 6a describes the situation in which there are only negative groups distributed in some fashion in the surface region of the solid, whereas Figure 6b represents the case where, in addition to the covalently bound anions, cations can also penetrate into the solid. Situation (a) is representative for solid polymers, impervious to counterions, whereas picture (b) rather applies to open, gel-like solids. In the second case the charge attributed by negative groups inside the solid σ_0 need not be lower than in the first. It can even be much higher, but the potential ϕ_0 is much less negative since the negative surface charge is to a large extent screened by cations. In electrostatic terms, such surfaces have a high capacitance, but in the solution this is not felt. Since such solids are pervious to charges, the ensuing double layers are called "*porous*" *double layers.* It must be stressed that porosity towards ions does not necessarily imply porosity to other molecules. It is quite possible that, for instance, N_2 adsorption isotherms show little or no capillary consensation, but that there is nevertheless a substantial ion uptake. Also, the converse situation may be observed: substances porous to gases but not to ions.

In the picture of Figure 6 it was assumed that the solution side obeys some Gouy-Stern distribution with the same ϕ_d and, because of Eq. (8), in case (b) a lower potential decay in the molecular condenser. The total double layer capacitance C now consists of three capacitances in series: if the one due to the double layer part inside the solid is very high, its contribution to C is negligible.

From Figure 6 we conclude the following:

(A) If one is interested in problems like adhesion, the outer part of the double layer counts, and methods to describe this part (such as electrokinetics) are suitable. It does not matter very much how the charge in the solid is distributed.

(B) If one is interested in the charge inside the solid, a combination of two techniques is mandatory: one type of experiment should give σ_0, the other type gives σ_0 plus the cationic charge inside the solid, their difference constituting the *effective* surface charge.

The various double layer models described above are of course not exhaustive, but as they encompass the main features, and as more specialized models have a more restricted validity range, we shall not describe more refined models.

4. DOUBLE LAYER THEORY

In this section we shall describe the mathematical implementation of the models of Section 3, thereby restricting ourselves to the most important equations. Details of derivations will not be given, but the main formulas and principles will be illustrated by some diagrams.

4.1. GOUY–CHAPMAN THEORY

The basic premises of Gouy-Chapman theory are: (a) the ions are point charges; (b) specific adsorption does not occur, i.e., the energy of adsorption at any x amounts to $z_i F\phi(x)$; (c) the aqueous phase is a structureless continuum characterized by its bulk dielectric permittivity $\varepsilon\varepsilon_0$; and (d) the average potential $\phi(x)$ is equal to the potential of the mean force.

The first three assumptions have been, either explicitly or not, mentioned before. Assumption (d) is a subtlety of statistical thermodynamics that is governed by the way in which rearrangements of the ions due to the introduction of an additional one have to be counted. It is beyond the scope of this review to deal with this problem, but it is perhaps not necessary since we shall apply diffuse double layer theory only for the conditions where it is valid (low potentials and ionic strength) and under these conditions assumption (d) is acceptable.

In the GC theory the distribution of ions is described according to Boltzmann

$$c_i(x) = c_i(\infty) \exp[-z_i F\phi(x)/RT] \tag{13}$$

From $c_i(x)$ for all ionic species i in the system $\rho(x)$ is obtained and inserted into Eq. (9) to give the *Poisson-Boltzmann* (PB) equation, which for flat plates and simple electrolytes can be integrated to give $\phi(x)$ and $\mathrm{d}\phi(x)/\mathrm{d}x$ [or, with Eq. (8), the charge]. The resulting equations for one symmetrical electrolyte read as follows:

$$\tanh(zF\phi(x)/4RT) = \tanh(zF\phi_d/4RT)\exp[-\kappa(x-d)] \tag{14}$$

where κ is the reciprocal Debye length,

$$\kappa = (2F^2cz^2/\varepsilon\varepsilon_0RT)^{1/2} \tag{15}$$

At room temperature, with the electrolyte concentration c expressed in moles,

$$\kappa \sim z(10^{15}c)^{1/2}\ \mathrm{cm}^{-1} \tag{16}$$

For instance, in 10^{-3} and $10^{-1} M$ solutions of NaCl, κ^{-1} amounts to 10 nm and 1 nm, respectively.

The physical meaning of the Debye length stems from the fact that at not too high potentials the hyperbolic tangents in Eq. (14) may be replaced by their arguments, leading to the more simple expression

$$\phi(x) = \phi_d \exp[-\kappa(x-d)] \tag{17}$$

according to which κ^{-1} may be identified as the distance over which the potential has decayed to e^{-1} of ϕ_d:

$$\phi(x = \kappa^{-1}) = \phi_d\, \mathrm{e}^{-1} = 0{\cdot}368\phi_d \tag{18}$$

As it is not possible to assign a discrete value to the thickness of a diffuse double layer since in principle $\phi \to 0$ only at $x \to \infty$ it is customary to refer to κ^{-1} as the *diffuse double layer thickness*. As d is of the order of 0.5 nm, we see that concentrations of monovalent electrolytes above $10^{-1} M$ are needed to compress the diffuse part sufficiently to make its thickness comparable with that of the Stern layer. In 10^{-3} *M* solutions the diffuse part is several times thicker.

The parameter κ can also be regarded as a measure for the screening of the surface charge: the higher κ, the more counterions accumulate in the inner parts of the diffuse double layer. In line with this, at given ϕ_d, σ_d increases with $c^{1/2}$, i.e., with κ [see Eq. (15)], and so does the diffuse layer capacitance.

The term "reciprocal Debye length" stems from the Debye–Hückel theory for the description of activity coefficients in strong electrolytes. This theory also uses a PB-distribution to describe the ionic atmospheres, and κ is again a parameter describing the extent of screening of (in this case) ionic charges by other ions.

Examples of $\phi(x)$ curves are drawn in Figure 7. First, it is noted that this graph is entirely dimensionless, so that it can be used for any z and any κ. At room temperature one unit of $e\phi/kT$ corresponds within a few percent to 25 mV. For a higher salt concentration and/or higher valency the screening is better, and the potential decays more rapidly with distance [in the figure: if κ is high at a given $(x - d)$ the potential is low]. It is customary to say that due to high salt concentration and/or high z the diffuse double layer is more *compressed*. For colloids this is a very important feature; it is the very basis for the interpretation of the strong influence of electrolytes on colloid stability: if κ is high, the double layers are strongly compressed, equally charged particles can then approach each other very

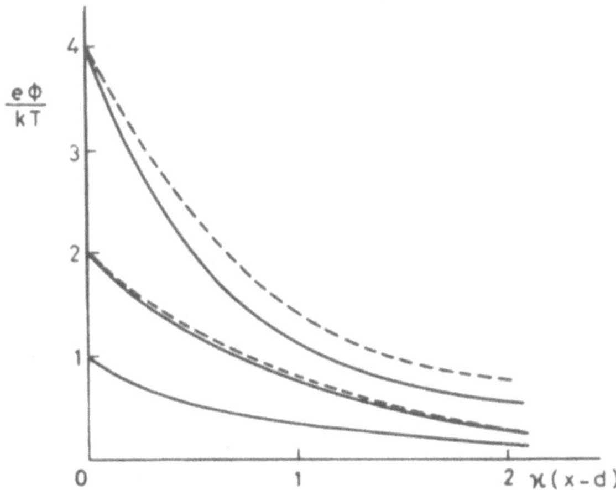

FIGURE 7. Potential distribution in a flat diffuse double layer. Drawn curve: complete expression [Eq. (14)]. Dashed curves: low-potential approximation [Eq. (17)].

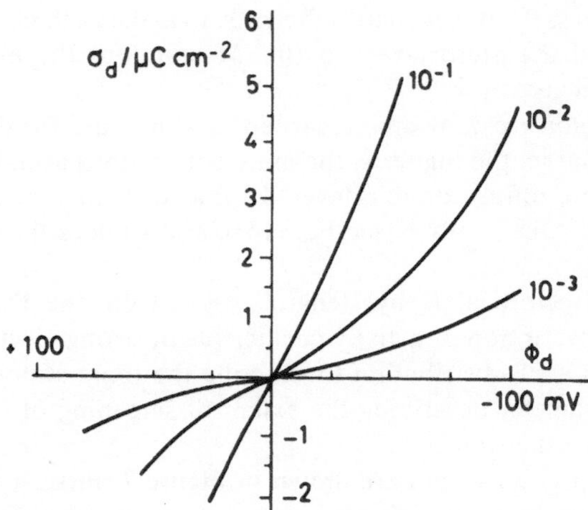

FIGURE 8. Surface charge density in a flat diffuse double layer at various (1–1) electrolyte concentrations. $T = 25°C$.

closely before electrostatic repulsion is felt, and then van der Waals attraction may be sufficiently strong to induce aggregation. Also for biopolymer adhesion similar reasoning applies. Note that, due to electrolytes, not only repulsion between like charges is reduced but also the attraction between oppositely charged particles.

Another typical feature of the diffuse double layers is the tendency of the $\phi(x)$ dependency to become progressively steeper with higher ϕ, i.e., more to the inner side of the layer. This is due to the Boltzmann principle, but as ion sizes are not accounted for, this strong increase is exaggerated in the theory: it would require counterion concentrations far above saturation. Hence, it is mainly this inner part that requires correction.

The charge in a diffuse double layer is represented in Figure 8. The basic equation reads for symmetrical (z–z) electrolytes

$$\sigma_d = -(8\varepsilon\varepsilon_0 cRT)^{1/2} \sinh (ze\phi_d/2kT) \tag{19}$$

amounting to $11.72c^{1/2} \sinh (ze\phi_d/2kT)\ \mu C\ cm^{-2}$ for aqueous solutions at 25°C with e in moles l^{-1}. Equation (19) follows from the PB equation by differentiating, using Eq. (8).

It is customary in electrochemistry to plot potentials with the negative values to the right. The charge in Eq. (19) has a sign opposite to that of the potential to conform with the situation in Figure 4, where $\sigma_d > 0$, but $\phi_d < 0$. It must be noted, however, that in literature it is more usual to plot σ_d and ϕ_d as having the same sign: then σ_d should not be read as the diffuse charge proper but as the compensating charge on the surface, i.e., as $-(\sigma_0 + \sigma_s)$. In the plot of Figure 8 the capacitance requires a minus sign: $C_d = -(d\sigma_d/d\phi_d)$. As a function of ϕ_d, σ_d is an uneven function: if ϕ_d reverses sign, so does σ_d. The capacitance, however, is an even function:

$$C_d = -(d\sigma_d/d\phi_d) = (2\varepsilon\varepsilon_0 cz^2F^2/RT)\cosh(ze\phi_d/2kT) \tag{20a}$$

$$C_d = \kappa\varepsilon\varepsilon_0 \cosh(ze\phi_d/2kT) \tag{20b}$$

At $\phi_d = 0$ it is minimal and equal to $\varepsilon\varepsilon_0\kappa = \varepsilon\varepsilon_0/\kappa^{-1}$; that is, it behaves as a condenser with plate distance equal to the double layer thickness κ^{-1} [compare Eq. (7)].

Both the capacitance and the charge increase rapidly with ϕ_d and κ. As said before, GC theory strongly overestimates this increase. In practice, usually not more than a few μC cm^{-2} are measured for the diffuse part. Stern theory corrects for this. In terms of capacitance, if κ and/or ϕ_d become high, C_d according to Eq. (20) also grows, but its contribution to the overall double layer properties diminishes in view of Eq. (11).

4.2. INNER LAYER THEORY

In addition to the fact that ion sizes have to be accounted for, specific adsorption must be introduced. In the Stern layer the adsorption free energy is not purely electrical but includes a "chemical" term that will be called Φ_i, the *specific adsorption free energy* of ion *i*. Hence, generally,

$$\Delta_{\text{ads}}G_{s,i} = z_iF\phi_s + \Phi_i \tag{21}$$

where the subscript *s* stands for "Stern layer" and $\Delta_{\text{ads}}G$ is the change in the free energy upon transporting an ion from $x = \infty$ to the place where it is adsorbed. The symbol ϕ_s generally stands for the potential at the point where the ion is adsorbed. In the simplified Stern picture of Figure 4 $\phi_s = \phi_d$, whereas in Grahame's picture $\phi_s = \phi(\text{iHp})$.

The very fact that in Eq. (21) $\Delta_{\text{ads}}G_{s,i}$ is the sum of an electrical and a chemical term and that breaking it up is thermodynamically inoperable and therefore requires a model has led to the development of two alternative theoretical approaches.

The first we call the *double layer approach.* Stern (or Grahame) layers are treated as molecular condensers in which, by way of complication, a chemical term is introduced. Electrostatics comes first; the chemical contribution is a correction.

The second is the *site binding approach.* Here chemistry comes first in that the binding of ions to the surface is formulated in terms of binding constants (pK's) that are corrected for the fact that the ions do not adsorb as individual entities but "feel" each other's presence through the double layer field.

The two treatments, if properly elaborated, should of course yield the same results, and it is a matter of taste rather than a matter of principle which one should be preferred. Also, this choice might depend on the problem under study. For instance, if one wants to describe the interaction between biomedical polymers and blood colloids, especially at not too short distances, the double layer approach is preferred; but if one is interested in ion binding to surfaces, the site binding picture is indicated.

Below we shall give some simple elaborations of each, thereby avoiding too much detail and disregarding intricacies such as the problem of how to describe double layers consisting of discrete rather than smeared-out changes. For more elaborate treatments the reader is referred to the pertinent literature.

The most simple Stern adsorption equation applies to non-superequivalent adsorption in the Stern plane ($x = d$) only, and reads as follows:

$$\frac{\theta}{1-\theta} = \frac{c_i}{55.5} \exp\left[-\frac{z_i F\phi_d + \Phi_i}{RT}\right] \tag{22}$$

with

$$\theta = -\sigma_s/\sigma_0 \tag{23}$$

It is an extension of the Langmuir equation which would result if in Eq. (22) $\phi_d \to 0$. The quotient $c_i/55.5$ is the mole fraction of ion i.

In two respects Eq. (22) is oversimplified. First, one should rather use ϕ(iHp) than ϕ_d, as explained above, and further it is somewhat illogical to assume specific adsorption to occur but not to allow for superequivalent adsorption. Nevertheless, experience has shown that Eq. (22) works satisfactorily for weak specific adsorption, for instance, in explaining lyotropic sequences for alkali ions in colloid stability.

If σ_0 is known by some tritation procledure (Section 6), Eq. (22), together with Eq. (23), Eq. (19), and the charge balance $\sigma_0 + \sigma_s + \sigma_d = 0$, allows the computation of Φ_i. However, as the equation is oversimplified, one usually finds Φ_i not to be a constant.

A better equation, but more difficult to solve, is Eq. (22) with ϕ(iHp) instead of ϕ_d. This increases the number of variables to three, viz. one must now know (i) the position of the iHp between $x = 0$ and $x = d$, (ii) the difference between ϕ(iHp) and ϕ_d, and (iii) the angle between the two linear sections of the potential-distance straight lines between ϕ_0 and ϕ(iHp), and between ϕ(iHp) and ϕ_d, respectively. Because of Eq. (8) this is determined by σ_s and the dielectric permittivities in these two layers which, as the layers are chemically different, need not be the same. In practice, some curve fitting is always involved to obtain the most suitable set of double layer parameters.

Stern himself[6] suggested another solution. He assumed all ions to adsorb at $x = d$, but allowed for superequivalent adsorption of two types of ions, a cation and an anion. Superequivalent adsorption must be invoked if shifts of the p.z.c. due to changes in the electrolyte concentration are to be accounted for.

Let us illustrate the introduction of Stern-type corrections by giving in Figure 9 a $\sigma_0(\phi_0)$ curve for a Gouy-Stern double layer without specific adsorption, that is, a Gouy-type double layer plus a charge-free inner layer (Figure 4, top). It is assumed that $C_s = 15\ \mu\text{F cm}^{-2}$ and is constant. Using $\sigma_0 = -\sigma_d$, Eqs. (12) and (20) allow $C(\phi_0)$ to be computed and integration gives $\sigma_0(\phi_0)$. The following features are observed:

(i) As now σ_0 is plotted as a function of ϕ_0, both have the same sign.
(ii) If $\phi_0 < 0$, the cation is the counterion, and as Ba^{2+} ions screen better than K^+ ions, the charge in dilute $Ba(NO_3)_2$ solutions is higher than that in an equivalent concentration of KNO_3.
(iii) For $\phi_0 > 0$ the NO_3^- ion is the counterion. It does not matter if it stems from $Ba(NO_3)_2$ or from KNO_3; the two 10^{-3} *N* curves are identical.
(iv) At high c_{salt}, $C \sim C_s$ and independent of the nature and concentration of the electrolyte.

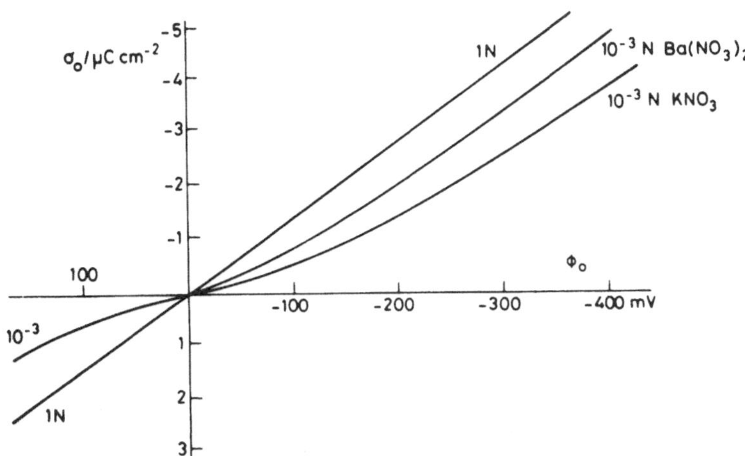

FIGURE 9. Surface charge density in a Gouy-Stern type double layer without specific adsorption in dilute and concentrated (1-1) and (2-1) electrolytes. $C_s = 15\ \mu\text{C cm}^{-2}$.

It is instructive to compare Figure 9 with Figure 8. The two graphs are identical only in the region of low c_{salt} and low potential. The main effect of the Stern correction is to eliminate the excessively high charges beyond this region.

By adjusting the value of C_s and by allowing it to be a variable and introducing specific adsorption, almost any experimental $\sigma_0(\phi_0)$ curve can be interpreted. For further reading References (1), (7)-(9) may be recommended. The main message for biomedical polymers, cells, proteins, and other blood colloids is that, although some important parameters like ϕ_0 are not known, theoretical interpretations are available that in other cases have proven to be viable. Their use may especially be relevant if electrokinetic and surface data (Sections 5 and 6) are to be compared.

In the site binding approach the biopolymer surface is envisaged as bearing a number of groups that can become charged by dissociation or association. Such groups can also take up a counterion. For instance, a surface carboxyl group, abbreviated as RCOOH, could dissociate in the following way:

$$\text{RCOOH} \rightleftarrows \text{RCOO}^- + \text{H}^+ \qquad K_a^s = \frac{[\text{H}^+][\text{RCOO}^-]}{[\text{RCOOH}]} \tag{24}$$

and an amino group which is positively charged at low pH could lose its proton at high pH according to

$$\text{RNH}_3^+ \rightleftarrows \text{RNH}_2 + \text{H}^+ \qquad K_b^s = \frac{[\text{H}^+][\text{RNH}_2]}{[\text{RNH}_3^+]} \tag{25}$$

Similar equations can be written for other functional groups. Also for specifically adsorbing ions such expressions may be formulated. For instance, for Ca^{2+} ions, forming a monodentate bond with a carboxyl group (an example of ion exchange),

$$\text{RCOOH} + \text{Ca}^{2+} \rightleftarrows \text{RCOOCa}^+ + \text{H}^+ \qquad K_{\text{Ca}}^s = \frac{[\text{H}^+][\text{RCOOCa}^+]}{[\text{RCOOH}][\text{Ca}^{2+}]} \tag{26}$$

Electrochemistry enters if the various K^s values are worked out. In Eq. (24), $[H^+]$ is not the bulk concentration of protons but the concentration in the plane of the head groups. These head groups are in the surface and feel an electrical potential ϕ_0. According to Boltzmann, therefore,

$$[H^+]_{\text{surface}} = [\text{H}^+]_{\text{bulk}} \exp(F\phi_0/RT) \tag{27}$$

so that

$$K_a^s = \frac{[H^+]_{bulk}[RCOO^-]}{[RCOOH]} \exp(F\phi_0/RT) = K_{a,intr} \exp(F\phi_0/RT) \quad (28)$$

where $K_{a,intr}$ is the purely chemical, or intrinsic, dissociation constant. Similar reasoning applies to Eq. (25). In Eq. (26) a difficulty arises in that Ca^{2+} ions do not adsorb at the head groups but remain outside of them. In our previous picture, they adsorb at the iHp where the potential is σ_s. Hence

$$K_{Ca}^s = K_{Ca,intr} \exp(F\phi_s/RT) \quad (29)$$

We now see that the site binding approach involves virtually the same parameters as the double layer treatment. The intrinsic binding constants are thermodynamically related to the free energy of transfer ($pK_{intr} = -\Delta G/RT$), where for a proton or a specifically adsorbing counterion ΔG is the chemical free energy of transferring the ion from $x = \infty$ to the side of binding; for the counterion this is nothing else but Φ_i [see Eq. (21)]. Furthermore, some model is needed to substitute values for ϕ_0 and ϕ_s. Over the past decades, progressively better double layer pictures have been used for that. It would be a poor approximation to substitute ϕ_d for ϕ_0 (see Figure 4); it would be a better procedure to substitute ϕ_d for ϕ_s (which is the Stern approximation). In equations like Eq. (29) the potentials can then be eliminated by introducing capacitances: from Eqs. (7) and (10) it follows that $\phi_0 - \phi_d = \sigma_0/C_s = \sigma_0 d/\varepsilon_s\varepsilon_0$. It is even better to distinguish also between ϕ_s and ϕ_d (Grahame). For a recent review in which this has been done see Reference (10). More information can be found in Reference (11). One of the gratifying results of such analyses is that, as a rule, the intrinsic pK's compare very well with the corresponding quantities measured in solution. The important consequence for the description of dissociation-association, ligand exchange, and specific binding equilibria on biomedical polymer surfaces is that, in the absence of further information, it appears to be a safe approximation to identify any wanted surface constant with the corresponding one in the bulk.

4.3. POROUS DOUBLE LAYERS

The models described in Sections 4.1 and 4.2 had in common that they all related to the solution side of the double layer. Let us now briefly discuss the charge distribution inside the polymer (Figure 6). For the sake of argument let the surface charge σ_0 be negative and let the co-absorbed countercharge inside the solid be σ_c. Hence, the effective charge on the solid amounts to:

$$\sigma_{\text{eff}} = \sigma_0 - \sigma_c \tag{30}$$

From the very onset it will be clear that the distribution of charge and potential inside the solid varies widely from system to system, so that it is impossible to give a comprehensive picture. Let us therefore restrict ourselves to a few general observations.

First, it is very unlikely that high space charges σ_0 develop inside the polymer, unless there is substantial co-uptake of compensating charge. The reason is, of course, that otherwise extremely high potentials would arise, ε being much lower for polymers than for aqueous solutions (section 2.3). The consequence is that σ_{eff} will never be very high. Experience with porous oxides like some silicas and glass have shown that at σ_0 values of over 100 $\mu C\ cm^{-2}$ the electrokinetic or diffuse charge is not more than a few $\mu C\ cm^{-2}$. The difference cannot be made up by assuming very high Stern charges, σ_s, because there is simply not enough space in this layer to accommodate the required number of ions.[12] Similar features have been observed for polyelecltrolyte coils. For polymeric matrices bearing even strongly dissociating acid groups (e.g., sulfonic acid groups) the result is that any of such groups residing in the interior of the polymer will be weakly dissociated or not dissociated at all, so that only the real surface groups are charge-determining.

With respect to the distribution of charge and potential inside the polymer, theories have been put forward assuming a certain distribution of the space charge $\rho(x_s)$ inside the solid, if x_s is the distance to the origin ($x = 0$ in Figure 6). This distribution depends on the properties of the polymer and, hence, on its preparation. If it is a structurally weak gel, $\rho(x_s)$ will be sensitive to pH, ionic strength, etc., and it is difficult to make predictions of general validity. On the other hand, theories can be developed if $\rho(x_s)$ is fixed, i.e., if the polymer is a solid matrix. The surface charge then follows from

$$\sigma_0 = \int_{x_s=0}^{\infty} \rho(x_s)\, dx_s \tag{31}$$

and σ_c depends on the affinity of the countercharges (H^+, cations) for the solid. The penetrability of these ions is included in this affinity: if a certain ion i binds strongly to the charged sites inside the solid but cannot reach them because it is too big to diffuse through the voids in the matrix, ion i is counted as having a low affinity.

Of the various models that have been elaborated we mention a Gaussian distribution (used for polyelectrolytes by Arnold and Overbeek[(13)]), an exponential one (Lyklema[(12)]), and a step function (Perram *et al.*[(14)]). By way of illustration, we select two typical results from Reference (12). This theory is based on an exponential decay of the space charge density of surface groups, $\rho(x_s) = \rho(0)\exp(-ax_s)$, and a Langmuir-Stern type of counterion uptake at any x_s, obeying Eq. (22). This equation is combined with Eq. (9) to give the Poisson-Langmuir equation, which is the equivalent of the Poisson-Boltzmann equation in the GC theory. Integration gives the charge and potential distribution.

Figure 10 represents the potential distribution plotted in a dimensionless way, where $\alpha = F^2zc(0)/\varepsilon\varepsilon_0 RTa^2$ is a dimensionless parameter, in which $c(0)$ is the concentration of charged sites at the surface ($x_s = 0$). We could say that $\alpha = \kappa_s^2/a^2$, where κ_s^{-1} is the diffuse double layer thickness in the solid and a^{-1} the distance into the solid where $\rho(0)$ had decayed to $\rho(0)/e$. It is a measure of the porosity. The second parameter $B = c_i \exp(-\Phi/RT)/55.5$ has already been encountered in Eq. (22); it is a measure of the affinity of counterions for the charged sites.

Figure 10 shows that at low potentials the decay is exponential, as it is in a diffuse layer (Figure 7), but if α (and ϕ) increase, the decay becomes linear. If $d\phi/dx_s$ = const. according to Poisson's law [Eq. (9)], no (additional) charge can be built in. A similar message is relayed by Figure 11: the co-uptake of compensating charge σ_c readily increases to such high fractions of σ_0 that σ_{eff} always remains low, the more so the deeper counterions can penetrate. Beyond a given σ_0 the curves become straight lines with a slope of 45°, meaning that each additional negative charge is fully compensated by a positive one.

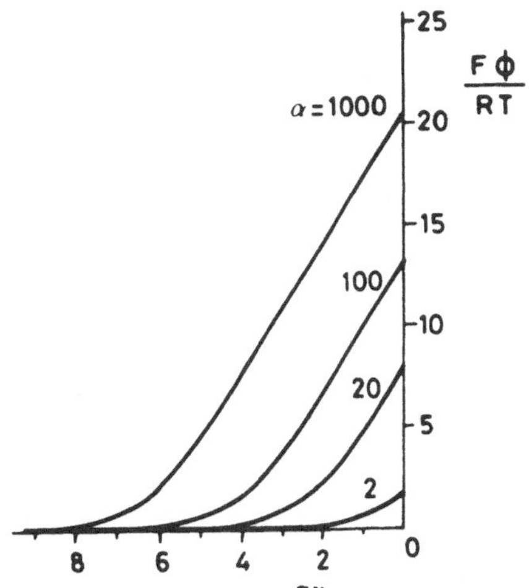

FIGURE 10. Potential distribution in a porous double layer. Poisson-Langmuir distribution and exponential decay of the surface site density. $B = 0.1$.

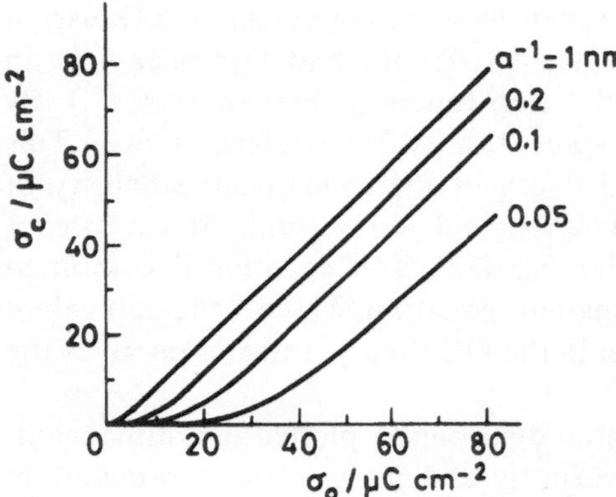

FIGURE 11. Compensation of the surface charge σ_0 by co-uptake of countercharge σ_c in a porous double layer. $B = 0.01$, $\varepsilon_s = 15$, $z = 1$. The penetration depth a^{-1} is indicated.

Although these figures are restricted by the necessity to assume some model, their qualitative features are of general validity.

5. INTERPRETATION OF ELECTROKINETIC POTENTIALS

Charges and potentials considered so far may be called "static" properties, because they are electrical magnitudes that are characteristic for double layers at rest. In addition, the important group of electrokinetic phenomena has been used as a technique for characterizing biomedical polymer surfaces. Typically, in these phenomena tangential motion of the liquid with respect to the solid takes place. Elaboration of existing formulas leads to an *electrokinetic potential or zeta* (ζ) *potential*, which is a kinetically determined quantity. What is at issue is how such potentials fit into the framework of double layers set up in the previous sections.

For the characterization of biomedical polymer surfaces the main electrokinetic techniques are (a) *streaming potential*, (b) *streaming current*, (c) *electroosmosis*, and (d) *electrophoresis*. Less relevant are sedimentation or centrifugation potentials and secondary electrokinetic phenomena like diffusiophoresis. Phenomena (a), (b) and (c) can be studied with the surface as such, whereas electrophoresis can only be done with a colloidal dispersion of it (a latex). Electrophoresis is perhaps the most familiar electrokinetic technique, but the difficulty involved is that one is not sure whether or not the surface properties of the (finely dispersed) latex are identical to those of large pieces of the same material.

All electrokinetic phenomena have in common that a moving and a stationary phase can be distinguished. Electrophoresis is perhaps the easiest to understand conceptually: simply stated it is the motion of charged colloid particles in an electric field. However, on closer inspection the situation is

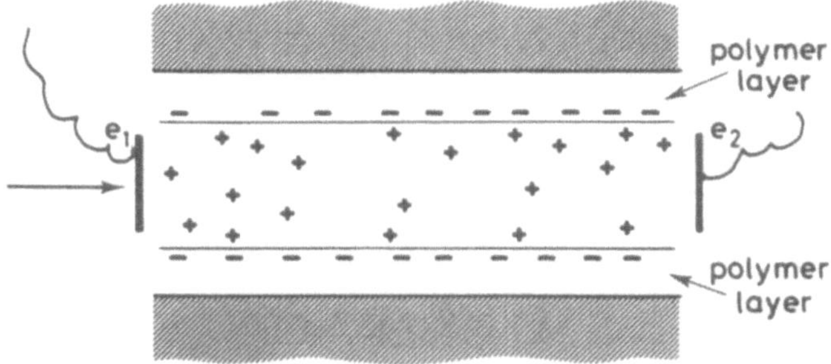

FIGURE 12. Principle of the creation of a streaming potential.

not so simple. If the particle is negative it has around it an equal compensating positive charge, so that particle + countercharge together are electroneutral. Nevertheless, such particles move to the anode, implying that in the electrokinetic process some separation occurs: part of the charge moves, whereas the remainder does not move. Somewhere in the double layer there must be a (sharp or unsharp) kinetic boundary.

Similar reasoning applies to the other electrokinetic phenomena. Let us consider the streaming potential in more detail. In Figure 12 a solid support capillary (hatched) is covered at its inner side with the polymer to be studied. Let us assume that the support is inert (it does not contribute to the flow of current) and that the polymer is negatively charged. An equal positive countercharge is in the solution. Now, in the direction of the arrow, liquid is pumped through the capillary. The result is a potential difference $\Delta\phi_{sp}$ between the two elecltrodes e_1 and e_2; this potential difference is the streaming potential. The right electrode is more positive than the left one. The explanation is as follows: due to the liquid flow, part of the positive countercharge is entrained. Accumulation of this positive charge near e_2 leads to back-flow of electricity by conduction. In the stationary state accumulation by liquid flow and conduction are in equilibrium with each other and then

$$\frac{\Delta\phi_{sp}}{\Delta P} = \frac{\varepsilon\varepsilon_0\zeta}{\eta\kappa_0} \tag{32}$$

Here ΔP is the applied pressure to which $\Delta\phi_{sp}$ is proportional, η is the liquid viscosity, κ_0 the specific conductivity of the solution, and ζ the electrokinetic potential. It is reasonable that η and κ_0 are in the denominator, since the viscosity of the liquid resists the transport of charges and since κ_0 is a measure of the back-flow through conduction. The dielectric permittivity enters through Poisson's law [Eq. (9)], which is, in turn, needed to balance the frictional (Newton) force with the electrical one. For characterization processes, ζ is relevant. As discussed in section 2.3, it is the potential of the slipping plane, i.e., the boundary between the moving (liquid) and the stationary (solid + a few layers of adhering liquid) phase. We defer the discussion of this slipping plane and its precise position until later in this section.

A variant of this procedure involves the same setup, but one having e_1 and e_2 short-circuited. Obviously, the stationary state is then characterized by an electrical current density i_{sc} through the circuit, for which this equation can be derived:

$$\frac{i_{sc}}{\Delta P}=\frac{\varepsilon\varepsilon_0\pi a^2\zeta}{\eta l} \tag{33}$$

Here a is the radius and l the length of the capillary. The quantity i_{sc} is the *streaming current.*

The counterpart of the streaming potential phenomenon is *electroosmosis.* An electrical field is now *applied* between e_1 and e_2. Due to the force exerted on the cations, these ions start to move, entraining liquid. The flow rate u is given through

$$\frac{u}{E}=\frac{\varepsilon\varepsilon_0\zeta}{\eta} \tag{34}$$

where E is the applied field strength.

Finally, we note that Eq. (34) is also the familiar Helmholtz-Smoluchowski equation for the *electrophoretic mobility* of large spheres ($\kappa a \gg 1$).

Let us first give some general comments on these expressions.

First, it must be noted that the equations given here are the most simple ones. In many cases complications arise, requiring modifications and/or extensions. For instance, in electrophoresis the lagging behind of the double layer leads to complex retardation phenomena that depend on κa. Also, for nonspherical particles other equations are needed. In streaming potential measurements, under some conditions surface conductance has to be taken into account, that is the fact that the specific conductivity in the double layer is higher than κ_0. For reviews on this matter see, e.g., References (15)-(17).

Second, it is striking that all the equations have much in common (they all contain $\varepsilon\varepsilon_0\zeta/\eta$), and so have the phenomena, in that the observed phenomena are always of a nature different from the applied force. For instance, in electrophoresis and electroosmosis an *electrical* field is applied, but a *mechanical* flow results (i.e., flow of a particle or of the liquid, respectively). With streaming potentials and streaming currents it is the reverse; a mechanical force now leads to an electrical phenomenon. Electrokinetic phenomena are typically secondary phenomena, in contradistinction to primary phenomena such as Poisseuille flow (mechanical flow due to mechanical force) and conduction (electric current due to electrical force). The facts that all the equations have much in common and that electrokinetic phenomena are second-order phenomena, are not accidental and find an explanation in the thermodynamics of irreversible processes. The similarity of the equations is a consequence of Onsager's reciprocal relations.

In terms of electrokinetic characterization of biomedical polymer surfaces this has one great advantage: *phenomenologically* it does not matter which electrokinetic technique is selected. One simply chooses the one that is experimentally and theoretically the simplest to handle. The only point left then is the interpretation of ζ.

The problem of how ζ fits into the double layer picture of section 4 is, in its generality, not yet solved. Experience has shown that ζ does not coincide with ϕ_0 but rather resembles ϕ_d. For instance, in the case of superequivalent adsorption (Figure 5), ζ and ϕ_0 have opposite signs. In the interpretation of the stability of hydrophobic colloids, which is governed by the diffuse part of the double layer, it appears often justified to identify ζ with ϕ_d.

Closer consideration of the physical meanings of ζ and ϕ_d shows that both quantities are abstractions of a complex reality. In tangential flow there will not be a sharp transition between the moving and the stationary parts of the liquid, but rather a gradual one.[18] Likewise, specific adsorption and ion size effects will diminish gradually with increasing distance from the surface, so that the transition from the molecular condenser to the diffuse part is also gradual. In both cases the gradual transitions have been replaced by step functions, and there is no way of telling if the "hydrodynamic step" (in electrokinetics) should coincide with the "electrostatic step" (in double layers). For a few simple systems such as surfactant monolayers and AgI sols it has been proven that within the resolution of present-day experiments the identification of ζ with ϕ_d is justified.[19] Very likely this applies also to simple flat polymer surfaces. A problem arises, however, if such surfaces are "hairy": oligomeric tails extending from the surface into the solution substantially modify the tangential flow pattern along the surface; if such hairs are present the slipping plane might well be much further out than the Stern plane. Actually, the ensuing difference between ζ and ϕ_d has been used to assess the (hydrodynamic) thickness of such adsorbates.[20,21] The problem is that, especially for polymeric surfaces, it is always hard to prove rigorously that hairs are absent.

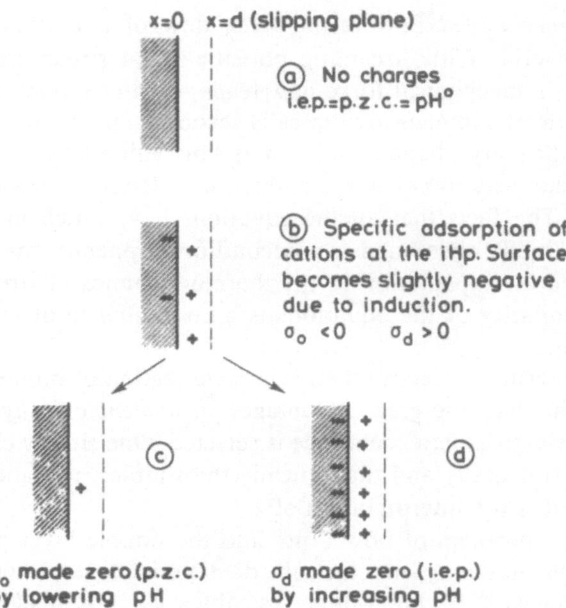

FIGURE 13. Illustration of the fact that if on amphoteric surfaces specific adsorption occurs, the p.z.c. and i.e.p. are no longer identical. Due to specific adsorption of cations the pH of the p.z.c. goes down, whereas that of the i.e.p. goes up. With anions the two trends are reversed.

The fact that ϕ_0 and ζ are different or, for that matter, the difference between σ_0 and $\sigma_{ek} \sim \sigma_d$ has a number of consequences, one of which is that the two quantities depend in a different fashion on experimental conditions, notably on c_{salt} and pH. At given ϕ_0 increasing c_{salt} reduces ζ. For amphoteric surfaces a pH can be identified where $\zeta = 0$. This is called the *isoelectric point* (i.e.p.) If no specific adsorption occurs, the i.e.p. is identical to the p.z.c., defined in section 2.1. However, if specific adsorption does occur, the two zero points differ, as Figure 13 illustrates. The difference between i.e.p. and p.z.c. is a measure of the extent of specific adsorption.

The conclusion for biomedical polymer surfaces is that methods yielding surface properties (especially σ_0) and electrokinetic measurements are not alternative but complementary pieces of information. For a full characterization of the electrical state of such a surface it is advisable to have both σ_0 and σ_{ek} available [$\sigma_{ek} \cong \sigma_d$, which follows from Eq. (19) after substituting ζ for ϕ_d], because both give information on an important, albeit very different part of the double layer. Subtraction of $\sigma_d = \sigma_{ek}$ from σ_0 yields σ_s, which is also an important double layer characteristic. Only if the problem under study requires information on just a part of the double layer may the system be taken to be sufficiently characterized by one of these two electrical parameters. For instance, if long-range interaction is at issue, it is sufficient to measure only the ζ potential.

6. EXPERIMENTAL

There are many reports spread throughout the literature involving the experimental characterization of the interfacial electrochemistry of biomedical polymers that can be used as prosthetic or perfusion devices. The materials on which these measurements are done, the experimental conditions (pH, ionic strength, sample pretreatment), and often also the techniques employed are disparate and not very systematical. It is beyond the scope of this section and, for that matter, it makes little sense to try to review all of this encyclopedically. Instead, against the background of the double layer pictures developed in the previous sections, we shall now discuss more systematically (a) the double layer parameters to be measured and the conditions under which the experiments should be executed, (b) the techniques available and their restrictions, and (c) the preparation of the samples. The discussion will be illustrated with some examples taken from literature.

At the very onset it should be noted that, although the present chapter is restricted to electrical properties, it can be expedient to consider simultaneously information obtainable from other sources. For instance, spectroscopic techniques can provide information on the nature of the surface groups that are involved in determining σ_0, X-ray fluorescence can give the (total) amount of sulfur-containing groups, and carboxyl groups can be detected by IR spectroscopy.

Electrostatically, for a complete characterization we need to establish: (i) the surface charge σ_0 *and* (ii) the electrokinetic charge σ_{ek} or electrokinetic potential ζ, because σ_0 is the *total* charge and the difference between σ_0 and σ_{ek} provides information about the charge *distribution.*

Accepting that for some purposes only one of these characteristics suffices, it is the rule that measurements of σ_0 and ζ make sense only if done at specified pH, ionic strength ω and material pretreatment, or rather, full characterization requires all of these measurements to be done over a wide range of pH and ω. Let us briefly repeat the arguments.

The surface charge is pH-dependent if it contains weak groups, and for most prosthetic materials this is the case. Also, the pH dependence is relevant in establishing the p.z.c. The influence of ω on σ_0 is less pronounced than that of pH, but it is still not negligible; it is due to screening. Figure 9 is a typical illustration.

Electrokinetic potentials depend for the same reason on pH, although not necessarily in a parallel fashion because of specific adsorption (section

5). However, ζ is extremely sensitive to ω because it is the potential at some distance from the surface and the potential there is reduced with increasing ω due to double layer compression and specific counterion adsorption. Consequently, ζ decays with ω, so that beyond $\sim 10^{-1}$ M very low values are found. Under physiological conditions $\zeta \rightarrow 0$. If superequivalent adsorption takes place ζ may reverse sign, but with further increase of ω, ζ will eventually reduce to zero in this case also. We may state, therefore, that electrokinetic measurements are not useful unless pH and ω are established.

6.1. THE SURFACE CHARGE

For the measurement of σ_0 *potentiometric* and *conductometric titration* are common techniques. They require little fancy instrumentation; the problems are rather in the preparation and definition of the sample.

Potentiometric proton titrations are in principle conducted in the same way as for dissolved acids or bases. The difference is that now a surface and a solution are titrated together. The consumption of H^+ and OH^- by the surface is calculated after subtration of the same in bulk which is determined from the e.m.f. of the cell in which the titration is done (after standardization). Proceeding this way, *relative* values of the proton charge are obtained, i.e., values at a certain pH are compared with those at another pH. A reference point is needed to make the values absolute. This can be a pH where all groups are known to be fully dissociated or not dissociated at all, or the p.z.c. can be established, e.g., as the point where addition of indifferent electrolyte does not affect the proton charge (see the intersection point in Figure 9). The resulting σ_0 is the proton charge, i.e., $\sigma_0 = F(\Gamma_{H^+} - \Gamma_{OH^-})$, where Γ_i is the surface excess of i. For strongly acidic surfaces no groups are titrated at moderate pH; for weak groups steep parts are observed at pH $\sim$ pK_a.

Conductometric titrations are exchange titrations where the conductivity is followed as a function of the amount of titrant. Suppose a latex in the acidic form (low pH) is titrated with NaOH. Protons are bound to OH^- to give nonconducting species, so that as far as the conductivity is concerned H^+ ions are replaced by Na^+ ions. As the mobility of the latter is lower, this exchange leads to a lowering of the conductivity, which continues till the equivalence point. Beyond this point the only effect of NaOH addition is to increase the conductivity in proportion to the amount added. Under ideal conditions a conductometric titration graph, therefore, consists of two straight lines intersecting at the equivalence point.

A typical example[22] is presented in Figure 14. In this case the sample is a latex. The surface area of most implant devices as such is too low to measure proton charges with sufficient accuracy; therefore, fine dispersions of the material in the aqueous dispersion medium under study are needed.

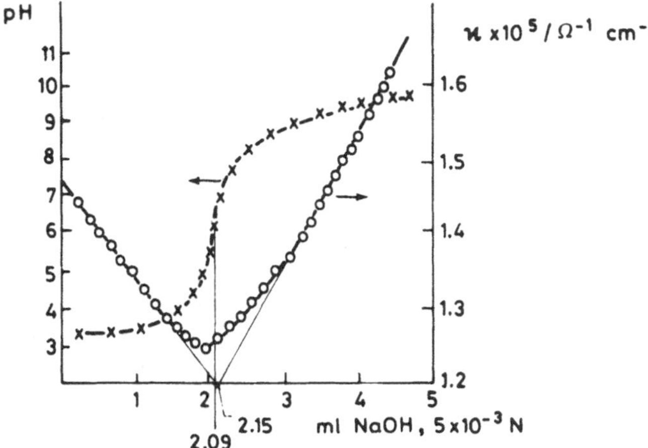

FIGURE 14. Potentiometric and conductometric titration of a polystyrene latex containing sulfate groups. (Taken from Furusawa *et al.*[22])

In this particular example the surface groups are strongly acidic sulfate groups, originating from $K_2S_2O_8$, used as the initiator in the polymerization. These groups accumulate at the surface because of their hydrophobicity. The discrepancy between the steep rise in the potentiometric curve and the (extrapolated) intersection point of the two conductometric branches is *ca.* 3%. The surface charge corresponding to this equivalence point is $-7.4\ \mu C\ cm^2$.

The sharpness of the equivalence points depends on such factors as the amount of titratable charges in the system, the concentration of the titrant, and the presence of other interfering groups. As upon standing or due to other treatments the nature of the surface groups might change, it is always advisable to repeat the measurements after some time.[23,24] Titrations with different bases can give rise to different shapes of the curves, reflecting differences with respect to the exchangeability of the proton. Thus, some insight into the binding of the protons, and perhaps into their distribution inside the polymer (Figure 6), is gained (it is imaginable that the titration of deeper situated protons is more affected by the radius of the exchanging cation than that of the more easily accessible surface groups). For instance, Vanderhoff and van den Hul[25] found for a given polystyrenesulfate latex a linear descending branch if titrated conductometrically with NaOH, but a curved one if titrated with $Ba(OH)_2$. The equivalence point, however, was the same, so that both titrants eventually "see" all surface groups. A similar observation was made by Furusawa and Kawai[26]. In other experiments evidence is found for the presence of two types of surface groups, e.g., carboxyl groups in addition to sulfate groups. Titration diagrams then have two equivalence points. Everett *et al.*[27]) and Stone-Masui and Watillon[28] described such curves and their variation with the mode of pretreatment of the latex. Labib and Robertson[29] subjected complex conductometric plots to a detailed analysis and recommended the inclusion of back-titration to avoid premature conclusons. Hen[30] reported that conductometry does not "see" deeply buried carboxyl groups occurring in latices prepared by copolymerization of styrene and itaconic acid; the fraction of these buried groups can be derived by subtracting the number of conductometrically measurable groups from the total number, obtained from the amount of acid incorporated during polymerization.

Although the majority of latices bear a negative charge, it is also possible to prepare them with positive σ_0[31] or with mixtures of positive and negative groups, so that they are amphoteric.[32] Such materials might be of interest to biomedical science; at any rate in many respects the adsorption of plasma proteins differs between positive and negative latices.[33]

In summary, although several problems remain to be solved, there are two techniques available that, if done sufficiently systematically, give good insight into the nature and magnitude of σ_0. The greatest problem remains that of the transferability of data: is it permissible to assume that the surface properties, established for finely dispersed polymers, are identical to those of the relatively coarse prosthetic materials? For this problem there is at present little answer other than trying to prepare the prosthetic polymers and the polymer colloids in as similar a manner as possible and subject both systems to exactly the same pretreatment.

6.2. THE ELECTROKINETIC CHARGE AND POTENTIAL

In a sense, the problem of transferability applies also to electrokinetics. In any case, electrophoresis has been a popular technique, and this can only be done with dispersed, preferably homodisperse, particles. Latex particles can be viewed in an (ultra-) microscope (dark field illumination), so that the motion can be directly observed in one of the commercially available electrophoresis apparatuses. For latices of which the refractive index does not differ enough from that of the ambient medium, the moving boundary (Tiselius) technique can be applied or the particles can be covered on a carrier, in which case it must be verified that the carrier is inert.[24]

For electroosmosis, streaming potentials, and streaming currents, one often works with porous plugs of the material under study or with capillaries drilled into it. It would be much more elegant to investigate the solids *in situ*, e.g., by covering an inert substrate at the inner side with the polymer as in Figure 12 and then do electrokinetic measurements. A good example of this approach has recently been elaborated[34] and applied to a great variety of polymers[35] by Van Wagenen *et al.* These authors studied stream-

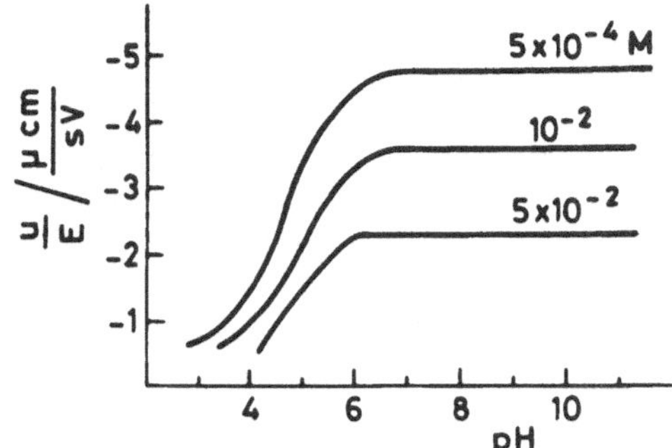

FIGURE 15. Electrophoretic mobility of carboxylate polystyrene latices as a function of pH. The electrolyte concentration (NaCl) is indicated. Particle radius, $a = 370$ nm. Results from Ottewill and Shaw[36].

ing potentials in a flat electrophoresis cell along the lines of Figure 12. Still, the main body of literature data involves electrophoresis, and since only on latices can comparisons between σ_{ek} and σ_0 be made, we shall pay some attention to it.

One of the more classical investigations is that by Ottewill and Shaw[36], from which Figures 15 and 16 are taken. The observed trends are rather typical for latices with weak acidic groups. Mobilities range between 0 and 5 cm μs^{-1} V^{-1}. The corresponding σ_{ek} never exceeds a few $\mu C\ cm^{-2}$, even if σ_0 is ten times as high. Above pH 6 all carboxyl groups are dissociated and the mobility is independent of pH. From pH 6 downward these groups start to associate and the mobility becomes negligibly small at low pH. The steepest decrease is around pH 4.6, which is $\sim pK_{a,\mathrm{intr}}$ of carboxyl groups, whose presence was confirmed by IR spectroscopy. For sulfate latices the mobility is independent of pH down to much lower values, although at very low and very high pH irregularities are observed which act as a pabulum for latex students, since they belong to the realm of system idiosyncracies,

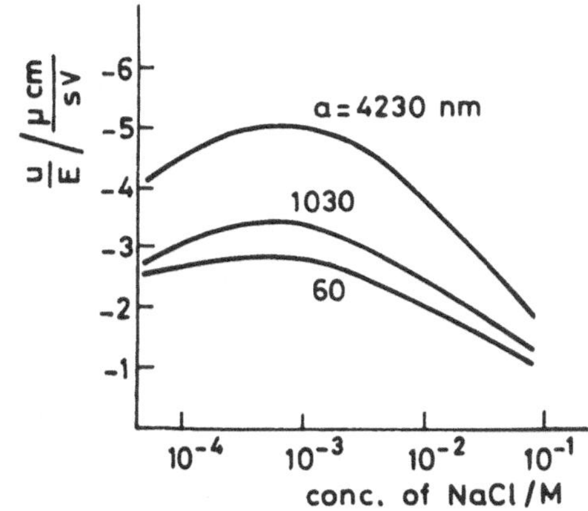

FIGURE 16. Dependence of the electrophoretic mobility on the ionic strength (NaCl). The particle radius is indicated. Data from Reference (36).

but at any rate for physiological conditions they play no role of any consequence.

In Figure 15 the role of indifferent electrolyte is as expected. It is also the case in Figure 16 for concentrations above about 5×10^{-2} *M*. However, the decrease of mobility with decreasing concentration at the lower branch is contrary to expectations. So is, at first sight, the fact that the mobility depends on *a*.

Part of the explanation can be sought and found in the theory of electrophoresis.[37] As discussed in section 5, electrophoresis involves intricate retardation phenomena due to the asymmetry of the double layer in the stationary state. It follows from theory, and can also be intuitively understood, that the extent of this retardation is a function of the double layer thickness relative to the particle radius. In formula terms, it depends on κa. Qualitatively this is the basis for the dependence on c_{salt} and a, but on close inspection electrophoresis theory cannot explain everything. For instance, Ottewill and Shaw observed that under certain conditions the particles moved faster than the maximum speed dictated by theory. In other cases, electrokinetic charges exceeding σ_0 have been derived under conditions where this would not be expected. For instance, for sulfate latices[38] a clear increase of σ_{ek} with increasing c_{NaCl} was reported in the very low concentration range (up to 10^{-3} *M*). As under these conditions σ_{ek} was more negative than σ_0, specific adsorption of Cl^- was put forward as one possible explanation, but this could not be independently confirmed. Hairiness of the particle, changing with c_{salt}, is an alternative possibility, or more precisely, the number of protruding chains is constant but the extent to which they are stretched into the solution might be affected by the nature and concentration of the ions.[39] For poly(vinyl toluene) latices, Bagchi *et al.*[24] report little difference between σ_0 and σ_{ek}, but their data involve only one electrolyte concentration (2×10^{-3} *M* KNO_3, which is not in the "dangerous" range), and it is rather disturbing that they could not confirm their microelectrophoretically obtained σ_{ek} values by moving boundary electrophoresis.

In connection with the use of electrokinetic methods to characterize surfaces of medical and biological significance, the behavior of u(pH) deserves some further comments. Figure 15 suggests that for pH > 7 at fixed ω the mobility is a good characteristic of the electrical state of the surface because it is then independent of pH. However, this is only partly correct. First, it should be realized that on surfaces where σ_0 continues to increase with increasing pH (for instance, with many oxidic surfaces)[10], the mobility also reaches a plateau. Also for materials for which σ_0 is insensitive to pH this trend is found, for example, long ago Troelstra and Kruyt[40] observed similar plateaus for the mobility of AgI particles as a function of pAg. The existence of such plateaus is apparently a general phenomenon. There are two possible theoretical reasons for this. First, there is the general trend that, if more and more charge is built into a double layer, this additional charge accumulates and is compensated by counterions in the very inner part of the double layer, so that at the slipping plane nothing is felt of this charge accumulation. The corollary of this in polyelectrolyte theory is the so-called counterion condensation.[41] Note also the fact that in diffuse double layer theory, according to the Boltzmann principle, $c_+(x)$ is proportional to $\exp(zF\phi(x)/RT)$, i.e., counterions accrue progressively in the range where the potential is highest. In the second place, there might well be a purely hydrodynamic reason. Hitherto the only existing quantitative picture of the slipping process views it as being governed by the increase of viscosity of adjacent water due to the electric field emanating from the surface.[18] According to this picture, increasing σ_0 leads to an increasing field strength [see Gauss's law, Eq. (8)], and hence to an outward displacement of the slipping layer, offsetting the effect of the increase of σ_0.

The conclusion is that the constancy of ζ or u/E is due to general double layer and/or hydrodynamic features, so that ζ is too insensitive to be considered a characteristic of the double layer. However, ζ *is* an acceptable characteristic for those experiments in which only the charge or potential in the outer double layer part counts, as in the case in problems like cell adhesion and long-distance interaction between the prosthetic materials and blood proteins.

In order to discriminate between hydrodynamic and counterion uptake it might well be advisable to measure the counterion adsorption directly. Radiotracer methods could be suitable for that. An example of such a study with latices is the work by Tabor *et al.*[42] By way of background information, the recent review by James on the electrophoretic characterization of bacterial cells may be mentioned.[43]

Electrokinetic measurements on the surfaces of prosthetic materials *in situ* are scarce and more studies are desirable. A good start has been made by Van Wagenen *et al.*,[35] using their streaming potential technique (Section 5). This work is not yet systematic in that it neither compares σ with σ_0, nor studies the influences of pH and ω, but it deserves attention because the applicability of the method is proved and many different polymers, spun-cast onto glass slides, are compared under identical conditions (pH 7,4; electrolyte 10^{-2} *M* KCl + 8×10^{-4} *M* Na_2HPO_4 + 2×10^{-4} *M* KH_2PO_4). Hence, the influence of the nature of the polymer on the streaming potential, that is: on ζ, is borne out. Most trends are more or less according to expectation. The majority of the polymers have naturally a negative ζ (discrimination between the various possible origins is not yet possible, since measurements of σ_0 and σ_s are not yet available). If more negative groups are incorporated into the polymer, ζ becomes more negative and the converse applies to positive groups. The trend is that it is not so easy to make $\zeta > 0$. The computed ζ potentials are all in the range of up to *ca.* -60 mV, which according to Eq. (19) would correspond to $|\sigma_{ek}| \lesssim 1.8\ \mu C\ cm^{-2}$, supporting our earlier conclusion that electrokinetic charges never exceed a few $\mu C\ cm^{-2}$.

7. CONCLUSIONS

The main purpose of this chapter was to review the interfacial electrochemistry of solid-solution interfaces, emphasizing biomedical polymers. Against this background, experimental procedures are discussed to characterize such surfaces electrochemically. In principle, two kinds of electrical techniques offer themselves: potentiometric and conductometric titrations to find σ_0 and electrokinetic procedures for σ_{ek}. If possible, such measurements should be done in combination with nonelectrochemical experiments, and effects of pH and ionic strength deserve due attention. Preferably, such measurements should be done with the biopolymer *in situ*. In addition to problems inherent to electrokinetics (e.g., surface conductivity, relaxation phenomena), attention should be paid to the pretreatment of the sample and the reversibility of the readings.

The art of electrochemical characterization of biopolymer surfaces is still in its infancy, but, in principle, theories and techniques are available to foster the development of this important bridge between colloid and biomedical science.

REFERENCES

1. J. Th. G. Overbeek, in: *Colloid Science* (H. R. Kruyt, ed.), Vol. I, Chapt. III, pp. 118–126, Elsevier, Amsterdam (1952).
2. J. Lyklema, *Medical Electron. Biol. Eng.* **2**, 265 (1964).
3. G. Gouy, *J. Phys.* (4) **9**, 457 (1910).
4. G. Gouy, *Ann. Phys.* (9) **7**, 129–184 (1917).
5. D. L. Chapman, A contribution to the theory of electrocapillarity, *Phil. Mag.* (6) **25**, 475–481 (1913).
6. O. Stern, *Z. Elektrochem.* **30**, 508–516 (1924).
7. D. C. Grahame, The electrical double layer and the theory of electrocapillarity, *Chem. Rev.* **41**, 441–501 (1947).
8. M. J. Sparnaay, *The Electrical Double Layer*, Pergamon Press, New York (1972).
9. J. O'M. Bockris, B. E. Conway, and E. Yeager, eds., *The Double Layer*, Vol. I of *Comprehensive Treatise of Electrochemistry*, Plenum Press, New York (1980).
10. R. O. James and G. A. Parks, in: *Surface and Colloid Science* (E. Matijevic, ed.), Vol. 12, pp. 119–216, Plenum Press, New York (1982).
11. W. Stumm and M. Morgan, in: *Aquatic Chemistry*, 2nd Edition, Chapt. 10, pp. 514–563, Wiley-Interscience, New York (1981).

12. J. Lyklema, The structure of the electrical double layer on porous surfaces, *J. Electroanal. Chem.* **18**, 341-348 (1968).
13. R. Arnold and J. Th. G. Overbeek. The dissociation and specific viscosity of polymethacrylic acid, *Rec. Trav. Chim.* **69**, 192 (1950).
14. J. W. Perram, R. J. Hunter, and H. J. L. Wright, The oxide-solution interface, *Austr. J. Chem.* **27**, 461-475 (1974).
15. J. Th. G. Overbeek, in: *Colloid Science* (H. R. Kruyt, ed.), Vol. I, Chapt. V, pp. 194-244, Elsevier, Amsterdam (1952).
16. S. S. Dukhin and B. V. Derjaguin, in: *Surface and Colloid Science* (E. Matijevic, ed.), Vol. 7, Chapt. 1-3, John Wiley and Sons, New York (1974).
17. R. J. Hunter, *The Zeta Potential in Colloid Science: Principles and Applications*, Academic Press, London (1981).
18. J. Lyklema and J. Th. G. Overbeek, On the interpretation of electrokinetic potentials, *J. Colloid Sci.* **16**, 501-512 (1961).
19. J. Lyklema, Water at interfaces: a colloid-chemical approach, *J. Colloid Interface Sci.* **58**, 242-250 (1977).
20. L. K. Koopal and J. Lyklema, Characterization of polymers in the adsorbed state by double layer measurements, *Faraday Disc. Chem. Soc.* **59**, 230-241 (1975).
21. J. Lyklema, Inference of polymer adsorption from electrical double layer measurements, *Pure Appl. Chem.* **46**, 149-156 (1976).
22. K. Furusawa, W. Norde, and J. Lyklema, *Kolloid-Z.Z. Polym.* **250**, 908-909 (1972).
23. D. E. Yates, R. H. Ottewill, and J. W. Goodwin, Purification of polymer latices, *J. Colloid Interface Sci.* **62**, 356-358 (1977).
24. P. Bagchi, B. V. Gray, and S. M. Birnbaum, Preparation of model poly(vinyl toluene) latices and characterization of their surface charge, *J. Colloid Interface Sci.* **69**, 502-528.
25. J. W. Vanderhoff and H. J. van den Hul, Well characterized monodisperse latexes as model colloids, *J. Macromol. Sci., Chem.* **A7(3)**, 677-707 (1973).
26. K. Furusawa and H. Kawai, Charge distribution in electric double layer on polystyrene latices, *Chem. Lett. Chem. Soc. Japan*, 693-696 (1979).
27. D. H. Everett, M. E. Gültepe, and M. C. Wilkinsin, Problems associated with the surface characterization of polystyrene latices, *J. Colloid Interface Sci.* **71**, 336-349 (1979).
28. J. Stone-Masui and A. Watillon, Characterization of surface charge on polystyrene latices, *J. Colloid Interface Sci.* **52**, 479-503 (1975).
29. M. E. Labib and A. Robertson, The conductometric titration of latices, *J. Colloid Interface Sci.* **77**, 151-161 (1980).
30. J. Hen, Determination of surface carboxyl groups in styrene/itaconic acid copolymer latexes, *J. Colloid Interface Sci.* **49**, 425-432 (1974).
31. J. W. Goodwin, R. H. Ottewill, and R. Pelton, Studies on the preparation and characterization of monodisperse polystyrene latices, *Colloid and Polymer Sci.* **257**, 61-69 (1969).
32. A. Homola and R. O. James, Preparation and characterization of amphoteric polystyrene latices, *J. Colloid Interface Sci.* **59**, 123-134 (1977).
33. P. Koutsoukos, C. A. Mumme-Young, W. Norde, and J. Lyklema, *Colloids and Surfaces* **5**, 93-104 (1982).
34. R. A. Van Wagenen and J. D. Andrade, Flat plate streaming potential investigations: hydrodynamics and electrokinetic equivalency, *J. Colloid Interface Sci.* **76**, 305-314 (1980).
35. R. A. Van Wagenen, D. L. Coleman, R. N. King, P. Triolo, L. Brostrom, L. M. Smith, D. E. Gregonis, and J. D. Andrade, Streaming potential investigations: polymer thin films, *J. Colloid Interface Sci.* **84**, 155-162 (1981).
36. R. H. Ottewill and J. N. Shaw, Electrophoretic studies polystyrene latices, *J. Electroanal. Chem.* **37**, 133-142 (1972).
37. P. H. Wiersema, A. Loeb, and J. Th. G. Overbeek, Calculation of the electrophoretic mobility of a spherical colloid particle, *J. Colloid Interface Sci.* **22**, 78-99 (1966).

38. A. E. J. Meyer, W. J. van Megen, and J. Lyklema, Pressure-induced coagulation of polystyrene latices, *J. Colloid Interface Sci.* **66**, 99–104 (1978).
39. A. G. van der Put and B. H. Bijsterbosch, *J. Colloid Interface Sci.* **92**, 499–507 (1983).
40. S. A. Troelstra and H. R. Kruyt, *Kolloid-Z.* **101**, 182–189 (1942).
41. G. S. Manning, The molecular theory of polyelectrolyte solutions with applications to the electrostatic properties of polynucleotides, *Quart. Rev. Biophys.* **11**, 179–246 (1978).
42. Z. Tabor, G. Deželic, and P. Strohal, *Croat. Chem. Acta* **50**, 219 (1977).
43. A. M. James, *Adv. Colloid Interface Sci.* **15**, 171 (1982).

9

Interface Acid–Base/Charge-Transfer Properties

Frederick M. Fowkes

Recent advances in the understanding of intermolecular interactions have led to new ideas about the interactions at interfaces between unlike substances, and these ideas are now being extended to interactions with biomaterials. Interactions at interfaces involve several kinds of intermolecular forces: dispersion forces, acid-base interactions, and the electrostatic attractions of ionic groups or of injected charges. This chapter reviews advances in: (1) acid-base interactions (including all hydrogen bonds) which had previously been labeled "polar"; and (2) the injection of electric charges upon contacting one material with another ("contact electrification") and how this effect influences interfaces.

1. THE ACID–BASE CHARACTER OF "POLAR" INTERACTIONS

For fifty years or so intermolecular interactions have been classified as "polar" or "non-polar," a result of the remarkable achievements in the 1920–28 period in the understanding of the nature of the attractive forces in gases. Van der Waals in 1873 had shown that in gases the attractive energies were inversely proportional to the sixth power of the intermolecular distance,[1] but the nature of the attractive forces were not understood until Keesom[2] and Debye[3] (in 1921) showed that the attractive energy between

---

Frederick M. Fowkes • Department of Chemistry, Building 6, Lehigh University, Pennsylvania 18015.

molecules 1 and 2 with permanent dipole moments (μ) could be quantitatively determined by:

$$U_{12}^{\mu\mu} = -2\mu_1^2\mu_2^2/3kTr_{12}^6(4\pi\varepsilon_0)^2 \tag{1}$$

$$U_{12}^{\mu\alpha} = -\mu_1^2\alpha_2/4\pi\varepsilon_0 r_{12}^6 \tag{2}$$

in which r_{12} is the center-to-center distance between the dipoles of Eq. (1) or between the dipole of molecule 1 and the center of polarizability α_2 of molecule 2, and ε_0 is the permittivity of free space (needed for SI units). The tour de force was completed in 1930 when London[4] showed that molecules without permanent dipoles attract each other because of mutual perturbation of their electron orbits, especially the outermost electrons whose binding energy is estimable from the experimental dispersion of light or from the ionization potential (I):

$$U_{12}^{d} = -3\alpha_1\alpha_2 I_1 I_2/2(I_1 + I_2)r_{12}^6 \tag{3}$$

London named this interaction the dispersion force attraction and proposed that the perturbation of orbiting electrons could be treated as a sum of fluctuating dipoles and quadrupoles; the above equation is the dominant term, due to interaction of fluctuating dipoles only.

In subsequent years other investigators tended to apply the conclusions of these gas studies to liquids and solids. It became customary to discuss the dipole-dipole interactions of Eq. (1) and the dipole-induced dipole interactions of Eq. (2) for molecules in the liquid or solid state, or at their interfaces. In this way the language of "polar" and "non-polar" interactions of molecules developed, without the realization that the extrapolation of these equations from the two-body interactions in gases to the multibody interactions in condensed phases might not be warranted. In contrast, the extension of Eq. (3) for dispersion forces from gases to condensed phases has been studied in much detail and with considerable success.[5-7]

Only very recently has the extrapolation from gases to condensed phases of the dipole interactions of Eqs. (1) and (2) been studied, and it is found so far that in condensed phases there is no evidence for any measurable dipole interaction energies of cohesion or adhesion. Instead it is found that the acid-base interactions of Lewis,[8] including the charge-transfer complexes of Mulliken,[9] occur between the so-called "polar" groups in liquids and solids, and that these interactions are quite independent of "polarity" as measured by dipole moment. Furthermore, it is found that the old rule of "like attracts like", etc., never applies to interactions between acids or between bases; instead we find "acids attract bases," This is particularly obvious in solubility or swelling studies with polymers, where acidic polymers interact strongly with basic solvents but not with acidic solvents.

The development of the modern theory of acids and bases started in 1923 when Bronsted,[10] Lowry,[11] and Lewis[8] made contributions which are still being exploited today. The Lewis concept of acids as electron acceptors and of bases as electron donors expanded the field of acid-base interactions to include nearly all molecules (except saturated hydrocarbons) as acidic, basic, or having both characteristics. It is now realized that most oxygens, nitrogens, sulfurs, and related elements tend to be the basic sites in molecules, that the π-electrons of olefins and aromatics, and all nucleophilic groups are basic sites, and that all anions tend to be bases. Similarly the "active" hydrogens of water, alcohols, "Bronsted" acids, and halogenated hydrocarbons are found to be acidic sites, as are all "electrophilic" sites, nitro groups, and all cations. The acidity or basicity of anions and cations is best correlated by the Lowry-Bronsted theory with later important contributions from many others, especially Schwarzenbach[12] and Pearson.[13] On the other hand, acidity or basicity of organic molecules is best correlated by the Drago correlations[14,15] of the enthalpies of acid-base interactions (ΔH^{ab}) in neutral solvents:

$$-\Delta H^{ab} = C_A C_B + E_A E_B \tag{4}$$

The two constants for the acid (C_A, E_A) and the two constants for the base (C_B, E_B) express the concepts of Schwarzenbach,[12] Pearson,[13] and others that the strength of interaction of a basic or acidic site depnds not only on the ability to donate or accept electrons, but also on the polarizability. Mulliken[9] introduced the terms "electrostatic" and "covalent" and Drago assigned E and C to these terms. Pearson introduced the terms "soft" and "hard" for acids and bases to indicate the relative ease of deformation of the outer electron orbitals by an electric field, with iodine featured as a soft acid and Bronsted acids as hard acids, or with sulfur or selenium featured as softer bases than oxygen. Drago proposed the ratio C/E as a measure of softness of an acid or base.

Edwards[16] actually preceded Drago in developing a four-constant equation for acid-base interactions, but he correlated the Gibbs free energies (equilibrium constants) instead of enthalpies. Much of the experimental data correlated in Drago's work comes from infrared spectra and represents therefore not actually an enthalpy change (ΔH) but an internal energy change (ΔU); however, in condensed systems the difference is trivial. The exact meaning of the C and E constants is not clear; although Drago had in mind the electrostatic and covalent terms of Mulliken, he arbitrarily set the values of C_A and E_A for iodine both equal to 1.00 $(\text{kcal/mol})^{1/2}$ and thereby fixed the actual value of C and E for all other species. If any other assignment were made, all values would change but the order of C/E ratios would not change. The original work was done with enthalpies in kilocalories per mole and nearly all studies to date have been reported in such units. Drago has determined the E and C constants for about eighty organic liquids,[14,15] many of which are representatives of large homologous series. He has also developed a correlation of ΔH^{ab} versus the "solvent shift" for the infrared OH stretching frequency of phenol when interacting with a variety of bases in dilute solutions in carbon tetrachloride.[14]

$$-\Delta H^{ab}\ (\text{kcal/mol}) = 3.08 + 0.0103\Delta\nu_{\text{OH}}(\text{cm}^{-1}) \tag{5}$$

Drago's C and E constants cover a wide range of acids and bases, mostly organic. Included are a wide range of hydrogen bonds, as is shown in Table 1. It can be seen that the strength of hydrogen bonds varies widely, from 2 kJ/mol to nearly 50 kJ/mol, but that the E and C equation gives very accurate predictions. This means that the basicity (C_B and E_B) of a proton acceptor site can be measured by its interaction with non-hydrogen acids such as iodine, antimony pentachloride, or an aluminum alkyl, and the acidity (C_A and E_A) of the proton donor can be measured by its ΔH with any bases, even sulfides or aromatics; yet one can take these C and E values and calculate accurately the ΔH of hydrogen bonding for nitrogen bases or oxygen bases. It is concluded, therefore, that hydrogen bonds are a subset of the general acid–base bonds, following all the principles established for Lewis acid–base interactions.

The effect of permanent dipoles on the enthalpy of acid–base interactions may be determined by comparing ΔH^{ab} values of acid–base pairs with large dipole moments versus those with small dipole moments. The basic data of Drago's tables come from calorimetric studies of the enthalpy

TABLE 1
Test of Equation (6) for Enthalpies of Hydrogen Bonds (kJ/mol)

Acid	Base	ΔH_{E+C}	ΔH_{Exp}
t-Butanol	Pyridine	18.0	18.0
	Dimethyl sulfoxide	15.1	15.1
$(CF_3)_2CHOH$	Pyridine	46.0	48.1
Phenol	Pyridine	33.0	33.5
	Trimethylamine	36.0	36.8
	Dibutyl ether	25.1	25.1
	Acetone	22.2	21.3
	Methyl acetate	19.2	20.1
Thiophenol	Pyridine	10.0	10.0
	Dimethylformamide	2.5	2.1

TABLE 2
Enthalpies (kJ/mol) of Acid–Base Interaction of *p*-Chlorophenol with Bases of Various Dipole Moments

Base	$\mu_1^2\mu_2^2[\mathrm{C\cdot m}]^4 \times 10^{94}$	ΔH_{E+C}	ΔH_{Exp}
Trimethylamine	2.1	39.7	40.2
Butyl ether	7.6	26.4	25.9
Ethyl sulfide	13.0	20.9	20.9
Ethyl acetate	17.3	20.9	21.3
Acetone	45.4	22.6	22.6

of mixing of acids and bases. If dipole–dipole interactions occur in these systems the enthalpy of mixing should include such an effect. Thus, if the C_B and E_B of a base with large dipole moment were to be determined by interactions with acids of zero dipole moment (such as iodine, boron trifluoride, etc.), and then one used these constants to predict the interaction with an acid of high dipole moment (such as *p*-chlorophenol), the predicted enthalpy should be less than observed by an amount equal to the enthalpy of the dipole–dipole interaction. The magnitude of such dipole–dipole interaction might be estimated from Eq. (1) for pair-wise interaction of one dipole with another, in which the energy of interaction is proportional to $\mu_1^2\mu_2^2$. Tables 2, 3, and 4 compare $\mu_1^2\mu_2^2$ for a series of acids and bases from Drago's tables, together with the measured enthalpies of mixing and those predicted by Eq. (4). It is apparent that there is little difference in predicted versus observed enthalpies, and in only one case (for acetonitrile) does the enthalpy of mixing for the high dipole moment pair exceed predictions. It is therefore concluded that present experimental results indicate that dipole–dipole interactions in liquids are negligibly small as compared with acid–base interactions or dispersion force interactions.

Why are dipole–dipole interactions so small in liquids when they are quite important in gases? One clue may be that in gases only two molecules interact with each other at the same time, and the Keesom equation [Eq.

TABLE 3
Enthalpies (kJ/mol) of Acid–Base Interaction of Isothiocyanic Acid with Bases of Various Dipole Moments

Base	$\mu_1^2\mu_2^2[\mathrm{C\cdot m}]^4 \times 10^{94}$	ΔH_{E+C}	ΔH_{Exp}
Butyl ether	5.0	27.2	26.8
Ethyl sulfide	8.6	14.6	14.6
Acetonitrile	57.3	19.2	20.9

TABLE 4
Enthalpies (kJ/mol) of Acid–Base Interaction of Acetone with Acids of Various Dipole Moments

Acid	$\mu_1^2\mu_2^2$[C·m(×10^46)]	ΔH_{E+C}	ΔH_{Exp}
Iodine	0	13.8	13.8
Trimethyl aluminum	2.6	83.7	84.9
Chloroform	10.6	15.1	15.1
Water	34.9	13.4	13.4
p-Chlorophenol	45.4	22.6	22.6

(1)] takes this into account, where the kT term allows for a Boltzmann average of orientations of the two molecules to maximize their dipole interactions. In a liquid where each molecule has many nearest neighbors, such pair-wise orientation is not possible, for the optimal orientation to one neighbor tends to oppose optimal orientation to others, ending up in actual repulsion between many adjacent molecules.

Many of the compounds usually referred to as "polar" have both acidic and basic sites, as in water, glycols, alcohols, phenol, nitrobenzene, and acetone. Often these have large dielectric constants resulting from a chaining-type of association in which the molecules self-associate through acid-base bonds to line up the dipole moments. A recent infrared study by Shurvell and Bulmer[(17)] show this polymer-type of association for phenol in some detail. Our recent studies with acetone indicate similar behavior. In both cases the enthalpy of self-association is about 12 kJ/mol and linear self-association results in alignment of the dipole moments, thereby resulting in high dielectric constants.

2. THE SURFACE ACIDITY OR BASICITY OF POLYMERS

The surface characteristics of solid polymers vary appreciably. Some polymers (such as polyethylene, polypropylene, and polytetrafluoroethylene) have rather nonpolar surfaces. Others are dominated by basic sites (polystyrene, polyesters, polyethers) or acidic sites (the polyimides and the polyvinyl or polyvinylidene chlorides or fluorides). In addition, there are polymer surfaces with both acidic and basic sites (polyamides, polycarbonates, nitrocellulose, and polyalcohols).

There are at least three current approaches to characterize the acidic or basic surface groups in polymers: contact angle studies with acidic or basic liquids,[(18)] inverse gas chromatography with acidic or basic gases,[(19)] and infrared absorption peak shifts of acid–base interactions between polymers and test gases or liquids of known acidity or basicity.[(20)] In all of these studies the measured interactions involve two kinds of interactions: the general attractions of the dispersion forces (d), and the specific attractions of acid-base interactions (ab).

2.1. INFRARED MEASUREMENTS OF ACID-BASE POLYMER INTERACTIONS

In the infrared studies it is found that absorption peaks of acidic or basic sites shift to lower frequencies upon complexation. Drago and coworkers determined the $\Delta\nu_{OH}$ shifts of the OH stretching frequency for phenol upon complexation with test bases in solutions of phenol in neutral solvents which were dilute enough to avoid the self-association of phenol; Eq. (5) evolved in these studies. In our current studies at Wright-Patterson Air Force Base[20] we are concentrating on the shift of the carbonyl stretching frequency of ketones and esters upon complexation with test acids and are extending the findings of Bellamy and Williams[21] to the much stronger acids of Drago's tables. We find that the $\Delta\nu_{C=O}$ shift on going from the vapor phase to solution depends on both dispersion force interactions and acid-base interactions:

$$\Delta\nu = \Delta\nu^{d} + \Delta\nu^{ab} \tag{6}$$

with a $\Delta\nu^{d}$ of about 12 cm^{-1} for cyclohexane. The value of $\Delta\nu^{ab}$ was often determined in cyclohexane solutions to avoid any contribution from $\Delta\nu^{d}$ values in the acid-base studies. Figure 1 shows the correlation of experimental ΔH^{ab} values with $\nu_{C=O}$ for ethyl acetate interacting with a wide range of nonassociated acids, resulting in:

$$\Delta H^{ab} = 1.00\,\Delta\nu_{C=O}\ (\text{kJ mol}^{-1}\ \text{cm}) \tag{7}$$

which was confirmed by the recent studies of Massat and Dubois[22] with acetone (shown in Figure 1). Figure 1 also shows our results with poly(methyl methacrylate) (PMMA), a weaker base than ethyl acetate.

The shifts of absorption peaks with acid-base complexation shown in Figure 1 may be used to determine the C and E constants as illustrated in Figure 2 where a rearranged form of the Drago equation is plotted for each ΔH^{ab} of the base with a test acid:

$$E'_B = -\Delta H^{ab}/E_A - C'_B(C_A/E_A) \tag{8}$$

in which E'_B and C'_B are trial values and the slope is the "softness index" for the test acid (C_A/E_A). The intersection of the lines for the different acids gives the best estimate for C_B and E_B of the base being evaluated. In

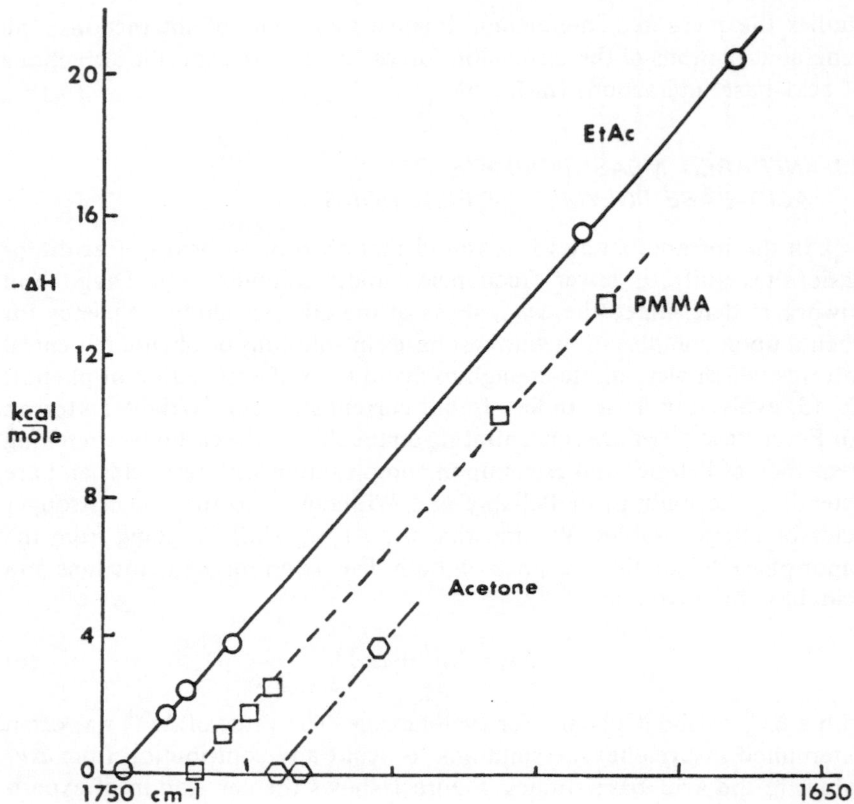

FIGURE 1. Correlation of calorimetric enthalpies of acid–base interaction (for ethyl acetate and acetone) with position of IR absorption peak for carbonyl stretching vibration. ΔH^{ab} values for poly(methyl methacylate) (PMMA) are predicted using the slope of these plots.

Figure 2a the two soft acids (iodine and antimony pentachloride) and two hard acids (chloroform and trimethylaluminum) have four significant intersections, yielding $C_B = 1.73 \pm 0.05$ and $E_B = 1.03 \pm 0.03$ (kcal/mol$^{1/2}$) for ethyl acetate. For PMMA (Figure 2b) and the same four acids the values obtained are $C_B = 0.95 \pm 10$ and $E_B = 0.65 \pm 07$ (kcal/mol)$^{1/2}$.

The spectral shift $\Delta\nu_{CO}$ with PMMA can be used to determine the relative acidity of solvents or of acidic polymers (Figure 3). The carbonyl shifts shown in Figure 3 are for transmission spectra of polymer blends in solution in CH_2Cl_2; films of acid–base polymer blends cast from these solutions showed essentially the same shifts by ATR spectroscopy.

Acidic polymers such as poly(vinyl chloride) (PVC), chlorinated poly(vinyl chloride) (CPVC), poly(vinyl fluoride) (PVF), poly(vinylidene fluoride) (PVF_2), or poly(vinyl butyral) (PVB) were found to form acid–base

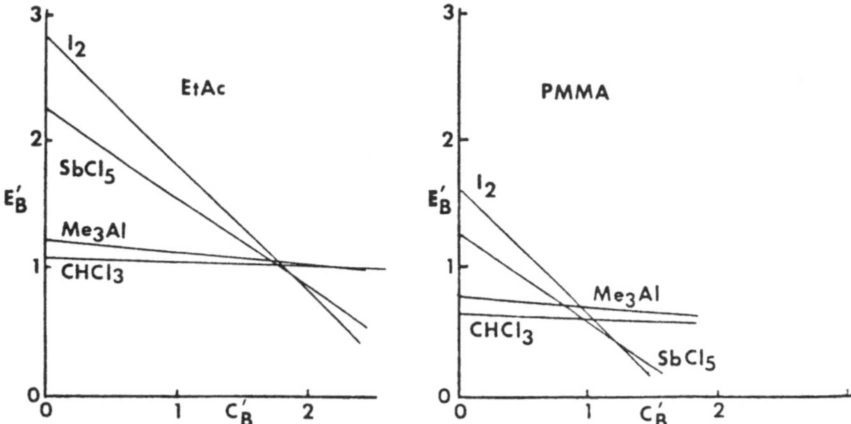

FIGURE 2. Graphical method for determining E_B and C_B values for ethyl acetate (EtAc) and poly(methyl methacrylate) (PMMA) with two hard acids (chloroform and trimethyl aluminum) and two soft acids (iodine and antimony pentachloride).

complexes having ΔH^{ab} values of −13 to −30 kJ/mol with esters, and larger enthalpies with nitrogen bases. The complexes formed readily in solution with ester-containing compounds, including monoester solvents, diester plasticizers, or high-molecular weight polyesters such as poly(methyl methacrylate) (PMMA). Acidic polymers such as CPVC were found to imbibe the vapors of a wide series of esters to form swollen complexes of exact stoichiometry, one ester group per four monomers; however, the solubility of CPVC at room temperature in a homologous series of esters is negligibly small for the lower esters (esters with solubility parameters between 8.8 and 10.2), but is quite appreciable (10–20%) for higher esters (hexyl, amyl, and butyl acetates, having solubility parameters of 8.6–8.7). These findings indicate that in ester solvents the solubility of CPVC is governed by the solubility parameter of the CPVC–ester complex (8.65) rather than by the solubility parameter of the polymer (about 10).

Similar behavior was observed with poly(vinyl butyral) (19% OH), an acidic polymer with a solubility parameter of about 9.5 which forms strong ester complexes ($\Delta H^{ab} = -4.6$ kcal/mol). In propyl acetate ($\delta = 8.8$) the PVB–ester complex is extremely soluble, but all of the other aliphatic esters ($\delta = 8.6$, 8.7, 8.9, 9.5, 9.9) are very poor solvents. Poly(vinyl butyral) is quite soluble in other basic solvents such as ketones (acetone, $\delta = 9.62$; methyl ethyl ketone, $\delta = 9.45$; and cyclohexanone, $\delta = 10.4$) and in nitrogen bases with higher solubility parameters (n-hexylamine, $\delta = 8.45$; n-butylamine, $\delta = 8.66$; or pyridine, $\delta = 10.62$) but not in nitrogen bases with lower solubility parameters (dibutylamine, $\delta = 8.15$; or tributylamine, $\delta = 7.76$).

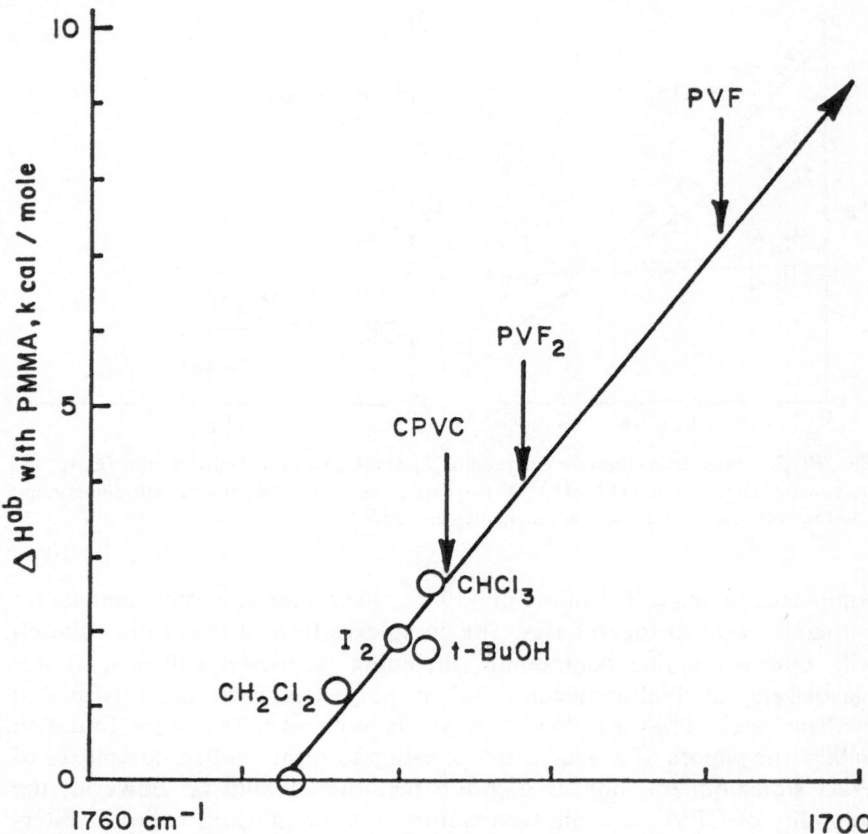

FIGURE 3. Predictions of enthalpies of acid-base complexation, ΔH^{ab}, of polymers from carbonyl stretching frequencies. Reprinted from *Physicochemical Aspects of Polymer Surfaces* K. Mittal, ed., Vol. 2, p. 583 (1983) by permission of Plenum Press.

These infrared studies give much insight into the acidic and basic sites of polymers. The studies are being extended to mulls of silica and other inorganic powders of high surface area, where the shift of absorption peaks upon adsorption of basic carbonyl groups (as in PMMA) gives a precise measurement of the surface acidity of interacting sites.

2.2. CONTACT ANGLE MEASUREMENTS OF SURFACE ACIDITY OR BASICITY

The surface free energy of acid-base interaction can be determined by contact angle measurements on polymer surfaces, and sometimes also on inorganic surfaces. Our initial studies were made with acidic or basic liquids from Drago's tables on a series of copolymers of ethylene with either vinyl acetate (to give basic surface sites) or with acrylic acid (to give acidic surface sites).

The Helmholtz free energy of interfacial interaction per unit area is $-W_{12}$, where W_{12} is the "work of adhesion." At the solid–liquid interface W_{SL} is given by:

$$W_{SL} = \gamma_L(1 + \cos\theta) + \pi_e \tag{9}$$

and fortunately on most polymer surfaces with finite contact angles π_e the "spreading pressure" of absorbed liquid vapor on the polymer–vapor interface is zero.[23] As in the spectral studies of intermolecular interaction the work of adhesion has contributions from dispersion forces (d) and acid–base interactions (ab):

$$W_{SL} = W_{SL}^d + W_{SL}^{ab} \tag{10}$$

From earlier dispersion force studies it is already known that

$$W_{SL}^d = 2(\gamma_S^d \gamma_L^d)^{1/2} \tag{11}$$

where γ_S^d and γ_L^d are the dispersion force contributions to Helmholtz surface free energies of solid S and liquid L. By determining the contact angle on polymers with a liquid having only dispersion force interactions and no π_e value, Eqs. (9) and (11) can be solved to determine γ_S^d for the polymer:

$$\gamma_S^d = \frac{\gamma_L^2(1 + \cos\theta)^2}{4\gamma_L^d} \tag{12}$$

We usually use methylene iodide (CH_2I_2) for this purpose because it has a high surface tension and no self-association ($\gamma_L = \gamma_L^d = 50.8$ mJ/m^2 at 20°). From its contact angle on polyethylene (52°) Eq. (12) predicts $\gamma_{PE}^d =$ 33.1 mJ/m^2 at 20°C. Polyethylene can be used in turn with various liquids to determine γ_L^d values. For liquids having both acidic and basic sites, self-association results in γ_L values in excess of γ_L^d, as is shown in Table 5.

TABLE 5
Surface Tensions of Test Liquids (mJ/m^2) at 20°C

Liquid	γ_L	γ_L^d
Water	72.8	22.0
CH_2I_2	50.8	(50.8)
Pyridine	38.0	37.2
Dimethylformamide	37.3	32.4
Dimethyl sulfoxide	43.5	34.9
Tricresylphosphate	40.7	36.2
Tricresylphosphate + 35% phenol	39.6	35.0
Tricresylphosphate + 48% phenol	39.5	34.8
Tricresylphosphate + 72% phenol	39.9	36.1

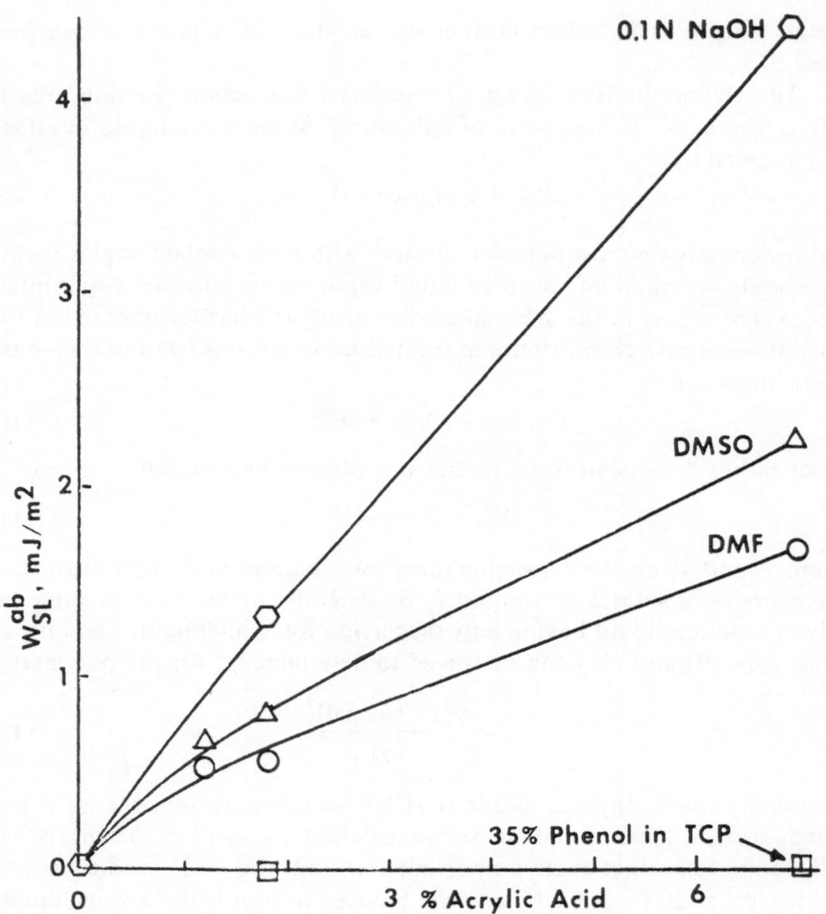

FIGURE 4. The acid–base contribution to the work of adhesion of acidic or basic liquids on copolymers of ethylene and acrylic acid. Reprinted from *Adhesion and Adsorption of Polymers A*, L.-H. Lee, ed., p. 43 (1980) by permission of Plenum press.

The contact angles and surface tensions of test liquids on polymer surfaces provide measured values of W_{SL}, and the γ_S^d values allow calculation of W_{SL}^{ab} (for systems where π_e is zero[(23)]):

$$W_{SL}^{ab} = W_{SL} - W_{SL}^d = \gamma_L (1 + \cos\theta) - 2(\gamma_S^d \gamma_L^d)^{1/2} \tag{13}$$

The results of such measurements are shown in Figure 4 in which W_{SL}^{ab} was determined with various basic liquids on a series of copolymers of ethylene and acrylic acid. With increasing acrylic acid content and with increasing basicity of the test liquid the magnitude of W_{SL}^{ab} is seen to increase. An

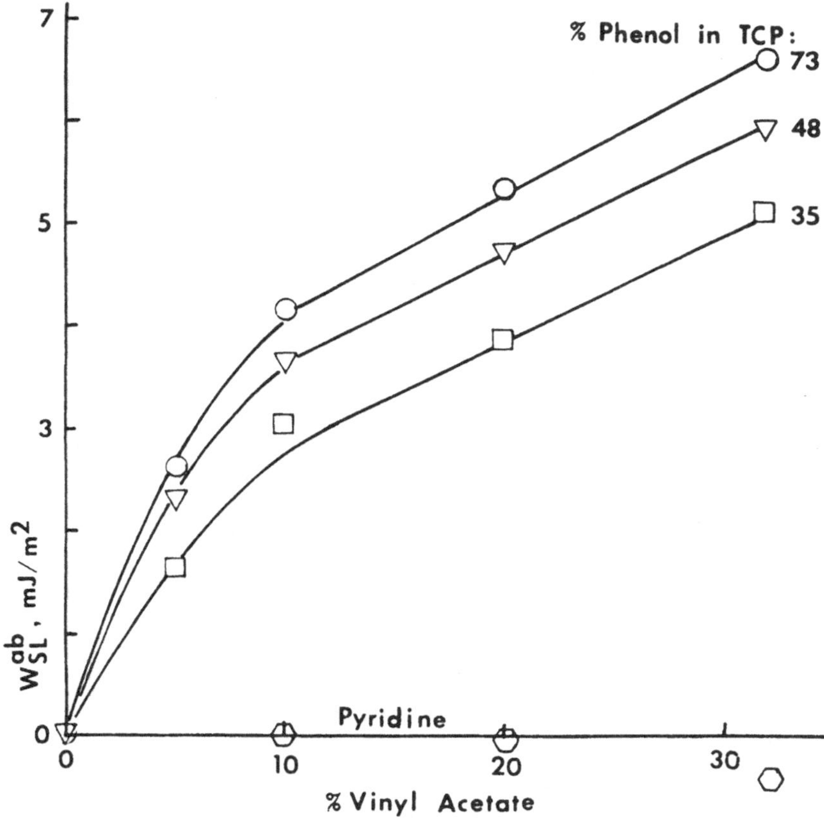

FIGURE 5. The acid-base contribution to the work of adhesion of acidic or basic liquids on ethylene-vinyl acetate copolymers. Reprinted from *Adhesion and Adsorption of Polymers A*, L.-H. Lee, ed., p. 43 (1980) by permission of Plenum Press.

acidic test liquid (35% phenol in tricresylphosphate) is seen to have no measurable value of W_{SL}^{ab}. At this time it is not surprising that there is no specific interaction between phenol and an acrylic acid site, but in earlier times we might have expected "polar" interactions between these dipoles. Figure 5 shows similar results for a copolymer of ethylene and vinyl acetate, where the specific interaction energy per unit area, W_{SL}^{ab}, increases with the surface concentration of acetate groups and with the acidity of the test liquid. A basic test liquid, pyridine, is seen to have no specific interaction with the basic acetate sites. These results show that two polar groups interact only when one is acidic and the other basic, and that in such cases the old adage that "like attracts like" is as wrong for acidic and basic sites as it is in electrostatics or magnetics.

TABLE 6
Acid-Base Surface Characteristics of Polymers[a]

Polymer	γ_s^d	DMSO	72% phenol in TCP
Polyethylene	33.15	0	0
Poly(ethylene, 32% vinyl acetate)	36.9	1.4	6.5
Poly(ethylene, 7% acrylic acid)	33.2	2.2	0
Poly(vinyl chloride)	45.5	4.9	0
Polyimide	45.3	7.0	0
Polycarbonate	45.4	3.5	0
Poly(vinylidene fluoride)	37.8	14.5 +	0

[a] All values in mJ/m^2.

Table 6 is a summary of W_{SL}^{ab} measurements on a series of commercial polymers. Such measurements are relatively easy and this table should be extended. We are a bit limited by the current test liquids, for their surface tensions are not high enough to give finite contact angles on many of the more interesting polymers. We are currently seeking to develop a series of test liquids of higher surface tensions, and are also extending this kind of measurement to inorganic solids.

3. THE SURFACE ACIDITY OR BASICITY OF INORGANIC SURFACES

The surface acidity of the SiOH surface groups on silica has long been recognized, and by the use of basic indicator dyes in benzene the surface acidity has been titrated with amines.[24,25] In the same way one can use acidic indicator dyes (such as Brom Phenol Magenta E, EK 6810) and titrate the surface basicity of glass powders with trifluoracetate acid.[26] The strong basicity of the surface silicate ions in glasses determines many of their surface properties. This is particularly obvious in the adhesion of polymers to glass surfaces. Films of a basic polymer such as poly(methyl methacrylate) (PMMA) when cast from CH_2Cl_2 onto ordinary glass are easily peeled off. However, films of an acidic polymer such as chlorinated poly(vinyl chloride) (CPVC) when cast from CH_2Cl_2 onto ordinary glass adhere so strongly that they can't be peeled off, even under water. On rinsing the glass with dilute HCl the surface silicates become silanols and now the PMMA cannot be peeled off and the CPVC peels easily.

Metal oxides vary appreciably in surface acidity and basicity. In general the oxides of the lower valance metals (alkali and alkaline earth metals) are basic and those of higher valence are acidic.[27] The surface of metal powders can be characterized by the adsorptivity of acidic or basic solutes from neutral organic solvents, and by using ellipsometry[28] or contact potential differences[29] one can do the same kind of study on flat metal surfaces. In these studies there is abundant evidence that aluminum surfaces have only weakly basic sites, that chromium surfaces have only acidic sites, and that iron surfaces have both kinds with the acidic sites predominant.

3.1. ADSORPTION OF POLYMERS ON INORGANIC SURFACES

Our initial studies of the forces which provide adhesion of polymers to inorganic surfaces were carried out by adsorption studies with carefully chosen acidic or basic polymers, acidic or basic inorganic solids, and acidic or basic solvents from Drago's tables.[30] The acidic polymer was a B. F. Goodrich post-chlorinated poly(vinyl chloride) (CPVC) and the basic polymer was polymethylmethacrylate (PMMA); the acidity and basicity of both have been verified by infrared peak shift ($\Delta\nu_{c=O}$) studies. The acidic solid was a high-surface area silica gel prepared by silane pyrolysis, and the basic solid was a precipitated calcium carbonate.

In the case of adsorption of PMMA onto silica, the cause of adsorption is the bonding of the basic methacrylate carbonyl groups to the acidic silanol surface groups of silica, proven many years ago by Fontana and Thomas, who measured $\Delta\nu_{C=O}$ with silica mulls in hydrocarbon solutions of poly-alkylmethacrylates.[31] They found that about 40% of the carbonyl absorption peaks were shifted and concluded that these polymers adsorb with very little looping from the surface into the solution. We found that PMMA adsorbed onto silica most strongly from carbon tetrachloride, our most neutral solvent (Figure 6). The amount adsorbed, 12×10^{-4} g/m^2 is the equivalent of about two geometric monolayers. However, only 2×10^{-5} g/m^2 of PMMA (4% of a monolayer) adsorbed from CCl_4 onto calcium carbonate, showing that basic polar polymers don't adsorb onto basic polar substrates from nonpolar solvents.

In the case of the acidic polymer CPVC no adsorption at all occurred on the acidic silica from any solvent, but strong adsorption of about one geometric monolayer (4×10^{-4} g/m^2) was observed to occur on calcium carbonate from CCl_4 and CH_2Cl_2 solutions.

The acidity or basicity of the solvent was found to have a profound influence on adsorption (Figure 6), for strong adsorption was observed only from the more neutral solvents. In the case of PMMA adsorption on silica, the more basic solvents (as measured by their ΔH^{ab} with butanol) preferentially solvated the acidic surface silanol groups of silica so that the less basic methacrylate groups could hardly compete. Similarly in more acidic solvents (as measured by their ΔH^{ab} with ethyl acetate) the basic methacrylate carbonyl sites of PMMA tended to bind to the acidic solvent instead of binding to the less acidic silanols of the silica particles.

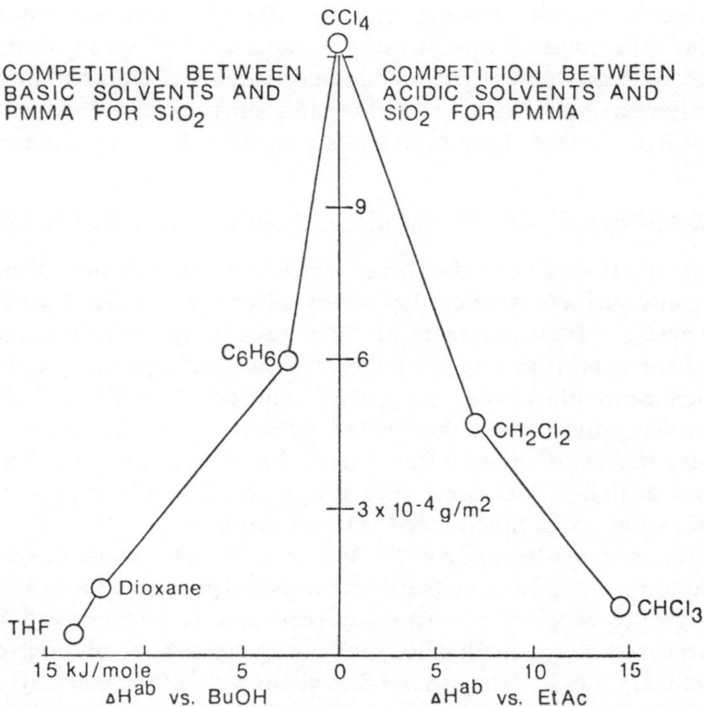

FIGURE 6. Effects of acidity or basicity of solvents on adsorption of poly(methyl methacrylate) (PMMA) on silica gel. Reprinted from *Ind. Eng. Chem., Prod. Res. Develop.* **17**, 3 (1978) by permission of the American Chemical Society.

Quite similar competition by the more acidic or basic solvents occurred in the adsorption of CPVC onto calcium carbonate. These findings appear to be very general, demonstrating the dominant role of acid-base interactions in the adsorption of polar solutes onto polar substrates from organic solvents.

The above studies provided a means of determining the acidity or basicity of unknown inorganic powders. Manson and Straume adsorbed CPVC from CH_2Cl_2 or PMMA from CCl_4 onto four candidate pigments with results as shown in Table 7. It can be seen that the first three were predominantly acidic and the fourth was predominantly basic, which was found to be in perfect agreement with dispersibility results.

3.2. HEATS OF ADSORPTION OF ORGANIC ACIDS AND BASES ON INORGANIC POWDERS

Currently we are studying the calorimetric heats of adsorption from neutral hydrocarbon solvents of acidic and basic solutes (from Drago's tables) onto powders of iron oxide, silica, titania, and glass. The heats of

TABLE 7
Ratio of Acidic to Basic Sites in Pigments

Pigment	1	2	3	4
PMMA adsorbed (mg/g)	30	97	35	6
CPVC adsorbed (mg/g)	0.09	1.7	8.5	61
%PMMA adsorbed	99.7	98	80	9
%CPVC adsorbed	0.3	2	20	91

adsorption are found to be independent of the dispersion force character (γ^d) of the solvent and are therefore believed to be equal to the ΔH^{ab} of acid-base interaction of the Drago equation. Thus we should be able to determine the C and E constants for acidic or basic sites on these surfaces. We are finding that the acidic sites on iron oxides have a distribution of acid strengths, so we expect to eventually find a distribution of C and E constants. We are using flow microcalorimetry coupled with an LC ultraviolet spectrometer and determine heats of adsorption per mole of acid or base at various degrees of coverage. Our initial estimate for the acid sites on iron, determined by the temperature coefficients of adsorption isotherms for triethylamine and pyridine on iron oxides,[32] is $E_A = 2.0$ and $C_A = 1.0$ $(\text{kcal/mol})^{1/2}$, showing iron oxide to be a weak acid, but nearly as soft as iodine. The softness, expressed as the C/E ratio,[14] indicates stronger bonding to sulfur bases than oxygen bases. The microcalorimetry results to date indicate a distribution of acid strengths rather than a single acid strength, however.

Infrared spectroscopy with modern FTIR instrumentation may become the best method for determining ΔH^{ab} for adsorption on inorganic solids. Initial tests with ethyl acetate adsorption on Cab-O-sil silica show the two populations of acid sites known to exist on this material. By adding a photoacoustic accessory to the FTIR, opaque materials such as iron oxides and carbon blacks can also be investigated.

The acidic or basic properties of metals, ceramics, or plastics used for medical purposes in the body may be of considerable importance in determining interactions with surrounding tissues or fluids. However, the chemistry of such surfaces has had remarkably little study to date. We are now on the threshold of knowing how to go about studying this aspect of biomaterials.

4. CHARGE TRANSFER BETWEEN PHASES

The generation of electrostatic charges at interfaces, often called "static electricity" or "triboelectricity," is a spontaneous contact phenomenon in which electrons or ions are transferred from one phase to another upon contact. The process is only partly understood at this time, for in some cases electrons are donated by basic materials (electron donors) to acidic materials (electron acceptors) and in other cases protons are donated by acidic materials to basic materials. However, the transfer of electrons into solid or liquid hydrocarbons is more of a puzzle.

4.1. CHARGE TRANSFER BETWEEN INORGANIC SOLIDS AND ORGANIC LIQUIDS

Finely divided dispersions of inorganic solids in organic liquids of low dielectric constant are often electrostatically charged and exhibit electrophoretic mobility when subjected to an electric field. The charged particles are surrounded by counterions in the surrounding organic liquid at a characteristic distance $1/\kappa$ (the Debye length). If the Debye length is greater than the radius (a) of the spherical particles, the charge per particle (q, in coulombs) can be estimated from the electrophoretic mobility (μ, in meters $\sec^{-1}$/volt meter^{-1}):

$$q = 6\pi a\eta\mu \tag{14}$$

where η is the viscosity. The zeta-potential (ζ, in volts) is therefore

$$\zeta = \frac{q}{a4\pi\varepsilon} = 1.5\eta\mu/\varepsilon \tag{15}$$

where ε is the permittivity of the liquid, equal to $\varepsilon_r\varepsilon_0$ where ε_r is the dielectric constant of the liquid and ε_0 is the permittivity of free space (8.854×10^{-12} coulomb V^{-1} m^{-1}). When the ratio of a to $1/\kappa$ is much less than unity:[(24)]

$$\zeta = \eta\mu/\varepsilon \tag{16}$$

Basic oil-soluble polymers or basic micellar sulfonates are found to give high negative zeta-potentials (−150 mV) to acidic carbon blacks in hydrocarbons,[(33)] while acidic oil-soluble polymers give positive zeta-potentials. The mechanism involves dynamic adsorption and desorption of the dispersing polymer, as illustrated in Figure 7. When the basic polymer adsorbs on the acidic surface, protons tend to transfer to the polymer, and then when the further adsorption of fresh polymer induces the charged polymer to desorb, it carries protons from the acid sites off into solution, leaving the particle positively charged. Experiments to test this mechanism

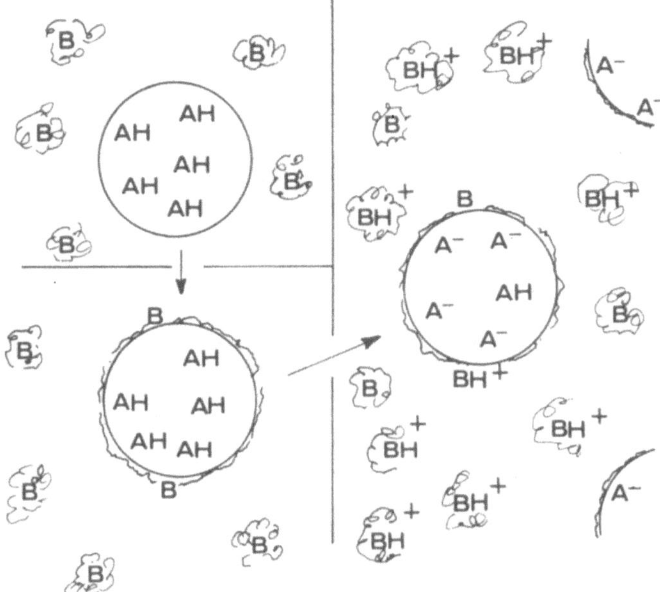

FIGURE 7. Mechanism of charging of acidic carbon blacks during adsorption and desorption of basic polymers in hydrocarbon media. Reprinted from ACS Symp. Ser. **200**, 307 (1982) by permission of the American Chemical Society.

involved the use of C-14 tagged polymers and tritium-tagged acidic sites on the carbon particles.[34,35] It was found that desorption times for adsorbed polymeric dispersant molecules were no more than ten minutes and that, in electrodeposition experiments, the carbon coated out on the anode, while the basic polymer with protons from the particle surface coated out on the clean cathode.

Similar studies were made by measuring zeta-potentials with a variety of inorganic powders of known acidity or basicity dispersed in acidic or basic organic solvents (from Drago's tables) in which polymers of known acidity or basicity were present over a range of concentrations.[35] Figure 8 illustrates the large positive zeta-potentials attained with the basic (electron-donating) solids calcium carbonate and calcium oxide dispersed in dioxane solutions of the acidic (electron-accepting) chlorinated poly(vinyl chloride) CPVC. Similarly, Figure 8 also illustrates the increasingly negative zeta-potential of acid-washed kaolin particles dispersed in dichloromethane with increasing concentrations of the basic Lexan polycarbonate.

These examples of charge transfer between organic liquids and inorganic solids are helpful to the understanding of some of the charge transfer phenomena occurring between solids.

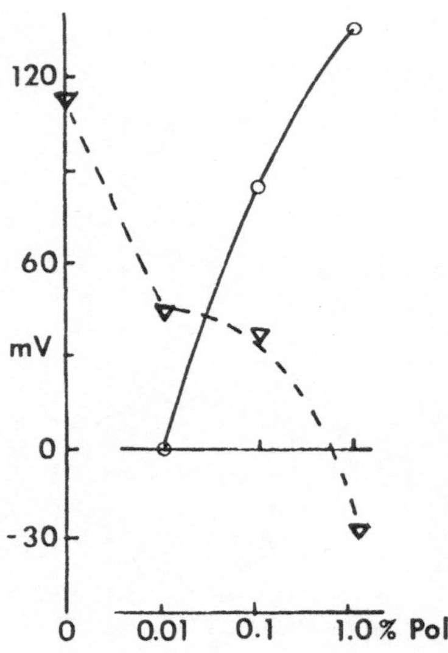

FIGURE 8. The effect of increasing concentration of interacting polymers on zeta-potential of dispersed inorganic powders in organic liquids. Solid line is for $CaCO_3$ in dioxane ($\varepsilon_r = 2.2$) with CPVC; dashed line is for acid-washed kaolin in dichloromethane ($\varepsilon_r = 9.1$) with Lexan polycarbonate.

4.2. CHARGE INJECTION INTO ORGANIC POLYMERS

The effects of charge injection into polymers are encountered daily in fabrics, film, fibers, and toners for copiers where "static charges" are an unwanted hazard or the basis for their usefulness.

A pioneering study of the role of charge injection in adhesion of polymers to metals was published by Skinner, Savage, and Rutzler, in 1953.[36] In this work the density of charge-trapping sites per unit volume in a polymer was shown to depend on the work function (ϕ, in electron volts) of the metal and on properties of the polymer. The electric discharge upon debonding of polymer–metal interfaces was measured and it was concluded that the electrostatic contribution to polymer–metal adhesion could be very appreciable in certain cases.

Several years later Deryagin and coworkers developed an improved method for measuring the interfacial density Q_s of charge injected by a substrate into a polymer, determined from measurements of the velocity of electrons ejected from the polymer when peeled in a vacuum.[37] In this work Deryagin showed that the familiar blue light emitted from the peeling zone when a polymer is peeled from a substrate is due to charge injection from the substrate into the polymer and that the act of peeling transforms mechanical energy into high-voltage electrons which are emitted across the air gap when the electric field reaches the breakdown level. In a vacuum the velocity of the emitted electrons was measured by electrostatic bending of a collimated beam and, assuming the kinetic energy of emitted electrons approximated their energy in the film at the time of emission, surface charge densities of 3×10^{-3} to 3×10^{-2} coulombs/m^2 were determined. These values of Q_s can be related to the electrostatic component of the tensile strength of the adhesive joint (F^e) by

$$F^e = Q_s^2/2\varepsilon_0, \tag{17}$$

giving values of F^e from 5 to 500 atmospheres. In these studies steel, glass, and rubber were found to inject electrons into polyethylene, poly(vinyl chloride), poly(vinyl butyral), cellulose nitrate, and a polyamide.

Many studies of charge injection in polymers have sought to determine a "triboelectric series" of polymers, but none of the published series are in agreement with each other.[38]

Much better agreement has been obtained in measurements of the density of injected charge in polymers after contacting metals, for the metals act as reservoirs of electrons of fixed free energy (at the Fermi level). Each metal has its own peculiar Fermi level (E'_F) which is related to the free energy of electrons in vacuum (E_V) by the work function $\phi' = E_V - E'_F$. Davies used direct surface potential (V_s) measurements of polymers in a vacuum chamber immediately after contacting metals, and showed that the injected charge Q_s was linearly dependent on work function ϕ, with positive values for metals of higher work function and negative values for metals of lower work function (Figure 9).[39] Davies referred to the work function of metals injecting zero charge into a given polymer as the "work function of the polymer," as is illustrated in Table 8. Similar measurements were published by von Harrach and Chapman in 1972 for Corning 7059 borosili-

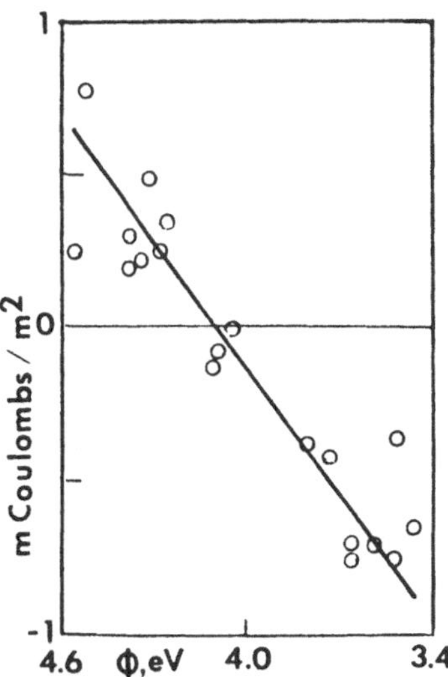

FIGURE 9. Charge injection into Nylon 66 from metals of a range of work functions (ϕ). Reprinted from D. K. Davies, *Brit. J. Appl. Phys.* **2**, 1533 (1969) by permission of the copyright owners.

TABLE 8
The "Work Functions" of Polymers[a]

Polymer	Work function of metal for zero charge injection (eV)
Poly(vinyl chloride)	4.85 ± 0.2
Polyimide	4.36 ± 0.06
Polycarbonate	4.26 ± 0.13
Polytetrafluororethylene	4.26 ± 0.05
Polyethyleneterephthalate	4.25 ± 0.10
Polystyrene	4.22 ± 0.07
Nylon 66	4.08 ± 0.06

[a] Reference (35).

cate glass,[(40)] for which V_s was zero with $\phi = 5.2$ eV. One can observe in Table 8 that the more acidic polymers are at the top of the list, and that they usually, but not always, act as electron acceptors. Similarly the more basic polymers are at the bottom of the list, and these usually, but not always, act as electron donors.

Studies of charge-trapping sites in polyethylene by Weaver[(41)] were made by conduction in thin films. These showed that the charge carriers were always positive and that these hopped from trap to trap by the Frenkel–Poole mechanism. Space charge-limited current characteristics showed a total trap density of $4 \times 10^{24}\ m^{-3}$, corresponding to about 100 Å between trapping sites.

More recent studies of trapping sites in polymers include an important series of papers by Fabish and Duke and coworkers at Xerox.[(42-48)] Figures 10 and 11 show their densities of electron and hole trapping sites (described as anion and cation states) in three polymers: polystyrene, poly(methyl methacrylate), and poly(vinylpyridine). Each polymer is shown to have a distribution of energy levels for the anion and for the cation states, determined experimentally by charge injection from a series of metals (shown in hatched rectangular bars). It was concluded in this study that the electrons at the Fermi level of a metal are in communication only with the trapping sites where electrons have the same energy (E_F) or down to 0.2 eV below E_F. In the series of diagrams of Figures 10 and 11 we see that nearly all of the test metals injected electrons into polystyrene, half of them injected electrons and half injected holes into PMMA, while nearly all metals injected holes into poly(vinylpyridine). The centroid of trap energies $\langle E \rangle$ lies between the energy levels for hole and electron traps and corresponds to the negative of the work function of the metals which inject no charge, the subject of Davies' work (Table 8). Poly(vinylpyridine) (the most basic polymer) has an $\langle E \rangle$ of -3.4 eV, PMMA has an $\langle E \rangle$ of -4.1 eV, and polystyrene (the least

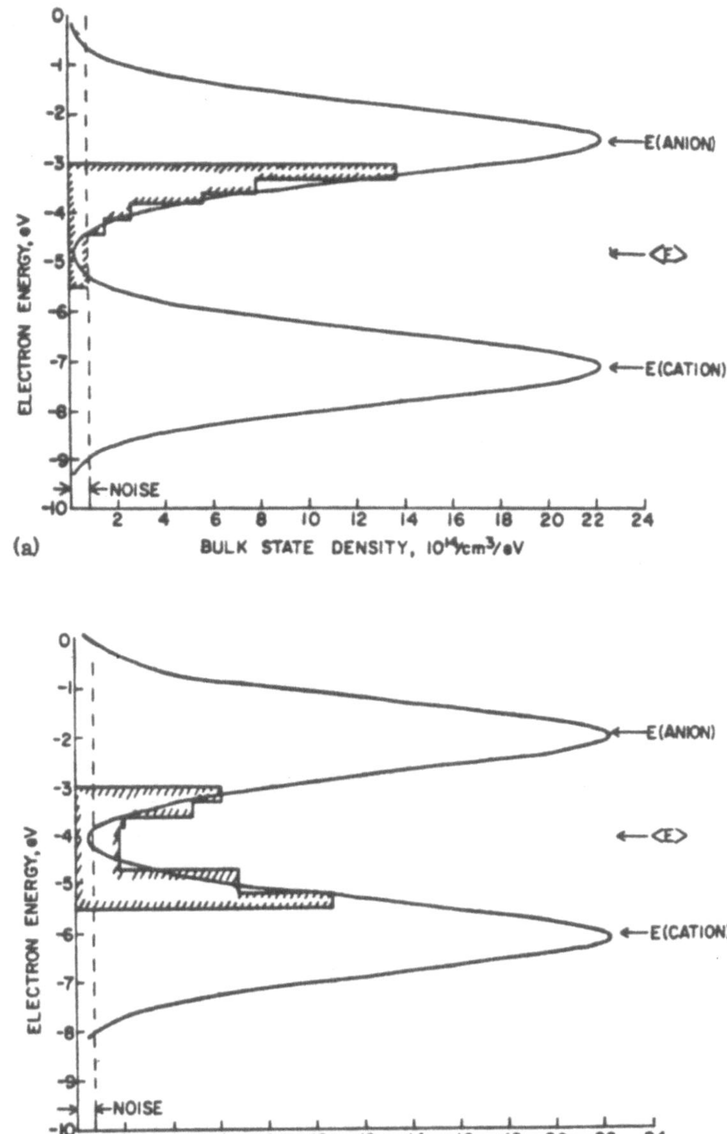

FIGURE 10. Gaussian representations of solid-state anion and cation states in polystyrene (top) and poly(methyl methacrylate) (bottom). Reprinted from T. J. Fabish and C. B. Duke, *J. Appl. Phys.* **48**, 4262 (1977) by permission of the copyright owner.

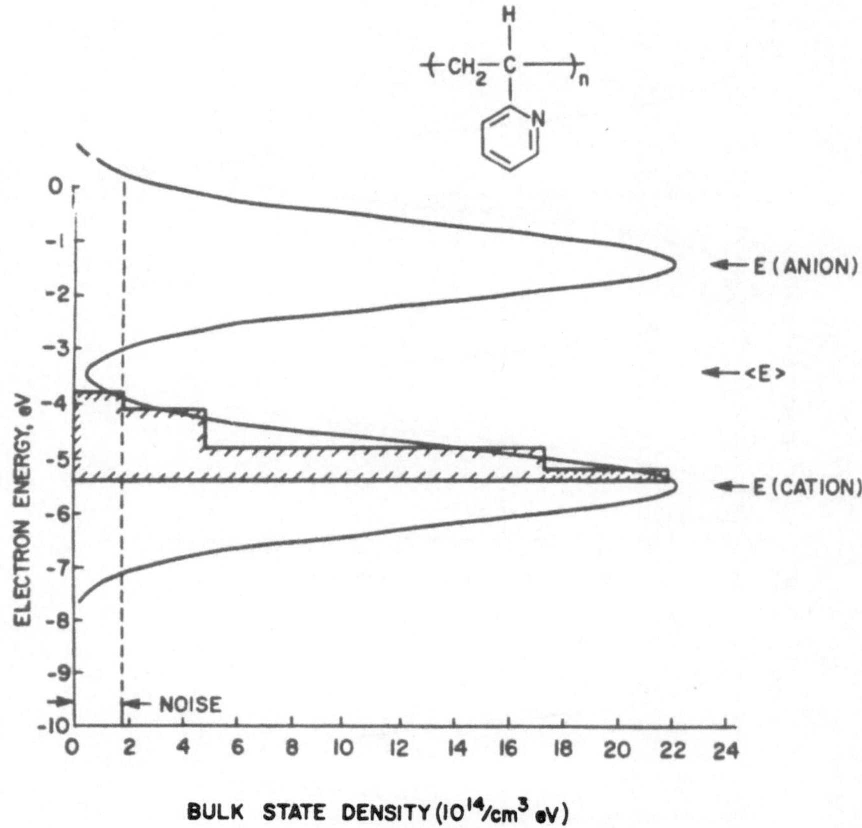

FIGURE 11. Measured (hatched) and fitted densities of molecular-ion states for poly(2-vinylpyridine). Reprinted from C. B. Duke *et al.*, *Phys. Rev.* **B18**, 5722 (1978) by permission of the copyright owner.

basic) has an $\langle E \rangle$ of -4.8 eV. An important part of this study was the successful prediction of the amount of electron injection by PMMA into polystyrene, based on the degree of overlap of the cation states of PMMA by the anion states of polystyrene.

Cottrell, Lowell, and Rose-Innes[49] have questioned the central postulate of the Duke and Fabish study, namely, that the electron states in the polymer which are in communication with a given metal have electron energies within 0.2 eV of the Fermi level. They suggest that the different metals contacted a different fraction of the surface and that is why a surface saturated by charge from one metal can pick up more charge from a second metal. A further exchange of comments followed.[50,51]

4.3. SEMICONDUCTOR DEVICES FOR SENSING CHARGE INJECTION INTO POLYMERS

Two research groups have independently developed laboratory techniques for sensing charge injection into polymers by field effect capacitors involving single crystal silicon as the grounded electrode, a thin polymer film as dielectric, and a metal (gold, chromium, or aluminum) as the biased electrode.[52–54] Figure 12 shows the capacitance–voltage (CV) diagram for this device made with n-type silicon. The dashed line represents a perfect dielectric with no injected charge and the solid line is the CV plot observed when the silicon injects positive charge into the polymer. The high capacitance observed when the metal is positive (C_0) results from the silicon becoming strongly n-type and highly conductive so that the dielectric film includes only the polymer. However, when the metal is negative the conduction electrons are driven deeper into the silicon, and the conductor-free region of the silicon also becomes part of the dielectric; the thicker dielectric layer results in lower capacitance. In the charge-free dielectric (dashed line) the capacitance at zero applied potential has an intermediate value calculable from the electron concentration in the silicon; this capacitance is called the "flat-band capacitance, C_{fb}." When positive charges are injected into the polymer from the silicon, the adjacent silicon is subject to a positive potential which enhances the electron concentration. In order to reduce the capacitance to C_{fb} the metal must now have a negative potential, as shown in Figure 12. If negative charges are introduced C_{fb} is attained when the metal has a positive potential.

Bui, Carchano, and Sanchez at Toulouse (France) made silicon–polymer–gold capacitors with thin films of plasma-polymerized styrene, and their CV plots (as in Figure 13) showed that positive charges are easily injected from the grounded silicon when the gold is at a negative potential, but that electrons are injected with greater difficulty (when the gold is at high positive potentials).[53,54] If charges are injected into the polymer, hysteresis loops result, as shown in Figure 13. For n-type silicon the hysteresis loops for charge injection from silicon are clockwise, whereas

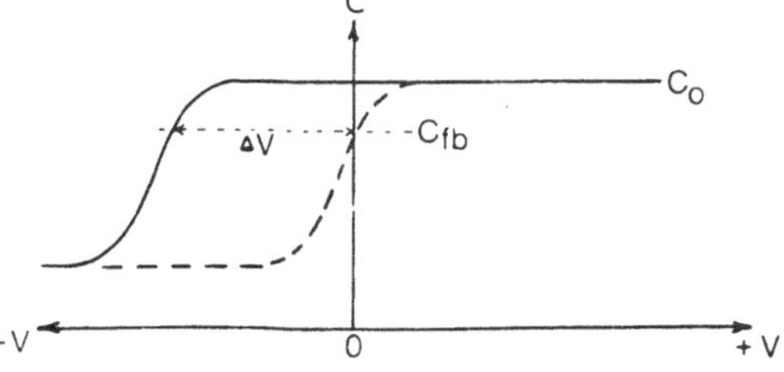

FIGURE 12. Capacitance vs. bias voltage measurements (CV) of n-type silicon–polymer metal capacitor. Dashed line is for no trapped charge in the polymer, solid line is for positive charge trapped in the polymer next to the silicon.

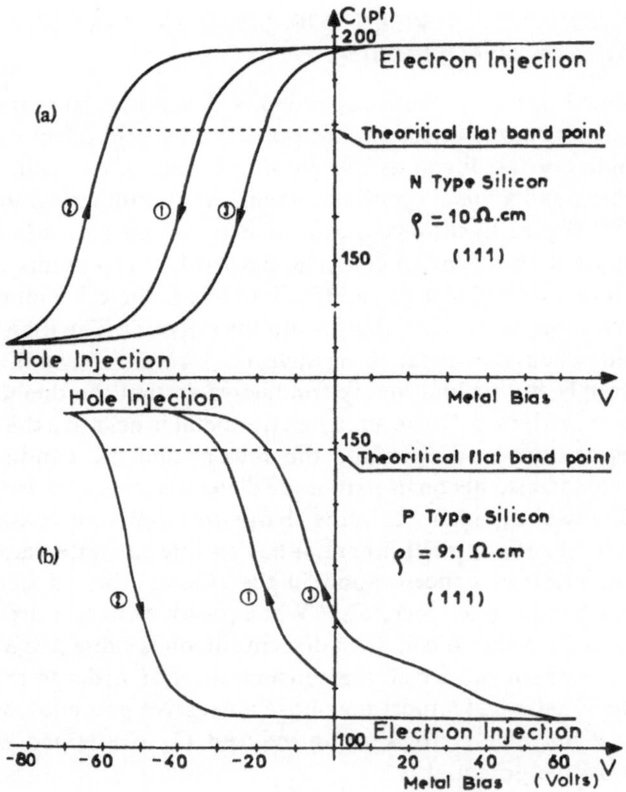

FIGURE 13. Demonstration by capacitance–voltage method that positive charges are injected into plasma-polymerized styrene from silicon. Reprinted from Bui, Carachano, and Sanchez, *J. Appl. Phys.* **43**, 3795 (1972) by permission of the copyright owner.

polarization of dipoles in the polymer (an earlier hypothesis) would give counterclockwise hysteresis loops.

At Lehigh University Hielscher made similar capacitors with solvent-cast films of a variety of polymers,[55] as shown in Table 9. With the exception of cellulose tributyrate, none of the polymers contained a net charge when no potential was applied to the metal electrode. Cellulose tributyrate showed a consistent initial positive built-in charge of 2×10^{12} electron charges per cm^2. Upon the application of a negative bias to the metal electrode, positive charges were injected into all polymers from the silicon interface. Charge accumulation was found to be a linear function of applied bias, approximately 10^{11} electron charges per cm^2 per volt, corresponding to $4 \times 10^{12}/cm^2$ near the breakdown voltage of 40 volts. The time for the charge to reach its final value upon application of bias was approximately five seconds. In

TABLE 9
Polymers Used in Hielscher's Study of Charge Injection and the Solutions from Which the Films Were Formed

Polymer/solution	Film thickness (Å)[a]
Group I (positive charges injected)	
Polystyrene—5% in xylene	1000
Polychlorostyrene—2% in benzene	1300
Poly(vinyl chloride)—2% in tetrahydrofuran	1200
Poly(vinyl acetate)—2% benzene	1000
Polycarbonate—1% in methylene chloride	1100
Cellulose tributyrate—2% in trichloroethylene	1200
Group II (positive and negative charges injected)	
Poly(*n*-octadecyl vinyl ether/maleic anhydride)—4% in tetrahydrofuran	1200
Chlorinated polyethylene—4% in benzene	1200
Chlorinated polypropylene—4% in benzene	1200
Poly(ethyl methacrylate)—2% in benzene	1200

[a] The film thickness was obtained by spinning at 13,000 rpm.

cellulose tributyrate the negative bias caused the positive charge to increase beyond the built-in charge of $2 \times 10^{12}/\text{cm}^2$, but the total charge was no greater than in the other polymers.

Application of a positive bias caused electron injection into only four of the ten polymer films which were studied (Group II of Table 9). In these four polymers the concentration of stored injected electrons was about equal to the concentration of stored positive charges, except for the copolymer with maleic anhydride where the negative charge storage was approximately 50% less than the positive charge storage.

The data on chlorinated polypropylene is representative of the extent of stored charge in the polymer films, and this is shown in Figure 14 as a function of bias of the metal electrode. For the six polymers of Group I, those which store only positive charge, the charge storage for positive metal bias is zero.

In a few cases, films consisting of two polymer layers were examined, where the thicker outer layer was a 1000-Å film of polystyrene formed from a 3% cyclohexane solution, and the thin (50 to 100 Å) inner layer was a different polymer. Polystyrene and other Group I polymers can store only positive charge near the silicon interface, and in fact when the inner layer was one of these polymers, only positive charge storage was observed. However, when the inner layer was a Group II polymer, such as chlorinated polypropylene, both positive and negative charge storage was obtained.

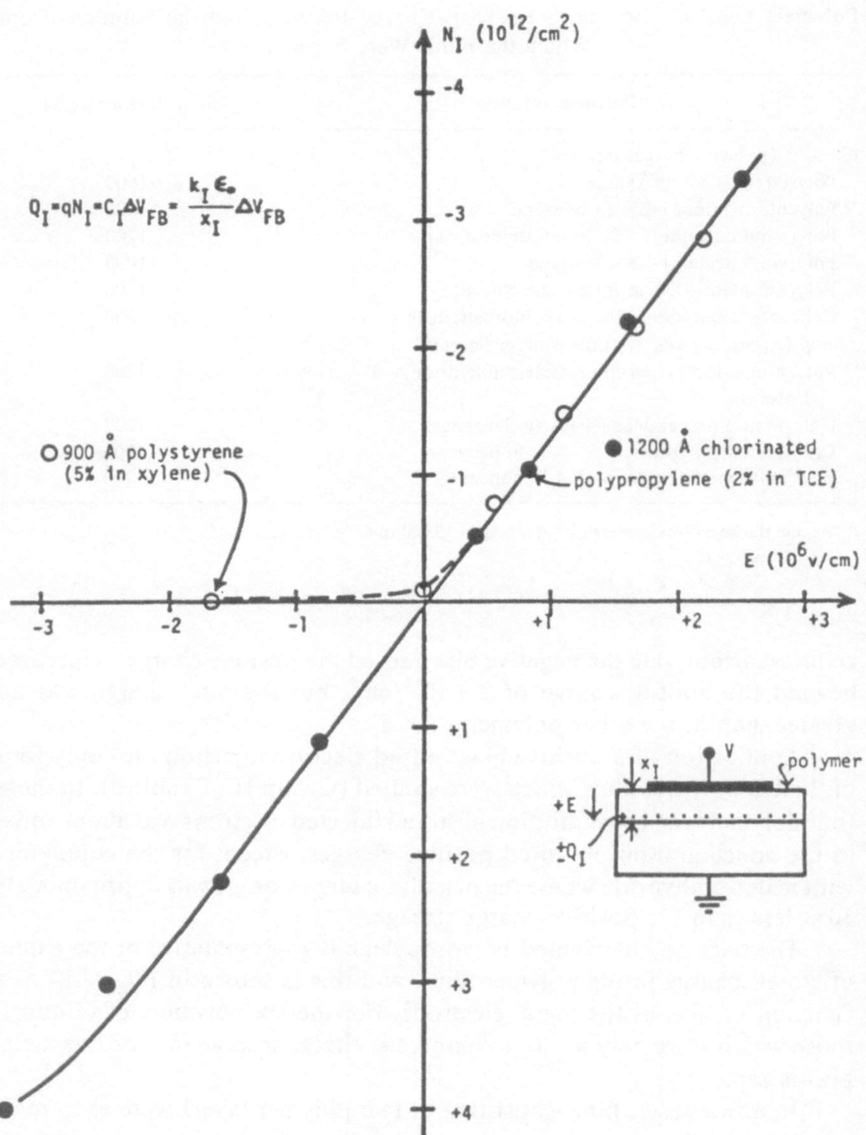

FIGURE 14. Charge injection into chlorinated polypropylene (a Type II polymer) from silicon as function of applied metal bias in a silicon-polymer-metal capacitor.

Thus, even though the inner layer was very thin, the charge storage was determined entirely, both in sign and magnitude, as if the whole composite polymer consisted of only the inner layer. These experiments suggest that the observed charge storage occurs within 100 Å of the silicon interface, and that the trap concentrations in these films are extremely high (greater than $10^{19}/cm^3$).

In the above experiments silicon was found to inject electrons or positive charges into the polymers of Group II, but only positive charges into the polymers of Group I. In subsequent joint experiments with H. R. Anderson of IBM (San Jose) the polymer was plasma-deposited tetrafluoroethylene and this film stored only injected electrons,[56] as illustrated in Figure 15. The electrons would be injected from the conduction band (with energy E_c) and the holes from the valence band of silicon (with energy E_v). If the centroid of trap energies $\langle E \rangle$ of the polymers fell between E_c and E_v for silicon one would expect both electron and hole injection, but if $\langle E \rangle$ were above E_c (as expected for the stronger electron-donor polymers) then only positive charges would be injected, and if $\langle E \rangle$ were below E_v (as expected for the stronger electron-acceptor polymers) then only electrons would be injected. Thus the Fabish and Duke model seems to fit these measurements, but our shallow depth of injection does not match their findings.

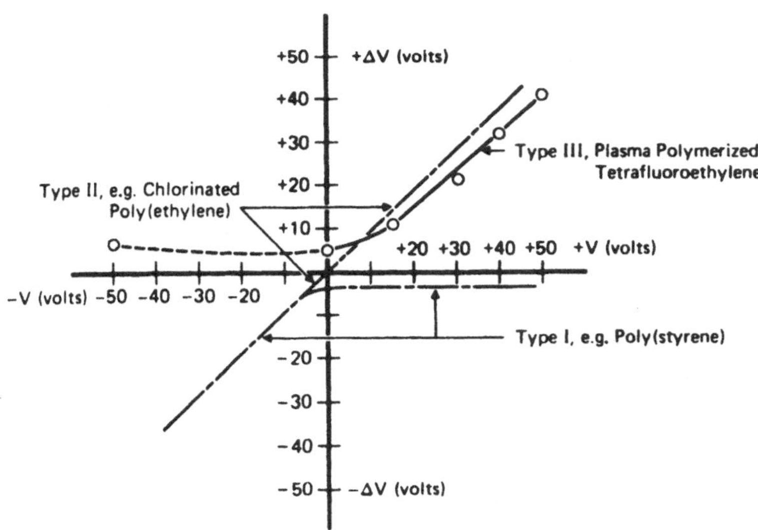

FIGURE 15. Charge injection into plasma-polymerized tetrafluoroethylene (a Type III polymer) versus metal bias in a silicon-polymer-metal capacitor. Reprinted from H. R. Anderson *et al.*, *J. Polym. Sci.*, *Phys.* **14**, 893 (1976) by permission of the copyright owner.

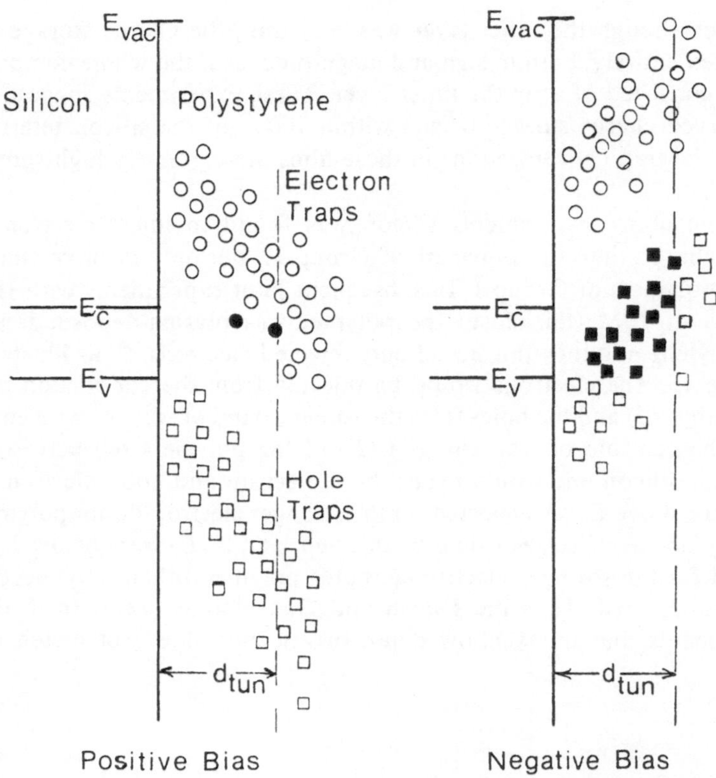

FIGURE 16. Energy band diagram of silicon–polymer–metal capacitor at +30V metal bias on a 1000 Å polymer film.

Figure 16 is a "band diagram" to illustrate the nature of the effect of applied field on the filling of traps in the polymer, showing that the applied field very appreciably changes the energy levels of the trapping sites in polymers with respect to the energy levels of electrons or holes in the silicon, bringing trapping sites within tunneling distance of the silicon.

Another type of semiconductor device for sensing charge injection in polymers, a floating gate field effect transistor, was devised by Fowkes and Hielscher to sense injection of charge into the surface of a polymer.[55,57] This structure, shown in Figure 17, is primarily a thin surface *p*-type channel over *n*-type silicon. The conductance through this surface channel is very sensitive to electric fields above it, increasing with negative potential and decreasing with positive potential. The *p*-type surface channel is made by epitaxial deposition of boron-doped silicon over *n*-type silicon (phosphorus-doped), and the electrical connections to it are made by indiffusing heavily

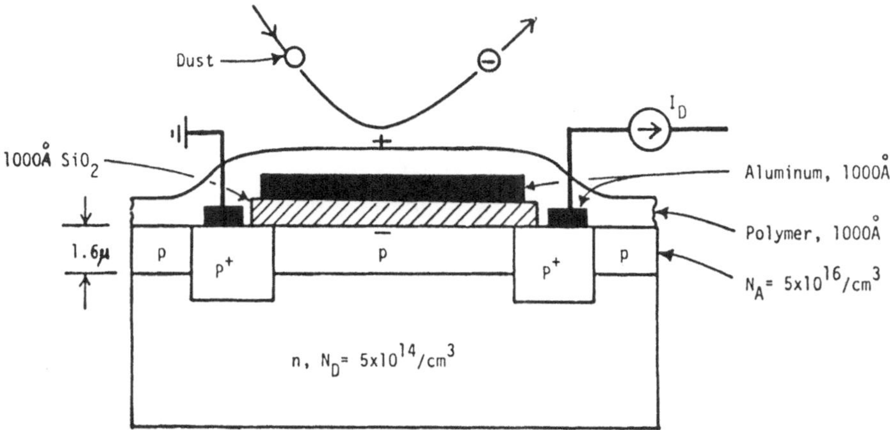

FIGURE 17. Floating gate field effect transistor (FET) for sensing charge injection in polymers.

boron-doped (p^+) "source" and "drain" electrode regions. Over the *p*-channel is grown a 500-Å silicon dioxide insulator and over this a 1000-Å chromium metal "gate" is deposited. Lastly a 1000-Å film of polymer is deposited by solvent casting or plasma deposition. Measurements are made by contacts made to the source and drain regions using a semiconductor device prober with electrolytically pointed contact probes and the current from source to drain was monitored with a Tektronix curve-tracer. Figure 18 shows the dependence of conductivity on charges injected into the polymer. For each experiment a probe is used to contact the gate metal and set its potential at a convenient level. The probe is withdrawn and the surface of the polymer is contacted with another material while the channel conductivity is measured.

The first experiments with the floating gate field effect transistor were done with polystyrene contacted with silica gel particles, blown so as to hit and bounce away. The *p*-channel conductance jumped to higher values in less than a second after the silica hit, showing that positive charges were injected instantly upon contact and that the silica particles carried off electrons from the polystyrene. Similar experiments were done with nine other polymers and with silica, carbon black, and coal particles. These results are shown in Table 10. It is known that all of these powders have surface acidity[58] but the coal and carbon black can also have surface basicity. It was expected that these particles would inject positive charges into most polymers, but that the coal or carbon black might inject electrons into acidic polymers. It can be seen that in 29 of 30 experiments positive charges were injected, but that coal powder injected electrons into a copolymer of *n*-octadecyl vinyl ether and maleic anhydride. It had been a goal of

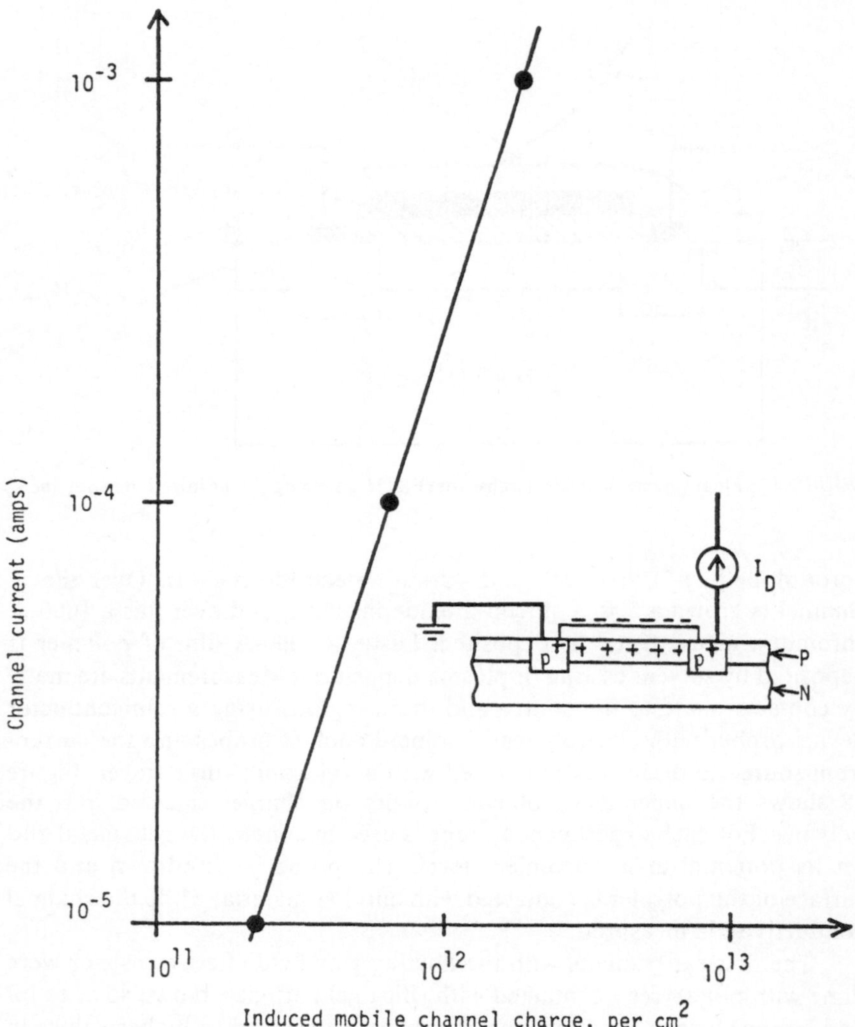

FIGURE 18. Transconductance of floating gate FET detector vs. injected charge density in polymer dielectric.

this study to develop a filter to selectively trap silica particles out of coal dust, and the findings suggest this is quite possible.

Another use for the device of Figure 17 is to detect charged injected by aqueous solutions into polymers. Figure 19 shows the results with polystyrene and with chlorinated polypropylene when contacted with an aqueous solution of a tin hydrosol at pH 2. It is seen that electrons are injected first and then deposition of the positive hydrosol neutralizes the

TABLE 10
Charge Injected into Polymer Films by Silica, Carbon Black, and Coal Particles[a]

Polymer	SiO_2	Carbon	Coal
Polystyrene	B	A	A
Polychlorostyrene	C	A	A
Poly(vinyl chloride)	C	B	B
Poly(vinyl acetate)	A	A	B
Polycarbonate	B	A	A
Cellulose tributyrate	C	A	C
Copolymer of maleic anhydride	C	A	*
Chlorinated polyethylene	A	A	A
Chlorinated polypropylene	B	B	B
Poly(ethyl methacrylate)	B	A	A

[a] A: Readily induces a positive charge greater than $10^{12}/cm^2$; B: readily induces a positive charge greater than $10^{11}/cm^2$; C: slowly induces a positive charge of $10^{11}/cm^2$; * readily induces a negative charge greater than $10^{11}/cm^2$.

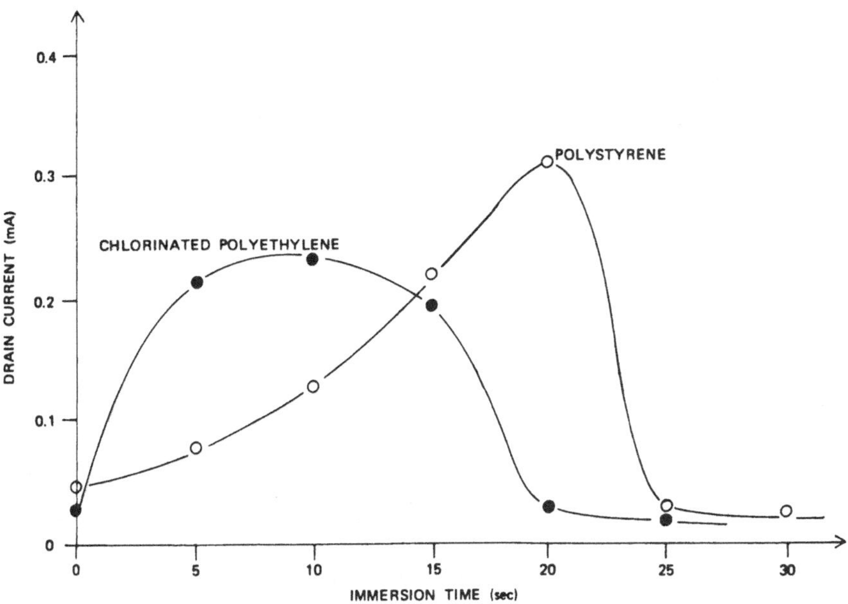

FIGURE 19. Surface density of charge injected into polymers from aqueous tin hydrosol with device of Figure 17.

polymer surface.[53] The injection of electrons into polystyrene by water is not surprising, for all hydrocarbons (solid or liquid) in contact with water are known to develop large negative zeta-potentials.[57] Figure 20 shows how the surface charge density of electrons injected into polystyrene, measured with the device of Figure 17, can establish the zeta-potentials of −50 to −60 mV which are observed with polystyrene latices specially prepared to be completely free of surface ionic groups.[57] The relationship of zeta-potential to surface charge density Q_s shown in Figure 20 was established long ago by Verwey and Overbeek.[60]

In conclusion it is hoped that the reader will realize that charge injection into polymers is an important everyday phenomenon which has been very much neglected in research, but that modern semiconductor sensing devices

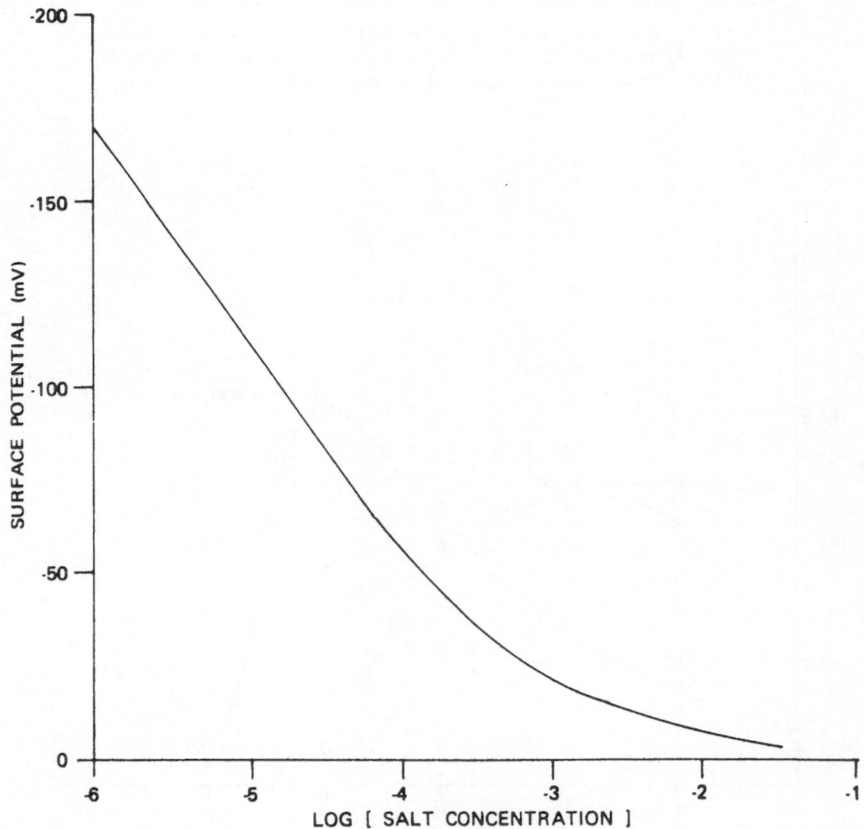

FIGURE 20. Zeta-potential of polystyrene immersed in aqueous solution which injected 10^{12} electrons per cm^2 into the surface region of the polymer, $\psi_0 = (2kT/ze)\sinh^{-1}[\sigma/(8\varepsilon_r\varepsilon_0 kTn_0)^{1/2}]$, $\sigma = 10^{12}$ electrons/cm^2.

and modern theories, such as those of Fabish and Duke, may give us the tools to explore these phenomena in much more depth. The performance of polymers in contact with blood or other body tissues could well depend on charge-injection phenomena.

ACKNOWLEDGMENT

The author thanks his coworkers at Lehigh University, especially Prof. Frank H. Hielscher, Sachio Maruchi, Dr. Mohammed Mostafa, Sara T. Joslin, and Lawrence A. Casper, and James A. Wolfe at the Air Force Materials Laboratory (Wright-Patterson AFB, OH). The support of the Office of Naval Research, the U.S. Bureau of Mines, IBM, Western Electric, and the Air Force Materials Laboratory is gratefully acknowledged.

REFERENCES

1. J. van der Waals, Thesis, Leiden University (1972).
2. W. H. Keesom, *Physik A* **22**, 129, 643 (1921).
3. P. Debye, *Physik Z.* **21**, 178 (1920); **22**, 302 (1921).
4. F. London, *Z. Phys. Chem.* **B11**, 222 (1930).
5. A. C. Hamaker, *Physica* **4**, 1058 (1937).
6. J. H. Hildebrand and R. L. Scott, *Solubility of Nonelectrolytes*, 3rd Edition, Reinhold, New York (1950).
7. F. M. Fowkes, *Ind. Eng. Chem.* **56**, 40 (1964).
8. G. N. Lewis, *Valence and the Structure of Atoms and Molecules*, Chem. Cat. Co., New York (1923).
9. R. S. Mulliken and W. B. Person, *Ann. Rev. Phys. Chem.* **13**, 107 (1962).
10. J. N. Bronsted, *J. Chem. Soc.* **119**, 574 (1921).
11. T. M. Lowry, *Chem. Ind.* **42**, 43 (1923).
12. G. Schwarzenbach, *Rec. trav. chim.* **75**, 562 (1956).
13. R. G. Pearson, *Hard and Soft Acids and Bases*, Dowden, Hutchinson, and Ross, Stroudsburg, Pennsylvania (1973).
14. R. S. Drago, G. C. Vogel, and T. E. Needham, *J. Am. Soc.* **93**, 6014 (1971).
15. R. S. Drago, L. B. Parr, and C. S. Chamberlain, *J. Am. Chem. Soc.* **99**, 3203 (1977).
16. J. O. Edwards, *J. Am. Chem. Soc.* **76**, 1540 (1954).
17. H. F. Shurvell and J. T. Bulmer, in: *Vibrational Spectra and Structure* (J. R. Durig, ed.), p. 91, Elsevier, New York (1977).
18. F. M. Fowkes and S. Maruchi, *Org. Coatings Plastics Chem.* **37**, 605 (1977).
19. H. P. Schreiber, C. Richard, and M. R. Wertheimer, in: *Physicochemical Aspects of Polymer Surfaces* (K. Mittal, ed.), Plenum, New York (1983).
20. F. M. Fowkes, D. O. Tischler, J. A. Wolfe, and M. J. Halliwell, *Org. Coatings Appl. Polym. Sci. Proc.* **46**, 1 (1982).
21. L. J. Bellamy and R. L. Williams, *Trans. Faraday Soc.* **55**, 14 (1959).
22. A. Massat and J. E. Dubois, *J. Mol. Struct.* **65**, 87 (1980).
23. F. M. Fowkes, D. C. McCarthy, and M. A. Mostafa, *J. Colloid Interface Sci.* **78**, 200 (1980).

24. H. A. Benesi, *J. Phys. Chem.* **61**, 970 (1957).
25. F. M. Fowkes, H. A. Benesi *et al.*, *J. Agr. Food Chem.* **8**, 203 (1960).
26. F. M. Fowkes and K. Knitter, unpublished results (1978).
27. J. C. Bolger and A. S. Michaels, in: *Interface Conversion Polymer Coatings* (P. Weiss, ed.), Elsevier, New York (1968).
28. L. A. Casper, Thesis, Lehigh University (1983).
29. F. M. Fowkes, *J. Phys. Chem.* **64**, 726 (1960).
30. F. M. Fowkes and M. A. Mostafa, *Ind. Eng. Chem., Prod. Res. Develop.* **17**, 3 (1978).
31. B. J. Fontana and J. R. Thomas, *J. Phys. Chem.* **65**, 480 (1961).
32. F. J. Fowkes, C.-Y. Sun, and S. T. Joslin, in: *Corrosion Control by Organic Coatings* (H. Leidheiser, Jr., ed.), p. 1 NACE, Houston (1981).
33. F. M. Fowkes, *Disc, Faraday Soc.* **42**, 246 (1966).
34. K. Tamaribuchi and M. L. Smith, *J. Colloid Interface Sci.* **22**, 404 (1966).
35. F. M. Fowkes, H. Jinnai, M. A. Mostafa, F. W. Anderson, and R. J. Moore, in: *Colloids and Surfaces in Reprographic Technology, ACS Symp. Ser.* **200**, 307 (1982).
36. S. M. Skinner, R. L. Savage, and J. E. Rutzler, *J. Appl. Phys.* **24**, 438 (1953).
37. B. V. Deryagin and V. P. Smilga, *Third Inter. Congr. Surface Activity* **2**, 349 (1960).
38. V. J. Webers, *J. Appl. Polym. Sci.* **7**, 1317 (1963).
39. D. K. Davies, *Brit. J. Appl. Phys.* **2**, 1533 (1969).
40. H. G. von Harrach and B. N. Chapman, *Thin Solid Films* **13**, 157 (1972).
41. C. Weaver, *Faraday Special Discussions* **2**, 1 (1972).
42. T. J. Fabish, H. M. Saltsburg, and M. L. Hair, *J. Appl. Phys.* **47**, 930–940 (1976).
43. C. B. Duke and T. J. Fabish, *Phys. Rev. Lett.* **37**, 1075 (1976).
44. T. J. Fabish and C. B. Duke, *J. Appl. Phys.* **48**, 4256 (1977).
45. G. B. Duke, T. J. Fabish, and A. Paton, *Chem. Phys. Lett.* **49**, 133 (1977).
46. C. B. Duke, *J. Vac. Sci. Technol.* **15**, 157 (1978).
47. C. B. Duke, *Phys. Rev.* ***B*18**, 5717 (1978).
48. C. B. Duke and T. J. Fabish, *J. Appl. Phys.* **49**, 315 (1978).
49. G. A. Cottrell, J. Lowell, and A. C. Rose-Innes, *J. Appl. Phys.* **50**, 374 (1979).
50. T. J. Fabish, C. B. Duke, M. L. Hair, and H. M. Saltsburg, *J. Appl. Phys.* **51**, 1247 (1980).
51. G. A. Cottrell, J. Lowell, and A. C. Rose-Innes, *J. Appl. Phys.* **51**, 1250 (1980).
52. F. M. Fowkes, F. H. Hielscher, and D. J. Kelly, *J. Colloid Interface, Sci.* **32**, 469 (1970).
53. A. Bui, H. Carchano, and D. Sanchez, *J. Appl. Phys.* **43**, 3794 (1972).
54. A. Bui, H. Carchano, and D. Sanchez, *Thin Solid Films* **13**, 207 (1972).
55. F. M. Fowkes and F. H. Hielscher, *Spontaneous Electrostatic Precipitation of Dust*, U.S. Bur. Mines Contract No.GO110269 Final Report (May 15, 1973).
56. H. R. Anderson, F. M. Fowkes, and F. H. Hielscher, *J. Polym. Sci., Phys.* **14**, 879 (1976).
57. F. M. Fowkes and F. H. Hielscher, *Org. Coatings Plastics Chem.* **42**, 169 (1980).
58. F. M. Fowkes, in: *Industrial Applications of Surface Analysis, ACS Symp. Ser.* **199**, 69 (1982).
59. F. M. Fowkes, F. H. Hielscher, and J. Emerson, to be published.
60. E. J. Verwey and J. Th. G. Overbeek, *Stability of Lyophobic Colloids*, p. 32, Elsevier, Amsterdam (1984).

10

Graft Copolymer and Block Copolymer Surfaces

Buddy D. Ratner

1. INTRODUCTION

Homopolymers and random copolymers are relatively homogeneous in composition throughout their bulk. At the outermost molecular layer of such materials, the surface chemistry might differ substantially from the average bulk chemistry due to orientation effects, oxidation, or contamination. Graft copolymers and block copolymers, on the other hand, often demonstrate large compositional differences between surface and bulk and these differences can be observed over many molecular layers extending from the surface into the bulk (see Figure 1). This review article will concentrate on the nature of the differences between the bulk and surface of graft and block copolymers and on methods which can be used to explore the surfaces of such systems.

It is useful to briefly review which type of systems fall into each category and consider synthetic methods used to prepare block and graft polymers. The structure of such polymers and related systems are schematically illustrated in Table 1. Table 2 outlines a few representative synthetic methods used to prepare these polymers. Preparative methods for block and graft copolymers are limited only by the imagination of the individuals performing the syntheses, and a comprehensive review of techniques would be unreasonably lengthy. A number of useful review articles and monographs on the preparation and bulk properties of these classes of materials have been published.[1-7]

Each of the structures described in Table 1 will generate surfaces which can differ substantially from the bulk in both chemistry and morphology.

Buddy D. Ratner • Department of Chemical Engineering and Center for Bioengineering, University of Washington, Seattle, Washington 98195.

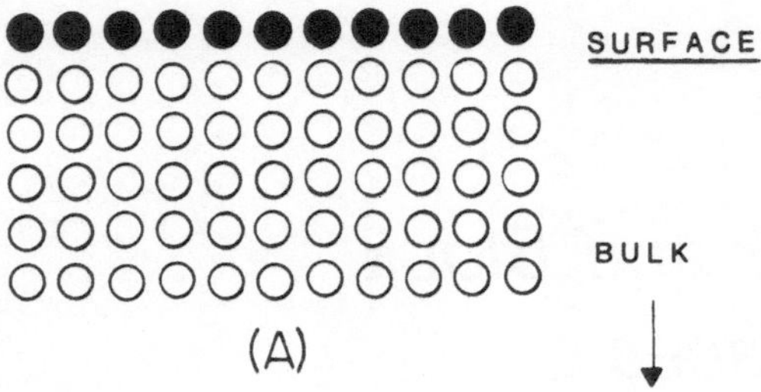

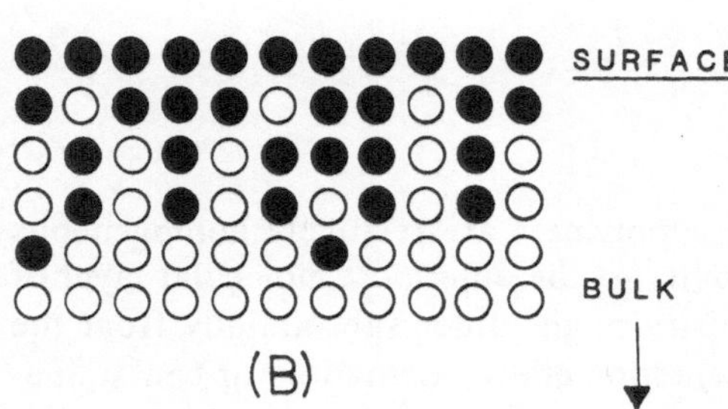

FIGURE 1. Typical surface/bulk compositional diagrams for (A) homopolymers and random copolymers and (B) block and graft copolymers. Homopolymers and random copolymers are often homogeneous throughout their bulk with only the outermost atomic layer(s) being of different composition. Graft and block copolymers typically have compositional gradients which extend many molecular layers from the surface into the bulk.

There are two primary factors which will govern the behavior of all these systems with regard to both surface/bulk differences and surface phase segregation. These are surface energetics and molecular mobility.

As discussed in Chapters 2 and 7, an interface will always attempt to achieve the lowest interfacial energy. If a polymer system contains more than one component, the component that, when localized at the interface will result in a minimum interfacial energy will, if possible, migrate to the surface. Thus, for a polymer/air interface, the lowest energy component (or functional groups) in the polymer will concentrate at the interface. For a polymer/water interface, it will be energetically desirable to have the most polar functional groups or most polar polymeric components oriented towards the interface.

The degree of molecular mobility exhibited by the polymer determines if the energetically desirable minimization of interfacial energy can indeed be achieved. Where molecular mobility is inhibited, energetically less favorable surface structures can be frozen into place. The processes of melting, dissolution in a solvent, raising the temperature above the glass transition, or plasticization can all serve to increase molecular mobility so that the most thermodynamically desirable states can prevail at the surface.

Table I

Schematic Representations of Polymer Systems

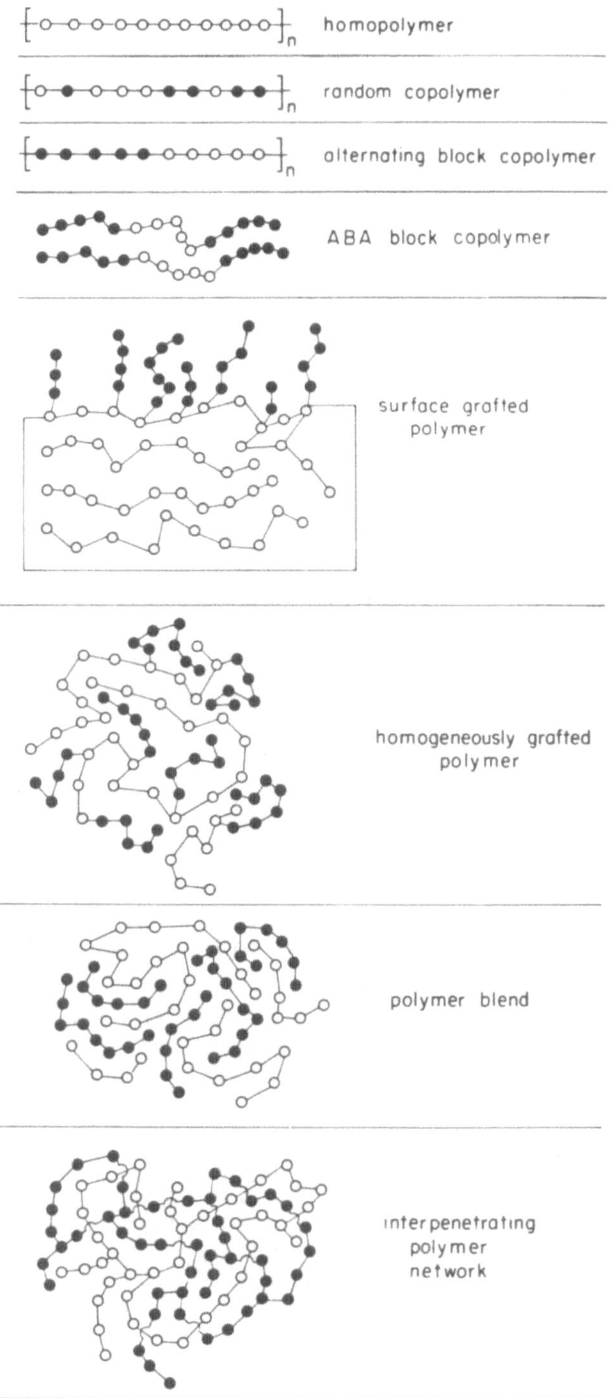

TABLE 2
Representative Methods for Generating Graft and Block Copolymers

Graft Copolymers:
- A. Free Radical Mechanism
  1. ionizing radiation (radiation grafting, UV grafting)
  2. chemical reactions (e.g., ceric ion reaction with —OH)
  3. peroxide formation (thermal or chemical activation)
  4. active vapor activation (RF or microwave plasma)
- B. Anionic (e.g., metal halogen formation on halogen-containing polymer)
- C. Cationic (e.g., aluminum alkyl-initiated graft of isobutylene onto PVC)

Block Copolymers:
- A. Anionic (e.g. living polymer technique using alkyllithium-reacted styrene)
- B. Cationic
- C. Cationic-Anionic (Ion Coupling Methods)
- D. Polycondensation (step growth) (e.g. segmented polyurethanes)
- E. Free Radical Methods
- F Ziegler-Natta

Two additional factors must also be considered in order to understand surface-bulk compositional differences. If polymer chains in a block or graft copolymer are highly cyrstallizable, they may sterically "exude" a second polymer component resulting in surface localization. Crystallization also impedes molecular motion and diffusion. Finally, the degree of incompatibility of two components will influence their tendency to phase separate, both in the bulk and at the surface.

2. GRAFT COPOLYMER SURFACES

Most surface studies on graft copolymer systems have concentrated primarily on ascertaining whether the graft is present at the surface. Such studies have been performed by infrared spectroscopy,[8-10] contact angle methods,[10,11] or x-ray photoelectron spectroscopy (XPS).[10,12]

For determining the presence of a surface graft and, in some cases, quantitating the amount of graft at the surface, XPS has proven to be a versatile tool. For poly(2-hydroxyethyl methacrylate) (pHEMA) grafts on

FIGURE 2. C-1*s* XPS spectra of polyethylene and HEMA radiation-grafted to polyethylene. (a) Ungrafted low-density polyethylene, C/O = 26.6; (b) polyethylene radiation-grafted at 0.25 Mrad with a 0.5% HEMA solution in ethanol/water solvent; graft level is unmeasurable by conventional gravimetric techniques, but higher binding energy C-1*s* species are observed (note peak at approx. 289 eV); C/O = 12.2; (c) polyethylene radiation-grafted at 0.25 Mrad with a 5% HEMA solution in ethanol/water solvent; graft level is 0.28 mg/cm^2 and the C-1*s* spectrum primarily reflects the HEMA graft; C/O = 2.8.

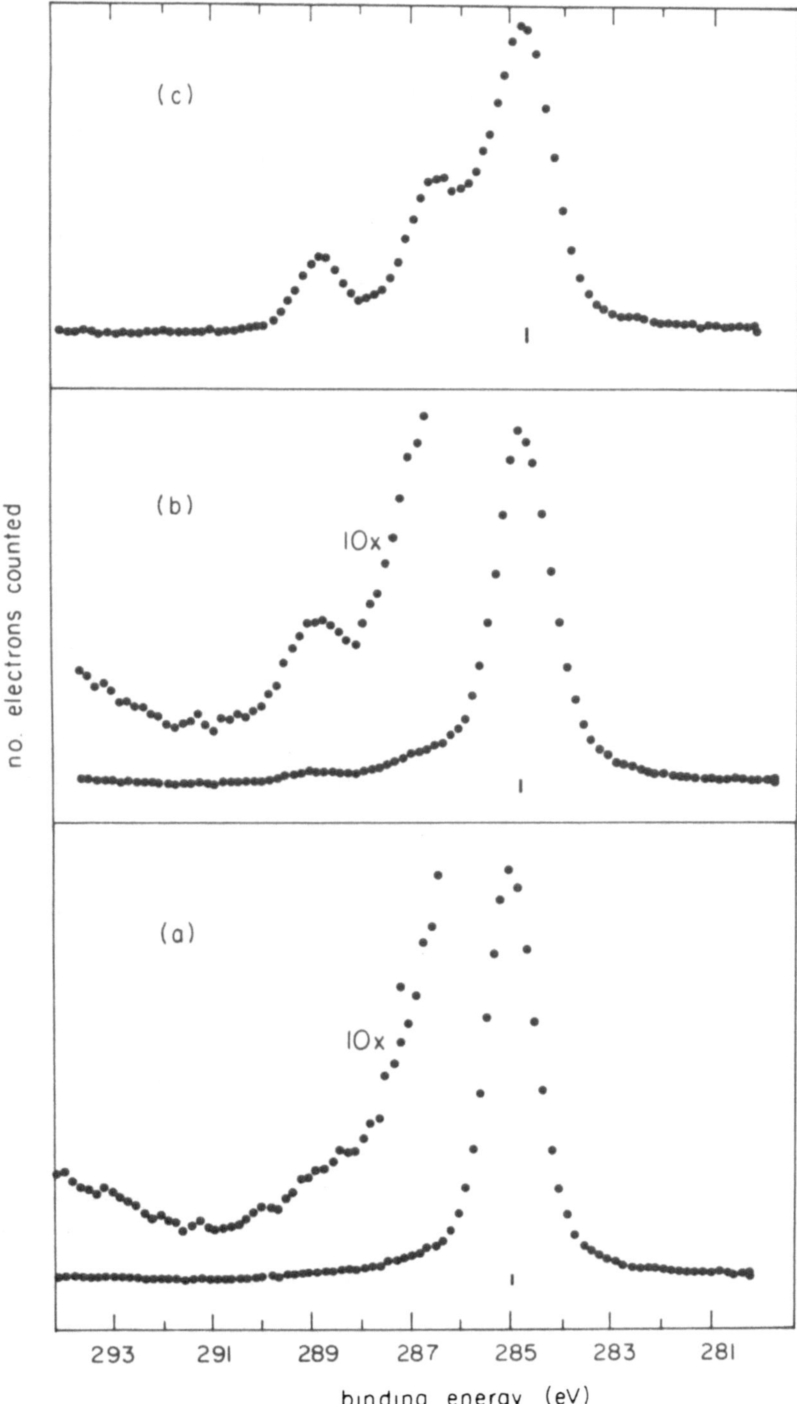
(c)
(b)
10x
(a)
10x
no. electrons counted
293
291
289
287
285
283
281
binding energy (eV)

low density polyethylene prepared in the author's laboratory, graft levels were too low to be accurately measured by gravimetric methods. However, XPS easily detects the presence of this graft (see Figure 2). A calibration curve can be constructed relating the fraction of the XPS signal primarily due to pHEMA to graft level (for graft levels high enough to be accurately measured gravimetrically). This calibration curve can then be extrapolated to zero graft level to estimate extremely low graft levels directly from the XPS signal. Other aspects of both the bulk and surface characterization of graft copolymers have been reviewed.[13,14]

The concept of interfacial energy minimization presented in the introduction to this article applies in a most significant way to graft copolymer systems. When radiation-grafted hydrogels on silicone rubber were examined by XPS in a hydrated state (at −160°C), appropriate graft spectra were observed along with some Si signal from the silicone rubber.[15] However, upon dehydrating these films (performed *in situ* in the XPS instrument by allowing the specimen to warm to ambient temperature), little or no hydrogel graft signal could be detected (see Figure 3 and Table 3). Since XPS will sample to the depths of approximately 100 Å, the surface of these materials must have restructured over many molecular layers. Two explanations for this behavior are possible, both based upon surface energy minimization arguments:

1. In order to reduce the interfacial free energy of the system, poly(HEMA) chains migrate into the silicone rubber matrix upon dehydration. Upon rehydration, again to reduce the interfacial energy of the system, the poly(HEMA) returns to the surface region to hydrogen bond with the aqueous medium.
2. In order to reduce the interfacial free energy of the system, low-molecular weight uncrosslinked silicone compounds which were not removed by the extraction migrate to the surface as the polymer is dehydrated. Rehydration results in a migration of these "oily" compounds back into the matrix.

TABLE 3
Elemental Ratios for Graft Polymers as Determined by ESCA

	C/Si		C/N	
Polymer	160° K[a]	303° K[a]	160° K[a]	303° K[a]
HEMA on silicone rubber	5.52	1.64	—	—
Acrylamide on silicone	7.70	2.05	3.41	∞

[a] 160° K = hydrated; 303° K = dehydrated.

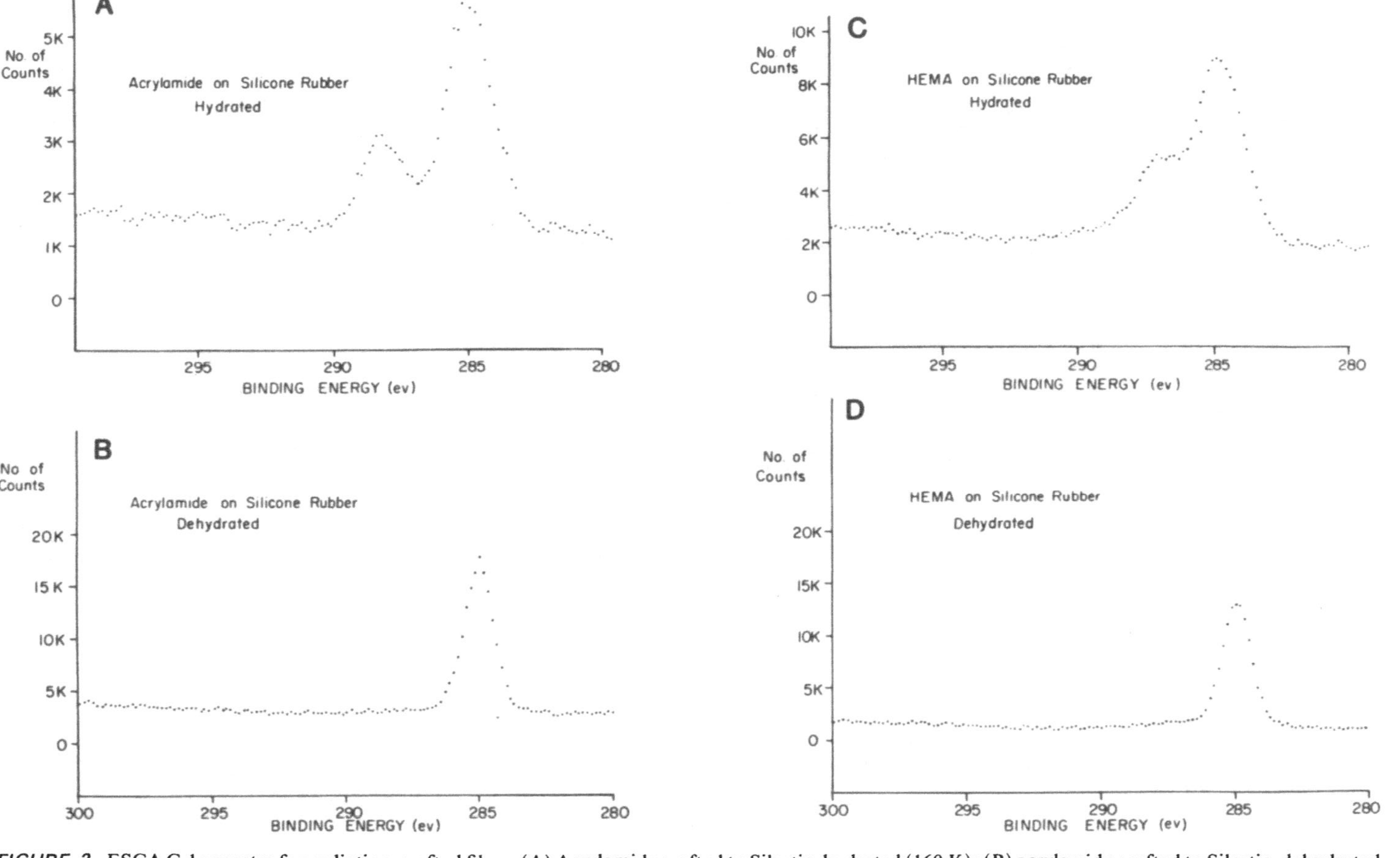

FIGURE 3. ESCA C-1*s* spectra for radiation-grafted films: (A) Acrylamide grafted to Silastic, hydrated (160 K); (B) acrylamide grafted to Silastic, dehydrated (303 K); (C) HEMA grafted to Silastic, hydrated (160 K); (D) HEMA grafted to Silastic, dehydrated (303 K).

Similar surface chemistry reversals have been noted with comblike graft copolymers containing hydrophilic and hydrophobic segments which have been synthesized and studied as additives for poly(methyl methacrylate).[16] Depending upon the surface against which these graft copolymer blends were annealed, the surface hydrophilicity/hydrophobicity can be strongly modified: annealing against Teflon or air produces a hydrophobic surface while annealing against a polar Mylar surface produces a more wettable surface. Surface chemistry reversals in response to the nature of the interface and its environment have been reported for many situations by many investigators.[15,17-21] In general, the driving force for such reversals would appear to be energetic—a reduction of interfacial energy. However, a recent study suggests that an entropically driven dilution of surface functional groups by the bulk may occur for certain systems exposed to solvents.[18] Thus, polyethylene exposed to an N_2 RF inductively coupled plasma was shown by XPS to undergo a decrease in surface concentration of nitrogen-containing functional groups in the presence of solvents which can disrupt interchain hydrogen bonds. Such non-covalent crosslinks, according to this model, reduce chain mobility and inhibit the diffusion of surface functional groups into the bulk.

Recent experiments involving radiation grafts formed by simultaneously grafting two monomers to a substrate have yielded some new insights into the nature of the surfaces of these complex systems.[22] In these experiments HEMA and ethyl methacrylate were radiation grafted to low-density polyethylene. Results obtained by XPS suggest that at various times during this grafting reaction, first one monomer, and then the other, strongly dominates the surface. Thus, the surface of the copolymer graft is not necessarily equivalent in composition either to that of a simple solution copolymer of the two monomers, or to the "average" composition of the copolymer graft on the substrate. A surface zone of unique composition is hypothesized based upon these experiments.

RF plasma-deposited surface grafts have undergone extensive surface characterization due to the interest in learning more about the unique chemistry of this type of polymer system. These unique chemistries are generated by the atomic rearrangements which occur when the monomer molecule is dissociated in the plasma environment to a variety of charged and free radical species. These species recombine and deposit upon any surface in the RF plasma reactor in a complex manner which is governed by such factors as the monomer concentration, the energy put into the plasma, the precise chemical species involved, the presence of other chemicals which can accelerate or inhibit reactions, and the ratio of surface etching rate to surface deposition rate. The preparation and properties of plasma-generated surface films have been reviewed recently in a number of publications.[23-25] The important property of plasma-deposited films as pertains to this review article is their sharp boundary between compositional regions. The graft–substrate interface, generally believed to consist of covalent bonds between the plasma-deposited film and substrate, is, for many systems, probably only a few molecular layers thick. This is in contrast to many "surface" grafted copolymers prepared using chemical or radiation techniques which can have gradients betweeen graft and bulk extending over hundreds or thousands of angstroms.[26,27]

Transmission electron microscope (TEM) and scanning electron microscope (SEM) studies can yield interesting information on the structure of graft copolymer systems. TEM images of microtomed cross sections of graft copolymers can be used to characterize the morphology of the interface between the graft and substrate and also reveal the morphology of the surface region. Examples of TEM cross sections of radiation grafts on low-density polyethylene are given in Figures 4 and 5. The potential for artifacts induced by cutting and staining is high in TEM sample preparation. In addition, for hydrogel grafts, additional artifacts can be introduced by examining them in the dry state. Therefore, control studies to attempt to assess the extent of artifact generation are always in order when a new type of sample is subjected to TEM study.

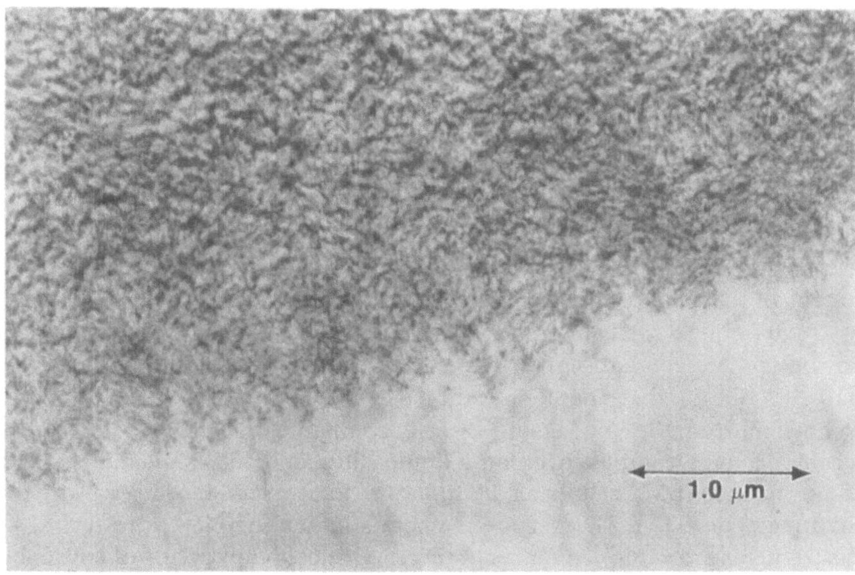

FIGURE 4. Transmission electron micrograph of a HEMA radiation graft on polyethylene (1.58 mg graft/cm^2) showing the HEMA/PE junction region. 30,000×, phosphotungstic acid stain.

FIGURE 5. Transmission electron micrograph of the surface of a HEMA radiation graft on polyethylene. 20,000×, stained by treating with cinnamoyl chloride and then osmium tetroxide.

For non-hydrogel graft specimens, SEM studies are considerably less subject to the generation of artifacts than TEM studies in which much more sample preparation is involved. For hydrogels, the dehydration process involved in SEM sample preparation usually leads to some shrinkage and therefore, potentially, to some distortion of morphology. Thus, positive conclusions about the morphology of the surface of all hydrogel grafts cannot be directly inferred by SEM images. Two techniques can help to establish the relationship between the surface morphology of wet and dry hydrogel grafts. First, hydrogel grafts can be examined under water with an optical stereomicroscope. If the features being observed are relatively large ($>10\text{m}\mu$), the general appearance of the images from the light microscope and from the SEM can be directly compared. Second, a wet hydrogel graft specimen can be frozen on a liquid nitrogen-cooled low temperature stage in the SEM, metalized *in situ,* and then examined directly in the hydrated condition. This second method is rather complex since few SEM instruments are equipped for such studies. Critical point drying of hydrogels can also be attempted to maintain the surface morphology of grafts. However, the solvents used to displace water from the hydrogel during the critical point drying process can potentially induce as much distortion as the dehydration which was being avoided, due to swelling or deswelling effects. In Figure 6 a series of SEM's of dry poly(hydroxyethyl methacrylate) radiation grafts on polyethylene are shown. Samples of this type had previously been examined under water using a light stereomicroscope. The existence of a similar surface morphology to that observed in Figure 6 was confirmed, thus establishing that dry and wet grafts have surface textures which differ primarily in magnitude ("bump size") rather than in general character. The surface texture is believed to be due to higher graft initiation rates at amorphous sites on the polyethylene surface. This is presently under investigation.

High-resolution SEM surface studies can further enhance the information which can be obtained about the morphology of polymer surfaces.[28] The development and commercialization of lanthanum boride (LaB_6) cathode electron guns have extended SEM resolution to 30 Å for such studies, However, particularly at high magnifications, the potential for beam damage is increased. Such beam damage can also induce artifacts. Short observation times are required to minimize beam damage.

3. BLOCK COPOLYMER SURFACES

The phenomenon of phase segregation (or domain formation) is well-documented for block copolymers. The driving force for such phase separation can be entropic and/or energetic. Various aspects of the bulk structure of these systems have been subject to thorough study and analysis in recent years.[1,2,29] In comparison, the surfaces of these systems have received relatively little attention.

In a number of studies dealing with the interactions of block copolymers with biological systems, attempts have been made to characterize the surfaces of these polymer systems using transmission electron microscopy (TEM). From TEM observations, conclusions were drawn concerning the effects of phase separation at the surface of the polymer on biological processes. TEM studies of stained thin polymer films can easily be misinterpreted when the surface region is under consideration. The possibilities for artifacts are illustrated by the four cases in Figure 7. However, useful morphologic information can be gleaned from both TEM and scanning electron microscopic (SEM) studies. Stained TEM specimens are particularly valuable for understanding the bulk morphology of block copolymers, especially where phase separation has occurred. Many publications have dealt with this subject[1,30] and no attempt will be made to review this material here. Replica methods for TEM specimen preparation can be useful, and, in at least one instance, this method has been used to corroborate surface structure evidence obtained using x-ray photoelectron spectroscopy.[31] Defocus electron microscopy can also be used to enhance contrast

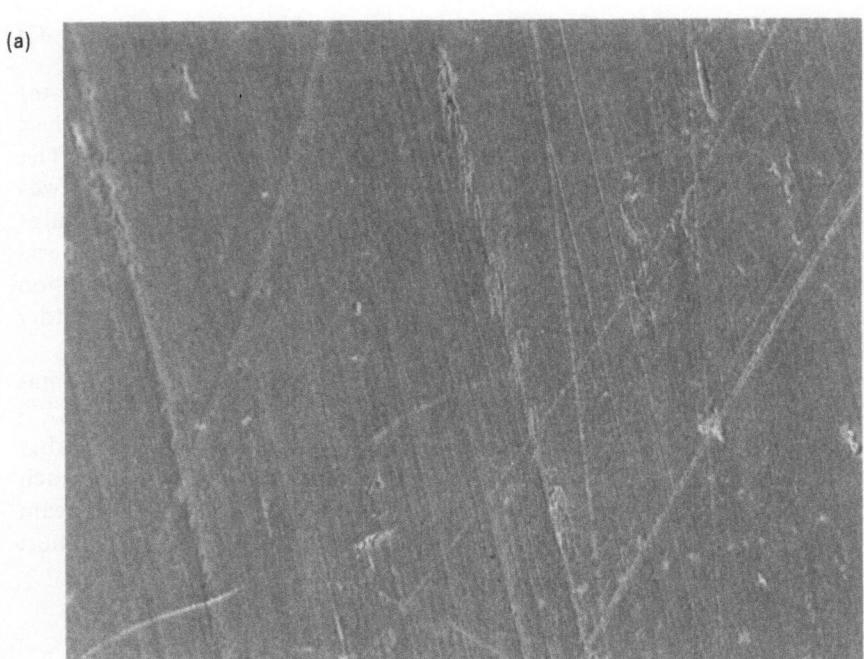

(a)

(b)

FIGURE 6. Scanning electron micrographs (200×) of HEMA radiation grafts on polyethylene: (a) Ungrafted, commercial grade, extruded low-density polyethylene film (control); (b) 0.5 mg/cm^2 HEMA graft on polyethylene; (c) 1.14 mg/cm^2 HEMA graft on polyethylene; (d) 3.36 mg/cm^2 HEMA graft on polyethylene.

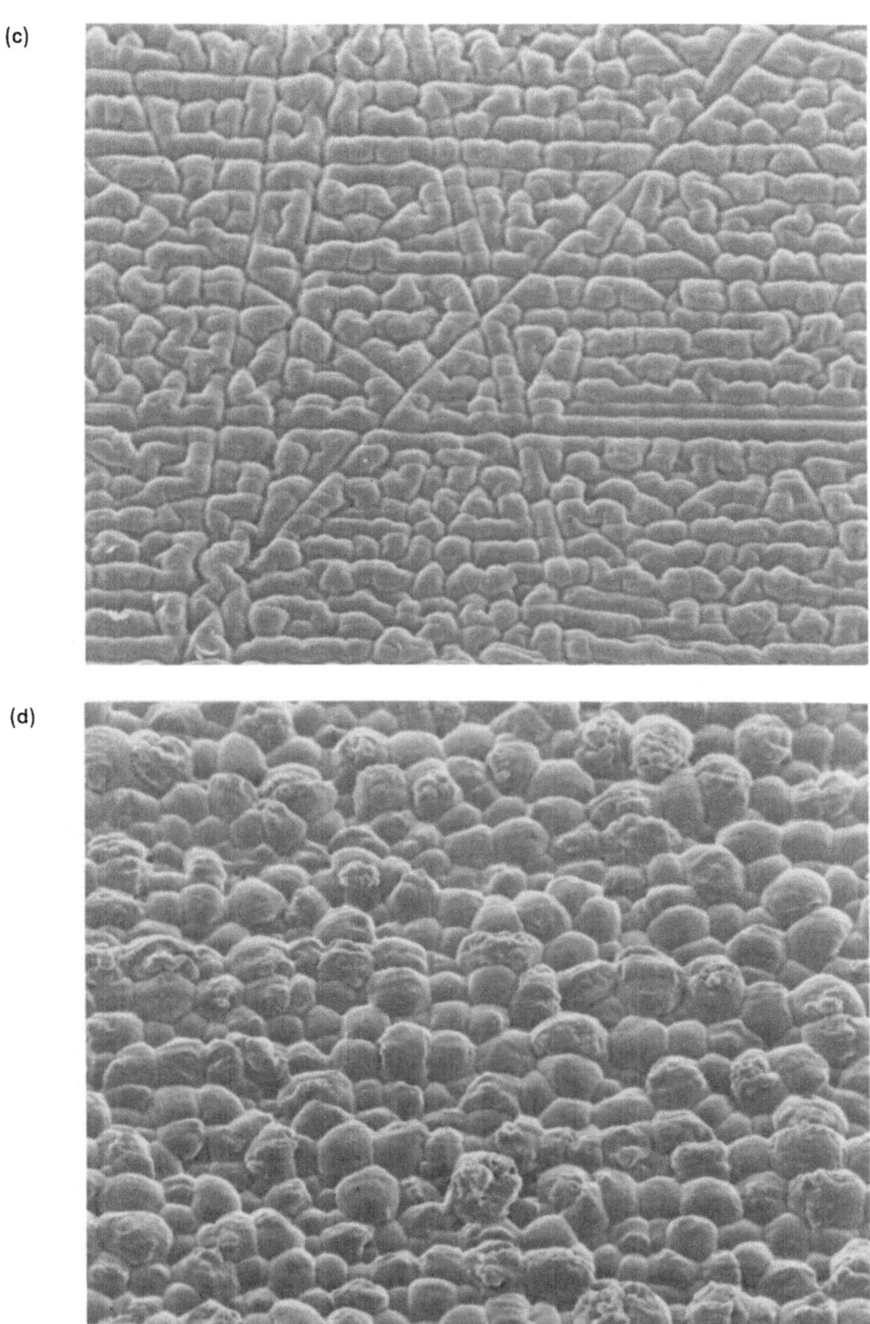

FIGURE 6 (continued)

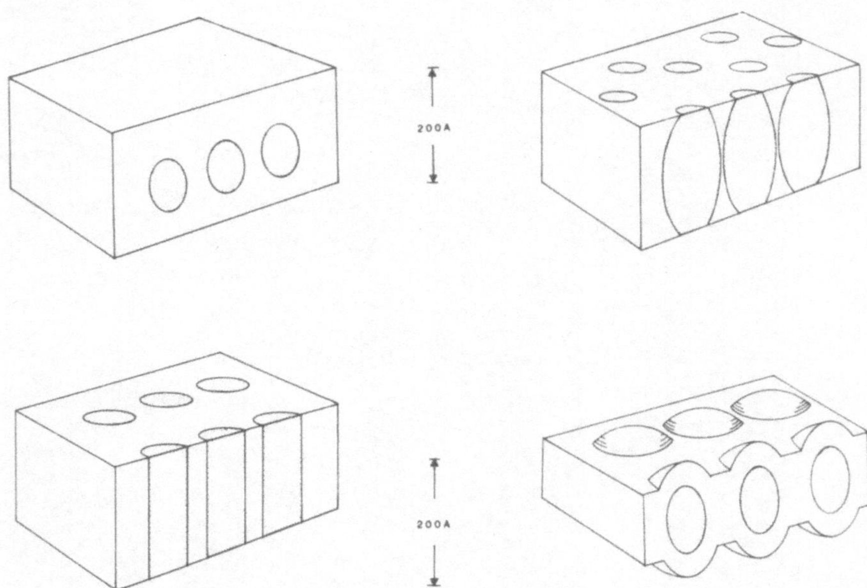

FIGURE 7. Models for domain structures in a 200-Å block copolymer film. Transmission electron micrographs taken through the 200-Å dimension of these four model films would all look identical.

in imaging multiphase polymeric systems, but great care must be exercised to avoid artifacts.[32]

SEM can be used with striking results to observe highly textured block copolymer surfaces and fracture planes.[33-35] A study of an osmium-stained polystyrene/polybutadiene blend using back-scattered electrons generated with accelerated electron beams of different energies has demonstrated that one can view domains in the surface region of an unsectioned bulk polymer.[36] In addition, the depth at which the domains are observed can be controlled. At 35-kV accelerating voltage, stereo pair images indicate that the depth of the polymer surface region penetrated and observed is greater than at 15 kV. Sampling depths are estimated to be as low as 100 Å. Although only blends were studied in this work, the method should be equally applicable to block copolymers.

XPS (also referred to as Electron Spectroscopy for Chemical Analysis, ESCA) has been found to be an unusually valuable tool for understanding the nature of block copolymer surfaces. The advantages of XPS for such studies are its surface nature (the top 10–100 Å are under study), the minimal specimen preparation required, the ability to nondestructively depth-profile materials, and the wealth of chemical information obtained.

One of the earliest studies on block copolymers by XPS examined poly(dimethylsiloxane)/polystyrene AB copolymer surfaces.[37] In all cases, the immediate surface was shown to consist of almost pure poly(dimethylsiloxane) (PDMS). By consideration of polystyrene shake-up peaks ($\pi \rightarrow \pi^*$ transitions) observed in the C-1*s* XPS spectra, the thickness of the poly(dimethylsiloxane) surface films were estimated to be ~13–40 Å. Casting solvent strongly influenced the thickness of this surface film.

In a later study using angular-dependent XPS techniques to observe similar block copolymers with lower-molecular weight poly(dimethylsiloxane) segments, surface concentration of the siloxane polymers was also noted.[38] However, in this case the surfaces were not dominated by overlayers of silicone as in the earlier study in which high-molecular weight PDMS blocks were considered. Polycarbonate/poly(dimethylsiloxane) copolymers were also examined by XPS and showed evidence of surface concentrations of the PDMS segment.[39] In addition, in this same study, polycarbonate/polysulfone block copolymers were observed. Results from both these polymer systems led the authors to conclude that the surface properties of block copolymers will be largely governed by the lower surface energy polymeric component.

It should be mentioned that other methods besides XPS can be used to observe the surface migration and concentration of PDMS blocks in block copolymers. Contact angle methods have been extensively employed for such analysis.[37,38,40,41] The pendant drop method has been used to study surface migration of a styrene–dimethylsiloxane block copolymer in toluene solutions containing polystyrene.[42] Surface tension measurements using a Rosano Wilhelmy plate surface tensiometer demonstrated the reduction in the surface tension of styrene solutions upon the addition of ABA polystyrene–polysiloxane–polystyrene block copolymers, presumably due to silicone block concentration at the solution surface.[41]

Polyether–polystyrene block copolymers have been extensively studied with regard to surface/bulk differentiation. Poly(tetrahydrofuran) (THF)/styrene (ST) block copolymers observed by TEM showed evidence that fibrils of poly(THF) coated with poly(ST) were formed after casting a THF-rich block copolymer from chloroform.[43] With increasing PS content in the copolymer, a lamellar morphology was observed with poly(THF) on the surface. The author explains this by considering that at high THF contents, crystallization of the poly(THF) units excludes PS units. At high PS contents, the surface activity of the THF units provides the driving force to coat the surface. The surface activity of the THF units was demonstrated using the pendant drop method on melts. Contact angle methods and inverse gas chromatography were also used to explore the complex surface segregations which occur in this system. Inverse gas chromatography was also used with great success to understand the surface properties of methyl methacrylate–stearyl methacrylate graft copolymers.[44]

Angular-dependent XPS studies have been applied to poly(ethylene oxide) (PEO)/polystyrene (PS) di- and tribock copolymers.[31,45] The data on both these systems suggest enrichment in the PS component at the surface, but also that the surface consists of discrete domains of PS and PEO. There is some evidence of electronic interaction and mixing at the surface between the PS and PEO. Changes in shape and intensity of the PS aromatic ring shake-up satellite, presumably due to interaction with ether oxygens, are interpreted as being indicative of this. Blends of PEO and PS were also considered by these authors and similar surface structures to those seen with the AB and ABA block copolymers were observed.[46]

Polystyrene/poly(2-vinylpyridine) block copolymers were subjected to XPS surface analysis.[47] These block copolymers were found to exhibit little electronic interaction between the two aromatic groups and phase separation at the surface. Enrichment of the surface in the vinylpyridine phase is suggested by the data. In contrast, a random copolymer of polystyrene and poly(2-vinylpyridine) shows strong evidence of electronic interaction. The authors presented a general hypothesis covering phase segregation and surface structure in such systems by considering aromatic ring steric accessibility (high between the styrene ring and vinylpyridine ring in the random copolymer, and low in the block copolymer) and the extent of electronic interaction (compatibility of the phases).

4. SEGMENTED POLYURETHANES

Segmented polyurethanes (SPU) materials have assumed great importance in biomedical applications (e.g., artificial hearts, pacemakers, catheters) and have generated much controversy with regard to the relationship between their surface structure and biological interactions. Therefore, they will be considered separately in this section. In actuality, they have much in common with the block copolymers described previously.

The most desirable property of SPU's for biomedical applications is their excellent tensile strength and resistance to failure under fatigue. Their blood compatibility, governed, to a large measure, by the surface chemistry of these polymers, has been the subject of much controversy.[48–50] For example, in a recent publication, it has been suggested that an increased proportion of hydrocarbon-type groups at the surface of polyurethanes will decrease platelet damage by the surface.[48] In another recent publication, increased polyether-type linkages at the surface have been related to reduced platelet interaction.[49,50] In any case, it has been clearly established that the nature of an SPU surface can strongly influence biological response.

Therefore, the factors governing the surface properties of this class of material should be elucidated.

Work by a number of investigators has established that the bulk and the surface of SPU's can be different. Most authors have suggested that a concentration of soft segment at the air interface of cast films occurs.[51,52] Localization of organic polymeric silicones at polyurethane surfaces has also been reported.[53,54] The nature of the casting surface during solvent casting has often been implicated as an important determinant of the surface composition of SPU's.[55]

Recent studies have demonstrated that SPU's contain a significant fraction of low-molecular weight material which can be compositionally different from the bulk of the polymer.[56-58] In particular, a polyether-rich low-molecular weight component has been extracted from these polymers. In order to understand the significance of such extractables with respect to the surface properties, a new technique has been developed in which models are constructed representing depth profiles of polymer surfaces—the XPS data expected from a material of similar surface structure to the model is

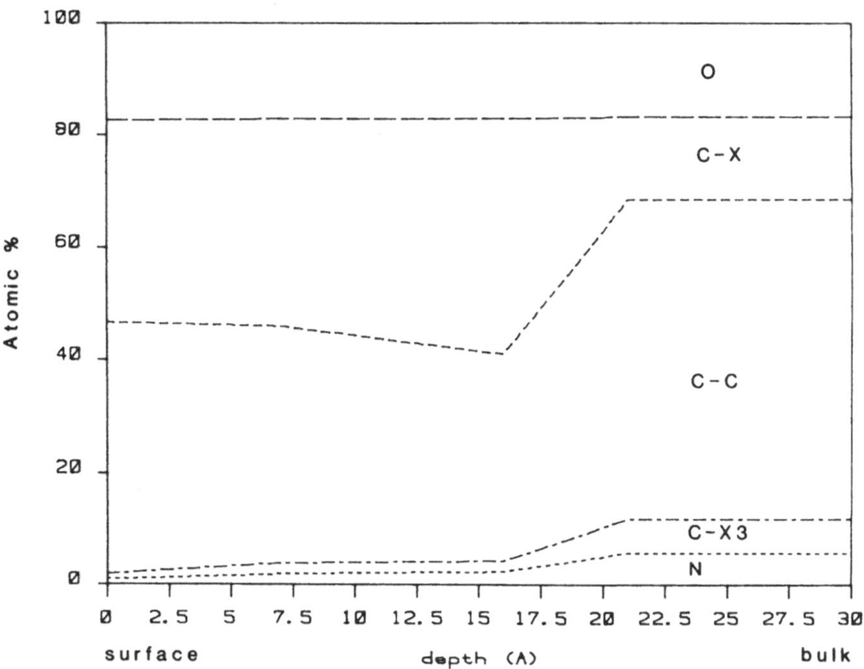

FIGURE 8. Depth profile diagram derived from XPS data for the air surface of a cast polyetherurethane film made with a 1000 MW poly(tetramethylene glycol) soft segment.

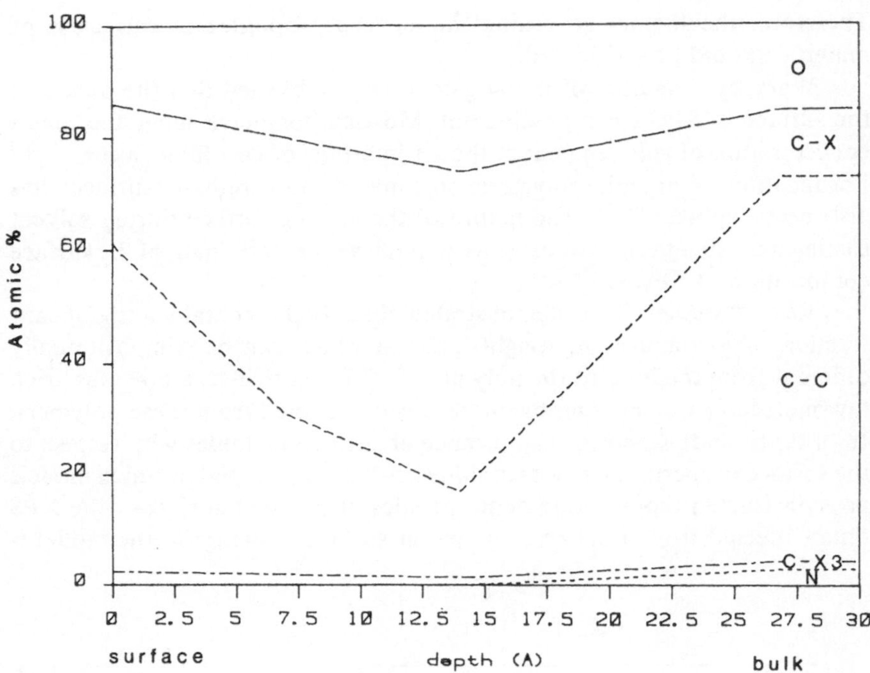

FIGURE 9. Depth profile diagram derived from XPS data for the air surface of a cast polyetherurethane film made with a 1000 MW poly(tetramethylene glycol) soft segment; free poly(tetramethylene glycol) was added to the casting solution.

then calculated and compared to experimentally obtained angular-dependent XPS data.[59,60] One such model, which shows close agreement with experimentally acquired angular-dependent XPS data, for a polyetherurethane with a 1000 MW poly(tetramethylene glycol) segment is shown in Figure 8. Upon the addition of a small amount of free polyether to the casting solution, a very different surface with a distinct polyether enrichment near the interface is obtained (Figure 9). Figure 9 also suggests the possibility of some hydrocarbon contamination at the air–polymer interface. No nitrogen is seen at the interface for either polymer. Thus, these models which best fit the XPS data would indicate that there may be no hard segment at the surface of SPU's. This is consistent with the concept that components of lower surface energy will migrate to the surface of a block copolymer to reduce interfacial energy.

Analysis of many SPU systems with and without low-molecular weight additives is currently in progress. The new algorithm for assisting in the interpretation of XPS data should contribute substantially to our abilities to understand and interpret the surface gradients occurring in all block and graft copolymer materials.

ACKNOWLEDGMENTS

The author would like to acknowledge the assistance of R. W. Paynter in calculating the surface gradient diagrams presented and to acknowledge J. Harris for the transmission electron micrographs. Some of the work presented in this article has been generated under NIH grants HL25951 and HL22163.

REFERENCES

1. B. R. M. Gallot, in: *Advances in Polymer Science* (H. J. Cantow, G. Dall'Asta, K. Dusek, J. D. Ferry, H. Fujita, M. Gordon, W. Kern, G. Natta, S. Okamura, C. G. Overberger, T. Saegusa, G. V. Schulz, W. P. Slichter, and J. K. Stille, eds.), Vol. 29, pp. 85-156, Springer-Verlag, Berlin (1978).
2. A. Noshay and J. E. McGrath, *Block Copolymers—Overview and Critical Survey*, Academic Press, New York (1977).
3. H. A. J. Battaerd and G. W. Tregear, *Graft Copolymers*, Interscience Publishers, New York (1967).
4. N. R. Legge, G. Holden, S. Davison, and E. DeLaMare, in: *Applied Polymer Science* (J. K. Craver and R. W. Tess, eds.), pp. 394-429, Organic Coatings and Plastics Chemistry Division of the American Chemical Society, Washington, D.C. (1975).
5. G. M. Estes, S. L. Cooper, and A. V. Tobolsky, Block polymers and related heterophase elastomers, *J. Macromol. Sci., Revs. Macromol. Chem.* **C4**, 313-366 (1970).
6. V. Stannett, Grafting, *Radiat. Phys. Chem.* **18**, 215-222 (1981).
7. S. L. Cooper and G. M. Estes (eds.), *Multiphase Polymers, ACS Advances in Chemistry Series*, American Chemical Society, Washington, D.C. (1979).
8. S. Yamakawa, Surface modification of polyethylene by radiation-induced grafting for adhesive bonding. I. Relationship between adhesive bond strength and surface composition, *J. Appl. Polym. Sci.* **20**, 3057-3072 (1976).
9. D. Lodesova, A. Pikler, M. Foldesova, and J. Tolgyessy, Contribution to the radiation-induced grafting of acrylonitrile and glycolmethacrylate to polypropylene, *Radiochem. Radioanal. Lett.* **32**, 327-336 (1978).
10. S. Tazuke and H. Kimura, Surface photografting, 2. Modification of polypropylene film surface by graft polymerization of acrylamide, *Makromol. Chem.* **179**, 2603-2612 (1978).
11. Y. Ikada, H. Iwata, F. Horlii, T. Matsunaga, M. Taniguchi, M. Suzuki, W. Taki, S. Yamagata, Y. Yonekawa, and H. Handa, Blood compatibility of hydrophilic polymers, *J. Biomed. Materials Res.* **15**, 697-718 (1981).
12. S. Yamakawa and F. Yamamoto, Surface grafting of polyethylene by mutual irradiation in methyl acrylate vapor. III. Quantitative surface analysis by x-ray photoelectron spectroscopy, *J. Polym. Sci., Polym. Phys. Ed.* **17**, 1581-1590 (1979).
13. B. D. Ratner, Characterization of graft polymers for biomedical applications, *J. Biomed. Materials Res.* **14**, 665-687 (1980).
14. Y. Ikada, in: *Advances in Polymer Science* (H. J. Cantow, G. Dall'Asta, K. Dusek, J. D. Ferry, H. Fujita, M. Gordon, W. Kern, G. Natta, S. Okamura, C. G. Overberger, T. Saegusa,

G. V. Schultz, W. P. Slichter, and J. K. Stille, eds.), Vol. 29, pp. 47-84, Springer-Verlag, Berlin (1978).

15. B. D. Ratner, P. K. Weathersby, A. S. Hoffman, M. A. Kelly, and L. H. Scharpen, Radiation-grafted hydrogels for biomedical applications as studied by the ESCA technique, *J. Appl. Polym. Sci.* **22**, 643-664 (1978).
16. Y. Yamashita and Y. Tsukahara, Control of polymer surface structure by tailored graft-copolymers, *ACS Org. Coatings Appl. Polym. Sci. Proc.* **46**, 75-78 (1982).
17. F. J. Holly and M. F. Refojo, in: *Hydrogels for Medical and Related Applications* (J. D. Andrade, ed.), *ACS Symp. Ser.* **31**, 252-266 (1976).
18. D. S. Everhart and C. N. Reilley, The effects of functional group mobility on quantitative ESCA of plasma modified polymer surfaces, *Surface Interface Anal.* **3**, 126-133 (1981).
19. R. N. King, J. D. Andrade, S. M. Ma, D. E. Gregonis, and L. R. Brostrom, Interfacial characterization of hydrogel-water interfaces, in: *Proceedings of the workshop on interfacial phenomena: Research needs and priorities*, University of Washington, February 15-16, 1979, pp. 458-502, National Science Foundation, Washington, D.C. (1979).
20. Y. C. Ko, B. D. Ratner, and A. S. Hoffman, Characterization of hydrophilic-hydrophobic polymeric surfaces by contact angle measurements, *J. Colloid Interface Sci.* **82**, 25-37 (1981).
21. R. G. Azrak, Surface property variations in melt-formed thermoplastics, *J. Colloid Interface Sci.* **47**, 779-794 (1974).
22. D. C. Cohn, A. S. Hoffman, and B. D. Ratner, Radiation grafted hydrogels for biomaterials applications: Synthesis, structure and composition of HEMA:EMA graft copolymers on LDPE film, Abstracts of the Fourth European Conference on Biomaterials, Belgium, August 31-September 2, 1983.
23. H. Yasuda, Glow discharge polymerization, *J. Polym. Sci., Macromol. Rev.* **16**, 199-293 (1981).
24. E. Kay and A. Dilks, Plasma polymerization of fluorocarbons in rf capacitively coupled diode system, *J. Vac. Sci. Technol.* **18**, 1-11 (1981).
25. H. V. Boenig, *Plasma Science and Technology*, Cornell University Press, Ithaca, New York (1982).
26. E. A. Hegazy, I. Ishigaki, A. Rabie, A. M. Dessouki, and J. Okamoto, Study on radiation grafting of acrylic acid onto fluorine-containing polymers. II. Properties of membrane obtained by preirradiation grafting onto poly(tetrafluoroethylene), *J. Appl. Sci.* **26**, 3871-3883 (1981).
27. A. Chapiro, Radiation induced grafting, *Radiat. Phys. Chem.* **9**, 55-67 (1977).
28. T. Tagawa, J. Mori, S. Aita, and K. Ogura, Application of the high resolution SEM to the fine structure study of polyethylene, *Micron* **9**, 215-221 (1978).
29. P. E. Gibson, M. A. Vallance, and S. L. Cooper, Morphology and properties of polyurethane block copolymers. *Dev. Block Copolym.* **1**, 217-259 (1982).
30. D. A. Thomas, Morphology characterization of multiphase polymers by electron microscopy, *J. Polym. Sci., Polym. Symp.* **60**, 189-200 (1977).
31. H. R. Thomas and J. J. O'Malley, Surface studies on multicomponent polymer systems by X-ray photoelectron spectroscopy. Polystyrene/poly(ethylene oxide) diblock copolymers, *Macromolecules* **12**, 323-329 (1979).
32. E. J. Roche and E. L. Thomas, Defocus electron microscopy of multiphase polymers: use and misuse, *Polymer* **22**, 333-341 (1980).
33. S. Y. Hobbs and V. H. Watkins, The use of chemical contrast in the SEM analysis of polymer blends, *J. Polym. Sci., Polym. Phys. Ed.* **20**, 651-658 (1982).
34. J. M. Short and R. G. Crystal, Morphology of block copolymers, *Appl. Polym. Symp.* **16**, 137-151 (1971).
35. J. N. Sultan, R. C. Laible, and F. J. McGarry, Microstructure of two-phase polymers, *Appl. Polym. Symp.* **16**, 127-136 (1971).

36. J. D. Andrade, D. L. Coleman, and D. E. Gregonis, Characterization of polymer surface morphology by scanning electron microscopy using backscattered electron imaging, *Makromol. Chem., Rapid Commun.* **1**, 101–104 (1980).
37. D. T. Clark, J. Peeling, and J. J. O'Malley, Application of ESCA to polymer chemistry. VIII. Surface structures of AB block copolymers of polydimethylsiloxane and polystyrene, *J. Polym. Sci., Polym. Chem. Ed.* **14**, 543–551 (1976).
38. D. Shuttleworth, J. G. VanDusen, J. J. O'Malley, and H. R. Thomas, An X-ray photoelectron study of low molecular weight polystyrene-polydimethyl siloxane block copolymers, *Polym. Prepr., Am. Chem. Soc., Div. Polym. Chem.* **20**, 499–502 (1979).
39. J. E. McGrath, D. W. Dwight, J. S. Riffle, T. F. Davidson, D. C. Webster, and R. Viswanathan, Bulk and surface segregation in polycarbonate-polysulfone and polycarbonate-polydimethylsiloxane block copolymers, *Polym. Prepr., Am. Chem. Soc., Div. Polym. Chem.* **20** (2), 528–530 (1979).
40. D. G. LeGrand and G. L. Gaines, Jr., Surface activity of block copolymers of dimethylsiloxane and bisphenol-A carbonate in polycarbonate, *Polym. Prepr., Am. Chem. Soc., Div. Polym. Chem.* **11**, 442–446 (1970).
41. M. J. Owen and T. C. Kendrick, Surface activity of polystyrene-polysiloxane-polystyrene ABA block copolymers, *Macromolecules* **3**, 458–461 (1970).
42. G. L. Gaines Jr., and G. W. Bender, Surface concentration of a styrene-dimethylsiloxane block copolymer in mixtures with polystyrene, *Macromolecules* **5**, 82–86 (1972).
43. Y. Yamashita, Surface properties of styrene-tetrahydrofuran block copolymers, *J. Macromol. Sci., Chem.* **13**, 401–413 (1979).
44. K. Ito, N. Usami, and Y. Yamashita, Syntheses of methyl methacrylate-stearyl methacrylate graft copolymers and characterization by inverse gas chromatography, *Macromolecules* **13**, 216–221 (1980).
45. J. J. O'Malley, H. R. Thomas, and G. M. Lee, Surface studies on multicomponent polymer systems by X-ray photoelectron spectroscopy. Polystyrene/poly(ethylene oxide) triblock copolymers, *Macromolecules* **12**, 996–1001 (1979).
46. H. R. Thomas and O'Malley, Surface studies on multicomponent polymer systems by X-ray photoelectron spectroscopy: Polystyrene/poly(ethylene oxide) homopolymer blends, *Macromolecules* **14**, 1316–1320 (1981).
47. T. J. Fabish and H. R. Thomas, Copolymer structure through charge injection and X-ray photoemission, *ACS Org. Coatings Plastics Chem. Prepr.* **42**, 406–411 (1980).
48. S. R. Hanson, L. A. Harker, B. D. Ratner, and A. S. Hoffman, in: *Biomaterials 1980* (G. D. Winter, D. F. Gibbons, and H. Plenk, Jr., eds.), pp. 519–530, John Wiley and Sons, Ltd., London (1982).
49. M. D. Lelah, L. K. Lambrecht, B. R. Young, and S. L. Cooper, Physiochemical characterization and *in vivo* blood tolerability of cast and extruded Biomer, *J. Biomed. Materials Res.* **17**, 1–22 (1983).
50. V. Sa Da Costa, D. Brier-Russell, E. W. Salzman, and E. W. Merrill, ESCA studies of polyurethanes: Blood platelet activation in relation to surface composition, *J. Colloid Interface Sci.* **80**, 445–452 (1981).
51. C. S. P. Sung and C. B. Hu, ESCA studies of surface chemical composition of segmented polyurethanes, *J. Biomed. Materials Res.* **13**, 161–171 (1979).
52. K. Knutson and D. J. Lyman, in: *Biomaterials: Interfacial Phenomena and Applications* (S. L. Cooper and N. A. Peppas, eds.), *Adv. Chem. Ser.* **199**, 109–132 (1982).
53. S. W. Graham and D. M. Hercules, Surface spectroscopic studies of Biomer, *J. Biomed. Materials Res.* **15**, 465–477 (1981).
54. E. Nyilas and R. S. Ward, Jr., in: *Science and Technology of Polymer Processing* (N. P. Suh and N. H. Sung, eds.), pp. 770–808, MIT Press, Cambridge, Massachusetts (1979).
55. S. I. Stupp, J. W. Kauffman, and S. H. Carr, Interactions between segmented polyurethane surfaces and the plasma protein fibrinogen. *J. Biomed. Materials Res.* **11**, 237–250 (1977).

56. B. D. Ratner, in: *Photon, Electron, and Ion Probes of Polymer Structure and Properties* (D. W. Dwight, T. J. Fabish, and H. R. Thomas, eds.), *ACS Symp. Ser.* **162**, 371-382 (1981).
57. B. D. Ratner, in: *Physicochemical Aspects of Polymer Surfaces* (K. L. Mittal, ed.), Vol. 2, pp. 969-983, Plenum Publishing Corp., New York (1983).
58. C. B. Hu and C. S. P. Sung, Surface chemical composition-depth profile of polyether polyurethaneureas as studied by FT-IR and ESCA, *Polym. Prepr., Am. Chem. Soc., Div. Polym. Chem.* **21** (1), 156-158 (1980).
59. R. W. Paynter, B. D. Ratner, and H. R. Thomas, Polyurethane surfaces—An XPS study, *Polym. Prepr., Am. Chem. Soc., Div. Polym. Chem.* **24** (1), 13-14 (1983).
60. B. D. Ratner, R. W. Paynter, and H. R. Thomas, Polyurethane surfaces—An XPS study, *Trans. Soc. Biomaterials* **6**, 21 (1983).

11

Interfacial Tensions at Amorphous High Polymer–Water Interfaces: Theory

Mu Shik Jhon and Youngie Oh

1. INTRODUCTION

The immiscibility of polymers is regarded as rather a rule than an exception in contrast with most small-molecule liquids. The interaction between unlike polymer constituent groups is less energetically favorable than the interaction between like ones in a relative sense. Both types of interactions are strongly attractive, as manifested by the fact that polymers form adhesive, highly incompressible phases. While the unfavorable interaction between unlike polymers tends to drive the two phases apart, an adhesive force acts strongly to prevent an actual gap from developing. The demixed polymer liquids thus form multiphase structures with stable but energetically unfavorable interfaces. Considerable attention has been given to interfacial properties of polymer melts, especially the interfacial tension, due to their importance in many technologies through their roles in the processes of polymer blends, wettability, and adhesion. Provided that one of the two demixed polymer phases is replaced by a water phase, a polymer-water interface will be formed. The polymer-water interfacial tension plays an important role in understanding interfacial interactions between the physiological environment and the surface of polymer materials. The polymer-water interface can be, however, investigated as an extension of the polymer-polymer interfacial system, and the interfacial tension of polymers is closely related to the surface tension, i.e., the excess free energy of polymer surface being exposed to atmosphere. Several reviews of the topic of surface

Mu Shik Jhon and Youngie Oh • Department of Chemistry, Korea Advanced Institute of Science and Technology, P.O. Box 150, Cheong Ryang Ri, Seoul, Korea.

and interfacial tensions of synthetic polymers, which compile accumulated experimental observations, are available.[1-5]

Surface tension, of course, is one of the fundamental properties of a liquid reflecting intermolecular interactions. Since the definitive works of Richards and Coombs[6] and Harkins and Humphery[7] reliable measurements of surface tension have been reported for hundreds of small-molecule liquids.[8-10] For polymeric liquids, the earliest report on the surface tension appears to have been made in 1939 by Gallaugher and Hibbert[11] who measured several ethylene glycol oligomers up to the heptamer. These materials are mobile fluids at room temperature and standard measurement techniques[12,13] yield reliable results. Most high polymers are, however, very viscous even at elevated temperature and exhibit non-Newtonian flow characteristics. Considerable care is, therefore, required to obtain surface tension values which can be accepted as valid. Accordingly, merely a few of all the standard techniques for the measurement of surface tension are satisfactory. Available methods for interfacial tension are more limited. Only the methods based on drop profiles are completely suitable for both surface and interfacial measurements; these include the pendant drop method,[14-26] the sessile bubble method,[17,18] and the rotating bubble method.[19,20]

As for solid polymers, the surface tension is indirectly determined by contact angle measurements, which consist of measuring the contact angle of a homologous series of liquids of known surface tension on plane solid surfaces.[21] Tilting plate methods[22,23] and the slide techniques[24] are also available. Certain relations between the contact angle and the solid surface tension are required in these indirect methods. The critical surface tension, which is an approximate characteristic of the surface, is defined as the extrapolated surface tension of liquid drops in the limit of zero contact angle. The critical surface tension varies with the liquid type used to determine it, and cannot be construed as a measure of the solid surface tension.[25] An equation of state, which defines a spectrum of critical surface tension when a series of testing liquids are used, has been proposed by Wu.[26] In this method the maximum value of the critical surface tension on that spectrum is taken as the solid surface tension. Extensive data on critical surface tensions of solid polymers are collected in the *Polymer Handbook*.[27] The surface tension of solid polymers can also be indirectly estimated by extrapolating the surface tension versus temperature plot of the melt to the solid range, or by extrapolating the surface tension versus molecular weight plot of the small-molecule homologous liquids to the solid range. The values which are obtained by the two extrapolations are consistent with each other.[2]

The temperature dependence of surface and interfacial tensions leads to an important quantity from which the surface and interfacial entropies are inferred. Most polymer liquids exhibit a linear dependence of both the surface and interfacial tension on temperature, like the small-molecule liquids far below the critical temperature. The observed temperature coefficients are, however, about one order of magnitude smaller than those for nonpolymeric liquids.[3] Since the absolute temperature coefficient is, essentially, the surface entropy, the smaller values for polymers are attributed to the conformational restrictions of long-chain molecules.[15,16]

The temperature coefficient of interfacial tension is usually less than 0.03 dyn/cm per degree, and is thus smaller than that of the surface tension, which is in the range of 0.05 to 0.08 dyn/cm per degree.[15,16,28,30] In order to explain this observation, Wu[1] has invoked MacLeod's equation[31] for the interface, noting that $d(\rho_1 - \rho_2)/dT$ is much smaller than $d\rho_1/dT$ or $d\rho_2/dT$, where the ρ_i's are the densities of the polymers and T is the temperature. This can be understood in the context of the free volume concept of simple liquids.[32] The extent of free volume for polymer units at the interface is definitely less than that of the surface due to the constraints of adjacent foreign polymer units. Since the entropy is logarithmically proportional to the extent of free volume, the temperature coefficient of interfacial tension is, thus, smaller than that of surface tension. More quantitative descriptions for the temperature coefficient of surface tension, which are based on the significant structure theory of liquids, have appeared,[33,34] The additional entropy of mixing in the interface is relatively immaterial, because it is only proportional to the number of polymer molecules rather than the number of monomeric units.

There has been great attention given to the molecular weight dependence of surface and interfacial properties. Bulk properties, such as glass transition temperature, heat capacity, specific volume, etc., of homologous series are known to depend linearly on the reciprocal of molecular weight.[35-40] The surface and interfacial tensions, however, do not obey such a linearity.[41-48] There are two alternative empirical equations which have been proposed to describe the molecular weight dependence of surface tension. Wu[2] proposed the use of

$$\gamma^{1/4} = \gamma_\infty^{1/4} - \frac{K_s}{M_n} \qquad (1)$$

where γ_∞ is the surface tension at infinite molecular weight, M_n is the number average molecular weight, and K_s is a positive constant. LeGrand and Gaines[47,48] have used

$$\gamma = \gamma_\infty - \frac{K_e}{M_n^{2/3}} \qquad (2)$$

where K_e is a constant which defines the extent of the molecular weight dependence. Both equations give comparable accuracy in fitting the experimental data for alkanes,[2] but both unfortunately lack accurate theoretical basis. The total surface energy, which is equal to $\gamma - T(\mathrm{d}\gamma/\mathrm{d}T)$, is nearly constant with respect to the molecular weight.[44,49] Moreover, for poly(ethylene glycols) and poly(propylene glycols) with —OH end groups, the surface tension is found to be substantially independent of molecular weight.[50,51] When the terminal hydroxyl groups are replaced by non-hydrogen bonding methyl ether or acetate groups, the surface tension then becomes dependent on the molecular weight, obeying Eq. (1) or Eq. (2). These facts suggest that the decrease of surface tension with decreasing chain length is associated largely with increasing surface entropy due to the larger proportion of the chain end groups relative to the repeating constituent groups. For interfacial tensions between polymer melts, a similar molecular weight dependence exists, which may be at least approximately expressed as[45,46]

$$\gamma_{12} = \gamma_{12}^{\infty} - \frac{K_1}{M_1^{2/3}} - \frac{K_2}{M_2^{2/3}} \tag{3}$$

Such a relation, of course, implies a criterion for miscibility between the two polymer fluids, since the interfacial tension becomes zero for molecular weights smaller than those that satisfy the equation

$$\frac{K_1}{M_1^{2/3}} + \frac{K_2}{M_2^{2/3}} = \gamma_{12}^{\infty} \tag{4}$$

Slow and Patterson[52] have compared the solubility parameter, parachor, and corresponding state approaches in regard to their ability to predict $\mathrm{d}\gamma/\mathrm{d}T$ and the $M^{-2/3}$ molecular weight dependence. They concluded that these concepts are closely related, in each case yielding temperature and molecular weight dependency through changes in free volumes. The solubility parameter relation does not predict either $\mathrm{d}\gamma/\mathrm{d}T$ or the molecular weight relation well. The parachor and corresponding states correlations, however, do yield fairly good predictions for the slopes of γ vs. T and $M^{-2/3}$, even though the absolute magnitudes of the calculated surface tensions are only approximately correct.

The interfacial tension may be reduced by the use of additives to improve the compatibility at the interface. A class of such additives is the block and graft copolymers with blocks or segments having the same chemical compositions as those of the two polymers which constitute the interface.[53] An AB block or graft copolymer will tend to bridge the interface between A and B, and thus reduce the interfacial tension and improve the compatibility.

The effect of polarity is one of the most important ingredients which determine the magnitude of the interfacial tension of organic polymers. The extent of nonpolar (dispersion) interaction between two phases does not vary greatly from system to system, but that of polar interactions can vary greatly. The polarity of a polymer can be estimated from the interfacial tension between this polymer and a nonpolar polymer such as polyethylene, employing a suitable equation which correlates the surface and interfacial tensions.[54] The polarity values thus obtained have been found to be independent of temperature within the investigated temperature range of 20–200°C.[54] It is due to the high polarity that poly(vinyl acetate) has an interfacial tension as high as 11.3 dyn/cm against polyethylene at 140°C. If poly(vinyl acetate) were nonpolar, the interfacial tension against polyethylene would be nearly zero. Theoretical aspects of the polar effect on interfacial tension have not yet been investigated thoroughly, despite its importance in practical applications.

The scope of this chapter is to compile present information on the interfacial tension of polymers from the theoretical viewpoint, and provide a proper perspective for the study of interfacial tension at polymer–water interfaces as an extension of polymer–polymer interfacial systems.

2. THE THEORY OF SURFACE AND INTERFACIAL TENSIONS OF AMORPHOUS HIGH POLYMERS

Theories of surface and interfacial tensions of small-molecule liquids have been well developed, and their successes and limitations in correlating with experimental data are also well-known; the major theories are those due to Kirkwood and Buff,[55] Cahn and Hilliard,[56] Ree *et al.*,[57] and Prigogine and Saraga.[58] Attempts have been made to apply to polymers most of the theories originally directed to understanding the surface and interfacial tensions of simple liquids. In this section we review the theories which have been proposed for the prediction of surface and interfacial tensions of amorphous high polymers.

2.1. LATTICE THEORIES

In view of the mathematical difficulties encountered in attempting to treat the liquid state rigorously, it has been natural that more or less intuitive approaches to this problem should also have been tried. The lattice theory has been outstanding in this regard. So-called cell and hole theories are included in the lattice theory. In cell theory the number of lattice sites is chosen equal to the number of molecules and each molecule is confined in a cell formed by nearest neighbors, while in a hole theory not all the lattice sites are occupied by molecules, some are vacant.

Lattice theories have a noble history in polymer science as well. Notable achievements are the Flory–Huggins theory[59] and to some extent Prigogine and coworkers' application of the cell model of simple liquids to polymer molecules.[60] Although shortcomings of the Flory–Huggins theory are well-known, its simplicity makes it extremely useful in interpreting and correlating large bodies of experimental data on polymeric systems. Employing the concepts of Flory–Huggins, Roe[61] and Helfand[62] have independently proposed lattice theories of interfacial tensions between polymers. In these theories one concentrates on the expression for the entropy for a polymer molecule placed in the interfacial region of nonuniform composition. Because of different assumptions used, Roe and Helfand obtained different results for the entropy expression and thus the interfacial tension. They have, nevertheless, embodied the conformational entropy of polymers at the diffuse interface, which is associated with the loss of conformational freedom resulting from molecules near the interface having to turn back from the opposite phase.

According to Roe's theory,[61] the interfacial system of polymers is defined in terms of the chain lengths of the two components, r_1 and r_2, and the Flory–Huggins interaction parameter, χ, which correlates the segmental pair interaction between unlike polymers. The chain lengths are defined as the ratios of the molecular volumes to the volume per lattice site. The lattice sites are assumed to be occupied by molecules with no vacancies, and to be arranged so as to form a collection of layers of sites, each parallel to the interface. The number of nearest neighbors around a site is equal to z, of which lz are in the same layer while mz are in each of the two adjacent layers ($l + 2m = 1$). The composition profile across the interface is given by the volume fraction of both components ψ_1^i and $\psi_2^i(\psi_1^i + \psi_2^i = 1)$ as a function of the lattice layer number i. The free energy of the system is then written as

$$F = (A/a)\sum_i f_i \tag{5}$$

where A is the total interfacial area, a the area per lattice site, and f_i the average contribution to the free energy by a segment located at layer i. The f_i comprise both energy and entropy terms. The energy term in f_i is readily expressed using the parameter χ. In deriving the entropy term, assumptions encountered are entirely analogous to the ones employed by Flory in deriving the entropy of mixing in polymer solutions.[59] One such example is the procedure of replacing the expectancy of finding a vacant site adjacent to a known vacant site with the average expectancy of vacancy of a site selected at random, which is equivalent in nature with the Bragg–Williams approximation of random mixing. But the random mixing of polymer segments is restricted within each lattice layer, which leads to the nonuniformity of composition in the interfacial region. The f_i are derived as follows,

$$\frac{f_i}{kT} = \chi\psi_1^i \sum_{\nu=1,0,+1} \lambda_\nu \psi_2^{i+\nu} + \sum_{K=1,2} \left[\psi_K^i \ln \psi_K^i - \left(1 - \frac{1}{\gamma_K}\right) \psi_K^i \ln \sum_{\nu=-1,0,+1} \lambda_2 \psi_K^{i+\nu} \right] \tag{6}$$

where $\lambda_0 \equiv l$ and $\lambda_1 = \lambda_{-1} \equiv m$. The first term in Eq. (6) arises from the nearest neighbor interaction between segments and the remainder represents the entropy, per lattice site, of mixing the molecules in accordance with the nonuniform concentration profile. For a homogeneous system, the entropy term is reduced to the Flory–Huggins form of entropy of mixing. The equilibrium concentration profile is given by the values of ψ_2^i which minimize F, subject to the condition of the conservation of mass

$$\sum_i \psi_2^i = \text{constant} \tag{7}$$

In the case that χ is small and $r_1 = r_2 \rightarrow \infty$, Roe obtained the following expression for interfacial tension

$$\gamma_{12} = 2m^{1/2}kT(2\chi)^{3/4}/(3a) \tag{8}$$

The composition profile ψ_2^i at equilibrium is obtained by numerical solution of a series of differential equations. Figure 1 shows the composition profile for the case of $m = 1/4$ (hexagonal close-packed lattice) and $r_1 = r_2 \rightarrow \infty$, calculated for varying values of χ by Roe's theory.

In Helfand's lattice theory, all the lattice sites are again occupied by molecules and assumed to be arranged to form a collection of layers parallel to the interface.[62] Assumptions encountered in this theory are entirely similar to the ones in Roe's theory described briefly in the preceding paragraph. However, Helfand used a more fundamental procedure of stochastic process for the expression of entropy of polymer chains at the interface where the chain lengths are taken as infinite. Also, he considered the anisotropy of bond directions of chain segments in order to express the loss of conformational entropy which is manifested by the nonuniformity of the interface together with an extra energy of mixing in the interphase. The loss of entropy δS^i per lattice site located in layer i is, in this theory, written as

$$\delta S^i/k = \sum_{K=1,2} \psi_K^i [m(g_{iK}^+ \ln g_{iK}^+ + g_{iK}^- \ln g_{iK}^-) + l g_{iK}^0 \ln g_{iK}^0] \tag{9}$$

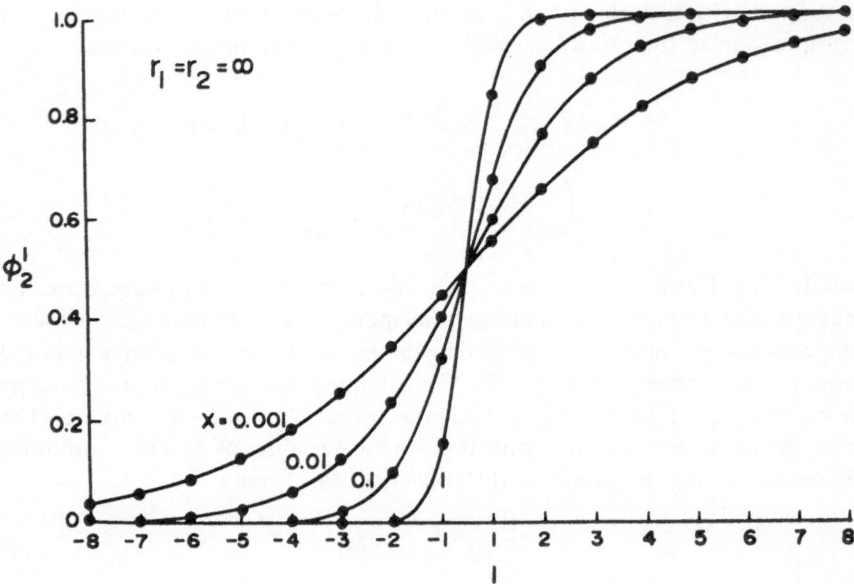

FIGURE 1. Composition profiles at the interface between two polymers of infinite molecular weight, calculated for the varying values of χ indicated (after R. J. Roe[61]).

where the quantities g_{iK}^{+}, g_{iK}^{-}, and g_{iK}° are anisotropy factors of bond direction, which reduce to unity under isotropic conditions while a value greater than unity indicates a greater than random tendency for bonds of polymer species K in layer i to be directed towards the upper, lower, or same layer relative to the layer i, respectively. For a homogeneous Flory–Huggins system, δS^i becomes zero evidently. Including the energy term, and employing some mathematical manipulations, Helfand obtained the expression for interfacial tension

$$\gamma_{12} = m^{1/2}kT\chi^{1/2}[1 + (1+\chi)\chi^{-1/2} \arctan(\chi^{1/2})]/(2a) \tag{10}$$

which is different from the one obtained by Roe's theory. The difference is caused not only by the difference in the qualitative nature of the assumptions used, but also by the different mathematical approximations employed in obtaining analytic forms of expression.

Numerical estimations of interfacial tensions by Roe's or Helfand's lattice theory are difficult, since the lattice parameters m and a which must be involved in the polymer density are uncertain. Figure 2 shows the variation of interfacial tension versus chain length for three selected values of χ, as calculated by Roe's theory. As the chain length increases beyond

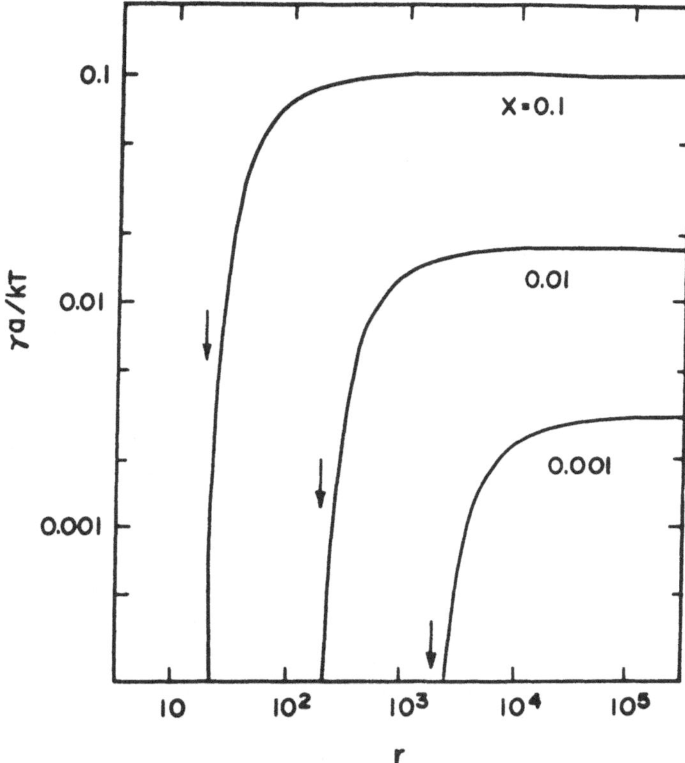

FIGURE 2. Interfacial tension of the polymer–polymer interface, in units of kT/a, plotted against $r(=r_1=r_2)$ for the three values of χ indicated. The arrows indicate the values of $r(=2/\chi)$ at which incipient phase separation is expected to occur (after R. J. Roe[61]).

the critical value for phase separation, the interfacial tension increases rapidly at first and then levels off to a plateau value.

A modification of the cell theory of simple liquids applicable to long-chain polymers was first proposed by Prigogine *et al.*;[58] and a variation (63) employing a square well potential was later found to give a reduced equation of state. Prigogine divided the $3r$ degrees of freedom of a r-mer molecule into the internal and external degrees of freedom. The internal degrees of freedom, corresponding to molecular vibrations, such as stretching and bending modes of the chain skeleton, are practically independent of intermolecular forces. The remaining external degrees of freedom corresponding to the translation, rotation, and internal rotations endow the chain constituent groups with lattice motions in the condensed phase. The parameter c which characterizes the external degrees of freedom has been introduced. Further work on the cell theory, and refinements of the theory, relaxing the lattice structure to introduce holes into the sites as carried out by Simha and coworkers[64-66] also include the parameter c, which is evaluated as an implicit measure of chain flexibility. Roe's early study on polymer surfaces is an application of the cell and hole concepts of Prigogine and Simha. Utilizing the parametric values of c evaluated by Nanda and Simha,[67] and employing the surface tension theory of Prigogine and Saraga,[58] Roe intended to calculate the surface tension, and its temperature coefficient, of polymers.[68] The works of Nose[69] and Patterson and Rastogi[70] utilize an analogous approach. These theories assume that molecular entities in a single surface layer exposed to the atmosphere are the origins of the surface energy and entropy. However, Nose suggested that the principle of corresponding states can be approximately satisfied for the reduced surface tension when the reduced variables are properly chosen.

Poser and Sanchez investigated the surface properties of polymer melts[71] using the lattice model of Sanchez and Lacombe[72] in conjunction with the Cahn–Hilliard theory of interface.[56] The lattice model of Sanchez and Lacombe is formally similar to the Flory–Huggins model except for an equilibrium number of vacant sites.[72] The configurational partition function is evaluated assuming random mixing of r-mers and vacancies. Minimizing the free energy of the system with respect to the fraction of vacant sites, an equation of state is derived. Since the model is not based on a cell concept but on the Ising lattice fluid, the question of how external and internal degrees of freedom are separated is not encountered, and the introduction of a "c" parameter is not required.[72]

Let us briefly summarize the Cahn–Hilliard theory of inhomogeneous small-molecule systems[56] before further discussing the treatment of Poser and Sanchez. The Cahn–Hilliard theory provides a means for relating a particular equation of state, based on a specific statistical mechanical model, to surface and interfacial properties. The local free energy per molecule, f, in a region of nonuniform composition will depend on the local composition as well as the composition of the immediate environment. f can be expressed in terms of an expansion in the local composition and local composition derivatives as follows

$$f(\psi_2, \nabla\psi_2, \nabla^2\psi_2 \cdots) = f_0(\psi_2) + \kappa_1\nabla^2\psi_2 + \kappa_2(\Delta\psi_2)^2 + \cdots \tag{11}$$

where $\kappa_1 \equiv [\partial f/\partial\nabla^2\psi_2]_0$, $\kappa_2 \equiv (1/2)[\partial^2 f/(\partial|\nabla\psi_2|^2)]_0$, and $f_0(\psi_2)$ is the free energy per molecule of a hypothetical mixture of uiniform composition ψ_2. Here, ψ_2 means the mole fraction of the second component molecules in the nonuniform interface, and the zero subscripts in the derivatives indicate that the derivatives are to be evaluated in the limit of $\nabla^2\psi_2$ and $\nabla\psi_2$ going to zero. If the composition gradient is sufficiently small, the free energy, F, of a system of volume V is given by

$$F = N_V \int_v [f_0(\psi_2) + \kappa(\nabla\psi_2)^2]\,\mathrm{d}v \tag{12}$$

where $\kappa \equiv -\mathrm{d}\kappa_1/\mathrm{d}\psi_2 + \kappa_2$, and N_V is the number of molecules per unit volume. Using these relations, and after a minimization procedure for the equilibrium composition profile, the interfacial tension for a planar interface can be written as

$$\gamma_{12} = 2N_V \int_{\psi_2^A}^{\psi_2^B} [\kappa\Delta f(\psi_2)]^{1/2}\,\mathrm{d}\psi_2 \tag{13}$$

in which ψ_2^A and ψ_2^B are the mole fractions of the second component in homogeneous phases A and B, respectively, which are far from the interface.

$\Delta f(\psi_2)$ is defined by

$$\Delta f(\psi_2) = f_0(\psi_2) - [\psi_2\mu_2(e) + (1-\psi_2)\mu_1(e)] \tag{14}$$

$$= \psi_2[\mu_2(\psi_2) - \mu_2(e)] + (1-\psi_2)[\mu_1(\psi_2) - \mu_1(e)] \tag{15}$$

where $\mu_1(e)$ and $\mu_2(e)$ are the chemical potentials per molecule [referred to the same standard state as $f_0(\psi_2)$] of species 1 and 2, respectively, in the A or B phases, $\mu_1(\psi_2)$ and $\mu_2(\psi_2)$ are the chemical potentials per molecule of the two molecular species in the interface at which the composition is characterized by ψ_2; and $\Delta f(\psi_2)$ is equal to $\chi(\mathrm{d}\psi_2/\mathrm{d}x)^2$, where x is the coordinate perpendicular to the flat interface. Equation (13) is a very general equation within the limit of slow variation of composition profile in the interface.

Poser and Sanchez applied Eq. (13) to the r-mer surface replacing the mole fraction ψ_2 with the density and the quantity $\Delta f(\psi_2)$ per molecule with the quantity per unit volume. The surface tension is written as

$$\gamma = 2\int_{\rho_l}^{\rho_g} [\kappa\Delta f(\rho)]^{1/2}\,\mathrm{d}\rho \tag{16}$$

and

$$\Delta f(\rho) = \kappa(\mathrm{d}\rho/\mathrm{d}x)^2 \tag{17}$$

where ρ_l and ρ_g are the bulk density of r-mers in the liquid and vapor phases, respectively. Poser and Sanchez evaluated the $\Delta f(\rho)$ utilizing the equation of state of their lattice model for r-mer liquids. If the value of κ

is known as a function of density, the density profile of the surface can be calculated by the integration of Eq. (17), and then the surface tension is estimated by using Eq. (16). There is no direct way to evaluate κ, but one can resort to an indirect method, regarding it as an adjustable parameter. The assumption of constancy of κ with respect to the density variation in the surface is essentially a neglect of the entropy effect due to the density gradient.[(56)] In order to extend the treatment of polymer surfaces to the polymer–polymer interfacial tension, Poser and Sanchez have attempted to construct a binary mixture model for polymers at interfaces.[(71)] Their work employs a number of assumptions in deriving an analytic form of interfacial tension, which makes it difficult to appraise the theory's applicability. They have, however, assessed the importance of compressibility effects on the mixture free energy for polymer pairs with relatively low interfacial tensions.

2.2. SIGNIFICANT STRUCTURE THEORY

The significant structure theory of the liquid state has been extensively applied to the thermodynamic properties of many liquids,[(74,75)] including amorphous and glassy polyethylene.[(76)] Stewart and von Frankenburg have applied the theory to the surface tension of molten polyethylene.[(33)]. More recently, a revised version of the surface tension theory for various amorphous polymers has been presented by Oh and Jhon,[(34)] in which a structural model for amorphous polymers has been proposed, considering the crystalline phase structure and the findings of neutron scattering experiments on the configuration of polymeric molecules in the amorphous phase. The scattering experiments have verified the absence of long-range bundle structure and the existence of an unperturbed random coil configuration of molecules much like in a dilute solution.[(77–81)]

In a crystalline linear polymer system, the crystallite sizes, lattice distortions, and the lengths of straight-chain segments follow certain distributions. When the temperature is evaluated above the glass transition temperature (T_g), a transition occurs, and long-range crystallinity is destroyed accompanied by a volume expansion of the system. In the theory of significant liquid structures, it is assumed that the volume expansion of an amorphous system is due to the introduction of fluidized vacancies into the lattice sites of the crystal. The chain segments adjacent to the vacancies translate from previous equilibrium (or nonequilibrium) positions to new ones much as in the gaseous state, and thus, the long-range crystallinity is destroyed. The wiggling movements of these segments do not necessarily perturb the overall configurations of molecules. The retention of straight-chain and short-range (5–10 Å) parallel alignments between neighboring chains has been assumed in this model. The assumption of spatial correlation seems to be reasonable in that the amorphous system recrystallizes as the temperature is lowered below T_g. Since the interaction between the chain constituent groups diminishes very rapidly with distance apart, one can evaluate the interaction energy by considering only the short-range spatial correlations of molecular entities, regardless of the overall configuration of the long molecular chain which follows. The theory identifies three nontrivial significant structures of amorphous polymers: (i) gas-like degrees of freedom which the fluidized vacancies confer on the neighboring segments, (ii) solid-like degrees of freedom of the remaining segments which are much like in the crystalline states, and (iii) possible positional degeneracy in the solid-like segments.

Regarding the monomer unit of polymers as a mass point, and ignoring end group effects, the theory assumes that each monomer unit independently contributes to the external modes with its one degree of freedom and contributes to the internal vibrations with its remaining two degrees of freedom. Then, the fluidized vacancies conver one-dimensional gas-like degrees of freedom on the adjacent monomer units. The configurational partition function for a solid-like monomer unit has been approximated by a one-dimensional cell partition function. The entropy effect on the surface tension has been included through the cell partition function of surface monomer units. In the assumed solid-like structure, the total potential energy of the system can be obtained by directly summing the interaction energy of all pairs of the solid-like monomer units in the parallel lattice excluding pairs belonging to the same chain. The long-range irregularity beyond 5–10 Å in the structure of the real system is unimportant for evaluating the potential energy because of the rapid spatial convergency of the interaction energies.The potential energy of a monomer unit in the force field of all its neighbors is written in terms of the Lennard-Jones (6–12) potential parameters, σ and ε, as follows

$$\psi(0) = 4\varepsilon\left(\frac{a}{\sigma_0}\right)\left[3.1633\left(\frac{\sigma}{a}\right)^{12} - 5.9969\left(\frac{\sigma}{a}\right)^{6}\right] \tag{18}$$

$$\psi'(0) = 4\varepsilon\left(\frac{a}{\sigma_0}\right)\left[2.3532\left(\frac{\sigma}{a}\right)^{12} - 4.2201\left(\frac{\sigma}{a}\right)^{6}\right] \tag{19}$$

where $\psi(0)$ and $\psi'(0)$ are the potential energy of a monomer unit in the bulk and surface state, respectively; a is the equilibrium distance between adjacent chains; σ_0 is the distance between adjacent monomer units in the same chain along the chain axis, which can be obtained from the bond lengths and angles of the chain skeleton. The potential parameters σ and ε can be estimated by employing the approximate method proposed by Davis.[82] This method correlates σ and ε with the polarizability α and diamagnetic susceptibility χ_D of a monomer unit through empirical formulas. The polarizabilities and diamagnetic susceptibilities are obtained indirectly by adding atomic or partial group contributons from tables given by Van Krevelen.[83] The estimated values of σ and ε have been tabulated for various polymers.[34,82]

The derivation of the surface tension equation in this theory is analogous to that of Ree *et al.*[57] for simple liquids, in which the surface energy and entropy are assumed to originate predominantly with the molecular entities of the single surface layer exposed to the atmosphere. The surface tension equation has been obtained as

$$\gamma = \beta^{-3/2}\frac{\sigma_0\sigma^3}{(V/N)^2}\left[\varepsilon\left(\frac{\sigma}{\sigma_0}\right)(3.5536\beta^{5/2} - 1.6162\beta^{11/2}) - \frac{1}{3}kT\ln\frac{1-0.875-\beta^{1/2}}{1-\beta^{1/2}}\right] \tag{20}$$

where V is the volume of amorphous polymers which consist of one mole of monomer units, N is Avogadro's number, and $\beta \equiv N\sigma_0\sigma^2/V_s$; V_s measures the molar volume occupied by solid-like monomer units. When the dimensionless parameter β is adjusted to 0.98, a good fit of experimental surface tensions of various polymers over a wide range of temperature is obtained. The temperature coefficient of surface tension, which is essentially the surface entropy, is especially satisfactory.

An extension of this theory to interfacial tensions between various polymer melts has been attempted,[84] employing the relation between the energy of cohesion and the energy of adhesion which was first proposed by Girifalco and Good.[85] In this extended work, the importance of polar force interactions between polymers having a permanent dipole in their constituent groups has been taken into account.

2.3. SCALED PARTICLE THEORY

The ideas[86] used in a radial distribution function theory of liquids consisting of rigid-sphere molecules have been extended by Reiss *et al.*[87] to develop an approximate treatment, which is called the scaled particle theory of liquids composed of attractive but impenetrable molecules. The following relationship for computing surface tensions was obtained from their treatment

$$\gamma = \frac{kI}{4\pi a^2}\left[\frac{12y}{(1-y)} + \frac{18y^2}{(1-y)^2}\right] - \frac{pa}{2} \tag{21}$$

where a is the hard-core diameter of the molecules, P is pressure, y is $4a^3N/6V$, N is Avogadro's number, and V is the molar volume. The term $pa/2$ generally is of negligible magnitude and can be ignored in the calculation. It can be seen that a knowledge of only the density of the liquid and of the hard-core diameter of the molecules is needed to calculate the value of the surface tension. Schonhorn[88] has employed the relation of Eq. (21) to develop an equation for the surface tension of polymers in terms of the cohesive energy density.

According to the scaled particle theory of liquids, the energy of vaporization per hard-core molecule is

$$\Delta E_v' = kT\frac{(1 - y^3)}{(1 - y)^3} \tag{22}$$

Using Eq. (22) and ignoring the term $pa/2$, Eq. (21) can be rewritten as

$$\gamma = \left(\frac{a}{4}\right)\left[\frac{(2 - y - y^2)}{(1 - y)^3}\right]\left(\frac{\Delta E_v}{V}\right) \tag{23}$$

where $\Delta E_v = N\Delta E_v'$. Equation (23) shows the relationship, previously introduced by Hilldebrand and Scott,[89] between the surface tension of liquids and the cohesive energy density. Knowing that $(4/3)\pi(a/2)^3 = V'/(nN)$, where V' is the hard-core volume of one mole of polymer and n is the number of identical groups in a molecule. Since $V' = yV$, then

$$\gamma = 0.310\left[\frac{y^{1/3}(2 - y - y^2)}{1 - y^3}\right]\left(\frac{V'}{nN}\right)^{1/3}\left(\frac{\Delta E_v}{V}\right) \tag{24}$$

Letting $m = y^{1/3}(2 - y - y^2)/(1 - y^3)$, Eq. (24) yields

$$\gamma = 0.310m\left(\frac{V}{nN}\right)^{1/3}\left(\frac{\Delta E_v}{V}\right) \tag{25}$$

Schonhorn has interpreted the value of m as the ratio of the internal pressure to the cohesive energy density, which is insensitive to structural variations in nonpolar species. However, the scaled particle theory approach of Schonborn hardly takes into account the connectedness of polymer segments.

2.4. SELF-CONSISTENT FIELD THEORY

Helfand and Tagami[90] formulated a statistical mechanical theory of the interface between immiscible polymers 1 and 2. The treatment of Helfand and Tagami is more fundamental than the other theories which we have discussed in preceding subsections, since it is based on Gaussian statistics. A self-consistent field approach is employed in this theory to estimate both the density profile of two immiscible polymers and the mean force field in the joining region, and then predict the interfacial tension. One of the weaknesses of the theory is that it has been developed only for the symmetrical case that polymers 1 and 2 are similar. Later, Helfand and Sapse extended it to be applicable to asymmetrical systems.[91]

In the extended treatment of this theory, a reduced density of each component $\bar{\rho}_K(\mathbf{r})$ is defined, $\bar{\rho}_K(\mathbf{r}) \equiv \rho_K(\mathbf{r})/\rho_{0K}$, where $K = 1$ or 2, $\rho_K(\mathbf{r})$ is the density of species K at a point $\mathbf{r}$ in the interfacial region, and ρ_{0K} is the density of pure K. The statistical segment length b_K is taken in such a way that the mean square end-to-end distance is equal to $Z_k b_K^2$, where Z_K is the degree of polymerization of K. In order to derive the local number density (or local volume fraction) of segments, Helfand and Tagami[90] have introduced a function q_K $(\mathbf{r}, t)$ which is proportional to the probability density that the end of a molecule, of type K and degree of polymerization t, is at $\mathbf{r}$. If the Gaussian statistics of segmental units in an effective force field U $(\mathbf{r})$ relative to the bulk phase are employed, a differential equation for q_K $(\mathbf{r}, t)$ can be obtained.

$$\frac{\partial q_K}{\partial t} = \left[\left(\frac{b_K^2}{6}\right)\nabla^2 - \frac{U_K(\mathbf{r})}{dT}\right]q_K \tag{26}$$

where ∇^2 is the Laplace operator, equal to $\partial^2/\partial x^2 + \partial^2/\partial y^2 + \partial^2/\partial z^2$, and an initial condition is $q_K(\mathbf{r}, 0) = 1$. For solving Eq. (26), the $U_K(\mathbf{r})$ is replaced by the work of bringing a segmental unit of K from bulk K to the point $\mathbf{r}$ in the interfacial region. This is a chemical potential-like quantity derivable from a free energy. Moreover, the theory clarifies the local and nonlocal aspects of the force fields. The former arises from the relatively unfavorable interaction of the two unlike species at the interface, and the latter is, essentially, the tendency of interdiffusion. Adapting the regular solution approach, the local free energy density of a hypothetical mixture of densities ρ_1 and ρ_2 has been taken as

$$\frac{\Delta f^*}{kT} = \alpha\bar{\rho}_1\bar{\rho}_2(\bar{\rho}_1 + \bar{\rho}_2) + (\bar{\rho}_1 + \bar{\rho}_2 - 1)^2/(2\kappa kT) \tag{27}$$

where α is a mixing parameter related to the Flory–Huggins interaction parameter χ, and κ is the compressibility of the polymers which has been assumed to be small and independent of composition. The nonlocal part, which tends to disfavor sharp boundaries, manifests itself in density gradient terms. Including the nonlocal part, in which higher-order terms than gradient squared are disregarded, the mean work of adding a segmental unit into the hypothetical mixture is then

$$W_K(x) = \rho_{0K}\frac{\partial(\Delta f)^*}{\partial \rho_K} + kT\left(\frac{\sigma^2\alpha}{\sigma}\right)\frac{\partial^2 \bar{\rho}_{K'}}{\partial x^2} \tag{28}$$

where $K \neq K'$, σ is the distance parameter, which measures the range of nonlocality, and x is the coordinate perpendicular to the interfacial plane.

The density of K at a point $\mathbf{r}$, $\rho_K(\mathbf{r})$, is expressed in terms of $q_K(\mathbf{r}, t)$ through the relation

$$\rho_K(\mathbf{r}) = \frac{\rho_{0K}}{Z_K}\int_0^{Z_K} q_K(\mathbf{v}, Z_K - t)q_K(\mathbf{V}, t)\,\mathrm{d}t \tag{29}$$

Equation (26), with W_K/ρ_{0K} replacing U_K, involves the unknown densities. Thus Eqs. (26) and (29) form a self-consistent set for the calculation of the density profile and the effective field simultaneously. For the case of quite large Z_K, Eqs. (26) and (29) are simplified and solvable without resorting to numerical methods. In this case, Helfand and Sapse[91] obtained the interfacial tension equation as follows

$$\gamma_{12} = kT\alpha^{1/2}\left[\frac{\beta_1 + \beta_2}{2} + \frac{1}{6}\frac{(\beta_1 - \beta^2)^2}{(\beta_1 + \beta_2)}\right] + \gamma_{12}^{(\text{nonlocal})} \tag{30}$$

and an expression for a characteristic thickness of the interface

$$a = 2[(\beta_1^2 + \beta_2^2)/(2\alpha)]^{1/2} \tag{31}$$

where $\beta_K = (\rho_{0K}b_K^2/6)^{1/2}$, and the nonlocal contribution to the interfacial tension is

$$\gamma_{12}^{(\text{nonlocal})} = \frac{\sigma^2\alpha}{18}\left[\frac{2}{(\beta_1 + \beta_2)} - \frac{2}{5}\frac{(\beta_1 - \beta_2)^2}{(\beta_1 + \beta^2)^3}\right] \tag{32}$$

With crude estimates of the parameter α, resorting to the use of the Hildebrand solubility parameter relation, $\alpha = (1/kT)(\delta_1 - \delta_2)^2$, and taking the values of δ_K adding up empirical group contributions, reasonable predictions of interfacial tension for various polymer pairs have been made. The nonlocal contributions are, in many cases, very small, and the characteristic thickness of interfacial regions are of the order of tens of angstroms.

2.5. THERMODYNAMIC THEORY

Kammer examined the interfacial phenomena of polymer melts from the thermodynamic point of view.[92] A system of thermodynamic equations has been derived to describe the temperature, pressure, and composition dependence of interfacial tension starting from the fundamental equation of Guggenheim,[93] and employing the well-known Gibbs–Duhem equation of intensive parameters. In order to obtain an expression for the interfacial tension, Kammer has adapted the Cahn–Hilliard formalism of the nonuniform system. Later, he proposed a more general thermodynamic theory for the interfacial tension[94] using an expression based on Prigogine's cell model of polymers for the free energy of the boundary layer. The derived expression for the interfacial tension contains the ratio of the chain lengths of the two polymer species, a parameter of chain flexibility, and an energy parameter. In terms of these parameters, geometric, structural, and energetic effects on the interfacial tension have been discussed, and a strong structural effect can be noted.

2.6. GEOMETRIC MEAN AND HARMONIC MEAN THEORIES

The interfacial tension can be expressed in terms of the free energy of adhesion per unit area for the interface between the phases 1 and 2, ΔF_{12}, and the free energy of cohesion per unit area of the two phases, ΔF_{11} and ΔF_{22},

$$\gamma_{12} = \frac{\Delta F_{11}}{2} + \frac{\Delta F_{22}}{2} - \Delta F_{12} \tag{33}$$

ΔF_{12} is defined as the work required to separate reversibly an interface, and ΔF_{11} and ΔF_{22} are identical with $2\gamma_1$ and $2\gamma_2$, respectively, where γ_1 and γ_2 are the surface tensions of the phases. A diagram of the process for thermodynamic definition of ΔF_{11}, ΔF_{22}, and ΔF_{12} is presented in Figure 3.

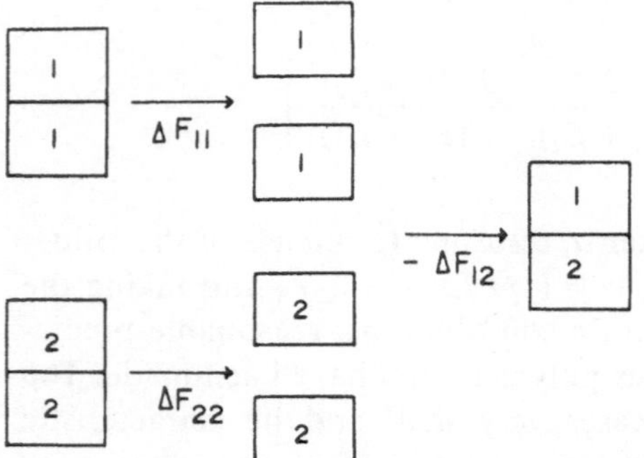

FIGURE 3. A diagrammatic definition of the free energy of cohesion and the free energy of adhesion.

If ΔF_{12} can be expressed as a function of ΔF_{11} and ΔF_{22}, then Eq. (33) provides a basis for relating the interfacial tension to the surface tension of the two phases.

It has been recognized by Girifalco and Good[85] that the free energy of adhesion is given by the geometric mean of the free energies of cohesion multiplied by a correction factor Φ_{12}.

$$\Delta F_{12} = \Phi_{12}(\Delta F_{11}\Delta F_{22})^{1/2} \tag{34}$$

They derived the expression for Φ_{12} using a quasi-continuum model of condensed phase.

$$\Phi_{12} = \frac{A_{12}}{(A_{11}A_{22})^{1/2}} \frac{d_{11}d_{22}}{d_{12}^2} \tag{35}$$

Here, A_{11} and A_{22} are the inverse sixth power dispersion coefficients for molecular species 1 and 2, respectively; A_{12} is the dispersion coefficient for the interaction of molecular species 1 and 2; d_{11}, d_{22}, and d_{12} are the equilibrium separation distances between the two semi-infinite bodies of the respective phases. Equation (33) is then

$$\gamma_{12} = \gamma_1 + \gamma_2 - 2\Phi_{12}(\gamma_1\gamma_2)^{1/2} \tag{36}$$

The calculated interfacial tension using Eq. (36) is very sensitive to the variation of the parametric quantity Φ_{12}. Therefore, one must be careful in predetermining the value of Φ_{12} in order to predict the interfacial tension. For high polymers, an expression of Φ_{12} in terms of the polarizability and the diamagnetic susceptibility of polymer constituent groups has been proposed by Davis on the basis of the Kirkwood-Müller relation for the potential parameters.[82]

A different approach to correlation of the surface and interfacial tensions has been proposed by Fowkes for the hydrocarbon-water interfacial system.[95] This approach employs a hypothesis of decomposing the surface tension of a liquid into the components of London dispersion force and polar or hydrogen bonding contributions according to the nature of the intermolecular forces in the liquid. A geometric mean expression for the free energy of cohesion between the hydrocarbon and water phases has been assumed such as

$$\Delta F_{12} = 2(\gamma_1\gamma_2^d)^{1/2} \tag{37}$$

where the subscripts 1 and 2 mean the hydrocarbon and water phases, respectively, and the superscript d designates the dispersion force contribution. Thus the interfacial tension becomes

$$\gamma_{12} = \gamma_1 + \gamma_2 - 2(\gamma_1 \gamma_2^d)^{1/2} \tag{38}$$

Since γ_1, γ_2, and γ_{12} are experimentally measurable, the γ_2^d can be obtained. For water at 20°C, γ^d is 22 dyn/cm, which amounts to about 30% of the surface tension. The concept of the dispersion force component of surface tension has been discussed in detail, and extended to various polar liquids including high polymers.[96] This is a simple and very useful method by which to correlate the experimental data on the interfacial tension of dipolar systems. But the use of the geometric mean relationship is based on certain assumptions and has certain limitations. A modification of this approach employing a harmonic mean equation has been proposed by Wu.[54] He used the equation

$$\gamma_{12} = \gamma_1 + \gamma_2 - \frac{4\gamma_1^d \gamma_2^d}{(\gamma_1^d + \gamma_2^d)} - 4\frac{\gamma_1^p \gamma_2^p}{(\gamma_1^p + \gamma_2^p)} \tag{39}$$

in order to correlate the experimental data on dipolar polymers, where the superscript p's are for the polar force contribution to surface tension.

2.7. REGULAR SOLUTION THEORY

The theory of Vrij[97,98] and later works of Nose[99] for the interface of polymer systems seem to be influenced by Debye's model of polymer solution[100] in the realm of the regular solution approach. The model pictures a polymer chain as a flexible string of monomers, and assumes that in each polymer coil the average density of monomers around the coil center of mass is spherically symmetric and thus can be represented by a Gaussian distribution. In Vrij's theory the density distribution of monomers in a polymer chain is assumed to be invariable even at the interface. However, Nose included the influence of chain deformation on the interfacial properties. Both theories derive equations for the interfacial tension and the quantities concerned with the interface profile on the basis of the theory of Cahn and Hilliard.[56] But the treatments of these theories are concentrated on the system of inhomogeneous polymer solutions near the consolute point, and thus an extension of them remains as further work.

3. THE THEORY OF INTERFACIAL TENSION AT POLYMER–WATER INTERFACES

3.1. SURFACE TENSION OF WATER

It is well-known that water exhibits a variety of unusual properties including its surface properties. A careful study of the variation of surface tension with temperature shows that for ordinary nonpolar liquids the average molar surface entropy is 24 J per degree, while for water this value is only 9.8 J per degree.[101] The entropy deficit of about 14 J per degree, or about 1.7 k per molecule, has been attributed to a preferred molecular orientation at the surface. Although orientational order is not the only factor responsible for the lowering of the slope of the $\gamma(T)$ characteristic, there are other reasons for expecting a higher degree of orientational order amongst the surface molecules with respect to those in the bulk.[102] It has been believed that the permanent quadrupole moment of the water molecule might be responsible for the preferred orientation at the interfacial boundary, and a specific surface polarization with associated surface field and surface potential will develop.[103] Numerous experimental and theoretical attempts have been made to determine the surface potential of liquid water, but they agree neither in sign nor magnitude, and the interfacial structure remains unsolved.[103]

Disregarding electrostatic depolarizing effects, a semiempirical approach for estimating the surface tension of water has been proposed by Fletcher,[104] bringing out the importance of hydrogen bonding, and regarding the water structure as an ice-like structure containing a considerable fraction of broken bonds, a good approximation for supercooled water. In order to obtain a formal expression for the difference in free energy between a partially oriented and a completely random water surface, Fletcher considered the enthalpy and entropy contributions. The empirical entropy deficit attributed to orientation, 1.7 k, has been incorporated in his formalism. Then the average free energy per surface oriented molecule is less than that of a random surface by 10^{-13} erg per surface molecule. Jhon *et al.*[105] have used their theory of the domain structure of water and improved Chang and coworkers' surface tension theory[106] to calculate the surface tension of liquid water. In their theory it has been assumed that an asymmetrical field of the surface makes those domains of water molecules grow which are favorably oriented with respect to the field at the expense of the less favorably oriented domains until a steady state is achieved. But the molecules other than those in the first surface layer are approximated by a symmetrical field, and thus, no such molecular orientation exists for them. Then the surface tension of water is mainly due to the orientation of molecules in the top layer which increases the orderliness at the surface, but is partly due also to the change in the density within a few molecular diameters of the interface, which can be calculated by using the method of Chang. Adjusting the quantity of field strength, Jhon *et al.* obtained good results for calculated surface tension and surface entropy.

Davis[107] has investigated the surface tension of water on the basis of the treatment of Kirkwood and Buff, for which, presumably, the long-range dipole-dipole interactions represent the principal anisotropic angular contribution to the pair potential. In this dipolar approximation Davis adopts a central Lennard-Jones interaction supplemented beyond some cutoff radius by a term for the dipole-dipole interaction. For water at 0°C, Davis calculated the dipolar component of surface tension $\gamma_{DD} = 114$, 60, and 38 dyn/cm for cutoff radii of $a = 2.85$, 4, and 5 Å, respectively (the nearest neighbor separation at the density of this temperature being 2.75 Å). The experimental value is 76 dyn/cm. Whilst it is clear that long-range dipolar interactions do make a contribution of substantial magnitude to the final value, it is virtually impossible to assess the effects of the various approximations employed in this theory. Fulton has also presented a similar approach for the surface tension of water.[108]

3.2. INTERFACIAL TENSION AT POLYMER–WATER INTERFACES

Because of incomplete knowledge of the surface structure of water, our understanding of polymer–water interfacial properties is not satisfactory, though there is substantial evidence for the existence of ordered water structures near aqueous/solid interfaces. Particular importance has been attached to the existence of thermal anomalies in the properties of interfacial water.[109] It has been suggested that these anomalies are evidence of high-order phase transition, such transitions only occurring in partially ordered structure units of a certain minimum size. However, the nature and extent of the ordering of the interfacial water molecules will depend upon the characteristics of the solid surface. The interfacial properties of polymer-water systems might exhibit some pecularities aside from the surface property of the polymer alone. An example is the surface tension of toluene solutions of ethyl acetate–vinyl acetate copolymer (γ_{sol}) and their interfacial tensions against water ($\gamma_{sol\text{-}w}$) plotted against the contents of vinyl acetate in the copolymer. γ_{sol} does not vary over the whole range of composition while $\gamma_{sol\text{-}w}$ changes drastically with composition, passing through a minimum and a maximum.[5] As for the interface between pure polymer and water, no relevant data are available. It is anticipated that knowledge of polymer–polymer interfacial properties, together with understanding of the surface anomaly of water molecules, will become a basis for modeling the interface of pure polymer–water systems.

REFERENCES

1. S. Wu, Interfacial and surface tensions of polymers, *J. Macromol. Sci., Revs. Macromol. Chem.* **C10**, 1-73 (1974).
2. S. Wu, in: *Polymer Blends* (D. R. Paul and S. Newman, eds.), Vol. 1, pp. 243-293, Academic Press, New York (1978).
3. G. L. Gaines, Jr., Surface and interfacial tension of polymer liquids, *Polym. Eng. Sci.*, **12**, 1-11 (1972).
4. H. L. Frisch, G. L. Gaines, Jr., and H. Schonhorn, in: *Treatise on Solid State Chemistry* (N. B. Hannay, ed.), Vol. 6B, pp. 343-412, Plenum Press, New York (1976).
5. T. Hata, *Polym. Sci. Technol.* **12A**, 15-41 (1980).
6. T. W. Richards and L. B. Coombs, The surface tensions of water, methyl, ethyl, and isobutyl alcohols, ethyl butyrate, benzene, and toluene, *J. Am. Chem. Soc.* **37**, 1656-1676 (1915).
7. W. D. Harkins and E. C. Humphery, The drop weight method for the determination of surface tension, *J. Am. Chem. Soc.* **48**, 228-236 (1916).
8. O. R. Quayle, The parachors of organic compounds, *Chem. Rev.* **53**, 439-589 (1953).
9. I. D. Sokolova and N. K. Voskresenskaya, *Russ. Chem. Rev.* **35**, 500-517 (1966).
10. H. L. Frisch and Z. W. Salsburg, *Simple Dense Fluids*, Academic Press, New York (1968).
11. A. F. Gallaugher and H. Hibbert, Studies on reactions relating to carbohydrates and polysaccharides, *J. Am. Chem. Soc.* **59**, 2514-2521 (1939).
12. J. F. Padday, in: *Surface and Colloid Science*, (E. Matijevic, ed.), Vol. 1, pp. 101-251, Wiley-Interscience, New York (1969).
13. A. W. Adamson, *Physical Chemistry of Surfaces*, 3rd ed., Wiley-Interscience, New York (1976).
14. R. J. Roe, V. L. Bachetta, and P. M. G. Wong, Refinement of pendant drop method for the measurement of surface tension of viscous liquid, *J. Phys. Chem.* **71**, 4190-4193 (1967).
15. R. J. Roe, Surface tension of polymer liquids, *J. Phys. Chem.* **72**, 2013-2017 (1968).
16. S. Wu, Surface and interfacial tensions of polymer melts, *J. Colloid Interface Sci.* **31**, 153-161 (1969).
17. T. Sakai, Surface tension of polyethylene melt, *Polymer* **6**, 659-661 (1965).
18. T. Hata, *Hyomen* **6**, 281-2898 (1968).
19. H. M. Princen, I. Y. Z. Zia, and S. G. Mason, Measurement of interfacial tension from the shape of a rotating drop, *J. Colloid Interface Sci.* **23**, 99-107 (1967).
20. H. T. Patterson, K. H. Hu, and T. H. Grindstaff, Measurement of interfacial and surface tensions in polymer systems, *J. Polym. Sci., C* **34**, 31-43 (1971).
21. W. A. Zisman, *Adv. Chem. Ser.* **43**, 1-51 (1964).
22. J. J. Bikerman, Sliding of drops from surfaces of different roughness, *J. Colloid Sci.* **5**, 349-359 (1950).
23. E. Baer and F. McLaughlin, The determination of surface properties of polymers from liquid drop stability on an inclined plane, *J. Appl. Polym. Sci.* **5**, 240-245 (1961).
24. W. Funke, G. E. H. Hellweg, and A. W. Neumann, *Angew. Makromol. Chem.* **8**, 185-193 (1969).
25. H. W. Fox and W. A. Zizman, The spreading of liquids on low energy surfaces, *J. Colloid Sci.* **7**, 605-609 (1979).
26. S. Wu, Surface tension of solids: an equation of state analysis, *J. Colloid Interface Sci.* **71**, 605-609 (1979).
27. E. G. Shafrin, in: *Polymer Handbook*, 2nd ed. (J. Brandrup *et al.*, eds.), pp. 111-221, Wiley-Interscience, New York (1975).
28. R. J. Roe, Interfacial tension between polymer liquids, *J. Colloid Interface Sci.* **31**, 228-235 (1969).

29. S. Wu, Surface and interfacial tensions of polymer melts, *J. Phys. Chem.* **74**, 632–638 (1970).
30. S. Wu, Calculation of interfacial tension in polymer systems, *J. Polym. Sci.*, C **34**, 19–30 (1971).
31. D. B. MacLeod, On a relation between surface tension and density, *Trans. Faraday Soc.* **19**, 38 (1923).
32. J. O. Hirschfelder, C. F. Curtiss, and R. B. Bird, *Molecular Theory of Gases and Liquids*, Wiley, New York (1964).
33. C. W. Stewart and C. A. von Frankenburg, Significant structure theory of the surface tension of polyethylene, *J. Polym. Sci., A-2* **6**, 1686–1688 (1968).
34. Y. Oh and M. S. Jhon, Theoretical estimation of surface tension of amorphous high polymers, *J. Colloid Interface Sci.* **73**, 467–474 (1980).
35. T. G. Fox and P. J. Flory, Second-order transition temperatures and related properties of polystyrene, *J. Appl. Phys.* **21**, 581–591 (1950).
36. T. G. Fox and P. J. Flory, The glass temperature and related properties of polystyrene, *J. Polym. Sci.* **14**, 315–319 (1954).
37. T. G. Fox and S. Loshaek, Influence of molecular weight and degree of crosslinking on the specific volume and glass temperature of polymers, *J. Polym. Sci.* **15**, 371–390 (1955).
38. K. Ueberreiter and G. Kanig, *Z. Naturforsch. A* **6**, 551–559 (1951).
39. R. F. Boyer, The relation of transition temperatures to chemical structure in high polymers, *Rubber Chem. Technol.* **36**, 1303–1421 (1963).
40. J. R. Martin, J. F. Johnson, and A. R. Cooper, *J. Macromol. Sci., Revs. Macromol. Chem.* **C8**, 57–199 (1972).
41. M. C. Phillips and A. C. Riddiford, The specific free surface energy of parafinic solids, *J. Colloid Interface Sci.* **22**, 149–157 (1966).
42. R. H. Dettre and R. E. Johnson, Jr., Surface tension, critical surface tension, and temperature coefficient of surface tension of polytetrafluoroethylene, *J. Phys. Chem.* **71**, 1529–1531 (1967).
43. T. Hata, *Kobunshi* **17**, 594–605 (1968).
44. H. Edwards, Surface tensions of liquid polyisobutylenes, *J. Appl. Polym. Sci.* **12**, 2213–2224 (1968).
45. D. G. LeGrand and G. L. Gaines, Jr., Immiscibility and interfacial tension between polymer liquids, *J. Colloid Interface Sci.* **50**, 272–279 (1975).
46. G. L. Gaines, Jr. and G. L. Gaines, III, The interfacial tension between *n*-alkanes and poly(ethylene glycols), *J. Colloid Interface Sci.* **63**, 394–398 (1978).
47. G. D. LeGrand and G. L. Gaines, Jr., The molecular weight dependence of polymer surface tensions, *J. Colloid Interface Sci.* **31**, 162–167 (1969).
48. D. G. LeGrand and G. L. Gaines, Jr., Surface tension of homologous series of liquids, *J. Colloid Interface Sci.* **42**, 181–184 (1973).
49. H. W. Starkweather, *SPE Trans.* **5**, 506 (1965).
50. G. W. Bender, D. G. LeGrand, and G. L. Gaines, Jr., Molecular weight dependence of surface tension and refractive index for some poly(ethylene oxide) derivatives, *Macromolecules* **2**, 681–682 (1969).
51. A. K. Rastogi and L. E. St. Pierre, Interfacial phenomena in macromolecular systems, *J. Colloid Interface Sci.* **35**, 16–22 (1971).
52. K. S. Slow and D. Patterson, The prediction of surface tensions of liquid polymers, *Macromolecules* **4**, 26–30 (1971).
53. N. G. Gaylord, *Adv. Chem. Ser.* **142**, 76–84 (1975).
54. S. Wu, Polar and nonpolar interactions in adhesion, *J. Adhesion* **5**, 39–55 (1973).
55. J. G. Kirkwood and F. P. Buff, The statistical mechanical theory of surface tension, *J. Chem. Phys.* **17**, 338–343 (1949).
56. J. W. Cahn and J. E. Hilliard, Free energy of a nonuniform system, *J. Chem. Phys.* **28**, 258–267 (1958).

57. T. S. Ree, T. Ree, and H. Eyring, Significant structure theory of surface tension, *J. Chem. Phys.* **41**, 524–530 (1964).
58. I. Prigogine and L. Saraga, *J. Chim. Phys.* **49**, 399–407 (1952).
59. P. J. Flory, *Principles of Polymer Chemistry*, Cornell University Press, Ithaca (1953).
60. I. Prigogine, N. Trappeniers, and V. Mathot, *Disc. Faraday Soc.* **15**, 93–107 (1953).
61. R. J. Roe, Theory of the interface between polymers and polymer solutions, *J. Chem. Phys.* **62**, 490–499 (1975).
62. E. Helfand, Theory of inhomogeneous polymers, *J. Chem. Phys.* **63**, 2192–2198 (1975).
63. I. Prigogine, A. Bellemans, and C. Naar-Colin, Theorem of corresponding states for polymers, *J. Chem. Phys.* **26**, 751–755 (1957).
64. R. Simha and A. J. Havlik, On the equation of state of oligomer and polymer liquids, *J. Am. Chem. Soc.* **86**, 197–204 (1964).
65. V. S. Nanda, R. Simha, and T. Somcynsky, Principle of corresponding states and equation of state of polymer liquids and glasses, *J. Polym. Sci., C* **12**, 277–295 (1966).
66. R. Simha and T. Somcynsky, On the statistical thermodynamics of spherical and chain molecule fluids, *Macromolcules* **2**, 342–350 (1969).
67. V. S. Nanda and R. Simha, Equation of state and related properties of polymer and oligomer liquids, *J. Phys. Chem.* **68**, 3158–3163 (1964).
68. R. J. Roe, Hole theory of surface tension of polymer liquids, *Proc. Nat. Acad. Sci. USA*, **56**, 819–824 (1966).
69. T. Nose, A hole theory of polymer liquids and glasses, *Polym. J.* **3**, 1–11 (1972).
70. D. Patterson and A. K. Rastogi, The surface tension of polyatomic liquids and the principle of corresponding states, *J. Phys. Chem.* **74**, 1067–1071 (1970).
71. C. I. Poser and I. C. Sanchez, Surface tension theory of pure liquids and polymer melts, *J. Colloid Interface Sci.* **69**, 539–548 (1979).
72. I. C. Sanchez and R. H. Lacombe, An elementary molecular theory of classical fluids, pure fluids, *J. Phys. Chem.* **80**, 2352–2362 (1976).
73. C. I. Poser and I. C. Sanchez, Interfacial tension theory of low and high molecular weight liquid mixtures, *Macromolecules* **14**, 361–370 (1981).
74. H. Eyring and M. S. Jhon, *Significant Liquid Structures*, Wiley, New York (1969).
75. M. S. Jhon and H. Eyring, in: *Theoretical Chemistry: Advances and Perspectives*, Vol. 3, pp. 55–141, Academic Press, New York (1978).
76. S. M. Ma, H. Eyring, and M. S. Jhon, The significant structure theory applied to amorphous and crystalline polyethylene, *Proc. Nat. Acad. Sci. USA* **71**, 3096–3100 (1974).
77. J. S. King, Neutron scattering from polymers, *J. Macromol. Sci., Phys.* **B12**, 13–25 (1976).
78. H. Benoit, Determination of polymer chain conformation in amorphous polymers, *J. Macromol. Sci., Phys.* **B12**, 27–40 (1976).
79. E. W. Fischer, J. H. Wendorff, M. Dettenmaier, and I. Voigt-Martin, Chain conformation and structure in amorphous polymers as revealed by X-ray neutron, light, and electron scattering, *J. Macromol. Sci., Phys.* **B12**, 41–59 (1976).
80. G. D. Patterson, Light scattering and the local structure of amorphous polymers, *J. Macromol. Sci., Phys.* **B12**, 61–74 (1967).
81. G. D. Wignall, D. G. H. Ballard, and J. Schelten, Chain conformation in molten and solid polystyrene and polyethylene by low angle neutron scattering, *J. Macromol. Sci., Phys.* **B12**, 75–98 (1976).
82. B. W. Davis, Estimation of surface free energies of polymeric materials, *J. Colloid Interface Sci.* **59**, 420–428 (1977).
83. D. W. Van Krevelen, *Properties of Polymers*, Elsevier, New York (1972).
84. Y. Oh, J. D. Andrade, and M. S. Jhon, Theoretical estimation of interfacial tension between molten polymers, *J. Korean Chem. Soc.* **23**, 210–216 (1979).
85. L. A. Girifalco and R. J. Good, A theory for the estimation of surface and interfacial energies, *J. Phys. Chem.* **61**, 904–909 (1957).

86. H. Reiss, H. L. Frisch, and J. L. Lebowitz, Statistical mechanics of rigid spheres, *J. Chem. Phys.* **31**, 369-380 (1950).
87. H. Reiss, H. L. Frisch, E. Helfand, and J. L. Lebowitz, Aspects of the statistical thermodynamics of real fluids, *J. Chem. Phys.* **32**, 119-124 (1960).
88. H. Schonborn, Theoretical relationship between surface tension and cohesive energy density, *J. Chem. Phys.* **43**, 2041-2043 (1965).
89. J. H. Hildebrand and R. L. Scott, *The Solubility of Nonelectrolytes*, Reinhold, New York (1950).
90. E. Helfand and Y. Tagami, Theory of the interface between immiscible polymers, *J. Chem. Phys.* **56**, 3592-3601 (1972).
91. E. Helfand and A. M. Sapse, Theory of unsymmetric polymer-polymer interfaces, *J. Chem. Phys.* **62**, 1327-1331 (1975).
92. H. W. Kammer, Surface and interfacial tension of polymer melts, *Z. Phys. Chemie, Leipzig*, **258**, 1149-1161 (1977).
93. E. A. Guggenheim, *Trans. Faraday Soc.* **36**, 1149-1161 (1977).
94. H. W. Kammer, *Z. phys. Chemie, Leipzig*, **261**, 519-528 (1980).
95. F. M. Fowkes, Determination of interfacial tensions, contact angles, and dispersion forces in surfaces, *J. Phys. Chem.* **66**, 381 (1962).
96. F. J. Fowkes, in: *Chemistry and Physics of Interface* (D. E. Gushee, ed.), Vol. 2, pp. 153-167, Am. Chem. Soc. Pub., Washington, D.C. (1971).
97. A. Vrij, Equation for the interfacial tension between demixed polymer solutions, *J. Polym. Sci., A-2* **6**, 1919-1932 (1968).
98. A. Vrig and G. J. Roebersen, Inhomogeneous polymer solutions, *J. Polym. Sci., Polym. Phys. Ed.* **15**, 109-125 (1977).
99. T. Nose, Theory of liquid-liquid interface of polymer systems, *Polym. J.* **8**, 96-113 (1976).
100. P. Debye, *J. Chem. Soc.* **31**, 680-687 (1959).
101. R. J. Good, Surface entropy and surface orientation of polar liquids, *J. Phys. Chem.* **61**, 810-813 (1957).
102. F. H. Stillinger and A. Ben-Naim, Liquid-vapor interface potential for water, *J. Chem. Phys.* **47**, 4431-4437 (1967).
103. C. A. Croxton, *Statistical Mechanics of Liquid Surface*, Wiley, New York (1980).
104. N. H. Fletcher, Surface structure of water and ice, *Phil. Mag.* **7**, 255-269 (1962).
105. M. S. Jhon, E. R. Van Artsdalen, J. Grosh, and H. Eyring, Further applications of the domain theory of liquid water, *J. Chem. Phys.* **47**, 2231-2234 (1967).
106. S. Chang, T. Ree, H. Eyring, and L. Matzner, in: *Progress in International Research on Thermodynamic and Transport Properties* (J. F. Masi and D. H. Tsai, eds.), pp. 88-92, Academic Press, New York (1962).
107. H. T. Davis, Statistical mechanics of interfacial properties of polyatomic fluids, *J. Chem. Phys.* **62**, 3412-3415 (1975).
108. R. L. Fulton, Contributions of long range polarization fluctuations to the surface tension of liquids, *J. Chem. Phys.* **64**, 1857-1858 (1976).
109. W. Drost-Hansen, in: *Chemistry and Physics of Interface* (D. E. Gushee, ed.), Vol. 2, pp. 204-241, Am. Chem. Soc. Pub., Washington, D.C. (1971).

12

Surface Raman Spectroscopy

W. M. Reichert and J. D. Andrade

1. INTRODUCTION

A large number of publications discuss the use of Raman spectroscopy as applied to the study of polymers in the bulk and in solution.[1-7] The principles of Raman scattering are thoroughly presented in modern textbooks.[7-10] This technique is extremely valuable in many areas of polymer characterization, including polymer synthesis, composition, conformation, chain orientation, stress, crosslinking, morphology, and impurity analysis. Only recently has Raman scattering been applied to the study of polymer surfaces and thin films.

The Raman effect (section 2) relies on the interaction of monochromatic light, generally the output of a laser, with vibrational/rotational modes of molecules to produce scattered light shifted in frequency away from the incident radiation. The magnitude of the Raman shift away from the laser line corresponds directly to the vibrational/rotational modes which produced the scattered light. An analysis of the frequency components present in the Raman scattered light can provide information about the structure, concentration, and identity of the molecules present in the scattering volume. While the Raman process is a weak effect in terms of conversion efficiency from incident to scattered light ($\cong 10^{-6}$ of the laser line intensity), molecular constituents in the low-ppm range can often be identified using existing lasers and detectors.

Systems designed for the detection of Raman-scattered laser light have four basic components: (1) a monochromatic laser source to excite scattering, (2) an optical system for focusing the laser beam onto the sample and for directing the Raman-scattered light to the spectrometer entrance slit, (3) a monochromator to disperse the scattered light into a frequency spec-

W. M. Reichert and J. D. Andrade ● Departments of Bioengineering and Materials Science and Engineering, College of Engineering, University of Utah, Salt Lake City, Utah 84112.

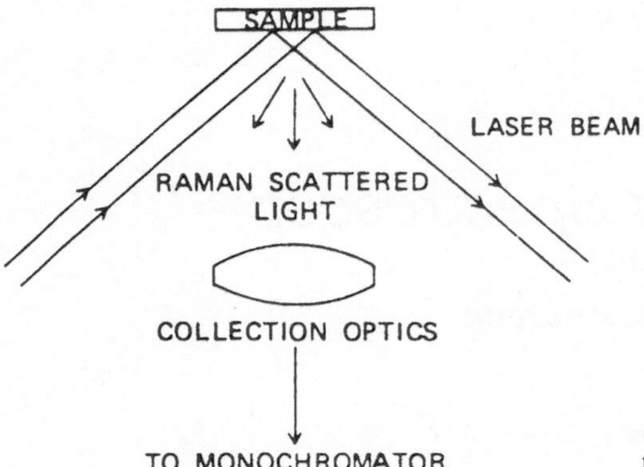

FIGURE 1. Schematic of 90° collection optics for a Raman system.

trum, and (4) a photon counter to measure the intensity of the scattered light dispersed by the monochromator. The four components are, in general, arranged into one of three collection geometries, the most common being the collection of scattered light at 90° to the sample surface (Figure 1).

The use of Raman scattered light for the characterization of surfaces and thin films is thus limited by the power of the laser light, the presence of a sufficient number of sample scattering sites within the scattering volume, and the efficiency with which the scattered light is collected.

2. RAMAN SCATTERING EFFECT

When monochromatic light of frequency ν_0 encounters a molecule the energy level of the molecule is increased by $h\nu_0$ where h is Plank's constant. This acquisition of energy causes the molecule to reach an excited virtual state. Upon relaxation the molecule can elastically return to an energy level equal to its original state, emitting light of energy $h\nu_0$, or the molecule can inelastically return to an energy level higher or lower than its original state emitting light of energy $h(\nu_0 - \nu_1)$ or $h(\nu_0 + \nu_1)$, respectively (Figure 2).

The vast majority of such molecular transitions emit light at the same frequency as the incident light (ν_0). This elastic emission of light is called Rayleigh scattering. The small number of inelastic transitions which cause light to be scattered at frequencies away from the Rayleigh line produce Raman-scattered light. The light scattered below ν_0 gives rise to Stokes lines ($\nu_0 - \nu_1$) and that scattered above ν_0 the anti-Stokes lines ($\nu_0 + \nu_1$) (Figures 2 and 3).

Molecular transitions resulting in scattered light occur when the oscillating electric field vector of the incident light exerts oppositely directed forces on the electrons and nuclei of the illuminated molecule, thus inducing

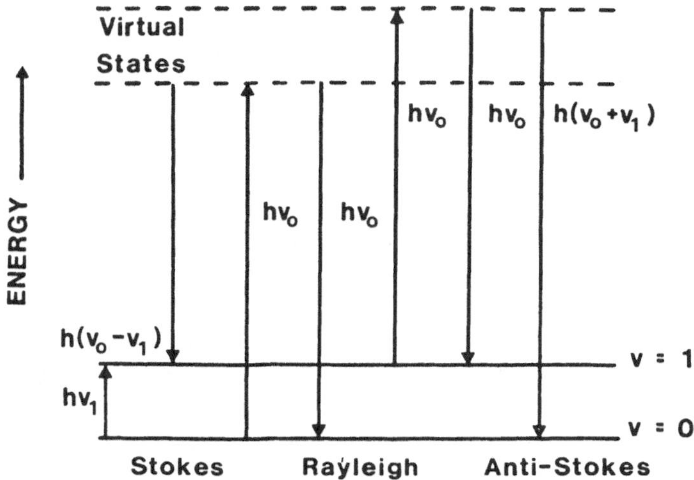

FIGURE 2. Energy level diagram showing elastic ($h\nu_0$) and inelastic [$h(\nu_0 - \nu_1)$ and $h(\nu_0 + \nu_1)$] molecular transitions.

an alternating dipole moment. The alternating dipole in turn emits the radiation detected as scattered light. If the polarizability of the molecule remains constant during illumination then the emitted radiation will have the same frequency as the incident light (Rayleigh scattering). However, if the molecule has a periodic change in polarizability (via vibrational or rotational modes) a small portion of the emitted dipole radiation will be

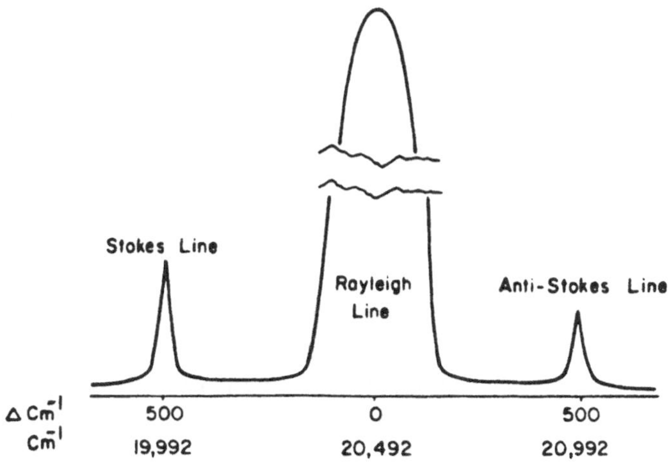

FIGURE 3. A hypothetical Raman spectrum displaying the Rayleigh line (ν_0), Stokes line ($\nu_0 - \nu_1$), and anti-Stokes line ($\nu_0 + \nu_1$). Note the relative intensities and symmetrical distribution. [Reprinted from Reference (9), p. 9, by permission.]

shifted away from the Rayleigh line. The frequency shift results from a loss $[h(\nu_0 - \nu_1)]$ or gain $[h(\nu_0 + \nu_1)]$ of energy equal to an energy level increase or decrease of the mode which altered the molecular polarizability. This altered dipole radiation is emitted as Raman-scattered light. Because an energy level increase and decrease are both probable the Raman-scattered light is shifted symmetrically with respect to the Rayleigh line (Figure 3). The magnitude of the frequency shift away from ν_0 is characteristic of specific modes. This fact makes Raman scattering a powerful characterization technique. An excellent overview of the Raman effect is given in Reference (9).

The ability of a molecule or molecular substituent to emit Raman scattered light can (in theory) be considered in terms of a polarizability interaction parameter called the Raman cross section.[(8)] This term measures the rate at which a given scattering site removes energy from the incident beam, relative to the intensity of the incident beam. Using this concept the Raman signal intensity, I_s, in counts per second is given by

$$I_s = \frac{N\Phi\Omega I_L \varepsilon}{E_p} \frac{d\sigma}{d\Omega} \tag{1}$$

where $N \equiv$ concentration of scatterers in number of scatterers per cm^3, $\Phi \equiv$ total scattering volume imaged onto the spectrometer entrance slit in cm^3, $\Omega \equiv$ solid angle of the collection optics in steradians, $I_L \equiv$ average intensity of the laser within the volume Φ in watts/cm^2, $\varepsilon \equiv$ overall efficiency of the Raman spectrometer including lens reflection losses, spectrometer throughput, and the quantum efficiency of the photodetector, $E_p \equiv$ energy of a laser photon in joules, and $d\sigma/d\Omega \equiv$ differential Raman cross section of a single scatterer per unit solid angle of collection in units of cm^2/steradians. The Raman cross section of many gases are readily available;[(11)] however, the values pertaining to solids and liquids have to be estimated from similar bands found in the gas literature.

3. SURFACE AND THIN FILM RAMAN SPECTROSCOPY

3.1. EXTERNAL REFLECTION

The majority of surface Raman studies have used an externally reflected laser beam (Figure 1) to excite Raman scattering from sample molecules adsorbed onto various substrates. The choice of substrate and sample is dependent upon: (1) the ability of the substrate to absorb a large number of sample molecules within the scattering volume of the reflected beam, and (2) on the ability of the sample to scatter light which is easily distinguished from the light scattered by the substrate. Generally the substrates were high-surface area oxides or silicates whose Raman spectra were sufficiently weak or simple to allow detection of the Raman lines arising from the sample. In most cases the spectrum of the adsorbed sample was compared to the bulk sample spectrum to detect shifts in the Raman bands that could give information about the adsorption process. These studies, their results, and problems have been reviewed in detail by several authors.[12-16]

The application of external reflection Raman to the study of surface-adsorbed polymers has been limited. Koenig and Shih[17-19] used this technique to study glass and silica fibers treated with vinylsilane and methyl methacrylate. In these studies it was found that treatment of the fibers with silane produced a chemically bound surface coat of poly(vinylsilane) (PVS), and that polymerization of methyl methacrylate (MMA) in the presence of the PVS-coated fibers resulted in a chemically bound PMMA/PVS copolymer on the fiber surface. Rives-Arnau and Sheppard[20] and Tsai *et al.*[21] used external reflection Raman to study the polymerization of acetylene on rutile (TiO_2) surfaces and its decomposition in the presence of oxygen.

Although the external reflection technique has been relatively fruitful, the adsorbed species produce inherently weak bands, usually $10^{-7}\ I_0$, where I_0 is the intensity of the reflecting laser beam.

In addition to increasing the number of sample scatterers present within the scattering volume, some methods exist for improving the collection of the light scattered by the sample. One approach is to use high laser power, high signal amplification, long photon counting times, or slow scanning speeds. The use of repeated scans for signal averaging purposes often produces high levels of background scattering which can swamp out the weak bands of the adsorbed molecules. Another option is the use of tunable dye lasers to obtain a resonance Raman effect which can enhance Raman bands of certain samples by a factor of up to 10^5.[10] Other investigators have attempted to improve the inefficient use of the incident beam intensity in generating scattered radiation.

Connel *et al.*[22] developed a system where the thickness of the sample and an underlying dielectric layer are adjusted such that the beam reflected from the sample is exactly cancelled by the light reflected at least once from a metallic reflector base upon which the sample and dielectric film rest (Figure 4). Furthermore, the conditions which allow the sample to have a zero reflectance are exactly those required to create an in-phase addition of the scattered radiation emitted from the sample, thus producing a stronger Raman signal.

Greenler and coworkers[23,24] used a multi-reflection technique to improve the exposure of the incident laser light by coating samples onto the surface of two parallel silver mirrors (Figure 5). Here a large standing wave is established at the mirror surface whose intensity is proportional to

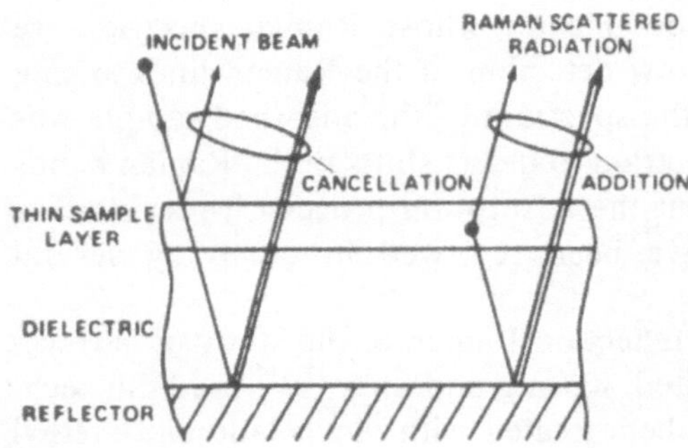

FIGURE 4. Schematic of in-phase addition geometry to enhance the Raman signal from thin film samples. [Reprinted from Reference (22), p. 31, by permission.]

the Raman scattering of the sample. This intensity is maximized through proper selection of the laser beam incident angle and the mirror separation. Additionally, the intensity of the collected scattered light could be maximized at a certain collection angle. This study[24] is of particular interest because the 1002 cm^{-1} Raman line of polystyrene was observed in films as thin as 50 and 200 Å (Figure 6). However, Greenler and coworkers did note that this technique is perhaps an order of magnitude less sensitive than that required for polymer surface and monolayer investigations.

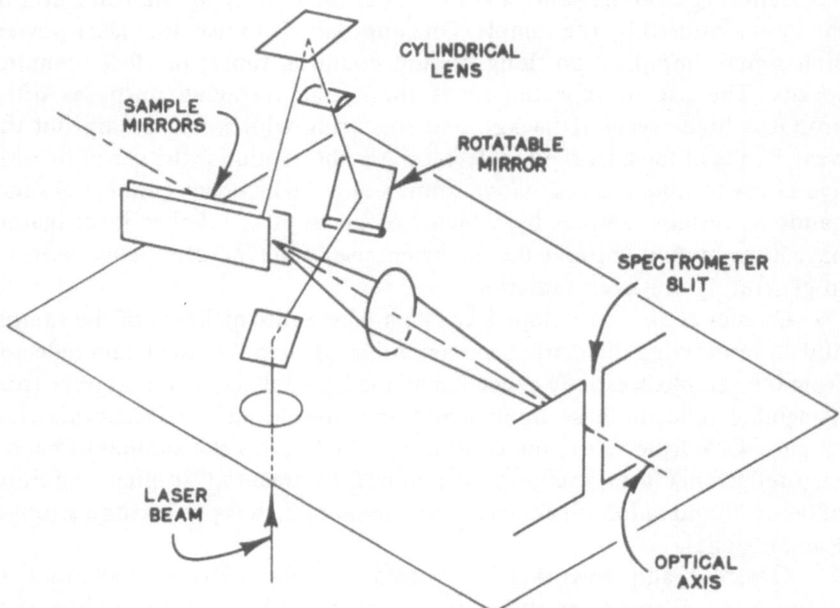

FIGURE 5. Schematic of the system used to record the Raman spectrum of polystyrene on silver mirrors. [Reprinted from Reference (24), p. 384, by permission.]

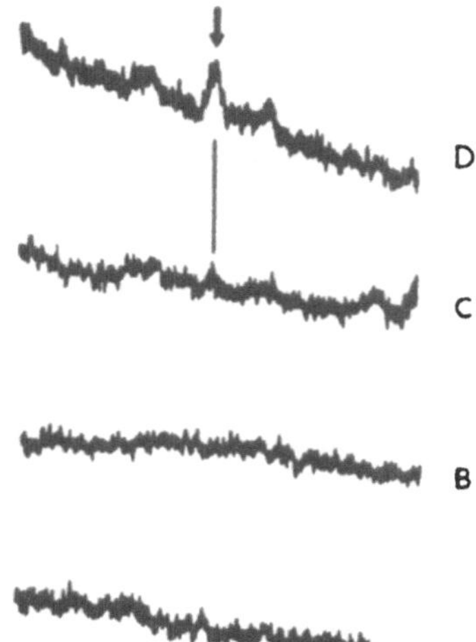

FIGURE 6. Raman spectra of polystyrene on silver mirrors. Band marked by the arrow is at 1005 cm^{-1}. (A) Spectrum of clean silver mirrors; (B) spectrum of silver mirrors rinsed with benzene; (C) spectrum of 50-Å layer of polystyrene on silver mirrors; (D) spectrum of 200-Å layer of polystyrene on silver mirrors. [Reprinted from Reference (24), p. 384, by permission.]

3.2. TOTAL INTERNAL REFLECTION (TIR)

Within the last five years or so several groups have been exploiting total internal reflection (TIR) optics to investigate the Raman scattering from a small region extending from the sample surface to a depth of about one wavelength of the incident laser beam. The basic sample configuration of a TIR Raman experiment consists of a three-phase system, $n_1 > n_2 \geq n_3$ (Figure 7), where n_i is the index of refraction of the three phases. In general, n_1 is the semi-infinite incident medium, usually an internal reflection element, n_2 is the thin film sample of thickness d_2, and n_3 is the semi-infinite final phase.

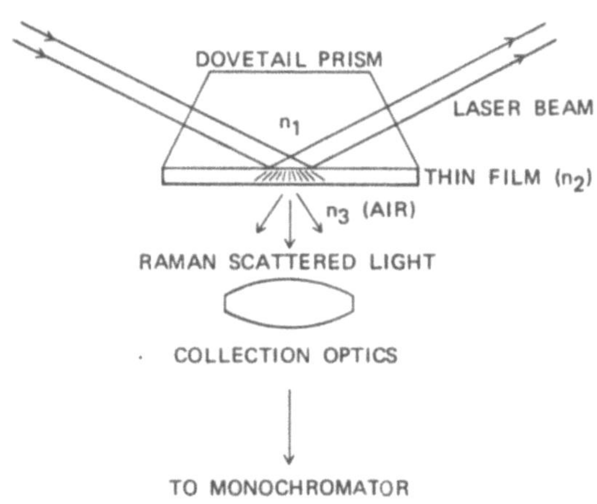

FIGURE 7. Schematic of basic three-phase sample configuration of a TIR Raman experiment.

TABLE 1
TIR Raman of Thin Films

Incident phase (n_1)	Sample film (n_2)[a]	Final phase (n_3)	Reference
Flint glass	CS_2	Air	26
Water	CO dye	Air	29
CCl_4	BRBS dye	Water	30, 31
Flint glass	Hemin, bilirubin, cobalamin	Air	32
Quartz	C_2H_5OH	Air	32
Fused silica	PVA	Water	33
Sapphire	Bovine albumin	Air	34
Sapphire	PS	Air	34
Sapphire	PS	PE	35, 36
Sapphire	PS	PC	36

[a] BRBS = Brilliant Red BS, CO = Cetyl Orange, PVA = poly(vinyl alcohol), PS = polystyrene, PE = polyethylene, PC = polycarbonate.

TIR Raman was apparently first proposed by Hirschfeld in 1970.[25] Another early study was that of Ikeshoji *et al.*[26] Detailed theoretical treatments of TIR Raman are available in the literature,[27,28] Table 1 lists the recently reported applications of TIR Raman to the study of thin films.

Briefly, the Raman scattering of thin film samples is excited by a partial penetration of the totally reflecting laser beam below the n_1/n_2 interface. This penetration produces an exponentially decaying, or evanescent, intensity field in n_2 and n_3. From theory, the fraction of the evanescent intensity residing within the film (I_2) and the final medium (I_3) is proportional to

$$I_2 \propto 1 - \frac{|T|^2}{|t|^2} e^{-2d_2/d_{p_3}} \tag{2}$$

and

$$I_3 \propto e^{-2d_2/d_{p_3}} \tag{3}$$

respectively. The terms $|t|$ and $|T|$ are the transmitted interfacial electric field amplitudes during TIR at the n_1/n_3 interface in the absence of a film ($d_2 = 0$) and at the n_2/n_3 interface in the presence of a film of thickness d_2, respectively. The value d_{p_3} is the depth of penetration of the final medium which defines the rate of the electric field decay from the n_2/n_3 interface into n_3. The light scattered by the sample is proportional to this field intensity. The electromagnetic considerations of TIR have been discussed in detail by Knutson and Reichert.[37]

Beyond such factors as collection optics and laser power, there are three basic considerations when designing a TIR Raman experiment: (1) the Raman spectra and background scattering of the incident and final phases, (2) the sample thickness, and (3) the angle of incidence of the laser beam at the n_1/n_2 interface (θ_1) which must be greater than the critical angle (θ_c) for total reflection [$\theta_1 > \theta_c = \sin^{-1}(n_2/n_1) \geq \sin^{-1}(n_3/n_1)$ where $n_1 > n_2 \geq n_3$].

The Raman bands from the incident and final phases mix with those arising from the sample film, thus presenting a problem when trying to isolate the sample spectrum. The optimum situation is one where n_1 and n_3 are colorless, transparent, possess no Raman bands within the region of the sample spectrum, and have no background scattering. For organic samples the incident phase is a high index material like flint glass or sapphire. Sapphire is particularly suitable because of its high purity (low background scattering) and lack of Raman bands above 800 cm^{-1} which is the fingerprint region of most organic materials.[36] The final medium is generally air or water, both of which are poor scatterers of light and have relatively low refractive indices.

The application of TIR Raman to the study of thin polymer films has essentially been limited to the work of Iwamoto *et al.*[33-36] This group has published the TIR Raman spectra of several polymers, most notably an investigation of polystyrene thin films (Figure 8).

The effect of sample thickness (d_2) and incident angle of the laser beam at the n_1/n_2 interface (θ_1) arises from their influence on the intensity distribution of the evanescent field and hence the intensity available to

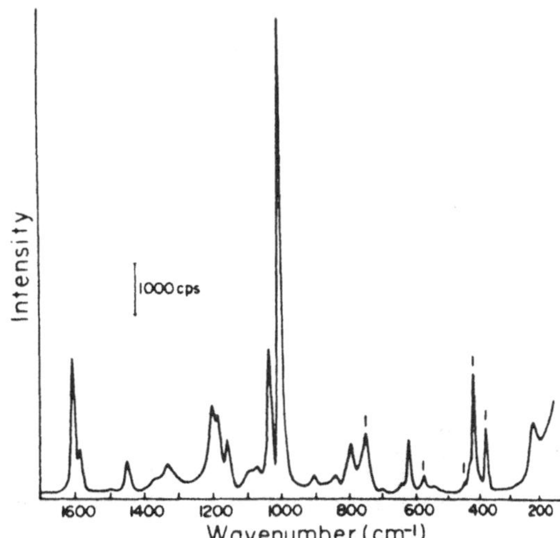

FIGURE 8. TIR Raman spectrum of a 0.7-μm thick polystyrene film at the critical angle. Marked peaks are due to the sapphire internal reflection element. [Reprinted from Reference (34), p. 586, by permission.]

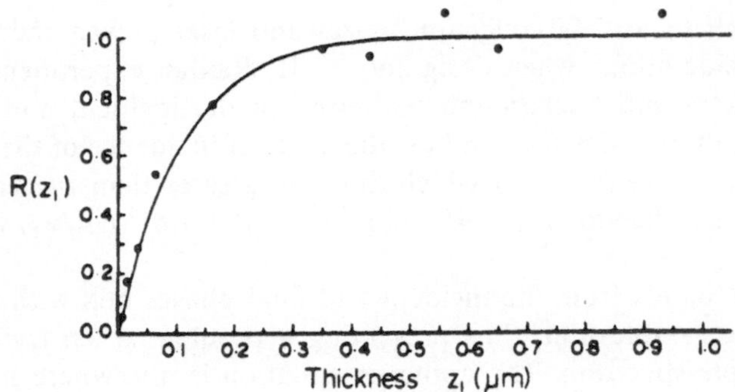

FIGURE 9. Relative TIR Raman intensity $R(z_1)$ of a polystyrene film as a function of film thickness z_1. Incident laser light is 488 nm. Solid curve: theoretical; circles: experimental. [Reprinted from Reference (36), p. 4786, by permission.]

excite Raman scattering within the film. The variables in Eqs. (2) and (3) are all functions of sample thickness and/or incident angle.

Figure 9 is a plot from the work of Iwamoto *et al.*[36] showing theoretical and experimental TIR Raman intensities. The 1002 cm^{-1} polystyrene peak is plotted as a function of sample film thickness for a system where n_1 is sapphire, n_2 is the polystyrene film, and n_3 is a polycarbonate substrate. The experimental Raman intensities, $R(z_1)$, were scaled relative to the intensity of a polystyrene sample much thicker than the distance penetrated by the evanescent field intensity. The experimental intensities were calculated from Eq. (2). Here a system was chosen where $n_2 = n_3$. This simplifies Eq. (2) to $1 - e^{-2d_2/d_p}$ where d_p is now the depth of penetration for both the polystyrene film and the polycarbonate final phase.

For films much thinner than the laser wavelength ($z_1 = d_2 \ll \lambda$) the Raman intensity increases linearly with film thickness (Figure 9). In this linear region the sample comprises less than half of the total evanescent field intensity in n_2 and n_3, thus making room for a considerable contribution from the final medium to the observed Raman spectrum. However, for film thickness on the order of a wavelength or greater ($z_1 = d_2 \gtrsim \lambda$) the Raman intensity of the film levels off, thus indicating that the polystyrene film thickness is reaching and eventually extending past the volume occupied by the evanescent field intensity.

Iwamoto *et al.*[36] also demonstrated the effect of incident angle, θ_1, on the measured Raman intensity of the polystyrene film (n_2) and the polycarbonate final phase (n_3). Figure 10 is an experimental and theoretical plot of the 890 cm^{-1} polycarbonate: 1002 cm^{-1} polystyrene band intensity ratio ($I_{PC}^{890}/I_{PS}^{1002}$) as a function of θ_1, where θ_1 is greater than the critical angle for total reflection (θ_c). The theoretical intensity ratio was calculated from the ratio of Eq. (2) to Eq. (3) where $n_2 = n_3$. Theory states the

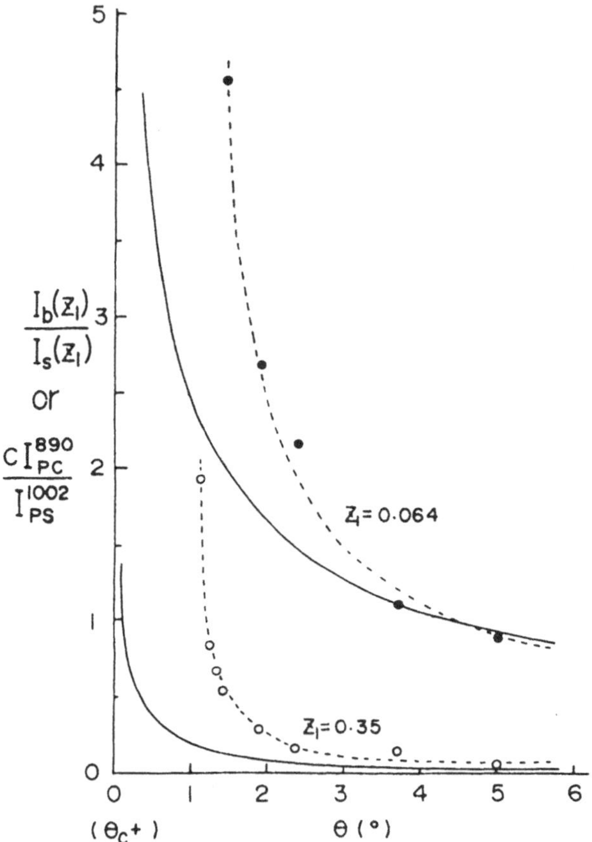

FIGURE 10. Dependence of relative TIR Raman intensities of the polycarbonate final phase (I_{PC}^{890}) to the polystyrene thin film (I_{PC}^{1002}) on the angle θ away from the critical angle θ_c ($\theta = \theta_1 - \theta_c$). Solid curve: theoretical; filled and open circles: experimental. Also note the effect of polystyrene film thickness. [Reprinted from Reference (36), p. 4789, by permission.]

evanescent field reaches its maximum scattering volume for $\theta_1 = \theta_c$, and decreases steadily as θ_1 increases away from θ_c.[37] This pattern is consistent with the data in Figure 10. With an increasing incident angle away from θ_c the scattering volume recedes away from the polycarbonate final phase into the polystyrene causing I_{PC}^{890} to decrease and the ratio of $I_{PC}^{890}/I_{PS}^{1002}$ to do likewise. Also note that in the thicker polystyrene film the intensity ratio goes to zero which indicates that at higher θ_1 values the evanescent field decays completely within the polystyrene film. The discrepancy between theory and experiment may be attributed to divergence of the laser beam and the fact that the Raman bands of polystyrene and polycarbonate have different Raman cross sections.[36]

The use of single reflection TIR Raman for the study of polymer surfaces is limited by the need for ultra-thin films (~50 Å) to increase the surface to volume ratio of the sample and hence the surface Raman signal. Unfortunately, this requirement produces a situation where the sample comprises a very small portion of the total evanescent scattering volume, thereby producing very weak Raman signals. Consequently, this technique is restricted to those transparent polymers with very strong scattering peaks. Polystyrene is a possible candidate for surface studies due to the identification of distinct Raman peaks in films as thin as 600 Å.[36] Another potential solution to this dilemma is a multiple reflection configuration to increase the number of reflection sites that contribute to the sample spectrum.

3.3. WAVEGUIDE METHODS

Swalen *et al.*[38-41] have described waveguide integrated optics and its application to determining polymer film thickness and refractive index. As shown by Rabolt *et al.*[44-49] and Levy *et al.*[50] the waveguide geometry provides an excellent technique to excite Raman scattering in micron and submicron polymer films (Figure 11a). The excitation source is the guided

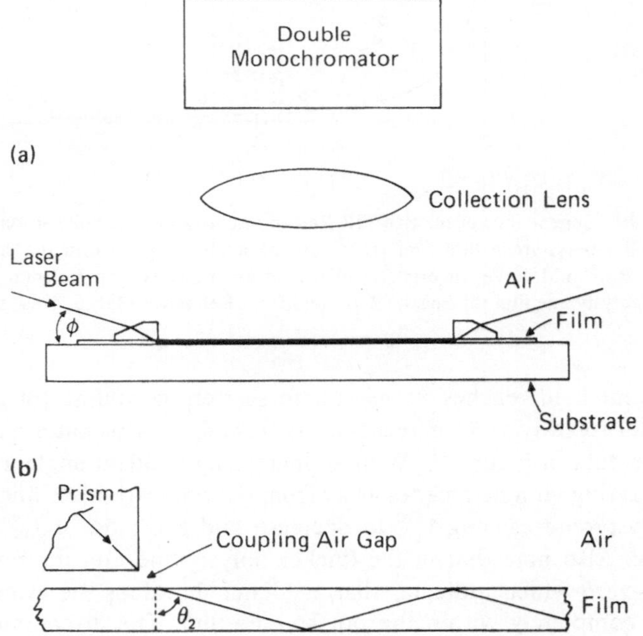

FIGURE 11. (a) Instrumental arrangement for recording the waveguide Raman spectrum of thin polymer films. (b) Laser beam coupled into a thin polymer film and undergoing multiple internal reflections. [Reprinted from Reference (44), p. 550, by permission.]

wave which propagates down the length of the polymer film by total internal reflection back and forth between the two film interfaces (Figure 11b). Here the refractive index of the film (n_2) must be greater than the refractive indices of the two adjacent media [n_1 (air) $<$ n_2 (polymer waveguide) $>$ n_3 (substrate)].

In addition to satisfying the critical conditions for total reflection at the n_1/n_2 and n_2/n_3 interfaces [$\sin^{-1}(n_1/n_2) < \theta_2 > \sin^{-1}(n_3/n_2)$] the angle of reflection within the film (θ_2) must equal one of the discrete angles associated with guided modes. The specific angles are required because the light wave in the film is not a single ray as shown in Figure 11b but a summation of light rays. The repeated reflection back and forth of these rays produces various interference patterns within the film depending upon the phase relationship between the reflecting rays. For a given set of n_1, n_2, n_3, film thickness d_2, and laser wavelength λ values there exists a finite set of θ_2 values which permit constructive interference patterns to develop, allowing the wave to propagate down the film. The propagating wave is called a guided mode.

More specifically, three events within the film alter the phase of the reflecting light rays: total reflection at the n_1/n_2 and n_2/n_3 interfaces, and traversing the film. If these phase changes sum to a multiple (m) of 2π then constructive interference results and a guided mode is launched. For a given mode number m, there are $(m + 1)$ field intensity modes across the film. Figure 12 shows the field intensity distribution across a poly(vinyl alcohol) wave guide (n_2) for the four guided modes which the waveguide will support. Also note the evanescent fields created in air (n_1) and the quartz substrate (n_3). Eigenvalue equations have been developed for calculating the mode-generating angles (θ_2) as a function of n_1, n_2, n_3, d_2, and λ. The theory of total internal reflection is discussed elsewhere.[(37)] The theory of optical waveguides has been presented by Marcuse[(42)] and Kapany and Burke.[(43)]

The waveguide system shown schematically in Figure 11 couples the laser light into the polymer film by an evanescent field created in an approximately 500-Å air gap between the polymer film and the high-index prism. The coupling angle ϕ is adjusted to obtain (via Snell's Law) the proper angle θ_2 that will allow the coupled beam to propagate down the film. The Raman scattered light is collected at an angle 90° to the direction of the guided mode. Through the principle of reciprocity, the beam can be coupled back out at the other end of the waveguide through another coupling prism.

In addition to collecting the Raman-scattered light directly from the polymer waveguide,[(44,46,50)] this geometry has been used to obtain the Raman spectra of ultra-thin polymer films spun cast on top of the waveguide[(45–58)] and the evanescently excited resonance Raman spectra of Langmuir–

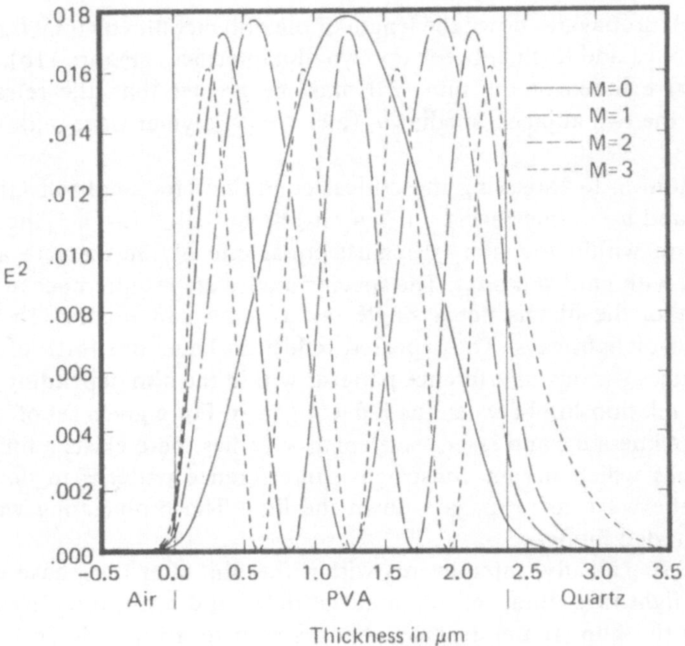

FIGURE 12. Mode intensity patterns for a 2.3-μm thick PVA waveguide. [Reprinted from Reference (41), p. 169, by permission.]

Blodgett dye monolayers deposited on the polymer waveguide surface.[47–49] These studies are summarized in Table 2.

The advantages of the waveguide Raman technique lie in the efficient manner in which guided modes excite Raman scattering within polymer films. Different mode intensity patterns emphasize different portions of the polymer waveguide (Figure 12). This feature is particularly effective in the study of polymer laminate structures.[45–48]

Typically, one observes a factor of 10 to 100 more reflections/cm with a guided mode than from an attenuated total reflection experiment. These multiply reflected beams produce a large scattering volume inside the thin films which appears macroscopically as a streak. The streak then can be focused directly onto the monochromator slit. Figure 13 compares two Raman spectra taken by these authors of the same 1-μm polystyrene film using the conventional back-scattering (external reflection) and the waveguide-enhanced techniques. Note the excellent signal enhancement (~55) and signal-to-noise ratio of the waveguide-enhanced spectrum. Finally, the evanescent streak at the waveguide surface can be used to excite the Raman spectra of very thin films deposited on a polymer surface (Figure 14).

TABLE 2
Waveguide Raman Spectroscopy of Polymers and Langmuir–Blodgett Dye Monolayers

Substrate	Waveguide[a]	Spun cast polymer film[a]	Dye monolayer[b]	Reference
Pyrex	PS (1-2 μm)			44
Pyrex	PMMA (1-6 μm)			50
PMMA	PVA (1-2 μm)			46, 47
Pyrex	PVA (1-2 μm)	PS (0.25-0.5 μm)		45, 46, 47
Pyrex	PVA (1-2 μm)	PMMA (0.25-0.5 μm)		46, 47
Pyrex	Corning glass (1 μm)	PS (0.05-0.5 μm)		48
Pyrex	Corning glass (1 μm)		CP dye, SQ dye	47, 48, 49
Pyrex	PVA (1-2 μm)		CP dye	47, 49

[a] PS = polystyrene, PVA = poly(vinyl alcohol), PMMA = poly(methyl methacrylate).
[b] CP = 1, 1-dioctadecyl-2,2-cyanine perchlorate, SQ = Squarylium.

3.4. PLASMON ENHANCEMENT

Because the Raman signal intensity is directly proportional to the average laser intensity [Eq. (1)] any technique or trick which concentrates or otherwise enhances the laser intensity in the scattering volume will increase or enhance the signal. A number of metal thin films, in particular, silver, can be used for this purpose. Since the volume we are interested in

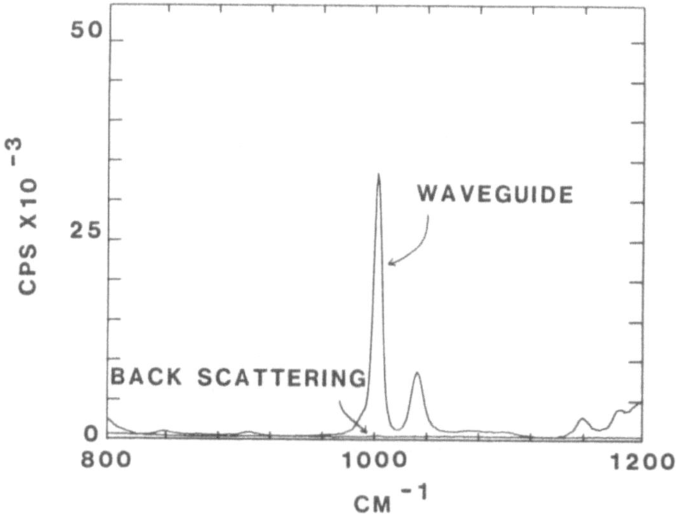

FIGURE 13. Comparison of waveguide and back-scattering Raman spectra of a 1-μm polystyrene film.

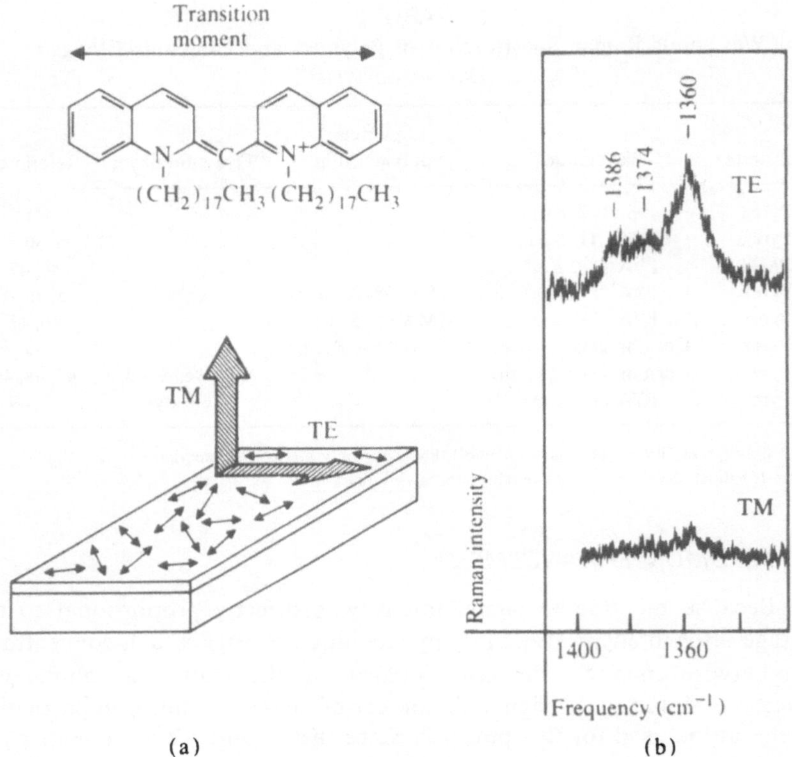

FIGURE 14. (a) Schematic of chromophore orientation on the waveguide surface and input laser polarization for TE and TM modes of propagation. (b) Resonant Raman spectra of the adsorbed chromophore in the 1320-1420 cm^{-1} range. [Reprinted from Reference (41), p. 7, by permission.]

is the interfacial volume, the trick is to couple or direct the laser beam energy only into the surface or interface of interest.

The optical and dielectric properties of matter are, to a large extent, determined by the interactions of the constituent atomic and molecular orbital electrons with the light. In the case of metals, the optical properties are dominated by the fact that one has a free electron gas in the bulk of the material. It can be shown that this free electron gas has its own resonance frequencies called plasma frequencies or plasmons. The plasma oscillation is a collective longitudinal excitation in the free electron gas. The energy associated with these plasma oscillations is directly measurable through electron energy loss measurements and is observed in XPS spectra (Chapter 5). These energies are in the range of roughly 4-15 eV for metals and 12-17 eV for various semiconductors.[51] Because of the different bonding environments in the vicinity of any surface, the electrons near the surface of the metal have different resonances or plasmon excitation energies from those in the bulk. Thus, surface plasmons are observed and can be readily measured by variable angle XPS (Chapter 5). Surface plasmons are generally several eV lower in energy than the bulk volume plasmons.

If one can direct a beam of light onto a metal surface or interface, such that the bulk or surface plasmons are excited, then this becomes a means of Raman signal enhancement. This is conveniently done in the case of silver films in the visible region by the use of total internal reflection optics which provide excitation of the bulk silver film plasmons through evanescent coupling.

The theory and mechanism of photon-excited bulk and surface plasmons has been extensively treated.[51-53] Chen *et al.* showed in 1976 that excitation of a 500-Å thick silver film, at the appropriate angle under total internal reflection conditions, resulted in the excitation of surface plasmons.[54] A specific well-defined angle of incidence is required in order to match the momentum components parallel to the surface for the photons and the surface plasmons. Under these conditions, the light intensity at the interface between the silver film and the ambient environment exceeds that of the incoming light up to two orders of magnitude, a substantial enhancement. Chen *et al.* suggested the application of this technique to Raman spectroscopy. In 1979 a number of groups demonstrated that this geometry can be used to obtain Raman and fluorescence spectra from thin films deposited on the silver support.[55-58] A particularly nice demonstration is the study by Benner *et al.*,[56] using a strontium titanate hemi-cylinder with silver deposited on one part and not on the other. It is convenient to use a hemi-cylinder which facilitates determination of the precise coupling angle.

Although the silver plasmon enhancement method has not been used for the study of polymer monolayers to any great extent, it obviously has considerable potential. It has been used for the study of phospholipids and related monolayers by Dupeyrat and his coworkers where Raman spectra from films as thin as 75 Å have been detectable.[57-58] A detailed treatment of the entire methodology has recently appeared.[59]

As long as the coupling requirements are met, the surface plasmon will be excited. This condition can be achieved through the use of appropriate surface geometries, such as by putting the silver film on a suitably prepared diffraction grating. This technique has been pioneered by the group at IBM Research Labs in San Jose.[60-61] Another approach is to use evanescent coupling via the so called "Otto-geometry" as opposed to the Kretschmann geometry which used classical ATR methods.[55]

It is generally accepted that a 500-Å uniform silver film deposited on the appropriate dielectric results in the optimum bulk and surface plasmon enhancement. Therefore, one can begin to consider the silver film as a sensor or probe. If one now places a polymer surface in contact with it, in principle, one would obtain a spectrum of that interfacial region. Clearly this technique has some potential for polymer surface analysis.

3.5. SURFACE-ENHANCED RAMAN SPECTROSCOPY (SERS)

During the last ten years, a technique has become available which utilizes external reflection from silver island films (as opposed to the continuous film discussed in the previous section) producing enhancements in the Raman signal of 10^3 to 10^6. This method, called surface-enhanced Raman scattering or SERS, is being studied as a means to greatly enhance the Raman scattering of monolayers and interfaces. The effect is believed to be due to at least two component mechanisms. One nonspecific component is due to the unique dielectric properties of very thin silver colloids which results in a local intensity enhancement of perhaps 10^3 to 10^4. The other component may be due to more specific silver surface effects. Whatever the exact nature of the mechanisms, thin silver island films are useful as probes. Submonolayers of various species can be deposited on these films and respectable Raman spectra obtained in short times. This method has been used in a very limited way to study polymer surfaces by Murray and Allara at Bell Laboratories[62] and has been used to study biomolecules by Cotton *et al.*[63] The method is described in detail by Cotton in Volume 2. A recent monograph is also available.[64] It certainly merits consideration as a means for the study of polymer thin films.

4. INSTRUMENTAL IMPROVEMENTS

There have been significant advances in the last several years in the components of Raman spectrometer systems. Most notable has been the development of optical multichannel analyzers (OMA), which allows a significant proportion of the total Raman spectrum to be detected in real time through the use of intensified photodiode array or vidicon type detectors.[65] These OMA's allow an increased throughput and therefore an increase in signal-to-noise by a factor of approximately 100 or more.[66] Advances have been made in the design of spectrometers and their gratings which allow more efficient throughput of the Raman scattered light and more efficient rejection of the Rayleigh scattered line.

The collection optics on the front end of the spectrometer have been optimized and improved considerably. The use of microscope optics has not only permitted the development of micro-area or Raman probe techniques, which have been widely applied to polymer analysis, but has improved the overall collection efficiency for many Raman applications.[59]

The continued extensive development in lasers and the recent development of truly simple trouble-free, easy to operate, tunable laser systems is making resonance Raman a useful routine analytical technique in many laboratories.

These many developments have permitted at least one group to obtain spectra of submonolayer quantities using conventional Raman, i.e., employing nonresonant methods and without any optical enhancements.[67]

5. CONCLUSION

Although Raman spectroscopy is certainly not a routine or well-known technique for the characterization of polymer surfaces today, the methods we have outlined for enhancing the Raman signal, coupled with continued advances in instrumentation, certainly lead us to anticipate that Raman spectroscopy of polymer surfaces will be widely applied in the near future and can be expected to be a common routine polymer surface characterization method in the not very distant future.

ACKNOWLEDGMENTS

We acknowledge discussions with R. E. Benner and R. A. Van Wagenen, University of Utah, on various aspects of Raman and fluorescence spectroscopy; J. D. Andrade acknowledges discussions with P. Barber, Clarkson College, and R. K. Chang, Yale University. We also acknowledge discussions and assistance of J. Swalen and J. Raybolt, IBM, San Jose. Preparation of the chapter was aided in part by NIH grants HL18519 and HL07520. We thank R. E. Benner for assistance in the theoretical aspects of Raman spectroscopy, silver plasmon enhancement, and SERS.

REFERENCES

1. J. G. Grasselli, M. K. Snavely, and B. J. Bulkin, Applications of Raman spectroscopy, *Physics Reports* **65**, 231-344 (1980).
2. D. J. Cutler, P. J. Hendra, and G. Frasier, Laser Raman spectroscopy of synthetic polymers, in: *Developments in Polymer Characterisation* (J. V. Dawkins, ed.), Vol. 2, pp. 71-143, Applied Science, London (1978).
3. P. J. Hendra, Raman spectroscopy, in: *Polymer Spectroscopy* (D. O. Hummel, ed.), pp. 151-188, Verlag Chemie, Deerfield Beach, Florida (1980).
4. P. J. Hendra, Laser Raman study of polymers, *Adv. Polym. Sci.* **6**, 151-169 (1969).
5. R. G. Snyder, Infrared and Raman spectra and polymers, in: *Polymers, Part A* (R. A. Fava, ed.), pp. 73-148, Academic Press, New York (1980).
6. B. Jasse and J. L. Koenig, Orientational measurements in polymers using vibrational spectroscopy, *J. Macromol. Sci., Rev.* **C17**, 61-135 (1979).

7. P. C. Painter, M. M. Coleman, and J. L. Koenig, *The Theory of Vibrational Spectroscopy and Its Applications to Polymeric Materials*, John Wiley and Sons, New York (1982).
8. D. A. Long, *Raman Spectroscopy*, McGraw-Hill, New York (1977).
9. A. T. Tu, *Raman Spectroscopy in Biology*, John Wiley and Sons, New York (1982).
10. P. R. Carey, *Biochemical Applications of Raman and Resonance Raman Spectroscopies*, Academic Press, New York (1982).
11. A. Weber, ed., *Raman Spectroscopy of Gases and Liquids*, Springer-Verlag, New York (1979).
12. T. Takenaka, Application of Raman spectroscopy to the study of surface chemistry, *Adv. Colloid Interface Sci.* **11**, 291-313 (1979).
13. R. P. Cooney, G. Curthoys, and N. T. Tam, Laser Raman spectroscopy and its application to the study of adsorbed species, *Adv. Catalysis* **24**, 293-342 (1975).
14. P. J. Hendra and M. Fleishman, Raman spectroscopy at surfaces, in: *Topics in Surface Chemistry* (E. Kay and P. S. Baggs, eds.), pp. 373-402, Plenum Press, New York (1978).
15. H. Yamada, Resonance Raman spectroscopy of adsorbed species on solid surfaces, *Appl. Spectrosc. Rev.* **17**, 227-277 (1981).
16. T. A. Egerhn and A. H. Hardin, The application of Raman spectroscopy to surface chemical studies, *Catal. Rev., Sci. Eng.* **11**, 1-40 (1975).
17. J. L. Koenig and P. T. K. Shih, Raman studies of the glass fiber-silane-resin interface, *J. Colloid Interface Sci.* **36**, 247-253 (1971).
18. P. T. K. Shih and J. L. Koenig, Raman studies of the hydrolysis of silane coupling agents, *Mater. Sci. Eng.* **20**, 137-143 (1975).
19. P. T. K. Shih and J. L. Koenig, Raman studies of silane coupling agents, *Mater. Sci. Eng.* **20**, 145-154 (1975).
20. V. Rives-Arnau and N. Sheppard, Raman spectroscopic study of the reaction of oxygen with polyacetylene adsorbed on rutile and of benzene adsorbed on rutile, *J. Chem. Soc., Faraday Trans. 1* **77**, 953-961 (1981).
21. P. Tsai, R. P. Cooney, J. Heaviside, and P. J. Hendra, Resonance Raman spectra of poly(acetylene) on zeolite and alumina surfaces, *Chem. Phys. Lett.* **59**, 510-513 (1978).
22. G. A. N. Connel, R. J. Nemanich, and C. C. Tsai, Interference enhanced Raman scattering from a very thin absorbing film, *Appl. Phys. Lett.* **36**, 31-33 (1980).
23. R. G. Greenler and T. L. Slager, Method for obtaining the Raman spectrum of a thin film on a metal surface, *Spectrochim. Acta.* **29A**, 193-201 (1973).
24. M. L. Howe, K. L. Walters, and R. G. Greenler, Investigation of thin surface films and adsorbed molecules using laser Raman spectroscopy, *J. Phys. Chem.* **80**, 382-385 (1976).
25. T. Hirschfeld, ATR sampling for Raman spectroscopy, Abstracts, 9th National Meeting, Soc. for Appl. Spectrosc. Oct., 1970, No. 108.
26. T. Ikeshoji, Y. Ono, and T. Mizuno, TIR Raman spectra, *Appl. Optics* **12**, 2236-2239 (1973).
27. L. D. Hooge, J. M. Vigreaux, and C. Menn, General theory of the Raman scattering close to a plane surface. Evanescent Raman spectra, *J. Chem. Phys.* **74**, 3639-3659 (1981).
28. L. D'Hooge and J. M. Vigoreux, Evanescent field excitation of homogeneous Raman scattering close to a dielectric, *Chem. Rev. Lett.* **65**, 500-506 (1979).
29. T. Takenaka and H. Fukuzaki, Resonance Raman spectra of insoluble monolayers spread on a water surface, *J. Raman Spectrosc.* **8**, 151-154 (1979).
30. T. Takenaka and T. Nakanaga, Resonance Raman spectra of monolayers adsorbed at the interface between carbon tetrachloride and an aqueous solution of surfactant and a dye, *J. Phys. Chem.* **80**, 475-480 (1976).
31. T. Nakanaga and T. Takenaka, Resonance Raman spectra of monolayers of surface active dye adsorbed at the oil-water interface, *J. Phys. Chem.* **81**, 645-649 (1977).
32. G. J. Muller, Spectroscopy with the evanescent wave in the visible region of the spectrum, in: *Multichannel Image Detectors* (Y. Talmi, ed.), ACS *Symp. Ser.* **102**, 239-262 (1979).

33. R. Iwamoto, Laser Raman study of adsorbed medium molecules without interference from the surrounding medium, *Appl. Spectrosc.* **33**, 55-56 (1979).
34. R. Iwamoto, K. Ohta, M. Miya, and S. Mima, Total internal reflection Raman spectroscopy at the critical angle for Raman measurements of thin films, *Appl. Spectrosc.* **35**, 584-587 (1981).
35. R. Iwamoto, M. Miya, K. Ohta, and S. Mima, Total internal reflection Raman spectroscopy as a new tool for surface analysis, *J. Am. Chem. Soc.* **102**, 1212-1213 (1981).
36. R. Iwamoto, M. Miya, K. Otah, and S. Mima, Total internal reflection Raman spectroscopy, *J. Chem. Phys.* **74**, 4780-4791 (1981).
37. K. Knutson and W. M. Reichert, *Total Internal Reflection Spectroscopy: Theory and Principles*, Plenum Press, New York, in press.
38. J. D. Swalen, M. Tacke, R. Santo, and J. Fischer, Determination of optical constants of polymeric thin films by integrated optical techniques, *Opt. Commun.* **18**, 387-390 (1976).
39. J. D. Swalen, Optical wave spectroscopy of molecules at a surface, *J. Phys. Chem.* **83**, 1438-1445 (1979).
40. J. D. Swalen, M. Tacke, R. Sauto, K. E. Rieckhofl, and J. Fisher, Spectra of organic molecules in thin films, *Helv. Chim. Acta.* **61**, 960-977 (1978).
41. J. D. Swalen, R. Santo, M. Tacke, and J. Fisher, Properties of polymeric thin films by integrated optical techniques, *IBM J. Res. Develop.* **21**, 168-175 (1977).
42. D. Marcuse, *Theory of Dielectric Optical Waveguides*, Academic Press, New York (1974).
43. N. S. Kapany and J. J. Burke, *Optical Waveguides*, Academic Press, New York (1972).
44. J. F. Rabolt, R. Santo, and J. D. Swalen, Raman spectroscopy of thin polymer films using integrated optical techniques, *Appl. Spectrosc.* **33**, 549-551 (1979).
45. J. D. Swalen, N. E. Schlotter, R. Santo, and J. F. Rabolt, Raman spectroscopy of laminated polymer films by integrated optical techniques, *J. Adhesion* **13**, 184-194 (1981).
46. J. F. Rabolt, N. E. Schlotter, and J. D. Swalen, Spectroscopic studies of thin film polymer laminates using Raman spectroscopy and integrated optics, *J. Phys. Chem.* **85**, 4141-4144 (1981).
47. J. F. Rabolt, R. Santo, N. E. Schlotter, and J. D. Swalen, Integrated optics and Raman scattering: Molecular orientation in thin polymer films and Langmuir-Blodgett monolayers, *IBM J. Res. Develop.* **26**, 209-216 (1982).
48. J. F. Rabolt, R. Santo, and J. D. Swalen, Raman measurements on thin polymer films and organic monolayers, *Appl. Opt.* **34**, 517-521 (1980).
49. J. F. Rabolt, N. E. Schlotter, J. D. Swalen, and R. Santo, Comparative Raman studies of monolayer interactions at a dye/polymer and a dye/glass interface, *J. Polym. Sci.* **12**, 1-9 (1983).
50. Y. Levy, C. Imbert, J. Cipriani, S. Racine, and R. Dupeyrat, Raman scattering of thin films as a waveguide, *Opt. Commun.* **11**, 66-69 (1974).
51. C. Kittel, *Introduction to Solid State Physics*, 5th ed., J. Wiley and Sons, New York (1976).
52. H. Raether, Surface plasma oscillations and their applications, *Physics Thin Films* **9**, 145-258 (1977).
53. E. Brustein, W. P. Chen, Y. J. Chen, and A. Hartstein, Surface polaritons—Propagating electromagnetic mechanisms at interfaces, *J. Vac. Sci. Technol.* **11**, 1004-1009 (1974).
54. Y. J. Chen, W. P. Chen, and E. Burstein, Surface electromagnetic wave enhanced Raman scattering, *Phys. Rev. Lett.* **36**, 1207-1210 (1976).
55. B. Pettinger, A. Tadjeddine, and D. M. Kolb, Enhancement in Raman intensity by use of Surface plasmons, *Chem. Rev. Lett.* **66**, 544-548 (1979).
56. R. E. Benner, R. Dornhaus, and R. K. Chang, Angular emission profiles of dye molecules excited by surface plasmon waves of metal surfaces, *Opt. Commun.* **30**, 145-149 (1979).
57. A. Aurengo, M. Masson, R. Dupeyrat, Y. Levy, H. Hasmonay, and J. Barbillat, Technical device for obtaining Raman spectra of ultrathin films of phospholipids, *Biochem. Biophys. Res. Commun.* **89**, 559-564 (1979).

58. M. Delhaye, M. Dupeyrat, R. Dupeyrat, and Y. Levy, An improvement in the Raman spectroscopy of very thin films, *J. Raman Spectrosc.* **8**, 351–353 (1979).
59. A. Aurengo, Y. Levy, and R. Dupeyrat, Optimization of ultrathin film nonresonant Raman spectra provided by combining a light-trapped device, a high N.A. objective and a spectrometer, *Appl. Opt.* **22**, 602–608 (1983).
60. A. Girlando, M. R. Philpott, D. Heitmann, J. D. Swalen, and R. Santo, Raman spectra of thin organic films enhanced by plasmon surface polaritons on holographic metal gratings, *J. Chem. Phys.* **72**, 5187–5191 (1980).
61. W. Kohl, M. R. Philpott, J. D. Swalen, and A. Girlando, Surface plasmon enhanced Raman spectra of monolayer assemblies, *J. Chem. Phys.* **77**, 2254–2260 (1982).
62. C. A. Murray and D. L. Allara, Measurement of the molecule-silver separation dependence of surface enhanced Raman scattering in multilayered structures, *J. Chem. Phys.* **76**, 1290–1303 (1982).
63. T. M. Cotton, S. G. Schultz, and R. R. Van Duyne, Surface-enhanced resonance Raman scattering from water-soluble porphyrins adsorbed on a silver electrode, *J. Am. Chem. Soc.* **104**, 6528–6533 (1982).
64. R. K. Chang and T. E. Furtak, *Surface Enhanced Raman Scattering*, Plenum Press, New York (1982).
65. Y. Talmi, ed., *Multichannel Image Detectors, ACS Symp. Ser.* **102** (1979).
66. J. J. Freeman, J. Heaviside, P. J. Hendra, J. Prior, and E. S. Reid, Raman spectroscopy with high sensitivity, *Appl. Opt.* **35**, 196–205 (1981).
67. A. Campin, J. K. Brown, and V. M. Orizzle, Surface Raman spectroscopy without enhancement, *Surface Sci.* **115**, 153–158 (1982).

13

Polymer Surface Analysis: Conclusions and Expectations

J. D. Andrade

1. INTRODUCTION

We have now examined twelve chapters and a wide variety of methods and techniques for probing polymer surfaces and their interfaces in various environments.

Chapter 1, "Introduction," introduced us to the literature, societies, and conferences relating to polymer surfaces. It also clearly indicated the limitations of this volume, particularly in the areas of electronic properties of polymer surfaces, the surface properties of solid polymers considerably above room temperature, polymer surface morphology, polymer monolayers at solution/air interfaces, polymer surface modification, and highly porous surfaces and polymer colloids. It was not that these topics were considered unimportant, only that this particular volume, due to space and time limitations, was unable to consider those subjects. In this chapter we consider methods and techniques not usually applied to polymers or not discussed earlier. Several ideas and totally new methods with considerable potential are also discussed.

2. MICROSCOPY

The very first surface characterization which should be performed on any surface is to look at it, noting color, reflectivity, texture, scratches or other defects, large particles, etc. The eye, with the aid of a small magnifying lens, can see surface features to less than 100 microns. Such features, if

J. D. Andrade ● Departments of Bioengineering and Materials Science and Engineering, College of Engineering, University of Utah, Salt Lake City, Utah 84112.

present, often dominate the surface and interfacial properties. Particles and deposits which may be present can have very significant influences on blood interactions; surface textures and roughnesses, in addition to being important in surface reactions and interactions in their own right, also influence subsequent surface characterizations, such as contact angle measurements (Chapter 7).

A number of reviews have carefully noted the importance of surface topography, texture, and general cleanliness for subsequent surface reactions and biomedical applications.[1,2]

The information obtained from optical and microscopic examination of polymer surfaces can fall into three categories:

(1) general cleanliness, for example, the presence of particulate matter, fibers, bacteria or other organisms, etc.
(2) surface roughness and topography—the intrinsic texture of the surface, machining or molding marks, extrusion lines, replication of the mold surface, etc.
(3) the intrinsic surface morphology of the polymer itself, i.e., phase distribution on the surface, presence of filler particles which may penetrate through the surface, etc.

Beyond the simple eyeball or low magnification identification, one goes to the stereo optical microscope which, with magnifications of up to 400 ×, allows one to see in the range of 1–10 microns. Unfortunately, at these higher magnifications in the optical microscope most depth of field is lost, which makes it difficult to analyze certain types of specimens. Nevertheless, the optical microscope with its various stages of magnification is the first and foremost surface characterization which should be done on any material.

There are detailed atlases and manuals for the optical analysis of materials, such as the McCrone[3] atlas. In addition, there are good examples in the literature of the use of these techniques for the analysis of biomaterial surfaces.[4]

The optical microscope has a number of features and accessories which enable more detailed examination.[5,6] For example, true surface topography is best observed in the reflected light mode rather than in the more common transmission illumination. The reflected light mode is sometimes enhanced by the deposition of a very thin reflective metallic film on the surface. Dark field imaging is particularly useful in picking up surface features, due to the scattering mode of imaging. Polarized light methods, particularly in transmission, are highly useful for the examination of thin films to study their optical and morphological properties, for example, the well-known method of cross polarizer analysis of polymer spherullites and related morphological features. In transmission the phase contrast technique produces a mode of contrast which helps with many types of specimens. Various interference contrast methods, particularly in reflection, can also be highly useful in monitoring surface topography and in the separation of morphological features.

A resolution significantly less than one micron or where increased depth of field is desirable requires the scanning electron microscope (SEM).[5,6] The SEM, using secondary electron imaging, provides an excellent way to characterize surface topography and related features, but provides almost no chemical information. Sample preparation is particularly important. Samples must be stable to the high-vacuum conditions. In addition, the samples should be noncharging, i.e., conductive; this is usually accomplished by depositing a very thin metal or carbon coating.

Although these methods are suitable for observing surface topography, they are not suitable for the measurement of surface composition or for detecting the presence of two or more different phases present at the surface, unless there is a topographical difference between them. If one assumes two or more phases are present on the surface, one would like to observe each directly. There are really only two approaches.

One is the technique of ion beam milling or etching which is discussed later in this chapter. Here we assume that one of the phases is preferentially etched by the ion beam, resulting in a surface topography. If one now coats the surface or takes a replica of the surface and then does conventional SEM or TEM (transmission electron microscopy), one observes topography. If one knows which phase etches at the more rapid rate, then one can deduce the surface distribution of the two phases. Because ion beam interactions with surfaces are highly energetic, extensive local heating can result which can destroy the very surface morphology which one is attempting to examine. For this reason the use of ion beam milling to develop surface texture or topography should only be done with deep frozen samples in order to freeze-in the appropriate texture which results.

The other approach is to use selective stains coupled with a different method of scanning electron microscopic imaging, called back-scattered electron detection.[5-9] In addition to the secondary electrons commonly used for imaging, the high-energy electrons from the primary beam may be scattered back into an appropriate detector, hence the term back-scattered electron imaging. It turns out that the scattered electrons are scattered more efficiently from high-mass atoms. Thus, if one has a phase which can selectively bind or interact with a reagent containing heavy atoms, then that phase will be a more efficient scatterer and will appear white in the scanning electron micrograph. This is analogous in many respects to the surface derivatization of functional groups for XPS analysis discussed in Chapter 5. A common reagent for this purpose has been osmium tetroxide, OsO_4, which reacts with unsaturated groups. Other useful reagents include iodine vapor, mercuric trifluoroacetate, ethanolic bromine solution, and ruthenium tetroxide. All of these methods have been reviewed in some detail by Hobbs[6] and Thomas.[7] A general review of all microscopic techniques for polymers, including replica techniques, has been given by Cope.[5]

In addition to the ion beam etching method, a variety of chemical etchants can be used. Many of these have been used for the study of the morphology of the surface of polymer single crystals and polymer spherullites.[6] Etching may also result from selective solvation. That is, if the two phases of interest are not crosslinked to one another or to each other, then by the use of an appropriate solvent for one of the phases, one can literally just dissolve it away, producing a surface topography which can then be observed directly or replicated and then observed.

Although most of these techniques are not in any way in routine use for the study of biomedical polymer surfaces, there clearly is a wealth of information available in the polymer microscopy literature which allows the investigator to seriously probe polymer surface topography and morphology. Using staining techniques and back-scattered electron detection in the SEM, one can easily obtain resolutions of about 1000 Å. Features smaller than 1000 Å usually require transmission electron microscopy. The use of thin sections means that one is not necessarily looking at the real surface. To accurately determine surface morphologies of bulk materials on the sub-1000-Å level requires indirect methods, such as variable angle X-ray photoelectron spectroscopy (Chapter 5), scanning Auger microscopy, or replica TEM.

One interesting characteristic of back-scattered electron detection using heavy metal stains is that by changing the electron beam voltage in the SEM one is, in essence, probing different depths into the polymer. Figure 1 shows some of our work[10] in which osmium tetroxide-stained blends containing polybutadiene were studied using a back-scattering detector and two different angles in the SEM to obtain a stereo pair. In this particular figure the voltages used were 15 KeV and 35 KeV. Note the different morphologies at the two voltages, because one is probing different depths into the sample. A series of stereo pairs of different voltages would enable extrapolation to zero voltage, corresponding to the true outermost surface of the material. To our knowledge this has not been done, although instrumentation is readily available which would permit such an extrapolation.

A variety of monographs and textbooks are now available on scanning electron microscopy, including the application of back-scattered electron detection to polymers. Perhaps the best series is *Scanning Electron Microscopy*, published on an annual basis as the Proceedings of the Annual SEM symposium.[11]

The reader is referred to the extensive reviews by Hobbs[6] and Cope[5] for further details.

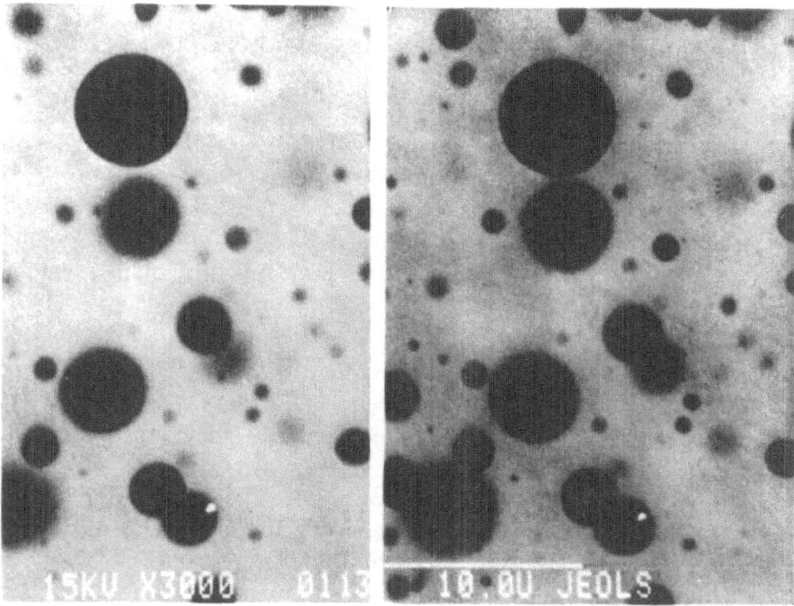

FIGURE 1. Stereo pairs obtained with a back-scattering detector in a JEOL-JSM-35C microscope at 15 (above) and 35 (below) KeV. The bar is 10 microns. The sample is a polystyrene/polybutadiene blend (weight ratio 1 : 3). Note the increased depth of the 35-KeV image. [Reprinted from Reference (10) with permission.]

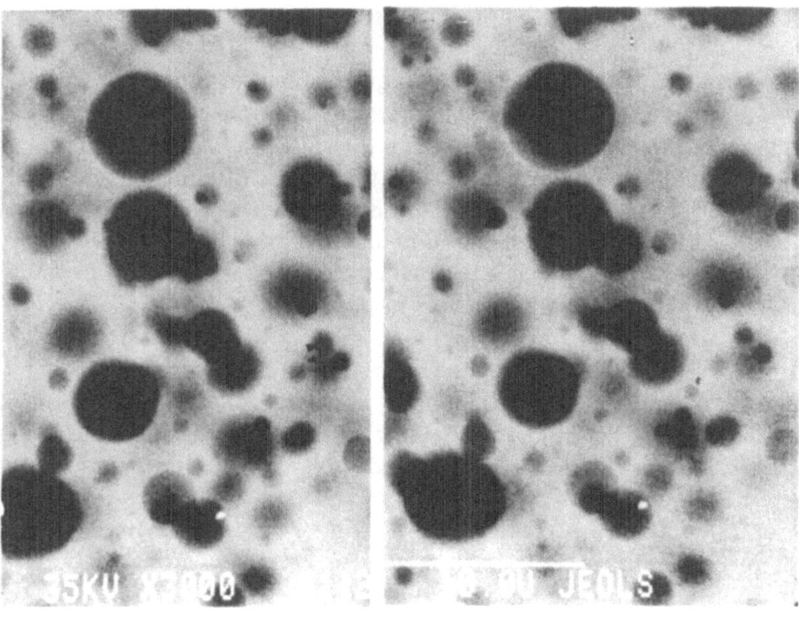

3. EXPECTATIONS

3.1. ION BEAM METHODS—SIMS AND ISS

Secondary Ion Mass Spectroscopy (SIMS) is a surface analysis technique wherein a focused ion beam is rastered over a sample surface. The ion beam interacts with the surface destructively, breaking the interatomic and intermolecular bonds in the surface region and producing fragments of the surface, either atomic ions or ionic molecular fragments, which are then directed into a mass spectrometer. A mass spectrometer detects the charge-to-mass ratio of the emitted particles, either positively charged or negatively charged ions. Neutral species can also be directed into an ionizing region and then detected as ionized species in the mass spectrometer. These methods are being widely studied and extensively developed for surface analysis.[12-15]

Since SIMS is a destructive technique, one is basically doing ion beam chemistry at the surface. There has been considerable question as to the application of such techniques to organic and polymer surfaces. Clearly, one is changing molecular weight, either by chain scission and degradation, by crosslinking, or both, during the course of such an analysis. Various pendant groups which are particularly ionizable may be preferentially removed, for example. The composition of the surface and certainly its bonding nature may change during the course of analysis. As these effects become better understood, applications to practical polymer analyses will expand.[12-22]

A problem with the SIMS technique applied to inorganic surfaces is that the sensitivity factors for different ions and ion clusters vary greatly, over several orders of magnitude.[23] It is thus difficult to obtain a quantitative elemental analysis. As these ionization factors are often matrix dependent, to obtain a rigorous quantification one needs standards which are, in essence, practically identical to the material being analyzed. In spite of these problems the SIMS technique holds considerable promise for surface analysis in general and for polymer surfaces in particular.[12-22]

One advantage of this technqiue is that it utilizes an ion beam which can be scanned and/or rastered. The ion beam can be as small as 0.5 μ in diameter, which allows one to obtain an image of the distribution of a particular mass spectral peak derived from the surface. Imaging SIMS has, of course, been available for over a decade in the form of the ion microprobe, where a 1-μ beam of oxygen ions interrogates the surface. Although ion microprobe techniques usually require a highly conductive sample and have been very difficult to apply to insulating or inorganic surfaces, there have been some very important applications to biological interfaces.[24,25] Ion microprobe methods use a high-energy and high-flux beam, which etches the surface at a fairly rapid rate. The technique, called dynamic SIMS, is often used to depth profile minerals, semiconductors, and related materials. Static SIMS refers to a technique utilizing a much lower beam flux which etches the surface at a much slower rate.[12] Static SIMS analyzes the topmost surface; there is no significant sputtering or profiling through the surface region.

There are several SIMS studies of polymers in the literature.[13-22] The two major studies are those by Gardella and Hercules[18,19] and by Briggs.[21,22,26] Gardella and Hercules compared SIMS, ISS, and XPS for a series of polymethacrylates using a 2-KeV low-intensity Ne^{20} ion beam and charge neutralization.[19] They were able to obtain static SIMS spectra on all the polymers. An analysis, based on bond-breaking events along the backbone and pendant side chains, showed that static SIMS can differentiate between small alkyl ester groups, including various isomers. It was difficult to analyze materials with long alkyl side chains from the static SIMS spectra alone.[19]

Campana *et al.* also used static SIMS to study poly(alkyl methacrylates),[20] concluding that the various homologues could be detected by mass spectral peak intensities. Briggs and Wootton[21] considered some of the problems involved in the practical application of SIMS to polymers, including the need for charge neutralization, uncertainty in surface potential, and electron-stimulated desorption of secondary ions.[13,14,21] They studied some short-chain hydrocarbons, polystyrene, and polytetrafluoroethylene films in the static SIMS mode. Briggs obtained fingerprint spectra[22] from a range of different polymers using electron beam charge neutralization and concluded that the spectra are highly characteristic and easily interpreted using fragmentation data from conventional electron impact mass spectrometry. Briggs has also used a 50-μ ion beam and demonstrated that secondary ion imaging of heterogeneous, insulating organic surfaces can be accomplished.[26] Features with dimensions of the order of 100 μ were successfully imaged within an area of several mm^2 using a static defocused electron beam to flood the whole sample for charge neutralization.

Ion Scattering Spectroscopy (ISS) is another ion beam technique wherein a beam of inert gas ions, helium, neon, or argon, is directed to the surface. This low-energy (1-5 keV) ion beam interacts with the surface atoms by elastic collisions, resulting in a loss of energy of the primary ions which depends on the mass of the atom in the surface with which they collided. Basically the process is a billiard ball problem. A high-mass ball hitting a low-mass ball transfers considerable energy to the low-mass ball. For the low-mass ball hitting a high-mass ball, little energy is transferred. The basic principles are outlined in a number of detailed reviews.[27-29]

ISS differs from SIMS in that the ion beam used to interrogate the surface is also the ion beam analyzed, the primary ions, whereas in SIMS the primary ion beam is used to produce ions characteristic of the surface species, secondary ions.

The ISS technique has the advantage that the primary ion beam samples only the outermost one or two atomic layers of the surface because of the low ion beam energy and flux. This is also a static technique, even though clearly some sputtering and etching is going on. ISS has the reputation of being the most surface sensitive of all the instrumental surface analytical techniques. Indeed, in our own experience an ISS spectrum of polytetrafluoroethylene shows only fluorine on the surface for the first few seconds of analysis. It is only after the first few seconds that one begins to see carbon in the spectrum. The exclusive surface sensitivity of ISS has been well demonstrated by looking at different crystal planes on inorganic crystals, clearly showing the dominance of the expected ions on the different crystal planes [Reference (15) and references therein].

Baun has reviewed the application of ISS to polymers.[(15)] which has been rather limited,[(30-33)] partially because of the charge neutralization problem,[(14,21)] and in part because of the unavailability of reliable ISS instrumentation, a problem which is hopefully now solved. ISS is also an imaging technique, in that the ion beam diameter is of the order of 100 μ, providing images on the 100–200 μ level.

The basic models for ion beam interaction with surfaces assume very high local heating. Based on those assumptions, one must expect very significant local side chain and main chain motions, which would scramble surface orientations and eliminate the surface sensitivity of the technique as applied to polymers. These problems can be alleviated if the sample is maintained at liquid notrogen or even liquid helium temperatures. Jonkman *et al.*[(34)] have shown that ISS and SIMS signals can be obtained from volatile organics condensed at cryogenic temperatures and entrapped in solid argon matrices. Studies similar to the elegant freeze etch XPS studies of Ratner *et al*[(35)] are in principle possible by both ISS and SIMS, given that suitable sample handling instrumentation is available. Such studies should be highly informative particularly with respect to the hypotheses presented in Chapter 2.

Because of the exquisite surface sensitivity of ISS, it should be possible to deduce information on surface dynamics and the surface orientation of pendant groups in various environments. The development of suitable sample mounting and handling systems which will permit the analysis of frozen, cryogenic samples and even fully hydrated, frozen samples would be very important for biomedical studies.

3.2. PHOTOELECTRON MICROSCOPY AND IMAGING XPS

It is usually stated that XPS is a large surface area technique and not capable of imaging. This is normally attributed to the fact that the X-ray beam can neither be focused nor rastered. Although this is in principle true, there are other methods of imaging.[43] For example, the CAMECA ion microprobe[25] produces submicron ion images of a surface without utilizing a rastered ion beam. It utilizes a 300-μ beam and, by well-designed and engineered ion optics, manages to produce images of high quality.

If one illuminates the surface uniformly with photons capable of exciting photoelectrons and directs those photoelectrons to a two-dimensional area detector, one has a photoelectron microscope. Such a technique, called photoelectron microscopy, has been under development for over ten years by Griffiths and coworkers in Oregon.[36,37] An analogous technique, called thermal emission microscopy, has been used by a small number of groups for many years, mainly in the metallurgical area, where the electrons emitted from the surface are imaged as a function of temperature.[38] In the Griffiths method all of the emitted photoelectrons are imaged with no attempt at energy discrimination. Since the photon source normally used is UV, the photoelectrons generated are all of molecular orbital or valence band nature (see Chapter 5). Therefore, elemental discrimination of the type normally performed by XPS is not possible. The images are highly useful however, because the intensity of photoelectrons is indeed related to the molecular characteristics of the surface.[36,37]

Beamson *et al.*[39] and Plummer *et al.*[40] use a method of analysis which allows them to generate an enlarged image of photoelectrons from a surface while preserving the original total energy distribution. The instrument is called a magnetic projection photoelectron microscope. The majority of the work has been done with 21.2-eV photons. They have clearly demonstrated the ability to obtain photoelectron spectra over a 0–16 eV range as a function of position in the specimen with lateral resolution in the micron range. It is expected that submicron resolution would be feasible with better instruments.[39] Figure 2 demonstrates some of the results of Beamson *et al.* A fragment of poppy anther is imaged using all electrons (0–16 eV) on the left. Electrons of energy from 8–16 eV produced the center image; electrons of energy greater than 11 eV produced the image on the right. These images can be correlated with the photoelectron energy spectra on a 0–16 eV scale rastered across the sample. Plummer *et al.* have used the same technique with X-ray photons and thereby provide an imaging XPS technique.[41]

These techniques have considerable advantages over Auger microscopy because the Auger technique uses a high current density electron beam, which produces radiation damage in most organic samples. Beamson *et al.* have argued[39] that the power used in the helium discharge is probably a factor of 1000 or more less than that of the Auger microscope for comparable magnification.

Gurker *et al.* proposed an imaging XPS instrument using the dispersion properties of a spherical condenser type spectrometer and a two-dimensional

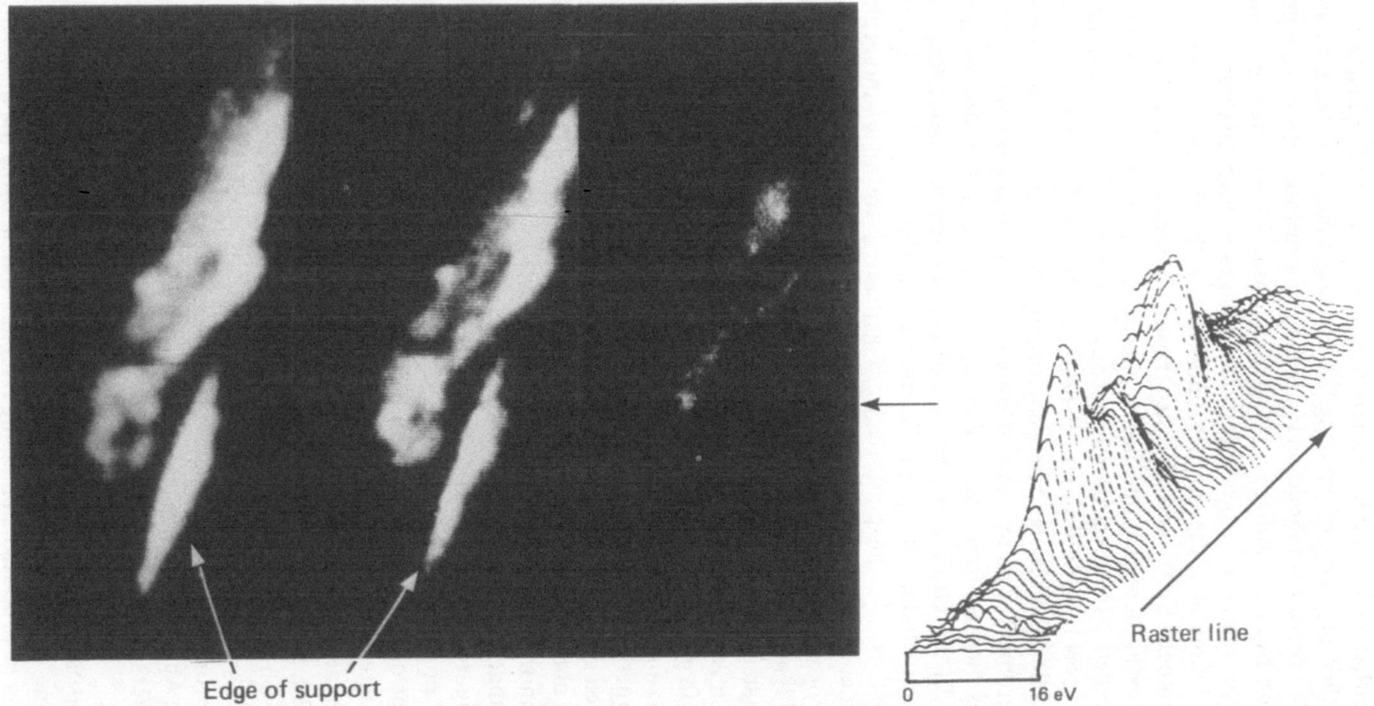

FIGURE 2. Photoelectron microscopy results of Beamson *et al.*[39] using poppy seed anther stimulated with HeI photons. On the far left all photoelectrons are detected (0–16 eV energy), the center image uses only 8–16 eV electrons, and the right image was produced with 11–16 eV electrons. The far right is a raster scan image showing the 0–16-eV photoelectron spectrum at different points along the line noted [from Reference (39) with permission.]

electron detector,[42] with mechanical scanning of the specimen to produce line distributions and area information. It was already indicated in Chapter 5 that a commercial XPS instrument is now available with a 150-μ X-ray beam and a two-dimensional detector. The movement of the sample with respect to the beam in this instrument should provide modest imaging.

It is clear that imaging XPS is, in principle, already available at the 100–200 micron level using conventional instruments. It is also clear that new methods of analysis, which permit magnification of the emitted photoelectrons by a means which maintains their relative positions on the surface, will permit the development of a photoelectron microscope with, in principle, submicron resolution. As these techniques are further developed and then commercially produced they will indeed revolutionize XPS and make it a legitimate micro-area imaging method. One can expect that within five and certainly within ten years this potential will have been achieved and commercial instruments will be available for routine XPS microscopy. Such instruments, fitted with the appropriate handling devices to permit the analysis of fully hydrated, frozen samples by the freeze etch technique previously mentioned will permit the probing of surface dynamics, surface orientations, surface degradation, etc.

It is doubtful that imaging XPS, Auger microscopy, or the ion beam methods will ever provide resolutions significantly below the submicron range. Therefore, such problems as surface distribution of the domains in block copolyether urethanes and related multiphase systems, where the phases are on the 30–100 Å level, will not be solvable by these methods in the near future. Such samples will have to continue to be studied by X-ray diffraction methods and by transmission electron microscopy, including transmitted electron energy loss techniques. The results, coupled with true surface information, will have to be extrapolated to the surface.

3.3. SURFACE FLUORESCENCE SPECTROSCOPY

Optical methods for the analysis of polymers are well-known. Light scattering is used for the determination of micromorphology. Polarized light methods are used to determine birefringence and strain in polymers. Fluorescence depolarization and fluorescence lifetimes have been used to measure molecular mobilities and other phenomena in polymers.[44,45] Raman scattering is widely used to characterize bulk polymer systems (Chapter 12). These are all bulk methods and cannot be considered surface sensitive.

There is considerable interest in total internal reflection methods[46] for the study of surfaces, exemplified by the TIR–IR technique for the study of polymer surfaces (Chapter 6), and the TIR technique applied to Raman spectroscopy of polymer thin films (Chapter 12). The use of total internal reflection fluorescence for the study of protein adsorption at surfaces is discussed in Volume 2. These approaches are all similar in that they use total internal reflection optics and the concept of the evanescent wave constrained to the solid/liquid interface. One can, in principle, use total internal reflection to excite fluorescence in a polymer thin film; by measuring fluorescence depolarization ratios one can obtain information on the molecular mobility in that thin film, using the methods now used for bulk polymer studies.[44,45] Although this technique has not yet been accomplished, the idea is relatively straightforward, using the principles outlined in Reference (46) and Volume 2.

Modern instrumentation with spectral subtraction and analysis capabilities should permit one to obtain data representing the outermost 50 to 100 Å of a polymer thin film. It is highly likely that interface fluorescence techniques will be developed in the near future to probe the surface and interfacial properties of polymer thin films, hopefully providing direct information on such subjects as polymer surface mobility and motions. It should even be possible, using two-dimensional vidicon and diode array detectors, to obtain a fluorescence image of a polymer surface in the total internal reflection mode. One can also envision using fluorescence probes, either bound to the polymer or designed to partition in the polymer phase, as a means to probe and to image that phase in the surface region of the polymer.

3.4. SURFACE MODIFICATION

Surface modification was intentionally excluded as a major subject in this volume. It is an extensive subject which is developing very rapidly. There is continual progress in organic chemistry in the development of reactions with which to derivatize and modify organic surfaces.[47,48] Gas phase techniques, particularly radio frequency glow discharge (RFGD) and plasma polymerization [see Reference (49), (50)], are now well developed. Plasma polymerization can produce thin films of widely different properties on widely different substrates, including polymer substrates.[49-52]

RFGD techniques, using a gas plasma to activate and produce a modified surface, are no longer nonspecific "sledgehammer" methods.[49,53] By careful attention to vacuum conditions, power, substrate conditions, and gas species, one has some control over specificity and the type of functional group which can be incorporated into the surface.[49,53] Indeed, this method has already been used to commercially produce cell culture substrates with unique cell binding and cell growth characteristics.[54] These gas phase methods, coupled with masking techniques commonly in use in the microcircuit industry, will permit the modification of polymer surfaces in appropriate patterns for specialty applications.

The development of monomers which can be oriented at air/solution interfaces by Langmuir monolayer techniques, transferred to solid surfaces by Langmuir–Blodgett methods, and then polymerized in place (see Chapter 4) permits the development of surfaces with well-defined orientation, functionality, and density of functional groups. Such surfaces can even be produced in two phases, each with different characteristics. A recent report indicates that such monomers can be adsorbed onto a surface, i.e., not requiring the more complex Langmuir–Blodgett procedures, and then polymerized in place to form a stable film.[55]

It is clear that polymer surfaces can be modified by a wide variety of methods for various applications, including biomedical.

3.5. REFERENCE SURFACES

One of the major problems of analytical surface chemistry is the availability of calibration or reference surfaces. Although this problem is far from solved, a start has been made with the development of the reference materials program of the Devices and Technology Branch of the National, Heart, Lung, and Blood Institute.[1] That group recognized that one of the problems limiting correlations between the surface properties of materials and their biomedical response, particularly blood interactions, was the fact that the surface properties varied widely for materials of apparently similar or even identical bulk composition.[1] They decided to embark on a program to provide reference materials with standard bulk and surface properties which investigators could use to calibrate their own techniques and against which to check the materials they produce. Only two reference materials are available: poly(dimethylsiloxane) and polyethylene. One would not expect such low-energy surfaces to be readily contaminated or modified in most situations. High-energy surfaces would be expected to adsorb material from the atmosphere and be readily contaminated and modified. It would, therefore, be desirable to have a series of high-energy surfaces of known properties available. Glass and quartz are commonly used by many investigators for this purpose.

3.6. SCANNING ELLIPSOMETRY

Ellipsometry is a technique that has been primarily used on reflective surfaces. One takes a beam of monochromatic plane-polarized light, allows it to externally reflect from the surface, and measures the change in polarization state of the reflected beam. By comparing the polarization state of the reflected and incident beams and using appropriate Fresnel optical equations, one can deduce the optical parameters of the surface.[56] If the surface is a homogeneous bulk material, then the complex refractive index is obtained, i.e., the real and imaginary part of the refractive index. The imaginary part is related to the absorption coefficient for that wavelength. If the surface contains a thin film and the substrate optical properties are known, then the thickness and refractive index of the thin film can be deduced.[(56)] In principle, this measurement can be made on an area of about 10 square microns. Such measurements are difficult for polymers because they are low-reflectivity surfaces and tend to be somewhat variable from point to point.

If one varies the wavelength the complex refractive index is found as a function of wavelength. Making the measurement as a function of wavelength basically gives an absorption spectrum of the surface of the material or of the thin film present. In principle, therefore, one can obtain a UV, visible, and IR absorption spectrum for only the surface region of material by the ellipsometry technique.[(57,58)] This can be done *in situ* with the polymer immersed in water or some other solution. There would, of course, be problems in the IR with this method in aqueous solutions because of the IR absorption of water. However, with carefully designed cells with pathlengths of only several microns, this method should be useful and should be competitive with total internal reflection IR techniques. Similar methods are already used to obtain external reflection IR spectra of metal electrodes in aqueous solutions.

The *in situ* scanning ellipsometry of polymer surfaces in aqueous solutions promises to be an important technique. Unfortunately, it appears that there is no commercial development of scanning ellipsometry instruments at present.[(58)]

3.7. OTHER

A micro-area analysis technique of considerable potential is laser induced mass spectrometry.[(59)] In this technique a focused laser beam is applied to the sample area of interest. The laser beam causes high local heating producing volatile species, which are then directed into a mass spectrometer for analysis. Although this technique is not surface sensitive, it is in many respects similar to high-flux dynamic SIMS for certain applications. One can envision optical tricks such as introducing the laser energy by total internal reflection means, which would constrain the excitation to the vicinity of an interface, thereby making the technique potentially surface sensitive. To our knowledge this has not been done; it certainly merits some consideration.

The Raman microprobe technique[(60–62)] is, in some respects, similar to the micro-laser mass spectroscopy method. Again, a focused laser beam is directed to the area of interest, but now the scattered Raman radiation is directed to a sensitive double monochromator for analysis. Again, this technique is not surface sensitive, but can, in principle, be made surface sensitive by application of total internal reflection and related optical techniques, discussed in Chapter 12. The same geometry and approach can, of course, be applied in the fluorescence mode (see Vol. 2).

High-flux, high-energy ion beams sputter material in such a manner that a unique surface texture may result.[63] Such surface texturing treatments are being evaluated for biomedical applications.[63]

A series of protocols for the surface analysis and characterization of biomedical polymers has become available. The recent publication of the NIH, on *Guidelines for Physicochemical Characterization of Biomaterials*,[1] includes a chapter on surface analysis and a set of recommendations as to levels of surface characterization for various medical applications. This protocol has now been revised and a second edition of this report is in press. Good examples of multi-technique surface analysis of polymers are available in Dwight and Riggs[64] as well as in the review by Andrade.[2]

4. CONCLUSIONS

Polymer surfaces can now be characterized by a wide variety of techniques, which provide information on the elemental surface composition, the organic functional groups present, the variation in composition and molecular character with depth, and to some extent the lateral distribution of the elemental and molecular character. In addition, the electronic properties of polymer surfaces can be determined. Although surface analysis and characterization at this time is not possible on the sub-100-μ level, it is expected that in the next five to ten years it will be possible at the submicron level. In addition to characterization, it is expected that the surface modification of polymers and other surfaces can be readily accomplished by both specific and nonspecific means, even at the micron level, using masking and rastering techniques commonly used in the microcircuit industry. The big limitation in polymer surface analysis is the lack of techniques by which one can probe the polymer–water or polymer–solution interface directly. Some progress has been made, mainly by external and total internal reflection methods, using IR, Raman, and fluorescence spectroscopies, though all three of these techniques have considerable problems. It is hoped and expected that they and other methods will be developed over the next few years, including external reflection techniques such as scanning ellipsometry.

REFERENCES

1. K. Keller, ed., *Guidelines for Physicochemical Characterization of Biomaterials*, NIH Publ. No. 80-2186, September, 1980. A revised version of this report is in press.

2. J. D. Andrade, Surface analysis of materials for medical devices and diagnostic products, *Med. Dev. Diag. Ind.* **2**, 22–23 (June, 1980).
3. W. C. McCrone and J. G. Delly, *The Particle Atlas*, 2nd ed., Vols. 1-3, Ann. Arbor Science Publ., Ann Arbor, Michigan (1973).
4. D. Coleman, J. Lawson, and W. J. Kolff, Scanning electron microscopic evaluation of the surfaces of artificial hearts, *Artificial Organs* **2**, 166–172 (1978).
5. B. C. Cope, Optical and Electron Microscopy, in: *Surface Analysis and Pretreatment of Plastics and Metals* (D. M. Brewis, ed.), pp. 95–119, Applied Publ., London (1982).
6. S. Y. Hobbs, Polymer Microscopy, *J. Macromol. Sci., Rev.* **C19**, 221–265 (1980).
7. D. A. Thomas, Morphology characterization of multiphase polymers by electron microscopy, *J. Polym. Sci. Symp.* **60**, 189–200 (1977).
8. S. Y. Hobbs and V. H. Watkins, Use of chemical contrast in the SEM analysis of polymer blends, *J. Polym. Sci. Phys.* **20**, 651–658 (1982).
9. J. S. Trent, J. I. Scheinbeim, and P. R. Couchman, Ruthenium tetraoxide staining of polymers for electron microscopy, *Macromolecules* **16**, 589–598 (1983).
10. J. D. Andrade, D. L. Coleman, and D. E. Gregonis, Characterization of polymer surface morphology by scanning electron spectroscopy using backscattered electron imaging, *Makromol. Chem., Rapid Commun.* **1**, 101–104 (1980).
11. O. Johari, ed., *Scanning Electron Microscopy*, SEM, Inc., P.O. Box 66507, AMF O'Hare (Chicago), Ill. 60666.
12. A. Benninghoven, Surface investigation of solids by static SIMS, *Surface Sci.* **35**, 427–457 (1973).
13. G. Muller, Surface analysis of insulating materials by SIMS, *Appl. Phys.* **10**, 317–324 (1976).
14. C. P. Hunt, C. T. H. Studdart, and M. P. Seah, Surface analysis of insulators by SIMS, *Surface Interface Analysis* **3**, 157–165 (1981).
15. W. L. Baun, Ion beam methods for the surface characterization of polymers, *Pure Appl. Chem.* **54**, 323–336 (1982).
16. G. D. Tantsyrev, M. I. Povolotskaya, and N. A. Kleimenov, Fluorine-containing copolymers by SIMS, *Polym. Sci. USSR* **19**, 2361–2370 (1977), and papers cited therein.
17. M. Gettings and A. J. Kenloch, Polysiloxane/metal oxide interfaces, *J. Mater. Sci.* **12**, 2511–2518 (1977).
18. J. A. Gardella and D. M. Hercules, Static SIMS of polymer systems, *Anal. Chem.* **52**, 226–232 (1980).
19. J. A. Gardella, Jr. and D. M. Hercules, Comparison of static SIMS, ISS, and XPS for surface analysis of acrylic polymers, *Anal. Chem.* **53**, 1879–1884 (1981).
20. J. E. Campana, J. J. de Corpo, and R. J. Colton, Characterization of polymeric thin films by low-damage SIMS, *Applic. Surface Sci.* **8**, 337–342 (1981).
21. D. Briggs and A. B. Wooton, Analysis of polymer surfaces by SIMS—1. Practical problems, *Surface Interface Analysis* **4**, 109–112 (1982).
22. D. Briggs, Analysis of polymer surfaces by SIMS—2. Fingerprint spectra, *Surface Interface Analysis* **4**, 151–155 (1982).
23. D. E. Newbury, Quantitative analysis by SIMS, in: *Quantitative Surface Analyses of Materials* (N. S. McIntyre, ed.), pp. 127–149, ASTM STP, 643 Amer. Soc. Testing and Materials (1978).
24. A. Benninghoven and W. K. Sichtermann, Biologically important compounds by SIMS, *Anal. Chem.* **50**, 1180–1184 (1978).
25. M. S. Burns, SIMS in biological research, *J. Microscopy* **127**, Part 3, 237–258 (1982).
26. D. Briggs, Analysis of polymer surfaces by SIMS—3, *Surface Interface Analysis* **5**, 113–118 (1983).
27. T. M. Buck, Low-energy ISS, in *Methods of Surface Analysis* (A. W. Czanderna, ed.), pp. 75–102, Elsevier Scientific, Amsterdam (1975).

28. E. Taglauer and W. Heiland, ISS, in: *Applied Surface Analysis* (T. L. Barr and L. E. Davis, eds.), pp. 111-124, ASTM STP 699, Amer. Soc. Testing and Materials, (1980).
29. W. L. Baun, ISS, *Surface Interface Analysis* **3**, 243-250 (1981).
30. A. C. Miller, A. W. Czanderna, H. H. G. Jellinek, and H. Kachi, ISS of polypropylene on CuO films, *J. Colloid Interface Sci.* **85**, 244-255 (1982).
31. A. T. DiBenedetto and D. A. Scola, S-glass/polysulfone adhesive failure Using ISS and SIMS, *J. Colloid Interface Sci.* **74**, 150-162 (1980).
32. A. T. DiBenedetto and D. A. Scola, S-glass/polymer interfaces using ISS and SIMS, *J. Colloid Interface Sci.* **64**, 480-500 (1978).
33. G. R. Sparrow and H. E. Mismash, Surface analysis of polymer and glass, in: *Quantitative Surface Analysis of Materials* (N. S. McIntyre, ed.), pp. 164-181, ASTM STP 643, Amer. Soc. Testing and Materials (1978).
34. H. T. Jonkman, J. Michl, R. N. King, and J. D. Andrade, Low-temperature positive SIMS of organic solids, *Anal. Chem.* **50**, 2078-2082 (1978).
35. B. D. Ratner, P. K. Weathersby, A. S. Hoffman, M. A. Kelly, and L. H. Scharpen, Radiation-grafted hydrogels studied by ESCA, *J. Appl. Polym. Sci.* **22**, 643-664 (1978).
36. O. H. Griffith, G. H. Lesch, G. F. Rempfer, G. B. Birrell, C. A. Burke, D. W. Schlosser, M. H. Mallon, G. B. Lee, R. G. Stafford, P. C. Jost, and T. B. Marriott, Photoelectron microscopy: A new approach to mapping organic and biological surfaces, *Proc. nat. Acad. Sci. USA* **69**, 561-565 (1972).
37. O. H. Griffith, G. F. Rempter, and G. M. Lesch, Photoelectron microscope for study of biological specimens, *Scanning Electron Microscopy/ 1981/II*, 123-130 (1981).
38. L. Wegmann, Photo-emission electron microscope, *J. Microscopy* **96**, Part 1, 1-23 (1972).
39. G. Beamson, H. Q. Porter, and D. W. Turner, Photoelectron spectromicroscopy, *Nature* **290**, 556-561 (1981).
40. I. R. Plummer, H. Q. Porter, and D. W. Turner, Photoelectron spectroscopy and microscopy, *J. Molec. Struct.* **79**, 146-162 (1982).
41. I. R. Plummer, H. Q. Porter, D. U. Turner, A. J. Dixon, K. Gehring, and M. Keenylside, Soft X-rays and fast atoms as image generators in photoelectron microscopy, *Nature* **303**, 599-601 (1983).
42. N. Gurker, M. F. Ebel, and H. Ebel, Imaging XPS principles, *Surface Interface Analysis* **5**, 12-19 (1983).
43. C. T. Hovland, Scanning ESCA, *Appl. Phys. Lett.* **30**, 274-275 (1977).
44. Y. Nishijima, Fluorescence methods in polymer science, *J. Polym. Sci.*, **31**, 353-373 (1970).
45. H. Morawetz, Applications of fluorimetry to synthetic polymer studies, *Science* **203**, 405-410 (1979).
46. K. Knutson and M. Reichert, *Total Internal Reflection Spectroscopy*, Plenum Press, in press (1985).
47. J. R. Rasmussen, E. R. Stedronsky, and G. M. Whitesides, Functional groups on the surface of low-density polyethylene, *J. Am. Chem. Soc.* **99**, 4736-4756 (1977).
48. D. E. Gregonis, R. Hsu, D. E. Buerger, L. M. Smith, and J. D. Andrade, Wettability of polymers and hydrogels, in: *Macromolecular Solutions* (R. B. Seymour and G. A. Stahl, eds.), pp. 120-133, Pergamon Press, New York (1982).
49. D. T. Clark and W. T. Feast, eds., *Polymer Surfaces*, Wiley, New York (1978).
50. D. H. Brewis, ed., *Surface Analysis and Pretreatment of Plastics and Metals*, Applied Science Publ. London (1982).
51. H. Yasuda and M. Gazicki, Biomedical applications of plasma polymerization and plasma treatment of polymer surfaces, *Biomaterials* **3**, 68-77 (1982).
52. M. Shen and A. T. Bell, eds., *Plasma Polymerization*, ACS Symp. Ser. **108**, (1979).
53. P. M. Triolo and J. D. Andrade, Surface modifications and evaluation of some commonly used catheter materials, *J. Biomed. Materials Res.* **17**, 129-147 (1983).

54. Anonymous, Modified culture plates improve cell growth, *Chem. Eng. News*, *p.* 43, December 20, 1982.
55. S. L. Regen, P. Kirszensztejn, and A. Singh, Polymer-supported membranes, *Macromolecules* **16**, 335–337 (1983).
56. R. H. Muller, Principles of ellipsometry, *Adv. Electrochem. Electrochem. Eng.* **9**, 167–222 (1973).
57. D. E. Aspnes, Spectroscopic ellipsometry of solids, in: *Optical Properties of Solids: New Developments* (B. O. Seraphin, ed.), pp. 800–876, North-Holland, Amsterdam (1976).
58. P. S. Hauge, Recent developments in instrumentation in ellipsometry, *Surface Sci.* **96**, 108–140 (1980).
59. LAMMA Symposium, *Fresenius, Z. Anal. Chem.* **308**(3), 193–315 (1981).
60. F. Adar, M. LeClercq, and R. E. Grayzel, Industrial applications of micro Raman analyses, *Am. Lab.*, pp. 56–65, March 1982.
61. G. J. Rosasco, Raman microprobe spectroscopy, *Adv. Infrared Raman Spectrosc.* **7**, 223–282 (1982).
62. P. Dhamelincourt, F. Wallart, M. Leclercq, A. T. N'Guyen, and D. O. Landon, Laser Raman molecular microprobe, *Anal. Chem.* **41**, 414A–421A (1979).
63. O. Auciello, Ion interaction with solids: Surface Texturing, *J. Vac. Sci. Technol.* **19**, 841–867 (1981).
64. D. W. Dwight and W. M. Riggs, Fluoropolymer surface studies, *J. Colloid Interface Sci.* **47**, 650–660 (1974).

Index